CAMBRIDGE STUDIES IN ADVANCED MATHEMATICS 159

STOCHASTIC ANALYSIS

Itô and Malliavin Calculus in Tandem

Thanks to the driving forces of the Itô calculus and the Malliavin calculus, stochastic analysis has expanded into numerous fields including partial differential equations, physics, and mathematical finance. This book is a compact, graduate-level text that develops the two calculi in tandem, laying out a balanced toolbox for researchers and students in mathematics and mathematical finance. The book explores foundations and applications of the two calculi, including stochastic integrals and stochastic differential equations, and the distribution theory on Wiener space developed by the Japanese school of probability. Uniquely, the book then delves into the possibilities that arise by using the two flavors of calculus together. Taking a distinctive, path-space-oriented approach, this book crystalizes modern day stochastic analysis into a single volume.

Hiroyuki Matsumoto is Professor of Mathematics at Aoyama Gakuin University. He graduated from Kyoto University in 1982 and received his Doctor of Science degree from Osaka University in 1989. His research focuses on stochastic analysis and its applications to spectral analysis of Schrödinger operations and Selberg's trace formula, and he has published several books in Japanese, including *Stochastic Calculus* and *Introduction to Probability and Statistics*. He is a member of the Mathematical Society of Japan and an editor of the *MSJ Memoirs*.

Setsuo Taniguchi is Professor of Mathematics at Kyushu University. He graduated from Osaka University in 1980 and received his Doctor of Science degree from Osaka University in 1989. His research interests include stochastic differential equations and Malliavin calculus. He has published several books in Japanese, including *Introduction to Stochastic Analysis for Mathematical Finance* and *Stochastic Calculus*. He is a member of the Mathematical Society of Japan and is an editor of the *Kyushu Journal of Mathematics*.

CAMBRIDGE STUDIES IN ADVANCED MATHEMATICS

All the titles listed below can be obtained from good booksellers or from Cambridge University Press. For a complete series listing visit: www.cambridge.org/mathematics.

Already published

119 C. Perez-Garcia & W. H. Schikhof *Locally convex spaces over non-Archimedean valued fields*
120 P. K. Friz & N. B. Victoir *Multidimensional stochastic processes as rough paths*
121 T. Ceccherini-Silberstein, F. Scarabotti & F. Tolli *Representation theory of the symmetric groups*
122 S. Kalikow & R. McCutcheon *An outline of ergodic theory*
123 G. F. Lawler & V. Limic *Random walk: A modern introduction*
124 K. Lux & H. Pahlings *Representations of groups*
125 K. S. Kedlaya *p-adic differential equations*
126 R. Beals & R. Wong *Special functions*
127 E. de Faria & W. de Melo *Mathematical aspects of quantum field theory*
128 A. Terras *Zeta functions of graphs*
129 D. Goldfeld & J. Hundley *Automorphic representations and L-functions for the general linear group, I*
130 D. Goldfeld & J. Hundley *Automorphic representations and L-functions for the general linear group, II*
131 D. A. Craven *The theory of fusion systems*
132 J. Väänänen *Models and games*
133 G. Malle & D. Testerman *Linear algebraic groups and finite groups of Lie type*
134 P. Li *Geometric analysis*
135 F. Maggi *Sets of finite perimeter and geometric variational problems*
136 M. Brodmann & R. Y. Sharp *Local cohomology (2nd Edition)*
137 C. Muscalu & W. Schlag *Classical and multilinear harmonic analysis, I*
138 C. Muscalu & W. Schlag *Classical and multilinear harmonic analysis, II*
139 B. Helffer *Spectral theory and its applications*
140 R. Pemantle & M. C. Wilson *Analytic combinatorics in several variables*
141 B. Branner & N. Fagella *Quasiconformal surgery in holomorphic dynamics*
142 R. M. Dudley *Uniform central limit theorems (2nd Edition)*
143 T. Leinster *Basic category theory*
144 I. Arzhantsev, U. Derenthal, J. Hausen & A. Laface *Cox rings*
145 M. Viana *Lectures on Lyapunov exponents*
146 J.-H. Evertse & K. Győry *Unit equations in Diophantine number theory*
147 A. Prasad *Representation theory*
148 S. R. Garcia, J. Mashreghi & W. T. Ross *Introduction to model spaces and their operators*
149 C. Godsil & K. Meagher *Erdős–Ko–Rado theorems: Algebraic approaches*
150 P. Mattila *Fourier analysis and Hausdorff dimension*
151 M. Viana & K. Oliveira *Foundations of ergodic theory*
152 V. I. Paulsen & M. Raghupathi *An introduction to the theory of reproducing kernel Hilbert spaces*
153 R. Beals & R. Wong *Special functions and orthogonal polynomials*
154 V. Jurdjevic *Optimal control and geometry: Integrable systems*
155 G. Pisier *Martingales in Banach spaces*
156 C. T. C. Wall *Differential topology*
157 J. C. Robinson, J. L. Rodrigo & W. Sadowski *The three-dimensional Navier–Stokes equations*
158 D. Huybrechts *Lectures on K3 surfaces*
159 H. Matsumoto & S. Taniguchi *Stochastic Analysis*

Stochastic Analysis

Itô and Malliavin Calculus in Tandem

Hiroyuki Matsumoto
Aoyama Gakuin University, Japan

Setsuo Taniguchi
Kyushu University, Japan

Translated and adapted from the Japanese edition

CAMBRIDGE
UNIVERSITY PRESS

32 Avenue of the Americas, New York NY 10013

One Liberty Plaza, 20th Floor, New York, NY 10006, USA

477 Williamstown Road, Port Melbourne, VIC 3207, Australia

4843/24, 2nd Floor, Ansari Road, Daryaganj, Delhi – 110002, India

79 Anson Road, #06–04/06, Singapore 079906

Cambridge University Press is part of the University of Cambridge.

It furthers the University's mission by disseminating knowledge in the pursuit of education, learning and research at the highest international levels of excellence.

www.cambridge.org
Information on this title: www.cambridge.org/9781107140516

Translated and adapted from the Japanese edition: *Kakuritsu Kaiseki*, Baifukan, 2013

First published 2017

Printed in the United States of America by Sheridan Books, Inc.

A catalog record for this publication is available from the British Library

ISBN 978-1-107-14051-6 Hardback

Contents

Preface

The aim of this book is to introduce stochastic analysis, keeping in mind the viewpoint of path space. The area covered by stochastic analysis is very wide, and we focus on the topics related to Brownian motions, especially the Itô calculus and the Malliavin calculus. As is widely known, a stochastic process is a mathematical model to describe a randomly developing phenomenon. Many continuous stochastic processes are driven by Brownian motions, while basic discontinuous ones are related to Poisson point processes.

The Itô calculus, named after K. Itô who introduced the calculus in 1942, is typified by stochastic integrals, Itô's formula, and stochastic differential equations. While Itô investigated those topics in terms of Brownian motions, they are now studied in the extended framework of martingales. One of the important applications of the calculus is a construction of diffusion processes through stochastic differential equations. The Malliavin calculus was introduced by P. Malliavin in the latter half of the 1970s and developed by many researchers. As he originally called it "a stochastic calculus of variation", it is exactly a differential calculation on a path space. It opened a way to take a purely probabilistic approach to transition densities of diffusion processes, which are fundamental objects in theory and are applied to many fields in mathematics and physics.

We made the book self-contained as much as possible. Several preliminary facts in analysis and probability theory are gathered in the Appendix. Moreover, a lot of examples are presented to help the reader to easily understand the assertions. This book is organized as follows. Chapter 1 starts with fundamental facts on stochastic processes. In particular, Brownian motions and martingales are introduced and basic properties associated with them are given. In the last three sections, investigations of path space type are made; the Cameron–Martin theorem, Schilder's theorem and an analogy with path integrals are presented.

Chapter 2 introduces stochastic integrals and Itô's formula, an associated chain rule. Although Itô originally discussed them with respect to Brownian motions, we formulate them with respect to martingales in the recent manner due to J. L. Doob, H. Kunita and S. Watanabe. Moreover, several facts on continuous martingales are discussed: for example, representations of them by time changes and those via stochastic integrals with respect to Brownian motions.

Chapter 3 presents several properties of Brownian motion. As direct applications of Itô's formula, problems in the theory of partial differential equations, like heat equations and Dirichlet problems, are studied. Although the Laplacian is only dealt with in this chapter, after reading Chapters 4 and 5, the reader will be easily convinced that the results in this chapter can be extended to second order differential operators on Euclidean spaces and Laplace–Beltrami operators on Riemannian manifolds.

Chapters 4 and 5 form the main portion of this book. Chapter 4 introduces stochastic differential equations and presents their properties and applications. Stochastic differential equations enable us to construct diffusion processes in a purely probabilistic manner. Namely, diffusion processes are realized as measures on a path space via solutions of stochastic differential equations. This is different from the analytical method by A. Kolmogorov, which uses the fundamental solution of the associated heat equation. It is also seen in the chapter that stochastic differential equations determine stochastic flows as ordinary differential equations. The flow property will be used in the next chapter.

The Malliavin calculus is developed in Chapter 5. The distribution theory on the Wiener space, which was structured by the Japanese school led by S. Watanabe, S. Kusuoka, and I. Shigekawa, is introduced. Moreover, the integration by parts formula and the change of variable formula on the Wiener space are presented. In the last two sections, the latter formula is applied to computing Laplace transforms of quadratic Wiener functionals.

Chapter 6 is a brief introduction to mathematical finance. In this chapter, we focus on the Black–Scholes model, the simplest model in mathematical finance. The existence and uniqueness of an equivalent martingale measure is shown and a pricing formula of European contingent claims is given. Moreover, as an application of the Malliavin calculus, we show ways to compute hedging portfolios and the Greeks, indices to measure sensitivity of prices with respect to parameters like initial price and volatilities.

Stochastic analysis is the analysis on path spaces, and it is deeply related to Feynman path integrals. It was M. Kac who gained an insight into this close relationship and achieved a lot of results. His achievements exerted great influence on not only probability theory but also other fields of mathematics.

Chapter 7 is intended to present results corresponding to such close relationship. It starts with an introduction of a Wiener space analog of the representation of a propagator by action integrals of classical paths, which was due to the physicist Van Vleck playing an active role in the early period of quantum mechanics. Next, applications of stochastic analysis to studies of eigenvalues of Schrödinger operators and Selberg's trace formula are presented. In these applications, probabilistic representations of a heat kernel with the aid of the Malliavin calculus provide a clear route to the results.

This book is based on the Japanese one *Kakuritsu Kaiseki* published by Baifukan in 2013. In this book, a section discussing the close conjunction of the Malliavin calculus and the rough path theory and a chapter on mathematical finance are newly added. During the writing of the Japanese book and this one, we have received much benefit from several representative monographs and books on stochastic analysis; these include:

- R. Durrett, *Brownian Motion and Martingales in Analysis*, Wadsworth, 1984;
- N. Ikeda and S. Watanabe, *Stochastic Differential Equations and Diffusion Processes*, 2nd edn., North Holland/Kodansha, 1989;
- I. Karatzas and S. E. Shreve, *Brownian Motion and Stochastic Calculus*, 2nd edn., Springer-Verlag, 1991;
- D. Revuz and M. Yor, *Continuous Martingales and Brownian Motion*, 3rd edn., Springer-Verlag, 1999.

As for the theory of martingales, we also gained some benefit from

- H. Kunita, *Estimation of Stochastic Processes* (in Japanese), Sangyou Tosho, 1976.

This book did not come into existence without the Japanese one. We deeply thank Professor Nobuyuki Ikeda, our supervisor, who recommended our writing the Japanese book. In writing the Japanese version, we received kind assistance and help from several people, whom we gratefully acknowledge. Professor Yoichiro Takahashi was on the editorial board of the series where our book appeared and encouraged our writing. He and Professors Masanori Hino, Yu Hariya, and Koji Yano read through the draft and gave us stimulating comments and kind help.

Hiroyuki Matsumoto
Setsuo Taniguchi

Frequently Used Notation

$\mathbb{N} \equiv \{1, 2, 3, \ldots\}$, $\mathbb{Z} \equiv \{0, \pm 1, \pm 2, \ldots\}$, $\mathbb{Z}_+ \equiv \{0, 1, 2, \ldots\}$
$a \wedge b = \min\{a, b\}$, $a \vee b = \max\{a, b\}$
$a^+ = \max\{a, 0\}$, $a^- = -\min\{a, 0\} = \max\{-a, 0\}$
$[a]$: the largest integer less than or equal to $a \in \mathbb{R}$
$[s]_n = 2^{-n}[2^n s] \in \{k2^{-n}\}_{k=0}^{\infty}$ $(s \geqq 0,\ n \in \mathbb{N})$
$\mathrm{sgn}(x) = -1$ $(x \leqq 0)$, $= 1$ $(x > 0)$
$\mathbf{1}_A$: the indicator function of a set A
$\sigma(\mathscr{C})$: the σ-field generated by a family $\mathscr{C}$ of subsets
$\sigma(X_1, \ldots, X_n)$: the σ-field generated by the random variables $X_1, \ldots, X_n$
$\mathscr{B}(S)$: the σ-field generated by open subsets of a topological space S
$C_0(S)$: the space of continuous functions on S with compact supports
$C^n(\mathbb{R}^d)$: the space of C^n-class functions on $\mathbb{R}^d$
$C_0^\infty(\mathbb{R}^d)$: the space of C^∞-functions on $\mathbb{R}^d$ with compact supports
$\mathscr{S}(\mathbb{R}^d)$: the space of rapidly decreasing C^∞ functions on $\mathbb{R}^d$
$\mathscr{S}'(\mathbb{R}^d)$: the dual space of $\mathscr{S}(\mathbb{R}^d)$, the space of tempered distributions
Δ : the Laplacian on $\mathbb{R}^d$, $\sum_{i=1}^d \left(\frac{\partial}{\partial x^i}\right)^2$
$g(t, x, y) = (2\pi t)^{-\frac{d}{2}} e^{-\frac{|y-x|^2}{2t}}$ $(t > 0,\ x, y \in \mathbb{R}^d)$: the Gauss kernel
$\mathbf{W} = \mathbf{W}^d = C([0, \infty) \to \mathbb{R}^d)$: the space of $\mathbb{R}^d$-valued continuous functions
$W = W^d = \{w \in \mathbf{W}^d;\ w(0) = 0\}$: the Wiener space
$W_T = W_T^d = C([0, T] \to \mathbb{R}^d)$: the Wiener space on $[0, T]$
$H_T = H_T^d$: the Cameron–Martin space of W_T^d
$\langle X \rangle$: the quadratic variation process of a semimartingale X
$\mathscr{M}^2$: the set of square-integrable martingales
$\mathscr{M}^2_{\mathrm{loc}}$: the set of square-integrable local martingales
$\mathscr{M}^2_{\mathrm{c,loc}}$: the set of square-integrable continuous local martingales

1

Fundamentals of Continuous Stochastic Processes

In this chapter fundamentals of continuous stochastic processes are mentioned, taking into account their applications to stochastic analysis.

1.1 Stochastic Processes

Let Ω be a set.

Definition 1.1.1 A family $\mathscr{F}$ of subsets of Ω is said to be a **σ-field** if

(i) $\Omega, \emptyset \in \mathscr{F}$,
(ii) if $A \in \mathscr{F}$, then $A^c := \{\omega \in \Omega; \omega \notin A\} \in \mathscr{F}$,
(iii) if $A_i \in \mathscr{F}$ $(i = 1, 2, \ldots)$, then $\bigcup_{i=1}^{\infty} A_i \in \mathscr{F}$.

The pair $(\Omega, \mathscr{F})$ is called a **measurable space**.

Definition 1.1.2 Let $(\Omega, \mathscr{F})$ be a measurable space. A set function $\mathbf{P} : \mathscr{F} \ni A \mapsto \mathbf{P}(A) \geqq 0$ is said to be a **probability measure** if

(i) $0 \leqq \mathbf{P}(A) \leqq 1$ for all $A \in \mathscr{F}$,
(ii) $\mathbf{P}(\Omega) = 1$,
(iii) for mutually disjoint subsets $A_i \in \mathscr{F}$ $(i = 1, 2, \ldots)$,

$$\mathbf{P}\Big(\bigcup_{i=1}^{\infty} A_i\Big) = \sum_{i=1}^{\infty} \mathbf{P}(A_i).$$

The triplet $(\Omega, \mathscr{F}, \mathbf{P})$ is called a **probability space**.

Throughout this book we denote by $\mathbf{E}$ or $\mathbf{E}^{\mathbf{P}}$ the expectation (integral) with respect to $\mathbf{P}$.

Given a family $\mathscr{C}$ of subsets of Ω, we denote the smallest σ-field including $\mathscr{C}$ by $\sigma(\mathscr{C})$:

$$\sigma(\mathscr{C}) = \bigcap_{\mathscr{G}} \mathscr{G},$$

where $\mathscr{G}$ runs over all σ-fields on Ω including $\mathscr{C}$.

If Ω is a topological space, $\mathscr{B}(\Omega)$ denotes the smallest σ-field containing all open subsets of Ω and is called the **Borel σ-field** on Ω.

Definition 1.1.3 Given a topological space E, a mapping $X : \Omega \to E$ is said to be $\mathscr{F}/\mathscr{B}(E)$-measurable if

$$X^{-1}(A) := \{\omega \in \Omega; X(\omega) \in A\} \in \mathscr{F}$$

holds for any $A \in \mathscr{B}(E)$. Such an X is called an E-valued **random variable**. The probability measure $\mathbf{P} \circ X^{-1}$ on E induced by X,

$$(\mathbf{P} \circ X^{-1})(A) = \mathbf{P}(X^{-1}(A)) \qquad (A \in \mathscr{B}(E)),$$

is called the **probability distribution** or **probability law** of X.

The purpose of this book is to study several kinds of analysis based on continuous stochastic processes. Here we introduce path spaces which play a fundamental role in such studies.

Definition 1.1.4 Let (E, d_E) be a complete separable metric space.
(1) For $T > 0$, $\mathbf{W}_T(E)$ stands for the set of E-valued continuous functions on $[0, T]$. $\mathbf{W}_T(E)$ is endowed with the topology of uniform convergence, or equivalently, with the distance function given by

$$d(w_1, w_2) = \max\{d_E(w_1(t), w_2(t)); 0 \leqq t \leqq T\},$$

which makes $\mathbf{W}_T(E)$ a complete separable metric space.
(2) The set of E-valued continuous functions on $[0, \infty)$ is denoted by $\mathbf{W}(E)$, and it is endowed with the topology of uniform convergence on compact sets, or equivalently, with the distance function given by

$$d(w_1, w_2) = \sum_{n=1}^{\infty} 2^{-n} \Big\{ \max_{0 \leqq t \leqq n} (d_E(w_1(t), w_2(t))) \wedge 1 \Big\},$$

by which $\mathbf{W}(E)$ is a complete separable metric space.

The Borel σ-fields with respect to the respective topologies are denoted by $\mathscr{B}(\mathbf{W}_T(E)), \mathscr{B}(\mathbf{W}(E))$.

Proposition 1.1.5 *Let $\mathscr{C}_T(E)$ be the totality of subsets of $\mathbf{W}_T(E)$ of the form*

$$\{w \in \mathbf{W}_T(E);\ w(t_1) \in A_1, w(t_2) \in A_2, \dots, w(t_n) \in A_n\} \tag{1.1.1}$$

with $0 < t_1 < t_2 < \cdots < t_n \leqq T$, $A_1, A_2, \dots, A_n \in \mathscr{B}(E)$. Then $\sigma(\mathscr{C}_T(E)) = \mathscr{B}(\mathbf{W}_T(E))$.

Proof For an open set G of E and for $t > 0$, $\{w \in \mathbf{W}_T(E); w(t) \in G\}$ is an open set of $\mathbf{W}_T(E)$. Hence $\sigma(\mathscr{C}_T(E)) \subset \mathscr{B}(\mathbf{W}_T(E))$.

To prove $\sigma(\mathscr{C}_T(E)) \supset \mathscr{B}(\mathbf{W}_T(E))$, it suffices to show

$$\Big\{w; \max_{0 \leqq t \leqq T} d_E(w(t), w_0(t)) \leqq \delta\Big\} \in \sigma(\mathscr{C}_T(E))$$

for any $w_0 \in \mathbf{W}_T(E)$ and $\delta > 0$. It is easily obtained from the following identity:

$$\Big\{w; \max_{0 \leqq t \leqq T} d_E(w(t), w_0(t)) \leqq \delta\Big\} = \bigcap_{0 \leqq r \leqq T,\, r \in \mathbb{Q}} \{w; d_E(w(r), w_0(r)) \leqq \delta\}. \qquad \square$$

We call a set of the form (1.1.1) a **Borel cylinder set**. We define the Borel cylinder sets of $\mathbf{W}(E)$ in the same way. Then, setting $\mathscr{C}(E) = \bigcup_{T>0} \mathscr{C}_T(E)$, we have $\sigma(\mathscr{C}(E)) = \mathscr{B}(\mathbf{W}(E))$.

Definition 1.1.6 Let $\mathbb{T}$ be $[0, T]$ $(T > 0)$ or $[0, \infty)$.
(1) A family $X = \{X(t)\}_{t \in \mathbb{T}}$ of E-valued random variables defined on a probability space $(\Omega, \mathscr{F}, \mathbf{P})$ is called an E-valued **stochastic process**. When $\mathbb{T} = [0, \infty)$, we write $X = \{X(t)\}_{t \geqq 0}$.
(2) A stochastic process X is said to be continuous if for almost all $\omega \in \Omega$ an E-valued function $X(\omega) : \mathbb{T} \ni t \mapsto X(t)(\omega) \in E$ on $\mathbb{T}$ is continuous.
(3) For each $w \in \Omega$, the function $X(\omega)$ is called a **sample path**.

Definition 1.1.7 Let $\mathbb{T}$ be the same as above and $X = \{X(t)\}_{t \in \mathbb{T}}$ and $X' = \{X'(t)\}_{t \in \mathbb{T}}$ be stochastic processes defined on a probability space $(\Omega, \mathscr{F}, \mathbf{P})$. If $\mathbf{P}(X(t) = X'(t)) = 1$ for all $t \in \mathbb{T}$, X' is called a **modification** of X.

Throughout this book, $\langle x, y \rangle$ denotes the standard inner product of $x = (x^1, x^2, \dots, x^d), y = (y^1, y^2, \dots, y^d) \in \mathbb{R}^d$ and $|x|$ denotes the norm of x:

$$\langle x, y \rangle = \sum_{i=1}^d x^i y^i, \qquad |x| = \Big\{\sum_{i=1}^d (x^i)^2\Big\}^{\frac{1}{2}}.$$

The next assertion is called **Kolmogorov's continuity theorem**.

Theorem 1.1.8 *Let $X = \{X(t)\}_{t \geqq 0}$ be an $\mathbb{R}^d$-valued stochastic process defined on a probability space $(\Omega, \mathscr{F}, \mathbf{P})$, and assume that for any $T > 0$ there exist positive constants α, β, C such that*

$$\mathbf{E}[|X(t) - X(s)|^\alpha] \leqq C(t-s)^{1+\beta} \qquad (0 \leqq s < t \leqq T).$$

Then, there exists a modification X' of X which satisfies

$$P\Big(\lim_{h \downarrow 0} \sup_{\substack{0 \leqq s < t \leqq T \\ t-s<h}} \frac{|X'(t) - X'(s)|}{(t-s)^\varepsilon} = 0\Big) = 1 \tag{1.1.2}$$

for any $\varepsilon \in (0, \frac{\beta}{\alpha})$. In particular, X' is continuous.

For a proof, we refer to [56, 62]. Kolmogorov's continuity theorem for random fields (Theorem A.5.1) should also be studied.

1.2 Wiener Space

We fix $T > 0$ and let $W_T = W_T(\mathbb{R}^d)$ be the set of $\mathbb{R}^d$-valued continuous functions w on $[0, T]$ with $w(0) = 0$:

$$W_T = \{w : [0, T] \to \mathbb{R}^d;\ w \text{ is continuous and } w(0) = 0\}.$$

As was mentioned in the previous section, the space W_T is endowed with the topology of uniform convergence. We denote the Borel σ-field with respect to this topology by $\mathscr{B}(W_T)$, omitting $\mathbb{R}^d$.

The next theorem due to Wiener is one of the starting points of modern probability theory.

Theorem 1.2.1 *There exists a unique probability measure μ_T on W_T satisfying*

$$\mu_T(\{w \in W_T;\ w(t_1) \in A_1, w(t_2) \in A_2, \ldots, w(t_n) \in A_n\})$$
$$= \int_{A_1} \mathrm{d}x_1 \int_{A_2} \mathrm{d}x_2 \cdots \int_{A_n} \mathrm{d}x_n \prod_{j=1}^n g(t_j - t_{j-1}, x_{j-1}, x_j) \tag{1.2.1}$$

for $0 = t_0 < t_1 < \cdots < t_n$ and $A_1, \ldots, A_n \in \mathscr{B}(\mathbb{R}^d)$. Here $x_0 = 0$ and the function $g(t, x, y)$ is given by

$$g(t, x, y) = \frac{1}{(2\pi t)^{\frac{d}{2}}} \mathrm{e}^{-\frac{|y-x|^2}{2t}}. \tag{1.2.2}$$

Definition 1.2.2 μ_T is called the d-dimensional **Wiener measure** and the probability space $(W_T, \mathscr{B}(W_T), \mu_T)$ is called the d-dimensional **Wiener space**

on $[0, T]$. Moreover, we call measurable functions (random variables) defined on the Wiener space **Wiener functionals**.

The function $g(t, x, y)$ is called the **Gauss kernel** or **heat kernel**. It is the fundamental solution for the heat equation. That is, it satisfies $\frac{\partial g}{\partial t} = \frac{1}{2}\Delta_x g$, Δ_x being the Laplacian acting on functions in x, and the function $u(t, x)$ defined by

$$u(t, x) = \int_{\mathbb{R}^d} g(t, x, y) f(y)\,\mathrm{d}y$$

for a bounded continuous function $f : \mathbb{R}^d \to \mathbb{R}$ satisfies

$$\frac{\partial u}{\partial t} = \frac{1}{2}\Delta_x u, \qquad \lim_{t\downarrow 0} u(t, x) = f(x).$$

Note that $u(t, x)$ is the expectation of the random variable $f(x + w(t))$ defined on $(W_T, \mathscr{B}(W_T), \mu_T)$. Moreover, $g(t, x, y)$ satisfies the equation

$$\int_{\mathbb{R}^d} g(t, x, y) g(s, y, z)\,\mathrm{d}y = g(t + s, x, z).$$

This relationship is known as the **Chapman–Kolmogorov equation**. We will discuss in detail the relationship between Brownian motions and Laplacians in Chapter 3.

When we need to write explicitly the dimension d of the Wiener space, we denote it by $(W_T^d, \mathscr{B}(W_T^d), \mu_T^d)$. We omit T when there is no fear of confusion.

The uniqueness of the Wiener measure follows from Proposition 1.1.5. We will mention a method to construct the Wiener measure after introducing a Hilbert space which plays an important role in stochastic analysis. But, before it, we show how to construct it via **random walks**.

It is sufficient to consider the case where $d = 1$ and $T = 1$. Let $\{\xi_i\}_{i=1}^\infty$ be a sequence of independent and identically distributed random variables taking values ± 1 with probability $\frac{1}{2}$. The discrete time stochastic process $\{S_k\}_{k=0}^\infty$ defined by $S_0 = 0$ and $S_k = \xi_1 + \cdots + \xi_k$ is a simple random walk. We define a sequence $\{X_n(t)\}_{0\leqq t\leqq 1}$ $(n = 1, 2, \ldots)$ of continuous stochastic processes by

$$X_n\Big(\frac{k}{n}\Big) = \frac{\xi_1 + \cdots + \xi_k}{\sqrt{n}}$$

for $t = \frac{k}{n}$ $(k = 1, 2, \ldots, n)$ and, for $t \in [\frac{k}{n}, \frac{k+1}{n}]$,

$$X_n(t) = X_n\Big(\frac{k}{n}\Big) + n\Big(t - \frac{k}{n}\Big)\Big(X_n\Big(\frac{k+1}{n}\Big) - X_n\Big(\frac{k}{n}\Big)\Big).$$

Then the probability measure on W_1 induced by $\{X_n(t)\}$ converges weakly to the Wiener measure μ_1. This is the so-called Donsker's invariance principle (see, for example, [56]).

Let H_T be the subset of the d-dimensional Wiener space W_T which consists of $h = (h_1, \ldots, h_d) \in W_T$ such that each component h_i of h is absolutely continuous and has square-integrable derivative $\dot{h}_i$ on $[0, T]$. We call H_T the **Cameron–Martin subspace**. We denote it also by H_T^d.

The inner product $\langle h_1, h_2 \rangle_{H_T}$ is defined by

$$\langle h_1, h_2 \rangle_{H_T} = \int_0^T \langle \dot{h}_1(t), \dot{h}_2(t) \rangle \, \mathrm{d}t \qquad (h_1, h_2 \in H_T),$$

where $\langle \cdot\,, \,\cdot \rangle$ on the right hand side is the Euclidean inner product. Then H_T is a real separable Hilbert space. We denote the norm on H_T by

$$\|h\|_{H_T} = (\langle h, h \rangle_{H_T})^{\frac{1}{2}}.$$

H_T is embedded in W_T densely and continuously. Hence, by identifying the dual space H_T^* [1] with H_T by the Riesz theorem, we have

$$W_T^* \subset H_T^* = H_T \subset W_T \qquad \text{(densely and continuously)}.$$

Proposition 1.2.3 *If $h \in H_T^*$ is of C^2-class, then $h \in W_T^*$ and, denoting by $\ddot{h}$ the second derivative, we have*

$$h(w) = \langle w(T), \dot{h}(T) \rangle - \int_0^T \langle w(t), \ddot{h}(t) \rangle \, \mathrm{d}t \qquad (w \in W_T). \tag{1.2.3}$$

Proof We define $\widetilde{h} \in W_T^*$ by the right hand side of (1.2.3). Then, for any $g \in H_T$,

$$\widetilde{h}(g) = \int_0^T \langle \dot{h}(t), \dot{g}(t) \rangle \, \mathrm{d}t = h(g),$$

which implies $\widetilde{h} = h$ in W_T^* because H_T is dense in W_T. □

Remark 1.2.4 If $h \in H_T^*$ is of piecewise C^2-class, that is, if there exists a sequence $0 = t_0 < t_1 < \cdots < t_N = T$ such that the restriction of h to (t_i, t_{i+1}) is of C^2-class, then $h \in W_T^*$ and

$$h(w) = \sum_{i=1}^N \Big\{ \langle w(t_i), \dot{h}(t_i - 0) \rangle - \langle w(t_{i-1}), \dot{h}(t_{i-1} + 0) \rangle - \int_{t_{i-1}}^{t_i} \langle w(t), \ddot{h}(t) \rangle \, \mathrm{d}t \Big\}.$$

Wiener constructed the Wiener measure by means of the Fourier expansion (see Example 1.2.6 below) and Lévy did so by the expansion using Schauder

[1] The set of continuous linear functionals on a linear topological space E is called the dual (or conjugate) space, which is denoted by E^* in this book.

functions (Example 1.2.7). We can understand these constructions in a unified manner by the celebrated Itô–Nisio theorem ([51]). We present it without proof.

Theorem 1.2.5 *Let $\{h_n\}_{n=1}^{\infty}$ be an orthonormal basis of the Cameron–Martin subspace H_T and $\{X_n\}_{n=1}^{\infty}$ be a sequence of independent and standard-normally distributed random variables defined on a probability space $(\Omega, \mathscr{F}, \mathbf{P})$. Then the sequence $\{S_n\}_{n=1}^{\infty}$ of W_T-valued random variables given by $S_n = \sum_{i=1}^{n} X_i h_i$ converges with probability* 1 *and the distribution of the limit (a probability measure on W_T) satisfies* (1.2.1).

Example 1.2.6 Let $d = 1$ and put

$$\psi_0(t) = \frac{t}{\sqrt{T}}, \quad \psi_n(t) = \frac{\sqrt{2T}}{n\pi} \sin\Big(\frac{n\pi t}{T}\Big) \quad (n \geqq 1).$$

Then $\{\psi_n\}_{n=0}^{\infty}$ is an orthonormal basis of H_T^1. Hence, letting $\{X_n\}_{n=0}^{\infty}$ be the same sequence as in Theorem 1.2.5 and setting

$$B(t) = \frac{X_0}{\sqrt{T}} t + \sum_{n=1}^{\infty} X_n \frac{\sqrt{2T}}{n\pi} \sin\Big(\frac{n\pi t}{T}\Big),$$

we see that the probability measure on W_T^1 induced by $\{B(t)\}_{0 \leqq t \leqq T}$ is the Wiener measure. This is called the Fourier expansion of Brownian motion.

Example 1.2.7 Letting $\boldsymbol{e}^1, \ldots, \boldsymbol{e}^d$ be the standard basis of $\mathbb{R}^d$, we define the $\mathbb{R}^d$-valued functions $\{h_j^i\}_{i=1,\ldots,d,\, j=0,1,\ldots}$ by $h_j^i(0) = 0$,

$$\dot{h}_0^i(t) = \frac{1}{\sqrt{T}} \boldsymbol{e}^i,$$

for $j = 0$ and

$$\dot{h}_{2^n+k}^i(t) = \begin{cases} \sqrt{\frac{2^n}{T}} \boldsymbol{e}^i & (\frac{k}{2^n} T \leqq t \leqq \frac{2k+1}{2^{n+1}} T) \\ -\sqrt{\frac{2^n}{T}} \boldsymbol{e}^i & (\frac{2k+1}{2^{n+1}} T \leqq t \leqq \frac{k+1}{2^n} T) \\ 0 & (\text{otherwise}) \end{cases}$$

for $j = 2^n + k$ $(k = 0, 1, \ldots, 2^n - 1,\ n = 0, 1, \ldots)$. $\{h_j^i\}$ forms an orthonormal basis of H_T^d. Lévy constructed the Wiener measure by using $\{h_j^i\}$.[2]

[2] $\dot{h}_j^i$ is called the **Haar function** and h_j^i is called the **Schauder function**.

The function $t \mapsto h^i_{2^n+k}(t)$ vanishes outside the interval $[\frac{k}{2^n}T, \frac{k+1}{2^n}T]$. It is piecewise linear on this interval and takes the maximum $\sqrt{\frac{T}{2^{n+2}}}$ at the midpoint. Moreover, since it is of piecewise C^2-class, it belongs to W_T^*.

We define $\ell_m(w) \in W_T$ by

$$\ell_m(w)(t) = \sum_{i=1}^{d} \sum_{j=0}^{2^m-1} \langle h^i_j, w\rangle h^i_j(t).$$

Then the function $t \mapsto \ell_m(w)(t)$ is piecewise linear and $\ell_m(w)(\frac{k}{2^m}T) = w(\frac{k}{2^m}T)$. Hence $\ell_m(w)$ is the piecewise linear function which connects $w(\frac{k}{2^m}T)$ $(k = 0, 1, \dots, 2^m)$ in order and Lévy's construction gives us the piecewise linear approximation of $w \in W$.

Next we consider the path space with the time interval $[0, \infty)$:

$$W = \{w : [0, \infty) \to \mathbb{R}^d;\ w \text{ is continuous and } w(0) = 0\}.$$

As mentioned in the previous section, we endow W with the topology of uniform convergence on compact sets. Note that a similar assertion to Proposition 1.1.5 holds. In this case we call the following space H the **Cameron–Martin subspace**:

$$H = \{h \in W;\ h \text{ is absolutely continuous and } \dot{h} \in L^2([0, \infty))\}.$$

For $T > 0$ we let $\varphi_T(w)$ be the restriction of $w \in W$ to $[0, T]$. Then, there exists a probability measure μ on W whose image measure under φ_T is the Wiener measure on W_T. We also call μ the Wiener measure and the probability space $(W, \mathscr{B}(W), \mu)$ the d-dimensional Wiener space on $[0, \infty)$.

We show some basic properties of the Wiener measure.

Theorem 1.2.8 (1) *Define the transforms* ψ_c $(c > 0)$, ϕ_s $(s > 0)$ *and* φ_Q *(*Q *is a* d*-dimensional orthogonal matrix) on* W *by*

$$\psi_c(w)(t) = \frac{1}{c}w(c^2t), \quad \phi_s(w)(t) = w(s+t) - w(s), \quad \varphi_Q(w)(t) = Qw(t).$$

Then μ *is invariant under* $\psi_c, \phi_s, \varphi_Q$.
(2) *Define the transform* Φ_T *on* W_T *by*

$$\Phi_T(w)(t) = w(T-t) - w(T) \quad (0 \leqq t \leqq T).$$

Then μ_T *is invariant under* Φ_T.

Proof To see the invariance under ψ_c, we have only to show

$$\mu\left(\left\{w;\ \frac{1}{c}w(c^2t_1) \in A_1, \frac{1}{c}w(c^2t_2) \in A_2, \ldots, \frac{1}{c}w(c^2t_n) \in A_n\right\}\right)$$
$$= \mu(\{w;\ w(t_1) \in A_1, w(t_2) \in A_2, \ldots, w(t_n) \in A_n\})$$

for $t_1 < t_2 < \cdots < t_n$, $A_1, A_2, \ldots, A_n \in \mathscr{B}(\mathbb{R}^d)$ by using the identity $g(t, x, y) = c^d \cdot g(c^2t, cx, cy)$ $(c > 0)$. The other assertions are shown similarly. □

Remark 1.2.9 Set $\Psi(w)(0) = 0$, $\Psi(w)(t) = tw(\frac{1}{t})$ $(t > 0)$. Then $\Psi(w)$ is continuous at $t = 0$ almost surely under the Wiener measure. Hence, we may regard Ψ as a transform of the Wiener space and can prove that the Wiener measure is invariant under Ψ in the same way as in Theorem 1.2.8.

Definition 1.2.10 A d-dimensional continuous stochastic process $\{X(t)\}_{t \geqq 0}$ defined on a probability space $(\Omega, \mathscr{F}, \mathbf{P})$ is called a d-dimensional **Brownian motion** (or **Wiener process**) starting from 0 if $X(0) = 0$ and its probability distribution on W is the Wiener measure.

1.3 Filtered Probability Space, Adapted Stochastic Process

Let $(W, \mathscr{B}(W), \mu)$ be the d-dimensional Wiener space on $[0, \infty)$ and define a function $\theta(s)$ $(s \geqq 0)$ on W by $\theta(s)(w) = w(s)$. Then the smallest σ-field which makes the behavior of each $w \in W$ up to time t measurable is

$$\mathscr{B}_t^0 := \sigma(\{\theta(s)^{-1}(A); A \in \mathscr{B}(\mathbb{R}^d),\ 0 \leqq s \leqq t\}) \tag{1.3.1}$$

and it forms an increasing sequence $\{\mathscr{B}_t^0\}_{t \geqq 0}$ of sub-σ-fields of $\mathscr{B}(W)$. Moreover, setting

$$\mathscr{B}_t = \bigcap_{u>t} \mathscr{B}_u^0,$$

we see that $\{\mathscr{B}_t\}_{t \geqq 0}$ is right-continuous, that is, $\mathscr{B}_{t+0} := \bigcap_{u>t} \mathscr{B}_u = \mathscr{B}_t$.[3]

The stochastic process $\{\theta(t)\}_{t \geqq 0}$ defined on the Wiener space W is called the **coordinate process**.

Definition 1.3.1 A quartet $(\Omega, \mathscr{F}, \mathbf{P}, \{\mathscr{F}_t\})$ of a probability space $(\Omega, \mathscr{F}, \mathbf{P})$ and an increasing sequence $\{\mathscr{F}_t\}_{t \geqq 0}$ of sub-σ-fields of $\mathscr{F}$ is called a **filtered probability space**.

[3] $\{\mathscr{B}_t^0\}$ is not right-continuous. For example, for a function ϕ on $[0, \infty)$, the random variable $\limsup_{u \downarrow t} \frac{w(u)-w(t)}{\phi(u-t)}$ is not $\mathscr{B}_t^0$-measurable, but it is $\mathscr{B}_t$-measurable.

$\{\mathscr{F}_t\}$ is called a **filtration**. If $\{\mathscr{F}_t\}$ is right-continuous and each $\mathscr{F}_t$ contains all **P**-null sets, $\{\mathscr{F}_t\}$ is said to satisfy the **usual condition**.

Next we mention the measurability of stochastic processes. While the purpose of this book is to develop analysis of continuous stochastic processes, we need to consider stochastic processes with discontinuous paths. Intuitive understanding is enough for our purpose and we do not discuss in detail but refer to [45, 56, 86, 114] and so on.

Let E be a complete separable metric space.

Definition 1.3.2 Let $X = \{X(t)\}_{t \geqq 0}$ be an E-valued stochastic process defined on a filtered probability space $(\Omega, \mathscr{F}, \mathbf{P}, \{\mathscr{F}_t\})$.
(1) X is called **measurable** if $X(t)(\omega)$ is $\mathscr{B}([0,\infty)) \times \mathscr{F}$-measurable as a function of (t, ω).
(2) X is $\{\mathscr{F}_t\}$-**adapted** if the E-valued random variable $X(t)$ is $\mathscr{F}_t$-measurable for each t.
(3) X is $\{\mathscr{F}_t\}$-**progressively measurable** if the map $[0,t] \times \Omega \ni (s, \omega) \mapsto X(s)(\omega) \in E$ is $\mathscr{B}([0,t]) \times \mathscr{F}_t$-measurable for each t.

We often write $X(s, \omega)$ for the value $X(s)(\omega)$ of the sample process at s, regarding it as a function in two variables (s, ω).

Proposition 1.3.3 *If an E-valued $\{\mathscr{F}_t\}$-adapted stochastic process X defined on a filtered probability space $(\Omega, \mathscr{F}, \mathbf{P}, \{\mathscr{F}_t\})$ has right-continuous paths, that is, if for any $\omega \in \Omega$ the map $t \mapsto X(t)(\omega)$ is right-continuous, then X is $\{\mathscr{F}_t\}$-progressively measurable.*

Proof Fix $t > 0$ and put for $n = 1, 2, \ldots$

$$X^{(n)}(0) = X(0),$$

$$X^{(n)}(s) = \sum_{j=0}^{\infty} X\Big(\frac{(j+1)t}{n}\Big) \mathbf{1}_{[\frac{jt}{n}, \frac{(j+1)t}{n})}(s) \qquad (s > 0).$$

Then, letting $j = j(s)$ ($s \in [0,t]$) be the integer such that $\frac{jt}{n} < s \leqq \frac{(j+1)t}{n}$, we see that both

$$[0,t] \times \Omega \ni (s, \omega) \mapsto X\Big(\frac{(j+1)t}{n}\Big)(\omega) \quad \text{and} \quad \mathbf{1}_{[\frac{jt}{n}, \frac{(j+1)t}{n})}(s)$$

are $\mathscr{B}([0,t]) \times \mathscr{F}_t$-measurable. Since $\lim_{n\to\infty} X^{(n)}(s)(\omega) = X(s)(\omega)$ for any (s, ω) by the right-continuity of $\{X(s)\}$, $X(s)(\omega)$ is $\mathscr{B}([0,t]) \times \mathscr{F}_t$-measurable. □

We end this section with an explanation on stopping times.

Definition 1.3.4 A $[0, \infty]$-valued random variable τ defined on a filtered probability space $(\Omega, \mathscr{F}, \mathbf{P}, \{\mathscr{F}_t\})$ is called an $\{\mathscr{F}_t\}$-**stopping time** if

$$\{\omega; \tau(\omega) \leqq t\} \in \mathscr{F}_t \quad (t \geqq 0).$$

Proposition 1.3.5 *For a filtered probability space $(\Omega, \mathscr{F}, \mathbf{P}, \{\mathscr{F}_t\})$ satisfying the usual condition, τ is an $\{\mathscr{F}_t\}$-stopping time if and only if $\{\omega; \tau(\omega) < t\} \in \mathscr{F}_t$ holds for any $t > 0$.*

Theorem 1.3.6 *For an E-valued $\{\mathscr{F}_t\}$-adapted continuous stochastic process $X = \{X(t)\}_{t \geqq 0}$ defined on a filtered probability space $(\Omega, \mathscr{F}, \mathbf{P}, \{\mathscr{F}_t\})$ which satisfies the usual condition, let τ_A be the* **first hitting time** *to a Borel set $A \in \mathscr{B}(E)$,*

$$\tau_A(\omega) = \inf\{t > 0; X(t)(\omega) \in A\},$$

where we put $\tau_A(\omega) = \infty$ if $X(t)(\omega) \notin A$ for all t. Then τ_A is an $\{\mathscr{F}_t\}$-stopping time.

We omit proofs of Proposition 1.3.5 and Theorem 1.3.6. See, for example, [56, 86].

1.4 Discrete Time Martingales

The notion of martingales was introduced by Ville and Lévy. The theory developed by Doob ([15]) plays a fundamental role in stochastic analysis. While we are mainly concerned with continuous time stochastic processes in this book, we mention in this section about discrete time martingales, the theory of which is a basis of that on continuous time martingales.

1.4.1 Conditional Expectation

We recall the conditional expectation and its properties. Let $(\Omega, \mathscr{F}, \mathbf{P})$ be a probability space and $\mathscr{G}$ be a sub-σ-field of $\mathscr{F}$. For an integrable random variable X, there exists a unique, up to the difference on $\mathbf{P}$-null sets, integrable $\mathscr{G}$-measurable random variable $\widehat{X}$ satisfying

$$\mathbf{E}[XY] = \mathbf{E}[\widehat{X}Y]$$

for any bounded $\mathscr{G}$-measurable random variable Y. We call $\widehat{X}$ the **conditional expectation** of X with respect to $\mathscr{G}$ and denote it by $\mathbf{E}[X|\mathscr{G}]$.

Theorem 1.4.1 *Let $X, X', X_1, X_2, \ldots$ be integrable random variables defined on a probability space $(\Omega, \mathscr{F}, \mathbf{P})$ and $\mathscr{G}$ be a sub-σ-field of $\mathscr{F}$.*
(1) *If X is $\mathscr{G}$-measurable, then* $\mathbf{E}[X|\mathscr{G}] = X$, $\mathbf{P}$-a.s. *where* $\mathbf{P}$-a.s. *means except for a set of* $\mathbf{P}$*-probability zero.*
(2) $\mathbf{E}[\mathbf{E}[X|\mathscr{G}]] = \mathbf{E}[X]$.
(3) [linearity] *For any $a, b \in \mathbb{R}$*

$$\mathbf{E}[aX + bX'|\mathscr{G}] = a\mathbf{E}[X|\mathscr{G}] + b\mathbf{E}[X'|\mathscr{G}], \quad \mathbf{P}\text{-a.s.}$$

(4) [positivity] *If* $X \geqq 0$, $\mathbf{P}$-a.s., *then* $\mathbf{E}[X|\mathscr{G}] \geqq 0$, $\mathbf{P}$-a.s.
(5) *If Y is $\mathscr{G}$-measurable and $XY \in L^1(\mathbf{P})$, then* $\mathbf{E}[XY|\mathscr{G}] = Y\mathbf{E}[X|\mathscr{G}]$, $\mathbf{P}$-a.s.
(6) [tower property] *If $\mathscr{H}$ is a sub-σ-field of $\mathscr{G}$, then*

$$\mathbf{E}[\mathbf{E}[X|\mathscr{G}]|\mathscr{H}] = \mathbf{E}[X|\mathscr{H}], \quad \mathbf{P}\text{-a.s.}$$

(7) [Jensen's inequality] *If $\varphi : \mathbb{R} \to \mathbb{R}$ is convex and $\varphi(X) \in L^1(\mathbf{P})$, then*

$$\varphi(\mathbf{E}[X|\mathscr{G}]) \leqq \mathbf{E}[\varphi(X)|\mathscr{G}], \quad \mathbf{P}\text{-a.s.}$$

In particular, if $X \in L^p(\mathbf{P})$, $p \geqq 1$, then

$$|\mathbf{E}[X|\mathscr{G}]|^p \leqq \mathbf{E}[|X|^p|\mathscr{G}], \quad \mathbf{P}\text{-a.s.}$$

(8) [Fatou's lemma] *If $X_n \geqq 0$ $(n = 1, 2, \ldots)$, then*

$$\mathbf{E}\Big[\liminf_{n\to\infty} X_n \Big| \mathscr{G}\Big] \leqq \liminf_{n\to\infty} \mathbf{E}[X_n|\mathscr{G}], \quad \mathbf{P}\text{-a.s.}$$

(9) [monotone convergence theorem] *If X_n $(n = 1, 2, \ldots)$ is monotone increasing in n and converges to a random variable X,* $\mathbf{P}$-a.s., *then* $\mathbf{E}[X_n|\mathscr{G}]$ *also converges to* $\mathbf{E}[X|\mathscr{G}]$ *almost surely.*
(10) [Lebesgue's convergence theorem] *If X_n converges to X,* $\mathbf{P}$-a.s. *and if there exists an integrable non-negative random variable Y such that $|X_n| \leqq Y$,* $\mathbf{P}$-a.s. $(n = 1, 2, \ldots)$, *then* $\mathbf{E}[X_n|\mathscr{G}]$ *converges to* $\mathbf{E}[X|\mathscr{G}]$, $\mathbf{P}$-a.s.
(11) *Let $p \geqq 1$. If X_n converges to X in L^p, then* $\mathbf{E}[X_n|\mathscr{G}]$ *converges to* $\mathbf{E}[X_n|\mathscr{G}]$ *in L^p.*
(12) *If X is independent of $\mathscr{G}$, then* $\mathbf{E}[X|\mathscr{G}] = \mathbf{E}[X]$, $\mathbf{P}$-a.s.

Proposition 1.4.2 *Let X, Y be random variables defined on a probability space $(\Omega, \mathscr{F}, \mathbf{P})$ and $\mathscr{G}$ be sub-σ-field of $\mathscr{F}$. Assume that X is independent of $\mathscr{G}$ and Y is $\mathscr{G}$-measurable. Then, setting* $\varphi(y) = \mathbf{E}[g(X, y)]$ $(y \in \mathbb{R})$ *for a bounded Borel-measurable function $g : \mathbb{R} \times \mathbb{R} \to \mathbb{R}$, we have*

$$\varphi(Y) = \mathbf{E}[g(X, y)|\mathscr{G}]\Big|_{y=Y} = \mathbf{E}[g(X, Y)|\mathscr{G}], \quad \mathbf{P}\text{-a.s.}$$

We refer the reader to [15, 45, 114, 126] and so on for more about the conditional expectation.

1.4.2 Martingales, Doob Decomposition

Let $\{\mathscr{F}_n\}_{n=0}^\infty$ be an increasing sequence of sub-σ-fields of $\mathscr{F}$.

Definition 1.4.3 Let $X = \{X_n\}_{n=0}^\infty$ be an $\mathbb{R}$-valued stochastic process and assume that it is $\{\mathscr{F}_n\}$-adapted, that is, X_n is $\mathscr{F}_n$-measurable for each n. Then, X is called an $\{\mathscr{F}_n\}$-**martingale** or simply a martingale, if X_n is integrable for each n and if

$$\mathbf{E}[X_{n+1}|\mathscr{F}_n] = X_n, \quad \mathbf{P}\text{-a.s.}$$

X is called a **submartingale** if the inequality $\mathbf{E}[X_{n+1}|\mathscr{F}_n] \geqq X_n$, $\mathbf{P}$-a.s. holds in place of the equality. X is called a **supermartingale** if the converse inequality holds.

If X is a submartingale, $\mathbf{E}[X_m|\mathscr{F}_n] \geqq X_n$, $\mathbf{P}$-a.s. for $m > n$. The following two propositions are easily proven.

Proposition 1.4.4 *If X is a submartingale, $\mathbf{E}[X_n]$ is monotone increasing in n. If X is a martingale, $\mathbf{E}[X_n]$ is a constant independent of n.*

Proposition 1.4.5 (1) *For a martingale X and a convex function φ, $\{\varphi(X_n)\}_{n=0}^\infty$ is a submartingale if $\varphi(X_n)$ is integrable for every n.*
(2) *For a submartingale X and a convex and monotone decreasing function φ, $\{\varphi(X_n)\}_{n=0}^\infty$ is a submartingale if $\varphi(X_n)$ is integrable for every n.*

Example 1.4.6 A simple random walk is a martingale. In fact, let $\{\xi_n\}_{n=1}^\infty$ be a sequence of independent identically distributed random variables such that $\mathbf{P}(\xi_n = \pm 1) = \frac{1}{2}$, and put $\mathscr{F}_0 = \{\emptyset, \Omega\}$ and $\mathscr{F}_n = \sigma\{\xi_1, \xi_2, \ldots, \xi_n\}$, the smallest σ-field which makes $\xi_1, \xi_2, \ldots, \xi_n$ measurable. Set $S_0 = 0$ and $S_n = \xi_1 + \xi_2 + \cdots + \xi_n$ $(n = 1, 2, \ldots)$. Then $\{S_n\}_{n=0}^\infty$ is an $\{\mathscr{F}_n\}$-martingale.

Moreover, set $f_1 = 1$ and

$$f_n = \begin{cases} 1 & (\text{if } \xi_{n-1} = 1), \\ 2^k & (\text{if } \xi_{n-1} = \xi_{n-2} = \cdots = \xi_{n-k} = -1,\ \xi_{n-k-1} = 1), \end{cases} \qquad (n = 2, 3, \ldots),$$

and define a stochastic process $Y = \{Y_n\}_{n=0}^\infty$ by

$$Y_0 = 0 \quad \text{and} \quad Y_n = f_1\xi_1 + f_2\xi_2 + \cdots + f_n\xi_n.$$

Then f_n is $\mathscr{F}_{n-1}$-measurable and Y is a martingale.

Imagine that we make a sequence of fair gambles and bet one dollar the first time. The martingale Y in Example 1.4.6 represents the gain (loss) at time n

when a gambler bets one dollar after a win and twice the previous one after a loss. Such a betting strategy is called a "martingale".

Definition 1.4.7 A stochastic process $f = \{f_n\}_{n=1}^{\infty}$is called **predictable** if f_n is $\mathscr{F}_{n-1}$-measurable for any $n = 1, 2, \ldots$ Setting $f_0 = 0$, we often consider $\{f_n\}_{n=0}^{\infty}$.

As in Example 1.4.6, we can construct a new martingale from a martingale and a predictable process. This gives an original form of the stochastic integrals.

Proposition 1.4.8 *Let $X = \{X_n\}_{n=0}^{\infty}$ be a martingale and $f = \{f_n\}_{n=1}^{\infty}$ be a predictable process. Then, if $f_n(X_n - X_{n-1})$ is integrable for each n, the stochastic process $Z = \{Z_n\}_{n=0}^{\infty}$ defined by*

$$Z_0 = 0 \quad and \quad Z_n = \sum_{k=1}^{n} f_k(X_k - X_{k-1}) \quad (n = 1, 2, \ldots) \tag{1.4.1}$$

is an $\{\mathscr{F}_n\}$-martingale.

Z is called the **martingale transform** of X by f. Together with the following proposition, we leave the proofs to the reader.

Proposition 1.4.9 *Let $X = \{X_n\}_{n=0}^{\infty}$ be a submartingale and $f = \{f_n\}_{n=1}^{\infty}$ be a non-negative, bounded predictable process. Then, the stochastic process $Z = \{Z_n\}_{n=0}^{\infty}$ defined by* (1.4.1) *is an $\{\mathscr{F}_n\}$-submartingale.*

Remark 1.4.10 Let $\{\mathscr{F}_n\}$ and $S = \{S_n\}_{n=1}^{\infty}$ be as in Example 1.4.6. Then, any $\{\mathscr{F}_n\}$-martingale $X = \{X_n\}_{n=1}^{\infty}$ is given by a martingale transform of S.

To see this, fix n. Since X_n is $\mathscr{F}_n$-measurable, there exists a function $\Phi_n : \{-1, 1\}^n \to \mathbb{R}$ such that $X_n = \Phi_n(\xi_1, \ldots, \xi_n)$. Then, by Proposition 1.4.2, we have that

$$\begin{aligned} X_n - X_{n-1} &= \Phi_n(\xi_1, \ldots, \xi_n) - \mathbf{E}[\Phi_n(x_1, \ldots, x_{n-1}, \xi_n)]\big|_{x_1=\xi_1,\ldots,x_{n-1}=\xi_{n-1}} \\ &= \frac{1}{2}\xi_n\{\Phi_n(\xi_1, \ldots, \xi_{n-1}, 1) - \Phi_n(\xi_1, \ldots, \xi_{n-1}, -1)\}, \end{aligned}$$

where we have used

$$\mathbf{E}[\Phi_n(x_1, \ldots, x_{n-1}, \xi_n)] = \frac{1}{2}\{\Phi_n(x_1, \ldots, x_{n-1}, 1) + \Phi_n(x_1, \ldots, x_{n-1}, -1)\}.$$

Thus, setting $f_n = \frac{1}{2}\{\Phi_n(x_1, \ldots, x_{n-1}, 1) - \Phi_n(x_1, \ldots, x_{n-1}, -1)\}$, we obtain the desired expression.

It should be remarked that, for every $\{\mathscr{F}_n\}$-martingale X, each X_n is bounded as seen in the above paragraph.

Let $\{M_n\}_{n=0}^{\infty}$ be an $\{\mathscr{F}_n\}$-martingale and $\{A_n\}_{n=0}^{\infty}$ be a predictable increasing process ($A_0 \leqq A_1 \leqq A_2 \leqq \cdots$) with $A_0 = 0$. Then, the stochastic process $X = \{X_n\}_{n=1}^{\infty}$ given by $X_n = M_n + A_n$ is an $\{\mathscr{F}_n\}$-submartingale.

The next theorem shows that any submartingale has such a decomposition, which is called the **Doob decomposition**.

Theorem 1.4.11 *Let $X = \{X_n\}_{n=0}^{\infty}$ be an $\{\mathscr{F}_n\}$-adapted integrable stochastic process. Then there exists an $\{\mathscr{F}_n\}$-martingale $M = \{M_n\}_{n=0}^{\infty}$ and a predictable process $A = \{A_n\}_{n=0}^{\infty}$ with $A_0 = 0$ such that $X_n = M_n + A_n$ and the decomposition is unique. In particular, if X is a submartingale, A is an increasing process.*

Proof At first we show the uniqueness. Let us assume that X_n has two decompositions

$$X_n = M_n + A_n = M_n' + A_n',$$

where $\{M_n\}, \{M_n'\}$ are martingales and $\{A_n\}, \{A_n'\}$ are predictable. Then, since $A_n - A_n' = M_n' - M_n$,

$$\mathbf{E}[A_n - A_n' | \mathscr{F}_{n-1}] = A_{n-1} - A_{n-1}'.$$

On the other hand, since $A_n - A_n'$ is $\mathscr{F}_{n-1}$-measurable, $A_n - A_n' = A_{n-1} - A_{n-1}'$. Hence, $A_n - A_n' = A_0 - A_0' = 0$ and $A_n = A_n'$, $M_n = M_n'$.

Next we show the existence of the decomposition. Put $M_0 = X_0$ and

$$M_n = \sum_{k=1}^{n} (X_k - \mathbf{E}[X_k | \mathscr{F}_{k-1}]) \qquad (n = 1, 2, \ldots).$$

Then $\{M_n\}_{n=0}^{\infty}$ is an $\{\mathscr{F}_n\}$-martingale. Moreover, since

$$A_n := X_n - M_n = \mathbf{E}[X_n | \mathscr{F}_{n-1}] - \sum_{k=1}^{n-1} (X_k - \mathbf{E}[X_k | \mathscr{F}_{k-1}]) \qquad (n = 1, 2, \ldots)$$

is $\mathscr{F}_{n-1}$-measurable, $X_n = M_n + A_n$ is the desired decomposition.

If X is a submartingale, since

$$A_n - A_{n-1} = \mathbf{E}[X_n | \mathscr{F}_{n-1}] - X_{n-1} \geqq 0 \qquad (n = 1, 2, \ldots),$$

$\{A_n\}$ is increasing. □

Example 1.4.12 Let $X = \{X_n\}_{n=0}^{\infty}$ be the martingale in Example 1.4.6. Then $\{X_n^2\}_{n=0}^{\infty}$ is a submartingale and $\{X_n^2 - n\}_{n=0}^{\infty}$ is a martingale.

1.4.3 Optional Stopping Theorem

We define stopping times also for stochastic processes with discrete time parameters.

Definition 1.4.13 An $\mathbb{N} \cup \{0, \infty\}$-valued random variable τ defined on a probability space $(\Omega, \mathscr{F}, \mathbf{P}, \{\mathscr{F}_n\})$ is called an $\{\mathscr{F}_n\}$-**stopping time** if $\{\tau \leqq n\} \in \mathscr{F}_n$ for any n.

Remark 1.4.14 The condition in the definition above is equivalent to that $\{\tau = n\} \in \mathscr{F}_n$ for any n.

Example 1.4.15 Let $X = \{X_n\}_{n=0}^{\infty}$ be an $\mathbb{R}$-valued, $\{\mathscr{F}_n\}$-adapted stochastic process and $G \in \mathscr{B}(\mathbb{R})$. Set $\tau_G = \min\{n; X_n \in G\}$, the first hitting time to G, where $\tau_G = \infty$ if $X_n \notin G$ for any n. Then, τ_G is an $\{\mathscr{F}_n\}$-stopping time.

The next theorem shows that a submartingale stopped at a stopping time is again a submartingale.

Theorem 1.4.16 (Optional stopping theorem) *Let $X = \{X_n\}_{n=0}^{\infty}$ be a submartingale. Then, for any stopping time τ, $X^\tau = \{X_{n\wedge\tau}\}_{n=0}^{\infty}$ is also a submartingale.*

Proof Set $f_n = \mathbf{1}_{\{\tau \geqq n\}}$. Then, since $\{f_n = 0\} = \{\tau < n\} = \{\tau \leqq n-1\} \in \mathscr{F}_{n-1}$, $\{f_n\}_{n=1}^{\infty}$ is a predictable process and we have

$$X_{n\wedge\tau} = X_0 + \sum_{k=1}^{n} f_k(X_k - X_{k-1}).$$

Hence, by Proposition 1.4.9, X^τ is a submartingale. □

We give a σ-field which represents information up to a stopping time.

Proposition 1.4.17 *If τ is an $\{\mathscr{F}_n\}$-stopping time, a family $\mathscr{F}_\tau$ of subsets in $\mathscr{F}$ given by*

$$\mathscr{F}_\tau = \{A \in \mathscr{F}; A \cap \{\tau \leqq n\} \in \mathscr{F}_n, n = 0, 1, 2, \ldots\}$$

is a σ-field and τ is $\mathscr{F}_\tau$-measurable.

Proposition 1.4.18 *Let σ and τ be $\{\mathscr{F}_n\}$-stopping times. Then,*
(1) $\{\tau < \sigma\}, \{\tau = \sigma\}$ and $\{\tau \leqq \sigma\}$ belong to both $\mathscr{F}_\sigma$ and $\mathscr{F}_\tau$.
(2) If $\tau \leqq \sigma$, then $\mathscr{F}_\tau \subset \mathscr{F}_\sigma$.

The proofs of the two propositions are easy and are omitted.

1.4.4 Convergence Theorem

We show that, if a submartingale $X = \{X_n\}_{n=0}^{\infty}$ satisfies $\sup_n \mathbf{E}[X_n^+] < \infty$, then X_n converges $\mathbf{P}$-a.s. as $n \to \infty$. Hence, if X is a non-positive submartingale or a non-negative supermartingale, it converges almost surely. For this purpose we define the number of upcrossings and prove an inequality (Theorem 1.4.20) due to Doob.

Let $X = \{X_n\}_{n=0}^{\infty}$ be an $\{\mathscr{F}_n\}$-adapted stochastic process. We consider a bounded closed interval $[a, b]$ $(a < b)$ and define a sequence of stopping times defined by

$$\sigma_1 = \inf\{n \geqq 0; X_n \leqq a\}, \qquad \tau_1 = \inf\{n \geqq \sigma_1; X_n \geqq b\},$$
$$\sigma_k = \inf\{n \geqq \tau_{k-1}; X_n \leqq a\}, \quad \tau_k = \inf\{n \geqq \sigma_k; X_n \geqq b\}, \quad (k = 2, 3, \ldots).$$

Here, if the set $\{\cdots\}$ is empty, $\inf\{\cdots\} = \infty$. We put

$$\beta(a, b) = \sup\{k; \tau_k < \infty\}$$

and call it the **upcrossing number** of X of $[a, b]$. By definition,

$$\text{if } \liminf_{n\to\infty} X_n < a < b < \limsup_{n\to\infty} X_n, \text{ then } \beta(a, b) = \infty,$$
$$\text{if } \beta(a, b) = \infty, \text{ then } \liminf_{n\to\infty} X_n \leqq a < b \leqq \limsup_{n\to\infty} X_n.$$

Hence, a necessary and sufficient condition that X_n converges or diverges to $\pm\infty$ as $n \to \infty$ is that $\beta(a, b) < \infty$ for any rational a and b.[4]

Proposition 1.4.19 *Let* $X = \{X_n\}_{n=0}^{\infty}$ *be a submartingale. Set* $\beta_N(a, b) = \sup\{k; \tau_k \leqq N\}$ *for* $N = 1, 2, \ldots$ *Then,*

$$(b - a)\mathbf{E}[\beta_N(a, b)] \leqq \mathbf{E}[(X_N - a)^+] - \mathbf{E}[(X_0 - a)^+], \tag{1.4.2}$$

where $x^+ = \max\{x, 0\}$ *for* $x \in \mathbb{R}$.

Proof At first we prove the proposition under the additional condition that $X_n \geqq 0$ and $a = 0$. Define $\{f_i\}_{i=1}^{\infty}$ by

$$f_i = \begin{cases} 1 & (i \in \bigcup_{k=1}^{\infty} (\sigma_k, \tau_k]) \\ 0 & (i \in \bigcup_{k=1}^{\infty} (\tau_k, \sigma_{k+1}]). \end{cases}$$

Then $\{f_i\}_{i=1}^{\infty}$ is predictable. Moreover, setting

$$Y_0 = 0 \quad \text{and} \quad Y_n = \sum_{i=1}^{n} f_i (X_i - X_{i-1}) \quad (n = 1, 2, \ldots),$$

[4] $\liminf_{n\to\infty} X_n = a$ or $\limsup_{n\to\infty} X_n = b$ does not imply $\beta(a, b) = \infty$.

we obtain

$$\mathbf{E}[Y_N] \geqq b\mathbf{E}[\beta_N(0,b)] \tag{1.4.3}$$

since $Y_N = \sum_{k=1}^{\infty}(X_{\tau_k \wedge N} - X_{\sigma_k \wedge N})$ and $Y_N \geqq b\beta_N(0,b)$.

On the other hand, since

$$X_n - X_0 - Y_n = \sum_{i=1}^{n}(1 - f_i)(X_i - X_{i-1}),$$

$\{X_n - X_0 - Y_n\}_{n=0}^{\infty}$ is a submartingale by Proposition 1.4.9. In particular, we have $\mathbf{E}[X_N - X_0 - Y_N] \geqq 0$ and, combining this with (1.4.3), we obtain

$$b\mathbf{E}[\beta_N(0,b)] \leqq \mathbf{E}[X_N] - \mathbf{E}[X_0]$$

and the assertion of the proposition.

In the general case, noting that $\{(X_n - a)^+\}_{n=0}^{\infty}$ is a submartingale and its upcrossing number up to time N of $[0, b-a]$ is equal to $\beta_N(a,b)$ and applying the above result, we obtain the assertion. □

Considering the upper bound in N of the right hand side of (1.4.2) and applying the monotone convergence theorem to the left hand side, we obtain the following.

Theorem 1.4.20 $(b-a)\mathbf{E}[\beta(a,b)] \leqq \sup_n \mathbf{E}[(X_n - a)^+] - \mathbf{E}[(X_0 - a)^+]$.

From Theorem 1.4.20, we get the convergence theorem for submartingales.

Theorem 1.4.21 *If a submartingale $X = \{X_n\}_{n=0}^{\infty}$ satisfies $\sup_n \mathbf{E}[X_n^+] < \infty$, then X_n converges to an integrable random variable* $\mathbf{P}$-a.s. *as $n \to \infty$.*

Proof Theorem 1.4.20 implies that $\beta(a,b) < \infty$ for any $a < b$ almost surely. Hence X_n converges almost surely. Moreover, since $|x| = 2x^+ - x$,

$$\mathbf{E}[|X_n|] = 2\mathbf{E}[X_n^+] - \mathbf{E}[X_n] \leqq 2\sup_n \mathbf{E}[X_n^+] - \mathbf{E}[X_0] < \infty.$$

Denote the limit of X_n by X_∞. Then, by Fatou's lemma,

$$\mathbf{E}[|X_\infty|] = \mathbf{E}\Big[\liminf_{n\to\infty} |X_n|\Big] \leqq \liminf_{n\to\infty} \mathbf{E}[|X_n|] < \infty. \qquad \square$$

Next we show that a uniformly integrable martingale (see Section A.3) converges in L^1 and almost surely. For this purpose we introduce the notion of closability, which is important also in the optional sampling theorem given in the next section.

Definition 1.4.22 A martingale $X = \{X_n\}_{n=0}^{\infty}$, for which there exists an $\mathscr{F}_\infty := \bigvee_n \mathscr{F}_n$-measurable and integrable random variable X_∞ satisfying

$$X_n = \mathbf{E}[X_\infty|\mathscr{F}_n], \qquad \mathbf{P}\text{-a.s.}\ (n = 0, 1, 2, \ldots),$$

is called a **closable martingale**.

When X is a submartingale, if there exists an $\mathscr{F}_\infty$-measurable and integrable random variable X_∞ satisfying

$$X_n \leqq \mathbf{E}[X_\infty|\mathscr{F}_n], \qquad \mathbf{P}\text{-a.s.}\ (n = 0, 1, 2, \ldots), \tag{1.4.4}$$

then X is called a **closable submartingale**.

As is mentioned in Theorem 1.4.24 below, if X is a closable martingale, then X_n converges to X_∞ as $n \to \infty$.

The Doob decomposition of a closable submartingale is given by the following.

Proposition 1.4.23 *A submartingale $X = \{X_n\}_{n=0}^{\infty}$ is closable if and only if X is a sum of a closable martingale and a non-positive submartingale.*

Proof Assume that X is a closable submartingale. Then, by the tower property,

$$\mathbf{E}[X_n - \mathbf{E}[X_\infty|\mathscr{F}_n]|\mathscr{F}_{n-1}] \geqq X_{n-1} - \mathbf{E}[X_\infty|\mathscr{F}_{n-1}].$$

Hence, X is a sum of the martingale $\{\mathbf{E}[X_\infty|\mathscr{F}_n]\}_{n=0}^{\infty}$ and the non-positive submartingale $\{X_n - \mathbf{E}[X_\infty|\mathscr{F}_n]\}_{n=0}^{\infty}$.

The converse may be easily shown. □

Theorem 1.4.24 *For a martingale $X = \{X_n\}_{n=0}^{\infty}$, the following conditions are equivalent:*

(i) *X is uniformly integrable,*
(ii) *X_n converges in L^1 as $n \to \infty$,*
(iii) *X is closable.*

Proof If X is uniformly integrable, $\sup_n \mathbf{E}[|X_n|] < \infty$ and $\sup_n \mathbf{E}[X_n^+] < \infty$. Hence, by Theorem 1.4.21, X_n converges almost surely as $n \to \infty$. Moreover, by Theorem A.3.7, X_n is also convergent in L^1.

Next we assume that X_n converges in L^1 and let X_∞ be the limit. Then, by Theorem 1.4.1(11),

$$\mathbf{E}[X_\infty|\mathscr{F}_n] = \lim_{m\to\infty} \mathbf{E}[X_m|\mathscr{F}_n]$$

in L^1. Since the right hand side is equal to X_n, $\mathbf{P}$-a.s., X is closable.

Finally we assume that X is closable. Then $|X_n| \leqq \mathbf{E}[|X_\infty||\mathscr{F}_n]$ and

$$\mathbf{E}[|X_n|\mathbf{1}_{\{|X_n|\geqq c\}}] \leqq \mathbf{E}[|X_\infty|\mathbf{1}_{\{|X_n|\geqq c\}}]. \tag{1.4.5}$$

Letting $c \to \infty$, we have

$$\sup_n \mathbf{P}(|X_n| \geqq c) \leqq \frac{1}{c} \sup_n \mathbf{E}[|X_n|] \leqq \frac{1}{c}\mathbf{E}[|X_\infty|] \to 0$$

and, combining this with (1.4.5), we obtain the uniform integrability of X by Proposition A.3.1. □

Theorem 1.4.25 *A submartingale $X = \{X_n\}_{n=0}^\infty$ is closable if and only if $\{X_n^+\}_{n=0}^\infty$ is uniformly integrable.*

Proof Assume that X is closable. Set $M_n = \mathbf{E}[X_\infty|\mathscr{F}_n]$. Then $\{M_n\}_{n=0}^\infty$ is a closable martingale and is uniformly integrable by Theorem 1.4.24. In particular, $\{M_n^+\}$ is uniformly integrable. Since $X_n \leqq M_n$ and $0 \leqq X_n^+ \leqq M_n^+$, $\{X_n^+\}$ is uniformly integrable.

Conversely, assume that $\{X_n^+\}$ is uniformly integrable. Then, since $\{X_n^+\}$ is a non-negative submartingale (Theorem 1.4.5), X_n^+ converges to an integrable random variable Y as $n \to \infty$ by Theorem 1.4.21. Moreover, since X_n^+ converges also in L^1 (Theorem A.3.7), we obtain

$$\mathbf{E}[Y|\mathscr{F}_n] = \lim_{m\to\infty} \mathbf{E}[X_m^+|\mathscr{F}_n] \geqq X_n^+, \quad \mathbf{P}\text{-a.s.},$$

which shows that $\{X_n - \mathbf{E}[Y|\mathscr{F}_n]\}_{n=0}^\infty$ is a non-positive submartingale. Since the martingale $\{\mathbf{E}[Y|\mathscr{F}_n]\}_{n=0}^\infty$ is closable, $\{X_n\}$ is a closable submartingale by Proposition 1.4.23. □

1.4.5 Optional Sampling Theorem

The following **optional sampling theorem** is useful in various situations.

Theorem 1.4.26 *Let $X = \{X_n\}_{n=0}^\infty$ be a closable submartingale. Then, for any stopping time τ, X_τ is integrable and, for another stopping time σ,*

$$\mathbf{E}[X_\tau|\mathscr{F}_\sigma] \geqq X_{\sigma\wedge\tau}, \ \mathbf{P}\text{-a.s.}$$

In particular, if X is a closable martingale, then

$$\mathbf{E}[X_\tau|\mathscr{F}_\sigma] = X_{\sigma\wedge\tau}, \ \mathbf{P}\text{-a.s.}$$

Proof Let X_∞ be the $\mathscr{F}_\infty$-measurable random variable satisfying (1.4.4). We can show $\mathbf{E}[X_\infty|\mathscr{F}_\sigma] \geqq X_\sigma$. In fact, for any $A \in \mathscr{F}_\sigma$, we have

$$\begin{aligned}\mathbf{E}[X_\infty \mathbf{1}_A] &= \sum_{k=0}^{\infty} \mathbf{E}[X_\infty \mathbf{1}_{A\cap\{\sigma=k\}}] \geqq \sum_{k=0}^{\infty} \mathbf{E}[X_k \mathbf{1}_{A\cap\{\sigma=k\}}] \\ &= \sum_{k=0}^{\infty} \mathbf{E}[X_\sigma \mathbf{1}_{A\cap\{\sigma=k\}}] = \mathbf{E}[X_\sigma \mathbf{1}_A].\end{aligned}$$

In particular, if X is a martingale, we have an equality.

Since, $X^\tau = \{X_{n\wedge\tau}\}_{n=0}^\infty$ is a submartingale by Theorem 1.4.16, the proof is completed once we have shown that X^τ is closable.

By Proposition 1.4.23, there exist a closable martingale $\{M_n\}_{n=0}^\infty$ and a non-positive submartingale $\{N_n\}_{n=0}^\infty$ such that $X_n = M_n + N_n$. Since $\{N_{n\wedge\tau}\}_{n=0}^\infty$ is closable by Theorem 1.4.21, it suffices to show the closability of $\{M_{n\wedge\tau}\}_{n=0}^\infty$. Denote the limit of M_n as $n \to \infty$ by M_∞. Since

$$\mathbf{E}[M_\infty|\mathscr{F}_{n\wedge\tau}] = M_{n\wedge\tau}$$

as is shown above,

$$\begin{aligned}\mathbf{E}[|M_{n\wedge\tau}|\mathbf{1}_{\{|M_{n\wedge\tau}|\geqq c\}}] &= \mathbf{E}\big[\big|\mathbf{E}[M_\infty|\mathscr{F}_{n\wedge\tau}]\big|\mathbf{1}_{\{|M_{n\wedge\tau}|\geqq c\}}\big] \\ &\leqq \mathbf{E}[|M_\infty|\mathbf{1}_{\{|M_{n\wedge\tau}|\geqq c\}}].\end{aligned}$$

Moreover, letting $c \to \infty$, we obtain

$$\sup_n \mathbf{P}(|M_{n\wedge\tau}| \geqq c) \leqq \frac{1}{c}\sup_n \mathbf{E}[|M_{n\wedge\tau}|] \leqq \frac{1}{c}\mathbf{E}[|M_\infty|] \to 0.$$

Hence $\{M_{n\wedge\tau}\}_{n=0}^\infty$ is uniformly integrable and $\{M_{n\wedge\tau}\}_{n=0}^\infty$ is closable by Theorem 1.4.24. □

Corollary 1.4.27 *Let $X = \{X_n\}_{n=0}^\infty$ be a submartingale and σ, τ be stopping times satisfying $\mathbf{P}(\sigma \leqq \tau \leqq N) = 1$ for some $N \in \mathbb{N}$. Then*

$$\mathbf{E}[X_\tau|\mathscr{F}_\sigma] \geqq X_\sigma, \quad \text{a.s.}$$

In particular, if X is a martingale, the equality holds.

Proof $X^N = \{X_{n\wedge N}\}_{n=0}^\infty$ is a closable submartingale. Hence, since $X_N^\tau = X_\tau$, Theorem 1.4.26 implies

$$\mathbf{E}[X_\tau|\mathscr{F}_\sigma] \geqq X_{\tau\wedge\sigma} = X_\sigma.$$ □

The next corollary is easily obtained from Theorem 1.4.26.

Corollary 1.4.28 *Let $X = \{X_n\}_{n=0}^{\infty}$ be an $\{\mathscr{F}_n\}$-closable submartingale and $\sigma_0, \sigma_1, \sigma_2, \dots$ be an increasing sequence of stopping times. Then, the stochastic process $\{Y_k\}_{k=0}^{\infty}$ defined by $Y_k = X_{\sigma_k}$ is an $\{\mathscr{F}_{\sigma_k}\}$-submartingale.*

Remark 1.4.29 If each stopping time is bounded, Corollary 1.4.28 is true also when X is not closable.

Remark 1.4.30 If X is not closable, the optional sampling theorem does not hold in general. For example, consider the martingale (simple random walk) $S = \{S_n\}_{n=0}^{\infty}$ given in Example 1.4.6. We have $\limsup_{n\to\infty} |S_n| = \infty$ almost surely. Hence, letting τ_a be the first hitting time to $a \in \mathbb{N}$, we have $\mathbf{P}(\tau_a < \infty) = 1$ and $\mathbf{E}[S_{\tau_a}] = a$. However, if we apply the optional sampling theorem, we should have $\mathbf{E}[S_{\tau_a}] = \mathbf{E}[S_0] = 0$ and a contradiction.

1.4.6 Doob's Inequality

We show **Doob's inequality** for the maximum of submartingales and its application.

Theorem 1.4.31 *Let $X = \{X_n\}_{n=0}^{\infty}$ be a submartingale and set $Y_n = \max\{X_k; 0 \leqq k \leqq n\}$. Then, for any $a \in \mathbb{R}$ and $N \in \mathbb{N}$,*

$$a\mathbf{P}(Y_N \geqq a) \leqq \mathbf{E}[X_N \mathbf{1}_{\{Y_N \geqq a\}}] \leqq \mathbf{E}[X_N^+].$$

Proof Let $A_k \in \mathscr{F}_k$ $(k = 0, 1, \dots, N)$ be a sequence of events given by $A_0 = \{X_0 \geqq a\}$ and

$$A_k = \{X_i < a\,(i = 0, \dots, k-1), X_k \geqq a\} \qquad (k = 1, 2, \dots).$$

Then

$$a\mathbf{P}(Y_N \geqq a) = \sum_{k=0}^{N} \mathbf{E}[a\mathbf{1}_{A_k}] \leqq \sum_{k=0}^{N} \mathbf{E}[X_k \mathbf{1}_{A_k}].$$

Since $\{X_k\}_{k=0}^{N}$ is a submartingale, $\mathbf{E}[X_k \mathbf{1}_{A_k}] \leqq \mathbf{E}[X_N \mathbf{1}_{A_k}]$ and

$$a\mathbf{P}(Y_N \geqq a) \leqq \sum_{k=0}^{N} \mathbf{E}[X_N \mathbf{1}_{A_k}] = \mathbf{E}[X_N \mathbf{1}_{\{Y_N \geqq a\}}].$$

The second inequality is obtained from

$$\mathbf{E}[X_N \mathbf{1}_{\{\tau_a \leqq N\}}] \leqq \mathbf{E}[X_N^+ \mathbf{1}_{\{\tau_a \leqq N\}}] \leqq \mathbf{E}[X_N^+].$$ □

The next inequality was first shown by Kolmogorov for a sum of independent random variables. It is obtained from Theorem 1.4.31 if we note that $\{X_n^2\}$ is a submartingale.

Corollary 1.4.32 *If $\{X_n\}_{n=0}^{\infty}$ is a square-integrable martingale, then for any $a > 0$ and $N \in \mathbb{N}$*

$$\mathbf{P}\Big(\max_{0 \leqq k \leqq N} |X_k| \geqq a\Big) \leqq \frac{1}{a^2}\mathbf{E}[|X_N|^2].$$

The next result enables us to extend the martingale convergence theorem to L^p- and almost sure convergences. We leave such extensions to the reader.

Theorem 1.4.33 *Let p, q be positive numbers such that $p^{-1} + q^{-1} = 1$ and $X = \{X_n\}_{n=0}^{\infty}$ be a non-negative p-th integrable submartingale.*
(1) Set $Y_n = \max\{X_k; 0 \leqq k \leqq n\}$. Then

$$\mathbf{E}[Y_n^p] \leqq q^p \mathbf{E}[X_n^p]. \tag{1.4.6}$$

(2) If $\sup_n \mathbf{E}[X_n^p] < \infty$, then $\sup_n X_n$ is also p-th integrable and

$$\mathbf{E}\Big[\sup_n X_n^p\Big] \leqq q^p \sup_n \mathbf{E}[X_n^p].$$

Proof (1) For $K > 0$, set $Y_n^K = K \wedge Y_n$. If $\lambda \leqq K$, then $\{Y_n^K \geqq \lambda\} = \{Y_n \geqq \lambda\}$ and, therefore,

$$\mathbf{P}(Y_n^K \geqq \lambda) = \mathbf{P}(Y_n \geqq \lambda) \leqq \frac{1}{\lambda}\mathbf{E}[X_n \mathbf{1}_{\{Y_n \geqq \lambda\}}] = \frac{1}{\lambda}\mathbf{E}[X_n \mathbf{1}_{\{Y_n^K \geqq \lambda\}}]$$

by Doob's inequality (Theorem 1.4.31). Since $\mathbf{P}(Y_n^K \geqq \lambda) = 0$ for $\lambda > K$,

$$\begin{aligned}\mathbf{E}[(Y_n^K)^p] &= \int_0^\infty p\lambda^{p-1}\mathbf{P}(Y_n^K \geqq \lambda)\,\mathrm{d}\lambda \leqq \int_0^\infty p\lambda^{p-1}\frac{1}{\lambda}\mathbf{E}[X_n \mathbf{1}_{\{Y_n^K \geqq \lambda\}}]\,\mathrm{d}\lambda \\ &= \mathbf{E}\Big[X_n \int_0^{Y_n^K} p\lambda^{p-2}\,\mathrm{d}\lambda\Big] = \frac{p}{p-1}\mathbf{E}[X_n (Y_n^K)^{p-1}].\end{aligned}$$

Hence, by Hölder's inequality, we obtain

$$\mathbf{E}[(Y_n^K)^p] \leqq q\mathbf{E}[(X_n)^p]^{\frac{1}{p}}\mathbf{E}[(Y_n^K)^p]^{\frac{1}{q}}$$

and, after a simple manipulation,

$$\mathbf{E}[(Y_n^K)^p] \leqq q^p \mathbf{E}[(X_n)^p].$$

Letting $K \to \infty$ on the left hand side, we obtain the assertion by the monotone convergence theorem.

(2) By the convergence theorem for submartingales (Theorem 1.4.21), both of $\lim X_n$ and $\sup X_n$ exist. Hence, taking the supremum in n on the right hand side of (1.4.6) and applying the monotone convergence theorem to its left hand side, we obtain the assertion. □

1.5 Continuous Time Martingale

1.5.1 Fundamentals

Let $(\Omega, \mathscr{F}, \mathbf{P}, \{\mathscr{F}_t\})$ be a filtered probability space.

Definition 1.5.1 (1) An $\mathbb{R}$-valued stochastic process $M = \{M(t)\}_{t \geqq 0}$ defined on $(\Omega, \mathscr{F}, \mathbf{P}, \{\mathscr{F}_t\})$ is called an $\{\mathscr{F}_t\}$-**martingale** if

(i) $\{M(t)\}_{t \geqq 0}$ is $\{\mathscr{F}_t\}$-adapted,
(ii) for any $t \geqq 0$, $\mathbf{E}[|M(t)|] < \infty$,
(iii) for $s \leqq t$,

$$\mathbf{E}[M(t)|\mathscr{F}_s] = M(s), \quad \mathbf{P}\text{-a.s.} \tag{1.5.1}$$

(2) If $\mathbf{E}[M(t)|\mathscr{F}_s] \geqq (\leqq) M(s)$, $\mathbf{P}$-a.s. holds instead of (1.5.1), M is called an $\{\mathscr{F}_t\}$-**submartingale** (**supermartingale**, respectively). Moreover, for $p > 1$, M is called an L^p-(sub, super)martingale if $\mathbf{E}[|M(t)|^p] < \infty$ $(t \geqq 0)$.

Proposition 1.5.2 (1) *If M is an $\{\mathscr{F}_t\}$-martingale and f is a convex continuous function such that $\mathbf{E}[|f(M(t))|] < \infty$ for any $t \geqq 0$, $\{f(M(t))\}_{t \geqq 0}$ is an $\{\mathscr{F}_t\}$-submartingale.*
(2) *If an $\{\mathscr{F}_t\}$-martingale M is continuous and satisfies $\mathbf{E}\big[\sup_{0 \leqq t \leqq T} |M(t)|\big] < \infty$ $(T > 0)$, the stochastic process $\{M_\eta(t)\}_{t \geqq 0}$ defined by*

$$M_\eta(t) = M(t)\eta(t) - \int_0^t M(s)\eta'(s)\,\mathrm{d}s$$

for $\eta \in C^1([0, \infty); \mathbb{R})$ is an $\{\mathscr{F}_t\}$-martingale.

Proof (1) For $s < t$, Jensen's inequality implies

$$\mathbf{E}[f(M(t))|\mathscr{F}_s] \geqq f(\mathbf{E}[M(t)|\mathscr{F}_s]) = f(M(s)), \quad \mathbf{P}\text{-a.s.}$$

(2) Let $s = t_0 < t_1 < \cdots < t_n = t$ be a partition of the interval $[s, t]$. Then, noting

$$
\begin{aligned}
M(t)\eta(t) - M(s)\eta(s) - \sum_{j=0}^{n-1} M(t_{j+1})(\eta(t_{j+1}) - \eta(t_j))& \\
= \sum_{j=0}^{n-1} \eta(t_j)(M(t_{j+1}) - M(t_j))&
\end{aligned}
$$

and taking the conditional expectation with respect to $\mathscr{F}_s$ of both sides, we obtain

$$
\mathbf{E}\Big[M(t)\eta(t) - M(s)\eta(s) - \sum_{j=0}^{n-1} M(t_{j+1})(\eta(t_{j+1}) - \eta(t_j))\Big|\mathscr{F}_s\Big] = 0.
$$

Letting the width $\max_{1\leqq j\leqq n}|t_j - t_{j-1}|$ of the partition tend to 0, we get

$$
\mathbf{E}\Big[M(t)\eta(t) - M(s)\eta(s) - \int_s^t M(u)\eta'(u)\,\mathrm{d}u\Big|\mathscr{F}_s\Big] = 0
$$

and the assertion. □

Remark 1.5.3 Let $\{X(t)\}_{t\geqq 0}$ be a d-dimensional Brownian motion and f be a subharmonic function on $\mathbb{R}^d$, that is, a function satisfying $\Delta f \geqq 0$. If in addition f satisfies an adequate growth condition, we can show that $\{f(X(t))\}_{t\geqq 0}$ is a submartingale. It is an easy application of Itô's formula given in the next chapter.

1.5.2 Examples on the Wiener Space

We give typical examples of martingales defined on the Wiener space.

Theorem 1.5.4 *Let $(W, \mathscr{B}(W), \mu)$ be the d-dimensional Wiener space, $\theta = \{\theta(t)\}_{t\geqq 0}$ be the coordinate process and $\{\mathscr{B}_t^0\}$ be the filtration given by* (1.3.1). *Then, the following stochastic processes are $\{\mathscr{B}_t^0\}$-martingales:*
(1) *$\{\theta^i(t)\}_{t\geqq 0}$ $(i = 1, 2, \ldots, d)$, where $\theta^i(t)$ is the i-th component of $\theta(t)$;*
(2) *$\{(\theta^i(t))^2 - t\}_{t\geqq 0}$ $(i = 1, 2, \ldots, d)$;*
(3) *$\{\exp(\langle\lambda, \theta(t)\rangle - \frac{|\lambda|^2 t}{2})\}_{t\geqq 0}$ for $\lambda \in \mathbb{R}^d$;*
(4) *$\{\exp(\mathrm{i}\langle\lambda, \theta(t)\rangle + \frac{|\lambda|^2 t}{2})\}_{t\geqq 0}$ for $\lambda \in \mathbb{R}^d$, where $\mathrm{i} = \sqrt{-1}$;*
(5) *$\{f(\theta(t)) - \frac{1}{2}\int_0^t (\Delta f)(\theta(s))\,\mathrm{d}s\}_{t\geqq 0}$ for a rapidly decreasing function f, where $\Delta = \sum_{i=1}^d (\frac{\partial}{\partial x^i})^2$ and f is said to be rapidly decreasing if, for any $\alpha_1, \ldots, \alpha_d \in \mathbb{Z}_+$ and $\beta > 0$, $(1 + |x|)^\beta \frac{\partial^{\alpha_1+\cdots+\alpha_d} f}{\partial^{\alpha_1} x_1 \cdots \partial^{\alpha_d} x_d}$ converges to 0 as $|x| \to \infty$.*

Remark 1.5.5 In (4), a $\mathbb{C}$-valued stochastic process is called a martingale if its real and imaginary parts are both martingales.

Proof By the definition of the Wiener measure,

$$\mathbf{E}[\varphi(\theta(t) - \theta(s))|\mathscr{B}_s^0] = \int_{\mathbb{R}^d} \varphi(x) g(t - s, 0, x)\, dx$$

for $s \leqq t$ and a continuous function φ with at most exponential growth, that is, there exist constants $C_1, C_2 \geqq 0$ such that $|\varphi(x)| \leqq C_1 e^{C_2|x|}$ ($x \in \mathbb{R}^d$). From this, the assertions (1)–(4) follow.

For (5) note that there exists a rapidly decreasing function $\widehat{f} : \mathbb{R}^d \to \mathbb{C}$ such that

$$f(x) = \int_{\mathbb{R}^d} \widehat{f}(\lambda) e^{i\langle \lambda, x\rangle} d\lambda \quad \text{and} \quad (\Delta f)(x) = -\int_{\mathbb{R}^d} \widehat{f}(\lambda)|\lambda|^2 e^{i\langle \lambda, x\rangle} d\lambda.$$

Then we obtain

$$\begin{aligned} M_f(t) &:= f(\theta(t)) - \frac{1}{2}\int_0^t (\Delta f)(\theta(s))\, ds \\ &= \int_{\mathbb{R}^d} \widehat{f}(\lambda)\Big(e^{i\langle \lambda, \theta(t)\rangle} + \frac{|\lambda|^2}{2}\int_0^t e^{i\langle \lambda, \theta(s)\rangle} ds\Big) d\lambda. \end{aligned}$$

Now let $\{M(t)\}_{t \geqq 0}$ be the martingale in (4) and set $\eta(t) = \exp(-\frac{|\lambda|^2 t}{2})$. Then, by Proposition 1.5.2, the stochastic process

$$e^{i\langle \lambda, \theta(t)\rangle} + \frac{|\lambda|^2}{2}\int_0^t e^{i\langle \lambda, \theta(s)\rangle} ds \qquad (t \geqq 0)$$

is a martingale. Hence, by Fubini's theorem, $\mathbf{E}[M_f(t)|\mathscr{B}_s^0] = M_f(s)$, $\mathbf{P}$-a.s. □

Conversely, the martingales discussed in the above theorem characterize the Wiener measure.

Theorem 1.5.6 *Let $X = \{X(t)\}_{t \geqq 0}$ be an $\mathbb{R}^d$-valued continuous $\{\mathscr{F}_t\}$-adapted stochastic process starting from 0 defined on a filtered probability space $(\Omega, \mathscr{F}, \{\mathscr{F}_t\}, \mathbf{P})$. For $\lambda \in \mathbb{R}^d$ and $f \in C_0^\infty(\mathbb{R}^d)$, set*

$$L_\lambda(t) = e^{\langle \lambda, X(t)\rangle - \frac{|\lambda|^2 t}{2}}, \qquad F_\lambda(t) = e^{i\langle \lambda, X(t)\rangle + \frac{|\lambda|^2 t}{2}},$$

$$M_f(t) = f(X(t)) - \frac{1}{2}\int_0^t (\Delta f)(X(s))\, ds.$$

Then, if one of the following conditions holds, the probability law of X, that is, the probability measure on W induced by X, is the Wiener measure:

(i) *for any $\lambda \in \mathbb{R}^d$, $\{L_\lambda(t)\}_{t \geqq 0}$ is an $\{\mathscr{F}_t\}$-martingale,*
(ii) *for any $\lambda \in \mathbb{R}^d$, $\{F_\lambda(t)\}_{t \geqq 0}$ is an $\{\mathscr{F}_t\}$-martingale,*
(iii) *for any $f \in C_0^\infty(\mathbb{R}^d)$, $\{M_f(t)\}_{t \geqq 0}$ is an $\{\mathscr{F}_t\}$-martingale.*

For the proof we prepare the following lemma.

Lemma 1.5.7 (1) *Let* $f : \mathbb{C}^n \times \Omega \to \mathbb{C}^d$ *satisfy that, for each* $\omega \in \Omega$, *the mapping* $\mathbb{C}^n \ni \zeta \mapsto f(\zeta, \omega) \in \mathbb{C}^d$ *is holomorphic*[5] *and, for any* $R > 0$, *there exists a non-negative integrable random variable* Φ_R *satisfying*

$$\left|\frac{\partial}{\partial \zeta^i} f(\zeta, \cdot)\right| \leqq \Phi_R \qquad (|\zeta| \leqq R,\ i = 1, \ldots, d).$$

Then, the mapping $\mathbb{C}^d \ni \zeta \mapsto \mathbf{E}[f(\zeta, \cdot)] \in \mathbb{C}$ *is holomorphic.*
(2) *If a* $\mathbb{C}^d$*-valued random variable* X *satisfies* $\mathrm{e}^{p|X|} \in L^1(\mathbf{P})$ *for any* $p > 1$, *then* $\mathbb{C}^d \ni \zeta \mapsto \mathbf{E}[\mathrm{e}^{\langle \zeta, X\rangle}]$ *is holomorphic.*

Proof (1) We only show the case when $n = 1$. Write $\zeta = x + \mathrm{i}y$ $(x, y \in \mathbb{R})$. Since $\frac{\partial}{\partial \zeta} = \frac{\partial}{\partial x} = -\mathrm{i}\frac{\partial}{\partial y}$ for holomorphic functions,

$$\left|\frac{\partial}{\partial x} f(\zeta, \cdot)\right| = \left|\frac{\partial}{\partial \zeta} f(\zeta, \cdot)\right| \leqq \Phi_R, \quad \left|\frac{\partial}{\partial y} f(\zeta, \cdot)\right| = \left|\frac{\partial}{\partial \zeta} f(\zeta, \cdot)\right| \leqq \Phi_R$$

by the assumption if $x^2 + y^2 < R^2$. Hence, by the Lebesgue convergence theorem, $\mathbf{E}[f(x + \mathrm{i}y, \cdot)]$ is of C^1-class in x, y and

$$\frac{\partial}{\partial x}\mathbf{E}[f(x + \mathrm{i}y, \cdot)] = \mathbf{E}\left[\frac{\partial f}{\partial x}(x + \mathrm{i}y, \cdot)\right],$$
$$\frac{\partial}{\partial y}\mathbf{E}[f(x + \mathrm{i}y, \cdot)] = \mathbf{E}\left[\frac{\partial f}{\partial y}(x + \mathrm{i}y, \cdot)\right].$$

Let $\frac{\partial}{\partial \bar{\zeta}} = \frac{\partial}{\partial x} + \mathrm{i}\frac{\partial}{\partial y}$. Then, since $\frac{\partial}{\partial \bar{\zeta}} f(\zeta, \cdot) = 0$,

$$\frac{\partial}{\partial \bar{\zeta}}\mathbf{E}[f(\zeta, \cdot)] = \mathbf{E}\left[\frac{\partial}{\partial \bar{\zeta}} f(\zeta, \cdot)\right] = 0.$$

This means the assertion (1).
(2) For $|\zeta| < R$,

$$|\mathrm{e}^{\langle \zeta, X\rangle}| \leqq \mathrm{e}^{|\zeta|\,|X|} \leqq \mathrm{e}^{R|X|} \quad \text{and} \quad \left|\frac{\partial}{\partial \zeta}\mathrm{e}^{\langle \zeta, X\rangle}\right| \leqq |X|\mathrm{e}^{|\zeta|\,|X|} \leqq \mathrm{e}^{(R+1)|X|}.$$

Hence (1) implies the assertion. □

Proof of Theorem 1.5.6. Assume (i). Then, for $s < t$, $A \in \mathscr{F}_s$ and $\lambda \in \mathbb{R}^d$,

$$\mathbf{E}[\mathrm{e}^{\langle \lambda, X(t)\rangle - \frac{|\lambda|^2 t}{2}} \mathbf{1}_A] = \mathbf{E}[\mathrm{e}^{\langle \lambda, X(s)\rangle - \frac{|\lambda|^2 s}{2}} \mathbf{1}_A]. \tag{1.5.2}$$

By Lemma 1.5.7, both sides are extended to holomorphic functions in λ and (1.5.2) holds for all $\lambda \in \mathbb{C}^d$ by the uniqueness theorem. In particular,

$$\mathbf{E}[\mathrm{e}^{\mathrm{i}\langle \lambda, X(t)\rangle + \frac{|\lambda|^2 t}{2}} \mathbf{1}_A] = \mathbf{E}[\mathrm{e}^{\mathrm{i}\langle \lambda, X(s)\rangle + \frac{|\lambda|^2 s}{2}} \mathbf{1}_A] \qquad (\lambda \in \mathbb{R}^d).$$

[5] A mapping $\mathbb{C}^d \ni \zeta \mapsto g(\zeta) \in \mathbb{C}^m$ is called holomorphic if each component $g^j(\zeta)$ $(j = 1, \ldots, m)$ of $g(\zeta)$ is holomorphic in each variable ζ^i $(i = 1, \ldots, d)$.

Thus (ii) is obtained.

Next assume (iii). For $\varphi(x) = \exp(\mathrm{i}\langle\lambda, x\rangle)$, there exists a sequence $\{f_n\} \subset C_0^\infty(\mathbb{R}^d)$ such that $f_n \to \varphi, \Delta f_n \to \Delta\varphi$ and $f_n, \Delta f_n$ are uniformly bounded. For example, let j_n be a function in $C_0^\infty(\mathbb{R}^d)$ such that $j_n(x) = 1$ if $|x| \leqq n$ and $j_n(x) = 0$ if $|x| \geqq n+1$. Then, $f_n = \varphi j_n$ is a desired function. Since $\{M_{f_n}(t)\}_{t\geqq 0}$ is a martingale by the assumption, we obtain (ii) by letting $n \to \infty$ and applying the bounded convergence theorem.

Finally we show, from (ii), that the probability distribution of X is the Wiener measure. By the assumption,

$$\mathbf{E}[\mathrm{e}^{\mathrm{i}\langle\lambda, X(t)-X(s)\rangle}|\mathscr{F}_s] = \mathrm{e}^{-\frac{|\lambda|^2}{2}(t-s)} = \int_{\mathbb{R}^d} \mathrm{e}^{\mathrm{i}\langle\lambda, x\rangle} g(t-s, 0, x)\,\mathrm{d}x$$

for any $\lambda \in \mathbb{R}^d$. From this identity we obtain

$$\mathbf{E}[f(X(t) - X(s))|\mathscr{F}_s] = \int_{\mathbb{R}^d} f(x) g(t-s, 0, x)\,\mathrm{d}x$$

for any $f \in C_0(\mathbb{R}^d)$ by a similar argument to the proof of Theorem 1.5.4(5). Hence $X(t) - X(s)$ is independent of $\mathscr{F}_s$ and obeys the d-dimensional normal distribution with mean 0 and covariance matrix $(t-s)I$, where I is the d-dimensional unit matrix. Therefore, for any $0 = t_0 < t_1 < \cdots < t_{n+1}, f_0, f_1, \ldots, f_n \in C_0^\infty(\mathbb{R}^d)$,

$$\mathbf{E}\Big[\prod_{i=0}^{n} f_i(X(t_{i+1}) - X(t_i))\Big] = \int_{\mathbb{R}^d} \cdots \int_{\mathbb{R}^d} \prod_{i=0}^{n} f_i(x_i) \prod_{i=0}^{n} g(t_{i+1} - t_i, 0, x_i)\,\mathrm{d}x_1 \cdots \mathrm{d}x_n,$$

which shows that the distribution of X is the Wiener measure. □

1.5.3 Optional Sampling Theorem, Doob's Inequality, Convergence Theorem

For continuous time martingales, the optional sampling theorem, Doob's inequality, and convergence theorems are proven in the same way as in the discrete time case. We begin with the definition of closability.

In this book, we always assume that a submartingale $X = \{X(t)\}_{t\geqq 0}$ is right-continuous, that is, the function $[0, \infty) \ni t \mapsto X(t, \omega) \in \mathbb{R}$ is right-continuous for all $\omega \in \Omega$. On a probability space $(\Omega, \mathscr{F}, \mathbf{P}, \{\mathscr{F}_t\})$ satisfying the usual condition, an $\{\mathscr{F}_t\}$-submartingale $\{X(t)\}_{t\geqq 0}$ whose expectation $\mathbf{E}[X(t)]$ is right-continuous in t has a modification whose sample path is right-continuous and has left limits (for details, see [98, Chapter 2]).

Definition 1.5.8 Let $X = \{X(t)\}_{t \geqq 0}$ be a submartingale defined on $(\Omega, \mathscr{F}, \mathbf{P}, \{\mathscr{F}_t\})$. If there exists an $\mathscr{F}_\infty := \bigvee_{t \geqq 0} \mathscr{F}_t$-measurable integrable random variable X_∞ such that

$$X(t) \leqq \mathbf{E}[X_\infty | \mathscr{F}_t], \quad \mathbf{P}\text{-a.s.} \quad \text{for every } t \geqq 0,$$

X is called **closable**.

If X is a martingale and if there exists an $\mathscr{F}_\infty$-measurable integrable random variable X_∞ such that

$$X(t) = \mathbf{E}[X_\infty | \mathscr{F}_t], \quad \mathbf{P}\text{-a.s.} \quad \text{for every } t \geqq 0,$$

X is called a **closable martingale**.

After showing that each stopping time (see Definition 1.3.4) is approximated by a decreasing sequence of discrete stopping times, we prove the optional sampling theorem. Before it, we define a σ-field which represents the information up to a stopping time in the same way as the discrete time case (Proposition 1.4.17).

Definition 1.5.9 For an $\{\mathscr{F}_t\}$-stopping time τ, the σ-field $\mathscr{F}_\tau$ is defined by

$$\mathscr{F}_\tau = \{A \in \mathscr{F} \; ; \; A \cap \{\tau \leqq t\} \in \mathscr{F}_t \text{ for each } t \geqq 0\}.$$

Lemma 1.5.10 *For an $\{\mathscr{F}_t\}$-stopping time τ, set*

$$\tau_n = \begin{cases} 0 & (\tau = 0) \\ \frac{k+1}{2^n} & (\frac{k}{2^n} < \tau \leqq \frac{k+1}{2^n}, \; k = 0, 1, 2, \ldots) \\ \infty & (\tau = \infty) \end{cases}$$

for $n = 1, 2, \ldots$. Then, for each n, τ_n is an $\{\mathscr{F}_t\}$-stopping time.

Proof For $t \geqq 0$, take $k \in \mathbb{Z}_+$ such that $\frac{k}{2^n} \leqq t < \frac{k+1}{2^n}$. Then,

$$\{\tau_n \leqq t\} = \left\{\tau_n \leqq \frac{k}{2^n}\right\} = \left\{\tau \leqq \frac{k}{2^n}\right\} \in \mathscr{F}_{\frac{k}{2^n}} \subset \mathscr{F}_t. \qquad \square$$

Theorem 1.5.11 *Let $X = \{X(t)\}_{t \geqq 0}$ be a closable submartingale. Then, for any stopping time σ and τ,*

$$\mathbf{E}[X(\tau) | \mathscr{F}_\sigma] \geqq X(\tau \wedge \sigma), \quad \mathbf{P}\text{-a.s.}$$

In particular, if X is a closable martingale, we have the equality.

Proof We show the first half. Let τ_n and σ_n be as in the above lemma, By the optional sampling theorem for discrete time submartingale (Theorem 1.4.26),

$$\mathbf{E}[X(\tau_n)|\mathscr{F}_{\sigma_n}] \geqq X(\tau_n \wedge \sigma_n), \quad \mathbf{P}\text{-a.s.}$$

and, hence, for any $A \in \mathscr{F}_\sigma \subset \mathscr{F}_{\sigma_n}$,

$$\mathbf{E}[X(\tau_n)\mathbf{1}_A] \geqq \mathbf{E}[X(\tau_n \wedge \sigma_n)\mathbf{1}_A]. \tag{1.5.3}$$

By the closability of X and the Lévy–Doob downward theorem ([126, Chapter 14]), $\{X(\tau_n)\}_{n=1}^\infty$, $\{X(\tau_n \wedge \sigma_n)\}_{n=1}^\infty$ are uniformly integrable and converge to $X(\tau)$, $X(\tau \wedge \sigma)$ in L^1, respectively. Therefore, letting $n \to \infty$ in (1.5.3), we obtain $\mathbf{E}[X(\tau)\mathbf{1}_A] \geqq \mathbf{E}[X(\tau \wedge \sigma)\mathbf{1}_A]$. □

Corollary 1.5.12 *Let $X = \{X(t)\}_{t\geqq 0}$ be an $\{\mathscr{F}_t\}$-submartingale. Then, for any stopping time τ, $X^\tau = \{X(t \wedge \tau)\}_{t\geqq 0}$ is also a submartingale.*

Proof For $s < t$, apply Theorem 1.5.11 by replacing τ by $t \wedge \tau$ and σ by s. Then we obtain $\mathbf{E}[X(t \wedge \tau)|\mathscr{F}_s] \geqq X(s \wedge \tau)$. □

Doob's inequality for a submartingale $\{X(t)\}_{t\geqq 0}$ follows from that for discrete time submartingales, since $\sup_{0\leqq s\leqq t} X(s) = \lim_{n\to\infty} \max_{0\leqq k\leqq n} X(\frac{kt}{n})$.

Theorem 1.5.13 *Let $X = \{X(t)\}_{t\geqq 0}$ be a submartingale. Then, for any $a, t > 0$,*

$$a\mathbf{P}\Big(\sup_{0\leqq s\leqq t} X(s) > a\Big) \leqq \mathbf{E}\Big[X(t)\ ;\ \sup_{0\leqq s\leqq t} X(s) > a\Big].$$

Theorem 1.5.14 *Let p, q be positive numbers such that $p^{-1} + q^{-1} = 1$ and $X = \{X(t)\}_{t\geqq 0}$ be a non-negative submartingale with $X(t) \in L^p$ $(t \geqq 0)$. Then $\sup_{0\leqq s\leqq t} X(s) \in L^p$ and*

$$\mathbf{E}\Big[\sup_{0\leqq s\leqq t} X(s)^p\Big] \leqq q^p\mathbf{E}[X(t)^p].$$

The following convergence theorem can also be proven in the same way as for discrete time submartingales.

Theorem 1.5.15 (1) *Let $\{X(t)\}_{t\geqq 0}$ be a submartingale defined on a probability space $(\Omega, \mathscr{F}, \mathbf{P}, \{\mathscr{F}_t\})$ which satisfies the usual condition. If $\sup_t \mathbf{E}[(X(t))^+] < \infty$, then $X(t)$ converges as $t \to \infty$ almost surely.*

(2) *A non-negative supermartingale defined on a probability space satisfying the usual condition converges almost surely as $t \to \infty$.*

The following theorem is used in various situations.

Theorem 1.5.16 *For an $\{\mathscr{F}_t\}$-martingale $\{X(t)\}_{t\geqq 0}$ defined on a probability space $(\Omega, \mathscr{F}, \mathbf{P}, \{\mathscr{F}_t\})$ satisfying the usual condition, the following three conditions are equivalent:*

(i) *M is closable,*
(ii) *$\{M(t)\}_{t\geqq 0}$ is uniformly integrable,*
(iii) *$\lim_{t\uparrow\infty} M(t)$ exists in L^1.*

Moreover, if $\sup_t \mathbf{E}[|M(t)|^p] < \infty$ for some $p > 1$, this is the case and $M(t)$ converges in L^p as $t \to \infty$.

Remark 1.5.17 When $p = 1$, the condition $\sup_t \mathbf{E}[|M(t)|] < \infty$ is not sufficient for the last assertion of the theorem to hold. For example, let $\{B(t)\}_{t\geqq 0}$ be a Brownian motion starting from 0 and set $M(t) = \exp(B(t) - \frac{t}{2})$. Then $\mathbf{E}[M(t)] = 1$, but $M(t) \to 0$ $(t \to 0)$.

Proof First we show (ii) from (i). Set $\Gamma_\alpha(t) = \{|M(t)| > \alpha\}$ for $\alpha > 0$. Then

$$\begin{aligned}\mathbf{E}[|M(t)|\mathbf{1}_{\Gamma_\alpha(t)}] &= \mathbf{E}\big[\big|\mathbf{E}[M_\infty|\mathscr{F}_t]\big|\mathbf{1}_{\Gamma_\alpha(t)}\big] \\ &\leqq \mathbf{E}\big[\mathbf{E}[|M_\infty|\big|\mathscr{F}_t]\mathbf{1}_{\Gamma_\alpha(t)}\big] = \mathbf{E}[|M_\infty|\mathbf{1}_{\Gamma_\alpha(t)}].\end{aligned}$$

By Chebyshev's inequality, we obtain

$$\mathbf{P}(\Gamma_\alpha(t)) \leqq \frac{1}{\alpha}\mathbf{E}[|M_\infty|],$$

and the uniform estimates in t for $\mathbf{P}(\Gamma_\alpha(t))$ and $\mathbf{E}[|M_\infty|\mathbf{1}_{\Gamma_\alpha(t)}]$.

Next, assume (ii). Then, since $\sup_t \mathbf{E}[(M(t))^+] < \infty$ by the assumption, $M(t)$ converges as $t \to \infty$ almost surely by Theorem 1.5.15. Hence, by the assumption again, $M(t)$ converges in L^1.

Finally we show (i) from (iii). Set $\lim_{t\to\infty} M(t) = M(\infty)$. Then

$$\begin{aligned}\mathbf{E}\big[\big|M(t) - \mathbf{E}[M(\infty)|\mathscr{F}_t]\big|\big] &= \mathbf{E}\big[\big|\mathbf{E}[M(t+s) - M(\infty)|\mathscr{F}_t]\big|\big] \\ &\leqq \mathbf{E}\big[\mathbf{E}[|M(t+s) - M(\infty)|\,|\mathscr{F}_t]\big] = \mathbf{E}[|M(t+s) - M(\infty)|]\end{aligned}$$

for $s > 0$. Since the right hand side converges to 0 as $s \to \infty$, M is closable.

When $\sup_t \mathbf{E}[|M(t)|^p] < \infty$, Doob's inequality implies $\sup_t |M(t)| \in L^p$. Hence $\{|M(t)|^p\}_{t\geqq 0}$ is uniformly integrable. □

1.5.4 Applications

We give applications of the optional sampling theorem and Doob's inequality. First we show, by using the optional sampling theorem, that non-trivial martingales do not have bounded variation.

Theorem 1.5.18 *If a continuous martingale $M = \{M(t)\}_{0 \leqq t \leqq T}$ $(T > 0)$ has bounded variation, then M is a constant.*

Proof We may assume $M(0) = 0$. For a partition $\Delta = \{0 = t_0 < t_1 < \cdots < t_n = T\}$ of $[0, T]$, set

$$V_\Delta^M(T) = \sum_{i=1}^n |M(t_i) - M(t_{i-1})|$$

and denote the supremum with respect to the partition (total variation) by $V^M(T) = \sup_\Delta V_\Delta^M(T)$. By the assumption, we have $\mathbf{P}(V^M(T) < \infty) = 1$.

Similarly, let $V^M(s)$ be the total variation of M in $[0, s]$ and set

$$\tau_K = \inf\{s \geqq 0; V^M(s) \geqq K\}$$

for $K > 0$. Then, τ_K is a stopping time and $M^K = \{M^K(t) = M(t \wedge \tau_K)\}_{t \geqq 0}$ is also a martingale. Hence,

$$\begin{aligned}
\mathbf{E}[(M^K(T))^2] &= \mathbf{E}\Big[\sum_{i=1}^n \{M^K(t_i) - M^K(t_{i-1})\}^2\Big] \\
&\leqq \mathbf{E}\Big[\max_i \{|M^K(t_i) - M^K(t_{i-1})|\} \sum_{i=1}^n |M^K(t_i) - M^K(t_{i-1})|\Big] \\
&\leqq K\mathbf{E}\Big[\max_i \{|M^K(t_i) - M^K(t_{i-1})|\}\Big].
\end{aligned}$$

Since $|M^K(t)| \leqq K$, the right hand side converges to 0 by the bounded convergence theorem as $|\Delta| = \max |t_i - t_{i-1}| \to 0$. Thus $M^K(T) = 0$, $\mathbf{P}$-a.s. Since $K > 0$ is arbitrary, $M(T) = 0$, $\mathbf{P}$-a.s. □

Next we give applications to Brownian motions.

Theorem 1.5.19 *Let $(W, \mathscr{B}(W), \mu)$ be the d-dimensional Wiener space and $\theta = \{\theta(t)\}_{t \geqq 0}$ be the coordinate process. Then, for $T > 0$ and $0 < \varepsilon < 1$,*

$$\mathbf{E}\left[\exp\left(\frac{\varepsilon}{2T} \max_{0 \leqq t \leqq T} |\theta(t)|^2\right)\right] \leqq \mathrm{e}(1 - \varepsilon)^{-\frac{d}{2}}.$$

In particular, for any $\lambda > 0$,

$$\mu\Big(\max_{0\leqq t\leqq T}|\theta(t)| > \lambda\Big) \leqq \mathrm{e}(1-\varepsilon)^{-\frac{d}{2}}\mathrm{e}^{-\frac{\varepsilon\lambda^2}{2T}}.$$

Proof The second assertion follows from the first one by Chebyshev's inequality. Hence we only show the first one. For $y \in \mathbb{R}^d$, consider the martingale given by $\{\exp(\langle y,\theta(t)\rangle - \frac{|y|^2 t}{2})\}_{t\geqq 0}$. By Doob's inequality, for $p > 1$,

$$\begin{aligned}
\mathbf{E}\Big[\max_{0\leqq t\leqq T}\mathrm{e}^{p\langle y,\theta(t)\rangle}\Big] &\leqq \mathrm{e}^{\frac{p|y|^2T}{2}}\mathbf{E}\Big[\max_{0\leqq t\leqq T}\mathrm{e}^{p(\langle y,\theta(t)\rangle-\frac{|y|^2t}{2})}\Big]\\
&\leqq \mathrm{e}^{\frac{p|y|^2T}{2}}\Big(\frac{p}{p-1}\Big)^p\mathbf{E}[\mathrm{e}^{p(\langle y,\theta(T)\rangle-\frac{|y|^2T}{2})}] = \Big(\frac{p}{p-1}\Big)^p\mathrm{e}^{\frac{p^2|y|^2T}{2}}.
\end{aligned}$$

Replace y by py to have

$$\mathbf{E}\Big[\max_{0\leqq t\leqq T}\mathrm{e}^{\langle y,\theta(t)\rangle}\Big] \leqq \Big(\frac{p}{p-1}\Big)^p\mathrm{e}^{\frac{|y|^2T}{2}}.$$

Since $\mathrm{e}^{\frac{|y|^2 s}{2}} = \int_{\mathbb{R}^d}\mathrm{e}^{\langle y,z\rangle}g(s,0,z)\,\mathrm{d}z$,

$$\begin{aligned}
\mathbf{E}\Big[\max_{0\leqq t\leqq T}\mathrm{e}^{\frac{\varepsilon|\theta(t)|^2}{2T}}\Big] &= \mathbf{E}\Big[\max_{0\leqq t\leqq T}\int_{\mathbb{R}^d}\mathrm{e}^{\langle\theta(t),z\rangle}g\Big(\frac{\varepsilon}{T},0,z\Big)\mathrm{d}z\Big]\\
&\leqq \Big(\frac{p}{p-1}\Big)^p\int_{\mathbb{R}^d}\mathrm{e}^{\frac{|z|^2T}{2}}g\Big(\frac{\varepsilon}{T},0,z\Big)\mathrm{d}z = \Big(\frac{p}{p-1}\Big)^p(1-\varepsilon)^{-\frac{d}{2}}.
\end{aligned}$$

Letting $p\to\infty$, we obtain the assertion. □

Theorem 1.5.20 *Let* $(W,\mathscr{B}(W),\mu)$ *be the one-dimensional Wiener space and* $\theta = \{\theta(t)\}_{t\geqq 0}$ *be the coordinate process, and set*

$$\tau_{-a} = \inf\{t>0;\theta(t)=-a\},\quad \tau_b = \inf\{t>0;\theta(t)=b\},\quad \tau=\tau_{-a}\wedge\tau_b$$

for $a,b>0$. *Then,*

$$\mu(\tau=\tau_{-a}) = \frac{b}{a+b}\quad \textit{and}\quad \mathbf{E}[\tau]=ab.$$

Proof Since $\{\theta(t\wedge\tau)\}_{t\geqq 0}$ is a martingale by the optional stopping time, $\mathbf{E}[\theta(t\wedge\tau)] = 0$. Since $-a\leqq\theta(t\wedge\tau)\leqq b$, by the bounded convergence theorem, we obtain, by letting $t\to\infty$,

$$0 = \mathbf{E}[\theta(\tau)] = -a\mu(\tau=\tau_{-a}) + b\mu(\tau=\tau_b).$$

Combine this identity with $\mu(\tau=\tau_{-a})+\mu(\tau=\tau_b)=1$. Then the first assertion follows.

For the second identity, note that $\{\theta(t \wedge \tau)^2 - (t \wedge \tau)\}_{t \geqq 0}$ is a martingale. Then

$$\mathbf{E}[\theta(t \wedge \tau)^2] = \mathbf{E}[t \wedge \tau].$$

By the bounded convergence theorem applied to the left hand side and the monotone convergence theorem to the right hand side, we obtain

$$\mathbf{E}[\tau] = \mathbf{E}[\theta(\tau)^2] = (-a)^2 \mu(\tau = \tau_{-a}) + b^2 \mu(\tau = \tau_b).$$

In conjunction with the first identity, the assertion follows. □

1.5.5 Doob–Meyer Decomposition, Quadratic Variation Process

As Doob's decomposition for discrete time submartingales (Theorem 1.4.11), a submartingale is decomposed into a sum of a martingale and an increasing process, which is called the Doob–Meyer decomposition. From this the quadratic variation processes of martingales are defined.

Let $(\Omega, \mathscr{F}, \mathbf{P}, \{\mathscr{F}_t\})$ be a probability space which satisfies the usual condition. Denote by $\mathbf{S}$ the set of $\{\mathscr{F}_t\}$-stopping times and by $\mathbf{S}_a$ $(a > 0)$ the subset of $\mathbf{S}$ which consists of elements τ satisfying $\tau \leqq a$ almost surely.

Definition 1.5.21 An $\{\mathscr{F}_t\}$-adapted stochastic process X is said to be of **class (D)** if the family of random variables $X(\tau)\mathbf{1}_{\{\tau<\infty\}}$ $(\tau \in \mathbf{S})$ is uniformly integrable, that is,

$$\lim_{\lambda \to \infty} \sup_{\tau \in \mathbf{S}} \mathbf{E}[|X(\tau)|\mathbf{1}_{\{\tau<\infty\}}; |X(\tau)| > \lambda] = 0.$$

If $X(\tau)$ $(\tau \in \mathbf{S}_a)$ is uniformly integrable for any $a > 0$, X is said to be of **class (DL)**.

Definition 1.5.22 A stochastic process $A = \{A(t)\}_{t \geqq 0}$ is called an **increasing process** if

(i) A is $\{\mathscr{F}_t\}$-adapted,
(ii) $A_0 = 0$, and $t \mapsto A(t)$ is right-continuous and non-decreasing,
(iii) for all $t > 0$, $\mathbf{E}[A(t)] < \infty$.

Definition 1.5.23 An increasing process $A = \{A(t)\}_{t \geqq 0}$ is said to be **natural** if, for any bounded martingale $M = \{M(t)\}_{t \geqq 0}$,

$$\mathbf{E}\Big[\int_0^t M(s)\,\mathrm{d}A(s)\Big] = \mathbf{E}\Big[\int_0^t M(s-)\,\mathrm{d}A(s)\Big] \qquad (t > 0),$$

where $M(s-) = \lim_{t \uparrow s} M(t)$.

It is known that, if an increasing process A is continuous, then A is natural. Moreover, an increasing process is natural if and only if it is predictable (see Definition 2.2.1). We omit the details and refer the reader to [13, 45].

Proposition 1.5.24 *All martingales and non-negative submartingales are of class (DL).*

Proof We show the result for martingales. A similar argument is applicable to submartingales. If $M = \{M(t)\}_{t\geqq 0}$ is a martingale, $\{|M_t|\}_{t\geqq 0}$ is a submartingale and, by the optional sampling theorem (Theorem 1.5.11),

$$\mathbf{E}[|M(\sigma)|\mathbf{1}_{\{|M(\sigma)|\geqq\lambda\}}] \leqq \mathbf{E}[|M(a)|\mathbf{1}_{\{|M(\sigma)|\geqq\lambda\}}]$$

for $\sigma \in \mathbf{S}_a$ and $\lambda > 0$. Moreover, we have

$$\sup_{\sigma\in\mathbf{S}_a} \mathbf{P}(|M(\sigma)| \geqq \lambda) \leqq \sup_{\sigma\in\mathbf{S}_a} \lambda^{-1}\mathbf{E}[|M(\sigma)|] \leqq \lambda^{-1}\mathbf{E}[|M(a)|],$$

which implies the uniform integrability

$$\lim_{\lambda\to\infty} \sup_{\sigma\in\mathbf{S}_a} \mathbf{E}[|M(\sigma)|\mathbf{1}_{\{|M(\sigma)|\geqq\lambda\}}] = 0.$$

□

The following is easily obtained.

Corollary 1.5.25 *Let M be a martingale and A be an increasing process. Then the stochastic process $\{X(t)\}_{t\geqq 0}$ given by $X(t) = M(t) + A(t)$ is of class (DL).*

The following theorem gives the converse of this corollary.

Theorem 1.5.26 (Doob–Meyer decomposition) *For a submartingale X of class (DL), there exist a martingale $M = \{M(t)\}_{t\geqq 0}$ and an increasing process $A = \{A(t)\}_{t\geqq 0}$ such that $X(t) = M(t) + A(t)$. We can find A from the natural ones and, in this case, the decomposition is unique.*

Proof We give only an outline of the proof. For details, see [45, 61, 95].

First we fix $T > 0$ and set

$$Y(t) = X(t) - \mathbf{E}[X(T)|\mathscr{F}_t] \qquad (0 \leqq t \leqq T).$$

Since

$$\begin{aligned}\mathbf{E}[Y(t)|\mathscr{F}_s] &= \mathbf{E}[X(t)|\mathscr{F}_s] - \mathbf{E}[\mathbf{E}[X(T)|\mathscr{F}_t]|\mathscr{F}_s] \\ &\geqq X(s) - \mathbf{E}[X(T)|\mathscr{F}_s] = Y(s)\end{aligned}$$

for $s < t$, $\{Y(t)\}_{0\leqq t\leqq T}$ is a submartingale and $Y(T) = 0$.

Second, set $t_j^{(n)} = jT2^{-n}$ and let Δ_n be a partition of $[0,T]$ given by $0 = t_0^{(n)} < t_1^{(n)} < \cdots < t_{2^n}^{(n)} = T$. Moreover, set

$$A^{(n)}(t_k^{(n)}) = \sum_{j=1}^{k} \{\mathbf{E}[Y(t_j^{(n)})|\mathscr{F}_{t_{j-1}}^{(n)}] - Y(t_{j-1}^{(n)})\} \qquad (k = 1, 2, \ldots, 2^n).$$

Then $\{A^{(n)}(T)\}_{n=1}^{\infty}$ is uniformly integrable. Hence, there exist a subsequence $\{n_\ell\}_{\ell=1}^{\infty}$ and an integrable random variable $A(T)$ such that $A^{(n_\ell)}(T)$ converges weakly to $A(T)$ in L^1. That is, for any bounded random variable U,

$$\lim_{\ell\to\infty} \mathbf{E}[A^{(n_\ell)}(T)U] = \mathbf{E}[A(T)U].$$

Now define $A = \{A(t)\}_{0 \leqq t \leqq T}$ by

$$A(t) = Y(t) + \mathbf{E}[A(T)|\mathscr{F}_t].$$

Then A is a natural increasing process and

$$X(t) = \mathbf{E}[X(T) - A(T)|\mathscr{F}_t] + A(t),$$

which is the desired decomposition. □

Corollary 1.5.27 (1) *Let M be a square-integrable $\{\mathscr{F}_t\}$-martingale with* $M(0) = 0$, **P**-a.s. *Then, there exists a natural increasing process $A = \{A(t)\}_{t\geqq 0}$ such that $\{M(t)^2 - A(t)\}_{t\geqq 0}$ is an $\{\mathscr{F}_t\}$-martingale.*
(2) *Let M, N be square-integrable $\{\mathscr{F}_t\}$-martingales with* $M(0) = N(0) = 0$, **P**-a.s. *Then there exists a stochastic process $A = \{A(t)\}$ which is represented as a difference of two natural increasing processes such that $\{M(t)N(t) - A(t)\}_{t\geqq 0}$ is an $\{\mathscr{F}_t\}$-martingale.*

Proof Since $\{M(t)^2\}_{t\geqq 0}$ is a submartingale, we can directly apply the Doob–Meyer decomposition to obtain (1). (2) is obtained by polarization. □

Definition 1.5.28 The process A in Corollary 1.5.27 (1) and (2) will be denoted by $\langle M\rangle = \{\langle M\rangle(t)\}_{t\geqq 0}$ and $\langle M, N\rangle = \{\langle M, N\rangle(t)\}_{t\geqq 0}$, respectively. $\langle M\rangle$ is called the **quadratic variation process** of M.

$\langle M, N\rangle$ is linear in M and N, and $\langle M, M\rangle = \langle M\rangle$.

Definition 1.5.29 The set of square-integrable $\{\mathscr{F}_t\}$-martingales defined on $(\Omega, \mathscr{F}, \mathbf{P}, \{\mathscr{F}_t\})$ is denoted by $\mathscr{M}^2(\{\mathscr{F}_t\})$ or $\mathscr{M}^2$:

$$\mathscr{M}^2 = \left\{ M = \{M(t)\}_{t\geqq 0}\ ;\ \begin{array}{l} M \text{ is an } \{\mathscr{F}_t\}\text{-martingale and} \\ \text{satisfies } \mathbf{E}[M(t)^2] < \infty \text{ for any } t \geqq 0. \end{array} \right\}.$$

The set of continuous elements in $\mathscr{M}^2$ is denoted by $\mathscr{M}_c^2$, and the set of the restrictions to the time interval $[0, T]$ of elements in $\mathscr{M}^2$ and $\mathscr{M}_c^2$ are denoted by $\mathscr{M}^2(T)$ and $\mathscr{M}_c^2(T)$, respectively.

Lemma 1.5.30 (1) *Identify two elements in $\mathscr{M}^2(T)$ or $\mathscr{M}_c^2(T)$ if one is a modification of the other. Then the set of equivalence classes $\overline{\mathscr{M}}^2(T)$ and $\overline{\mathscr{M}}_c^2(T)$ are Hilbert spaces with the inner product $\langle\langle M, M'\rangle\rangle_T := \mathbf{E}[M_T M'_T]$. Moreover, $\overline{\mathscr{M}}_c^2(T)$ is a closed subspace of $\overline{\mathscr{M}}^2(T)$.*
(2) *$\overline{\mathscr{M}}^2$ and $\overline{\mathscr{M}}_c^2$ are complete metric spaces with metric*

$$d(M, N) := \sum_{n=1}^{\infty} 2^{-n} \min\{\|M - N\|_n, 1\},$$

where $\|M\|_n = \sqrt{\langle\langle M, M\rangle\rangle_n}$.

The lemma is a straightforward consequence of Doob's inequality.

1.6 Adapted Brownian Motion

Let $(\Omega, \mathscr{F}, \mathbf{P}, \{\mathscr{F}_t\})$ be a filtered probability space.

Definition 1.6.1 Let $\mathbb{T}$ be $[0, T]\,(T > 0)$ or $[0, \infty)$. A stochastic process $B = \{B(t)\}_{t\in\mathbb{T}}$ defined on $(\Omega, \mathscr{F}, \mathbf{P}, \{\mathscr{F}_t\})$ is called a d-dimensional **$\{\mathscr{F}_t\}$-Brownian motion** if the following two conditions are satisfied:

(i) B is an $\{\mathscr{F}_t\}$-adapted $\mathbb{R}^d$-valued continuous stochastic process,
(ii) for any $0 \leqq s \leqq t, \lambda \in \mathbb{R}^d$,

$$\mathbf{E}[\mathrm{e}^{\mathrm{i}\langle\lambda, B(t)-B(s)\rangle}|\mathscr{F}_s] = \mathrm{e}^{-\frac{|\lambda|^2(t-s)}{2}}.$$

Moreover, if $\mathbf{P}(B(0) = x) = 1$ for $x \in \mathbb{R}^d$, we call B a d-dimensional $\{\mathscr{F}_t\}$-Brownian motion starting from x.

Let $B^0 = \{B^0(t)\}_{t\geqq 0}$ be the continuous stochastic process given by $B^0(t) = B(t) - B(0)$ for a d-dimensional $\{\mathscr{F}_t\}$-Brownian motion $B = \{B(t)\}_{t\geqq 0}$. The probability measure on the path space W induced by B^0 is the Wiener measure μ. In particular, for any $0 = t_0 < t_1 < \cdots < t_n$, $B(t_1) - B(t_0), \ldots, B(t_n) - B(t_{n-1})$ are independent and the distribution of $B(t_i) - B(t_{i-1})$ is the d-dimensional Gaussian distribution with mean 0 and covariance matrix $(t_i - t_{i-1})I$.

Conversely, let $\{\mathscr{B}_t^0\}$ be a filtration on the d-dimensional Wiener space $(W, \mathscr{B}(W), \mu)$ on $[0,\infty)$ given by

$$\mathscr{B}_t^0 = \sigma(\{\theta(s), s \leqq t\}) \qquad (t \geqq 0),$$

where $\{\theta(t)\}_{t\geqq 0}$ is the coordinate process on W (see (1.3.1)). Then $\{\theta(t)\}_{t\geqq 0}$ is a $\{\mathscr{B}_t^0\}$-Brownian motion starting from 0.

We mention several properties of the paths of Brownian motions. Since the components of a general dimensional Brownian motion are one-dimensional Brownian motions and are independent, we only consider the one-dimensional case.

Let $\{B(t)\}_{t\geqq 0}$ be a one-dimensional $\{\mathscr{F}_t\}$-Brownian motion on $[0,\infty)$ starting from 0 defined on $(\Omega, \mathscr{F}, \mathbf{P}, \{\mathscr{F}_t\})$. The next proposition is shown by an elementary computation.

Proposition 1.6.2 *Let $s < t$. Then, for $n = 1, 2, \ldots,$*

$$\mathbf{E}[|B(t) - B(s)|^{2n}] = (2n-1)(2n-3)\cdots 3\cdot 1(t-s)^n = \frac{(2n)!}{2^n \cdot n!}(t-s)^n.$$

In stochastic analysis, the following property on the variations of Brownian motion is fundamental.

Theorem 1.6.3 (1) *For $T > 0$ let $\Delta_n = \{0 = t_0^n < \cdots < t_{m_n}^n = T\}$ be a sequence of partitions of $[0,T]$ which satisfies $\Delta_n \subset \Delta_{n+1}$ $(n = 1, 2, \ldots)$ and $|\Delta_n| := \max(t_j^n - t_{j-1}^n) \to 0$ $(n \to \infty)$. Then*

$$\mathbf{P}\Big(\lim_{n\to\infty} \sum_{j=1}^{m_n} |B(t_j^n) - B(t_{j-1}^n)| = \infty\Big) = 1.$$

(2) *For a sequence $\{\Delta_n\}$ of partitions in (1),*

$$\lim_{n\to\infty} \mathbf{E}\Big[\Big(\sum_{j=1}^{m_n} |B(t_j^n) - B(t_{j-1}^n)|^2 - T\Big)^2\Big] = 0.$$

Proof While (1) can be shown from (2) by contradiction, we give a direct proof. Set $V_n = \sum_{j=1}^{m_n} |B(t_j^n) - B(t_{j-1}^n)|$. Then, since $\Delta_n \subset \Delta_{n+1}$, $\{V_n\}_{n=1}^{\infty}$ is non-decreasing and has a limit V as $n \to \infty$, admitting $V = \infty$. Hence, by the bounded convergence theorem, one has

$$\mathbf{E}[\mathrm{e}^{-V}] = \lim_{n\to\infty} \mathbf{E}[\mathrm{e}^{-V_n}] = \lim_{n\to\infty} \prod_{j=1}^{m_n} \int_{\mathbb{R}} \mathrm{e}^{-|x|} g(t_j^n - t_{j-1}^n, 0, x)\,\mathrm{d}x.$$

By the inequality $e^{-x} \leqq 1 - x + \frac{x^2}{2}$ $(x > 0)$,

$$\int_{\mathbb{R}} e^{-|x|} \frac{1}{\sqrt{2\pi s}} e^{-\frac{x^2}{2s}} dx \leqq 2 \int_0^{\infty} \frac{1}{\sqrt{2\pi s}} \Big(1 - x + \frac{x^2}{2}\Big) e^{-\frac{x^2}{2s}} dx = 1 - \sqrt{\frac{2s}{\pi}} + \frac{s}{2}$$

and

$$\mathbf{E}[e^{-V}] \leqq \limsup_{n\to\infty} \prod_{j=1}^{m_n} \Big\{1 - \sqrt{\frac{2}{\pi}} (t_j^n - t_{j-1}^n)^{\frac{1}{2}} + \frac{1}{2} (t_j^n - t_{j-1}^n)\Big\}.$$

Since $\sum_{j=1}^{m_n} (t_j^n - t_{j-1}^n)^{\frac{1}{2}} \to \infty$ $(n \to \infty)$, $\mathbf{E}[e^{-V}] = 0$ and $V = \infty$, $\mathbf{P}$-a.s.
(2) Set $D_{n,j} = |B(t_j^n) - B(t_{j-1}^n)|^2 - (t_j^n - t_{j-1}^n)$ $(j = 1, 2, \ldots, m_n)$. For each n, $\{D_{n,j}\}_{j=1}^{m_n}$ is an independent sequence of random variables. $D_{n,j}$ has mean 0 and variance $2(t_j^n - t_{j-1}^n)^2$ by Proposition 1.6.2. Hence,

$$\begin{aligned} \mathbf{E}\Big[\Big(\sum_{j=1}^{m_n} |B(t_j^n) - B(t_{j-1}^n)|^2 - T\Big)^2\Big] &= \mathbf{E}\Big[\Big(\sum_{j=1}^{m_n} D_{n,j}\Big)^2\Big] \\ &= \sum_{j=1}^{m_n} \mathbf{E}[(D_{n,j})^2] + 2 \sum_{i<j}^{m_n} \mathbf{E}[D_{n,i} D_{n,j}] \\ &= \sum_{j=1}^{m_n} 2(t_j^n - t_{j-1}^n)^2 \leqq 2|\Delta_n| \sum_{j=1}^{m_n} (t_j^n - t_{j-1}^n) \to 0. \end{aligned}$$

□

Theorem 1.6.3 shows that the paths of Brownian motions are not smooth. In fact, the following is well known.

Theorem 1.6.4 *Almost every path of Brownian motions is nowhere differentiable.*

On the continuity and uniform continuity of paths, the following is known.

Theorem 1.6.5 (Khinchin's law of iterated logarithm) *Let B be a Brownian motion with $B(0) = 0$. Then, almost surely,*

$$\limsup_{t\downarrow 0} \frac{B(t)}{\sqrt{2t \log\log(\frac{1}{t})}} = 1 \quad \textit{and} \quad \liminf_{t\downarrow 0} \frac{B(t)}{\sqrt{2t \log\log(\frac{1}{t})}} = -1.$$

Theorem 1.6.6 (Lévy's modulus of continuity) *For any $T > 0$*

$$\mathbf{P}\Big(\limsup_{\delta\downarrow 0} \sup_{\substack{0\leqq s<t\leqq T \\ t-s<\delta}} \frac{|B(t) - B(s)|}{\sqrt{2\delta \log(\frac{1}{\delta})}} = \sqrt{T}\Big) = 1.$$

We omit the proofs. For the proofs and more properties, see [56].

Remark 1.6.7 Theorem 1.6.5 and Remark 1.2.9 imply

$$\limsup_{t\to\infty} \frac{B(t)}{\sqrt{2t\log\log t}} = 1, \quad \liminf_{t\to\infty} \frac{B(t)}{\sqrt{2t\log\log t}} = -1.$$

We discuss the asymptotic behavior of Brownian motion as $t \to \infty$ in Section 3.2.

1.7 Cameron–Martin Theorem

Let $(W_T, \mathscr{B}(W_T), \mu_T)$ be the d-dimensional Wiener space on $[0,T]$. The Wiener measure μ_T is determined by the Cameron–Martin subspace H_T by the Itô–Nisio theorem (Theorem 1.2.5). Moreover, if we define a transform T_h ($h \in W_T$) of the Wiener space W_T by $T_h(w) = w + h$, the induced measure $\mu_T \circ T_h^{-1}$ and μ_T is mutually absolutely continuous if and only if $h \in H_T$. The purpose of this section is to prove this celebrated Cameron–Martin theorem.[6]

We now introduce the Wiener integral for $h \in H_T$. While the integral is a special case of the stochastic integral (Section 2.2), we introduce it in another way. For $\ell \in W_T^*$, let $\mathscr{I}(\ell)$ be the random variable given by $\mathscr{I}(\ell)(w) = \ell(w), w \in W_T$.

Lemma 1.7.1 *The distribution of $\mathscr{I}(\ell)$ is Gaussian with mean 0 and variance $\|\ell\|_{H_T}^2$, where we have used the inclusion that $W_T^* \subset H_T^* = H_T$.*

By the lemma,

$$\mathbf{E}[\mathscr{I}(\ell)^2] =: \|\mathscr{I}(\ell)\|_{L^2(\mu_T)}^2 = \|\ell\|_{H_T}^2.$$

Thus the mapping $\mathscr{I} : H_T \supset W_T^* \ni \ell \mapsto \mathscr{I}(\ell) \in L^2(\mu_T)$ extends to an isometry from H_T to $L^2(\mu_T)$, say $\mathscr{I}$ again. The random variable $\mathscr{I}(h), h \in H_T$, is called the **Wiener integral** of $h \in H_T$.

Proof Since $W_T = (W_T^1)^d$, it suffices to show the assertion when $d = 1$.

Let $\ell \in W_T^*$. We first compute the norm $\|\ell\|_{H_T}$. By the Riesz–Markov–Saks–Kakutani theorem ([97, Theorem 9.5.1]), there exists a bounded variation function $F : [0,T] \to \mathbb{R}$ such that

$$\ell(w) = \int_0^T w(t)\mathrm{d}F(t) \qquad (w \in W_T).$$

[6] The result in this section will not be used until Chapter 4 and the reader can skip this section.

By the integration by parts formula,

$$\ell(h) = \int_0^T \dot{h}(t)(F(T) - F(t))\mathrm{d}t \quad (h \in H_T).$$

Hence

$$\|\ell\|_{H_T}^2 = \int_0^T (F(T) - F(t))^2 \mathrm{d}t. \tag{1.7.1}$$

We now show that $\mathscr{I}(\ell)$ is a Gaussian random variable. Put

$$\mathscr{I}^{(n)}(\ell)(w) = \sum_{m=0}^{n-1} w\Big(\frac{mT}{n}\Big)\Big\{F\Big(\frac{(m+1)T}{n}\Big) - F\Big(\frac{mT}{n}\Big)\Big\}.$$

By the very definition of the Lebesgue–Stieltjes integral,

$$\mathscr{I}(\ell)(w) = \lim_{n\to\infty} \mathscr{I}^{(n)}(\ell)(w).$$

Moreover, by a simple algebraic manipulation, we have

$$\mathscr{I}^{(n)}(\ell)(w) = \sum_{m=0}^{n-1}\Big\{w\Big(\frac{(m-1)T}{n}\Big) - w\Big(\frac{mT}{n}\Big)\Big\}\Big\{F\Big(\frac{mT}{n}\Big) - F(T)\Big\}.$$

Since the distribution of $w(\frac{mT}{n}) - w(\frac{(m-1)T}{n})$ is Gaussian with mean 0 and variance $\frac{T}{n}$, $\mathscr{I}^{(n)}(h)$ is also a Gaussian random variable with mean 0 and variance $\frac{T}{n}\sum_{m=1}^{n-1}\{F(\frac{mT}{n}) - F(T)\}^2$. Therefore, by the bounded convergence theorem and (1.7.1),

$$\mathbf{E}[\mathrm{e}^{\mathrm{i}\lambda\mathscr{I}(\ell)}] = \lim_{n\to\infty}\mathbf{E}[\mathrm{e}^{\mathrm{i}\lambda\mathscr{I}^{(n)}(\ell)}] = \mathrm{e}^{-\frac{1}{2}\|\ell\|_{H_T}^2\lambda^2} \qquad (\lambda \in \mathbb{R}).$$

Hence $\mathscr{I}(\ell)$ is a Gaussian random variable with mean 0 and variance $\|\ell\|_{H_T}^2$. □

Theorem 1.7.2 (1) *If $h \in H_T$, then the distribution $\mu_T \circ T_h^{-1}$ of $T_h(w)$ is mutually absolutely continuous with respect to μ_T and the associated density function is* $\exp(\mathscr{I}(h)(w) - \frac{\|h\|_H^2}{2})$. *That is, the expectation of $F(w+h)$, F being a bounded continuous function on W_T, is given by*

$$\begin{aligned}\int_{W_T} F(w)(\mu_T \circ T_h^{-1})(\mathrm{d}w) &= \int_{W_T} F(w+h)\mu_T(\mathrm{d}w)\\ &= \int_{W_T} F(w)\mathrm{e}^{\mathscr{I}(h)(w) - \frac{\|h\|_H^2}{2}}\mu_T(\mathrm{d}w).\end{aligned} \tag{1.7.2}$$

(2) *If $h \notin H_T$, $\mu_T \circ T_h^{-1}$ is singular to μ_T.*

Proof (1) From the linearity of the Wiener integral, we have

$$\mathbf{E}[\mathrm{e}^{\xi\mathscr{I}(h_1)+\eta\mathscr{I}(h_2)}] = \exp\left(\frac{\xi^2\|h_1\|_{H_T}^2}{2} + \xi\eta\langle h_1, h_2\rangle_{H_T} + \frac{\eta^2\|h_2\|_{H_T}^2}{2}\right) \tag{1.7.3}$$

for any $\xi, \eta \in \mathbb{R}$, $h_1, h_2 \in H_T$. Moreover, this identity holds for all $\xi, \eta \in \mathbb{C}$ by Lemma 1.5.7.

Let $\ell \in W_T^*$ and $h \in H_T$. Since $\mathscr{I}(\ell)(w+h) = \mathscr{I}(\ell)(w) + \langle \ell, h\rangle_{H_T}$ $(w \in W_T)$, (1.7.3) implies

$$\int_{W_T} \mathrm{e}^{\mathrm{i}\mathscr{I}(\ell)}\mathrm{d}(\mu_T \circ T_h^{-1}) = \mathrm{e}^{-\frac{\|\ell\|_{H_T}^2}{2}+\mathrm{i}\langle \ell, h\rangle_{H_T}} = \int_{W_T} \mathrm{e}^{\mathrm{i}\mathscr{I}(\ell)}\mathrm{e}^{\mathscr{I}(h)-\frac{\|h\|_{H_T}^2}{2}}\mathrm{d}\mu_T.$$

Substituting $\ell = \sum_{i=1}^n \langle \xi_i, w(t_i)\rangle_{\mathbb{R}^d}$ into this identity, where $0 < t_1 < \cdots < t_n \leqq T$ and $\xi_1, \ldots, \xi_n \in \mathbb{R}^d$, we obtain

$$\begin{aligned}&\int_{W_T} \mathrm{e}^{\mathrm{i}\sum_{i=1}^n\langle \xi_i, w(t_i)\rangle_{\mathbb{R}^d}}(\mu_T \circ T_h^{-1})(\mathrm{d}w)\\ &\qquad = \int_{W_T} \mathrm{e}^{\mathrm{i}\sum_{i=1}^n\langle \xi_i, w(t_i)\rangle_{\mathbb{R}^d}}\mathrm{e}^{\mathscr{I}(h)-\frac{\|h\|_{H_T}^2}{2}}\mu_T(\mathrm{d}w).\end{aligned}$$

This implies (1.7.2).

(2) For $h \notin H_T$ we have

$$\sup\{|\varphi(h)|;\ \varphi \in W_T^*,\ \|\varphi\|_{H_T^*} = \|\varphi\|_{H_T} = 1\} = \infty.$$

Since W_T^* is dense in H_T, there exists a sequence $\{f_j\}_{j=1}^\infty$ in W_T^* such that

$$\langle f_i, f_j\rangle_{H_T} = \delta_{ij}, \qquad f_j(h) \geqq 2j \qquad (j = 1, 2, \ldots).$$

By definition, $\mathscr{I}(f_j) = f_j$, $j = 1, 2, \ldots$, and hence $\{f_j\}_{j=1}^\infty$ forms a sequence of independent and standard normally distributed random variables.

Set $A_j = \{f_j \leqq j\}$ and $A = \bigcup_{n=1}^\infty \bigcap_{j\geqq n} A_j = \liminf_{n\to\infty} A_n$. We shall show that $\mu_T(A) = 1$ and $(\mu_T \circ T_h^{-1})(A) = 0$, which implies the assertion.

Observe first that

$$\mu_T(A_j) = 1 - \int_j^\infty \frac{1}{\sqrt{2\pi}}\mathrm{e}^{-\frac{x^2}{2}}\mathrm{d}x \geqq 1 - \int_j^\infty \frac{x}{j}\frac{1}{\sqrt{2\pi}}\mathrm{e}^{-\frac{x^2}{2}}\mathrm{d}x = 1 - \frac{1}{\sqrt{2\pi}j}\mathrm{e}^{-\frac{j^2}{2}}.$$

Since $\sum_j j^{-1}\mathrm{e}^{-\frac{j^2}{2}} < \infty$, this yields that

$$\mu_T\Big(\bigcap_{j\geqq n} A_j\Big) \geqq \prod_{j\geqq n}\Big(1 - \frac{1}{\sqrt{2\pi}j}\mathrm{e}^{-\frac{j^2}{2}}\Big) \to 1 \qquad (n \to \infty)$$

and hence $\mu_T(A) = 1$.

Next note that

$$(\mu_T \circ T_h^{-1})(A_j) = \mu_T(f_j + f_j(h) \leqq j) \leqq \mu_T(f_j \leqq -j) = \int_{-\infty}^{-j} \frac{1}{\sqrt{2\pi}} \mathrm{e}^{-\frac{x^2}{2}} \mathrm{d}x.$$

This implies $(\mu_T \circ T_h^{-1})\Big(\bigcap_{j \geqq n} A_j\Big) = 0$ and $(\mu_T \circ T_h^{-1})(A) = 0$. □

Theorem 1.7.3 *If $h \in H_T \setminus W_T^*$, then $\mathscr{I}(h)$ is not continuous on W_T.*

Proof Let $h \in H_T \setminus W_T^*$ and suppose that $\mathscr{I}(h) : W_T \to \mathbb{R}$ is continuous on W_T. Take a sequence $\{\ell_n\} \subset W_T^*$ with $\|\ell_n - h\|_{H_T} \to 0$. Since the random variable $\mathscr{I}(\ell_n)$ converges to $\mathscr{I}(h)$ in L^2, we may assume that it converges almost surely, taking a subsequence if necessary. For each $k \in H_T$, by the Cameron–Martin theorem, $\mathscr{I}(\ell_n)(\cdot + k)$ converges to $\mathscr{I}(h)(\cdot + k)$ almost surely. Recalling that

$$\mathscr{I}(\ell_n)(w + k) = \mathscr{I}(\ell_n)(w) + \langle \ell_n, k\rangle_{H_T} \qquad (w \in W_T)$$

and then letting $n \to \infty$, we obtain

$$\mathscr{I}(h)(\cdot + k) = \mathscr{I}(h)(\cdot) + \langle h, k\rangle_H, \quad \mu_T\text{-a.e.}$$

Now let $\{k_n\}_{n=1}^\infty \subset H_T$ be a dense countable subset of W_T. By the observation above, there exists an $N \in \mathscr{B}(W_T)$ such that $\mu_T(N) = 0$ and

$$\langle h, k_n\rangle_{H_T} = \mathscr{I}(h)(w + k_n) - \mathscr{I}(h)(w) \qquad (w \notin N,\ n = 1, 2, \ldots).$$

In particular, for a fixed $w_0 \notin N$,

$$\langle h, k_n\rangle_{H_T} = \mathscr{I}(h)(w_0 + k_n) - \mathscr{I}(h)(w_0) \qquad (n = 1, 2, \ldots).$$

Since $\{k_n\}$ is dense in W_T and $\mathscr{I}(h) : W_T \to \mathbb{R}$ is continuous, $\langle h, \cdot\rangle_{H_T}$ extends to a continuous linear operator on W_T. This means $h \in W_T^*$, which is a contradiction. □

By the definition and the above theorem, a Wiener integral $\mathscr{I}(h)$ is continuous on W_T if and only if $h \in W_T^*$. This discontinuity of Wiener integrals is an obstacle to carrying out analysis on W_T. Such a difficulty will be overcome by the Malliavin calculus.

1.8 Schilder's Theorem

On the d-dimensional Wiener space $(W_T, \mathscr{B}(W_T), \mu_T)$ on $[0, T]$, we define a transform X_ε ($\varepsilon > 0$) on W_T by

$$X_\varepsilon(w)(t) = \sqrt{\varepsilon}\, w(t) \qquad (0 \leqq t \leqq T,\ w \in W_T).$$

As $\varepsilon \to 0$, $X_\varepsilon(w)$ converges to the constant path 0 for each w and the distribution $\mu_T^\varepsilon := \mu_T \circ X_\varepsilon^{-1}$ of $X_\varepsilon(w)$ converges to the Dirac measure concentrated at this path. The following theorem due to Schilder ([102]) asserts the exponential decay of the μ_T^ε measure of a subset of W_T which does not contain 0. The theorem is important from the theoretical and applied point of view; for example, it is applicable to prove the law of the iterated logarithm on the Wiener space. Moreover, it is also important to indicate the close relationship between Wiener integrals and Feynman integrals, as is shown in the next section. For more details, see [14].

Theorem 1.8.1 *Define the function J on W_T by*

$$J(w) = \begin{cases} \frac{1}{2}\|w\|_{H_T}^2 & (w \in H_T), \\ \infty & (w \in W_T \setminus H_T). \end{cases}$$

(1) *For any open set $G \subset W_T$,*

$$\liminf_{\varepsilon \downarrow 0} \varepsilon \log \mu_T^\varepsilon(G) \geqq - \inf_{w \in G} J(w). \tag{1.8.1}$$

(2) *For any closed set $F \subset W_T$,*

$$\limsup_{\varepsilon \downarrow 0} \varepsilon \log \mu_T^\varepsilon(F) \leqq - \inf_{w \in F} J(w). \tag{1.8.2}$$

Proof (1) We show

$$\liminf_{\varepsilon \downarrow 0} \varepsilon \log \mu_T^\varepsilon(G) \geqq -J(h)$$

for any $h \in G \cap H_T$. For any $\eta > 0$ there exists $g \in G \cap H_T$ of C^2-class such that

$$J(g) \leqq J(h) + \eta. \tag{1.8.3}$$

Since G is an open set, we can choose $\delta > 0$ such that

$$B(g, \delta) := \{w \in W_T; \|w - g\| < \delta\} \subset G,$$

where $\|w\| = \max_{t \in [0,T]} |w(t)|$. Then we have

$$\mu_T^\varepsilon(G) \geqq \mu_T^\varepsilon(B(g, \delta)) = \mu_T(\varepsilon^{\frac{1}{2}}(w - \varepsilon^{-\frac{1}{2}} g) \in B(0, \delta)).$$

By the Cameron–Martin theorem (Theorem 1.7.2),

$$\begin{aligned} \mu_T^\varepsilon(G) &\geqq \int_{W_T} \mathbf{1}_{B(0,\delta)}(\sqrt{\varepsilon} w) \exp\Big(-\frac{1}{\sqrt{\varepsilon}} \mathscr{I}(g)(w) - \frac{1}{\varepsilon} J(g)\Big) \mu_T(\mathrm{d}w) \\ &\geqq \int_{W_T} \mathbf{1}_{B(0,\delta)}(w) \exp\Big(-\frac{1}{\varepsilon} \mathscr{I}(g)(w) - \frac{1}{\varepsilon} J(g)\Big) \mu_T^\varepsilon(\mathrm{d}w) \end{aligned}$$

$$= \int_{B(0,\delta)} \exp\Big(\frac{1}{\varepsilon}\int_0^T \langle w(t), \ddot{g}(t)\rangle \mathrm{d}t - \frac{1}{\varepsilon}\langle w(T), \dot{g}(T)\rangle - \frac{1}{\varepsilon}J(g)\Big)\mu_T^\varepsilon(\mathrm{d}w).$$

Since $|w(t)| < \delta$ ($t \in [0,T]$) for any $w \in B(0,\delta)$, in conjunction with (1.8.3), this yields

$$\log(\mu_T^\varepsilon(G)) \geqq \log(\mu_T^\varepsilon(B(0,\delta))) - \frac{1}{\varepsilon}\delta T\|\ddot{g}\| - \frac{1}{\varepsilon}\delta\|\dot{g}\| - \frac{1}{\varepsilon}(J(h)+\eta).$$

Hence,

$$\liminf_{\varepsilon\downarrow 0} \varepsilon \log \mu_T^\varepsilon(G) \geqq -\delta T\|\ddot{g}\| - \delta\|\dot{g}\| - J(h) - \eta$$

because $\mu_\varepsilon(B(0,\delta)) \to 1$ as $\varepsilon \downarrow 0$. Letting $\eta, \delta \to 0$, we arrive at (1.8.1).
(2) First let $\delta > 0$ and set

$$F^{(\delta)} = \bigcup_{w\in F} B(w,\delta).$$

Using the piecewise linear approximation $\ell_m(w)$ of $w \in W_T$ given in Example 1.2.7, we have

$$\mu_T^\varepsilon(F) \leqq \mu_T^\varepsilon(\{w; \ell_m(w) \in F^{(\delta)}\}) + \mu_T^\varepsilon(\{w; \|\ell_m(w) - w\| \geqq \delta\}).$$

Set

$$\alpha_\delta = \inf_{w\in F^{(\delta)}} J(w) \geqq 0.$$

We shall complete the proof of the upper estimate (1.8.2) by showing the following three assertions.
(I) For any $\delta > 0$

$$\limsup_{\varepsilon\downarrow 0} \varepsilon \log \mu_T^\varepsilon(\{w; \ell_m(w) \in F^{(\delta)}\}) \leqq -\alpha_\delta.$$

(II) For $m = 1, 2, \ldots$

$$\limsup_{\varepsilon\downarrow 0} \varepsilon \log \mu_T^\varepsilon(\{w; \|\ell_m(w) - w\| \geqq \delta\}) \leqq -\frac{2^{m-4}\delta^2}{T}.$$

(III) As $\delta \downarrow 0$, $\alpha_\delta \to \inf_{w\in F} J(w)$.
From (I) and (II), we obtain

$$\limsup_{\varepsilon\downarrow 0} \varepsilon \log \mu_T^\varepsilon(F) \leqq -\min\Big\{\alpha_\delta, \frac{2^{m-4}\delta^2}{T}\Big\} = -\alpha_\delta$$

for large m. This, combined with (III), implies the upper estimate.
We prove the three claims in order. By the definition of α_δ,

$$\mu_T^\varepsilon(\{w; \ell_m(w) \in F^{(\delta)}\}) \leqq \mu_T^\varepsilon(\{w; J(\ell_m(w)) \geqq \alpha_\delta\}).$$

Using the same notations as in Example 1.2.7, we have

$$J(\ell_m(w)) = \frac{1}{2}\sum_{i=1}^{d}\sum_{j=0}^{2^m-1}(\mathscr{I}(h^i_j)(w))^2.$$

Since $\mathscr{I}(h^i_j)$ $(i = 1, \dots, d,\ j = 0, \dots, 2^m-1)$ are independent and obey the standard normal distribution under the Wiener measure, $2J(\ell_m(w))$ is a χ^2 random variable with $d2^m$ degrees of freedom. Therefore,

$$\begin{aligned}\mu^{\varepsilon}_T(\{w; J(\ell_m(w)) \geqq \alpha_\delta\}) &= \mu_T(\{w; \varepsilon J(\ell_m(w)) \geqq \alpha_\delta\}) \\ &= \int_{2\alpha_{\frac{\delta}{\varepsilon}}}^{\infty} \frac{1}{2^{d2^{m-1}}\Gamma(d2^{m-1})} x^{d2^{m-1}-1}\mathrm{e}^{-\frac{x}{2}}\,\mathrm{d}x.\end{aligned}$$

By an elementary inequality

$$\begin{aligned}\int_a^\infty x^{p-1}\mathrm{e}^{-\frac{x}{2}}\mathrm{d}x &= \int_a^\infty x^{p+1}\mathrm{e}^{-\frac{x}{2}}\frac{\mathrm{d}x}{x^2} \\ &\leqq \exp\Big(\sup_{x\geqq a}\Big(-\frac{x}{2} + (p+1)\log x\Big)\Big)\int_a^\infty \frac{\mathrm{d}x}{x^2} \\ &\leqq \frac{1}{a}\exp\Big(-\frac{a}{2} + (p+1)\log a\Big)\end{aligned}$$

for $a, p \in \mathbb{R}$ with $a \geqq 2(p+1)$ $(p > 0)$, we obtain (I) by setting $a = \frac{2\alpha_\delta}{\varepsilon}$.

In order to show (II), note that

$$\begin{aligned}&\mu^{\varepsilon}_T\Big(\Big\{w; \|\ell_m(w) - w\| \geqq \delta\Big\}\Big) \\ &\leqq \sum_{j=0}^{2^m-1}\mu^{\varepsilon}_T\Big(\Big\{w; \max_{\frac{jT}{2^m}\leqq t\leqq\frac{(j+1)T}{2^m}} |\ell_m(w)(t) - w(t)| \geqq \delta\Big\}\Big) \\ &\leqq 2^m\mu^{\varepsilon}_T\Big(\Big\{w; \max_{0\leqq t\leqq\frac{T}{2^m}} |\ell_0(w)(t) - w(t)| \geqq \delta\Big\}\Big)\end{aligned}$$

and, if $\max_{0\leqq t\leqq\frac{T}{2^m}} |w(t)| \leqq \frac{c}{2}$,

$$\max_{0\leqq t\leqq\frac{T}{2^m}} |\ell_0(w)(t) - w(t)| \leqq \max_{0\leqq t\leqq\frac{T}{2^m}} |\ell_0(w)(t)| + \max_{0\leqq t\leqq\frac{T}{2^m}} |w(t)| < c.$$

Then, by using Theorem 1.5.19, we obtain

$$\begin{aligned}\mu^{\varepsilon}_T(\{w; \|\ell_m(w) - w\| \geqq \delta\}) &\leqq 2^m\mu_T\Big(\Big\{w; \max_{0\leqq t\leqq\frac{T}{2^m}} |w(t)| \geqq \frac{\delta}{2\sqrt{\varepsilon}}\Big\}\Big) \\ &\leqq \mathrm{e}\,2^{m+\frac{d}{2}}\exp\Big(-\frac{2^m\delta^2}{16\varepsilon T}\Big).\end{aligned}$$

This implies the assertion (II).

Finally, in order to show (Ⅲ), we note that $\delta \mapsto \alpha_\delta$ is non-increasing and $\lim_{\delta\to 0}\alpha_\delta$ exists, admitting that it is ∞. When the limit is ∞,

$$\alpha_\delta \leqq \inf_{w\in F} J(w) \tag{1.8.4}$$

implies (Ⅲ) in the sense that both sides are ∞.

Assume that $\lim_{\delta\to 0}\alpha_\delta$ is finite. Then, $\alpha_{\frac{1}{n}} \leqq \lim_{\delta\to 0}\alpha_\delta$ $(n = 1, 2, \ldots)$ and we can choose $f_n \in F^{(\frac{1}{n})}$ such that

$$\alpha_{\frac{1}{n}} \leqq J(f_n) \leqq \alpha_{\frac{1}{n}} + \frac{1}{n}.$$

Since $\{f_n\}$ is a bounded set in the Hilbert space H_T, we may assume that $\{f_n\}$ converges weakly to f, taking a subsequence if necessary. Then

$$J(f) \leqq \liminf_{n\to\infty} J(f_n) = \lim_{\delta\to 0}\alpha_\delta.$$

In general, for $h \in H_T$ and $s, t \in [0, T]$,

$$|h(t) - h(s)|^2 = \left|\int_s^t \dot{h}(u)\,\mathrm{d}u\right|^2 \leqq \left(\int_0^T |\dot{h}(u)|^2 \mathrm{d}u\right)|t-s|$$

and $|h(t) - h(s)| \leqq \|h\|_{H_T}|t-s|$. Hence, $\{f_n\}$ is equi-continuous and has a subsequence $\{f_{n_j}\}$ which is convergent in W_T by the Ascoli–Arzelà theorem. Let $\bar{f}$ be the limit. Then, since F is a closed set, $\bar{f} \in F$ and

$$\ell(\bar{f}) = \lim_{j\to\infty}\ell(f_{n_j}) = \lim_{j\to\infty}\langle \ell, f_{n_j}\rangle = \langle \ell, f\rangle = \ell(f)$$

for any $\ell \in W_T^*$. Since W_T^* is dense in H_T, $f = \bar{f}$ and

$$\inf_{w\in F} J(w) \leqq J(f) = \lim_{\delta\to 0}\alpha_\delta.$$

This, combined with (1.8.4), implies (Ⅲ). □

Schilder's theorem is one of the most important results in the theory of **large deviations**. The next theorem, due to Donsker and Varadhan, the originators of the theory, shows that the Laplace method, which is fundamental in the theory of asymptotics of the integrals on finite dimensional spaces, also works on the Wiener space with suitable modifications.

Theorem 1.8.2 *Let F be a bounded continuous function on the Wiener space $(W_T, \mathscr{B}(W_T), \mu_T)$. Then,*

$$\lim_{\varepsilon\downarrow 0}\left\{\varepsilon\log\left(\int_{W_T} \mathrm{e}^{\frac{F(w)}{\varepsilon}}\mu_T^\varepsilon(\mathrm{d}w)\right)\right\} = \sup_{w\in W_T}\{F(w) - J(w)\}.$$

Proof Let $E_{n,k}$ $(k \in \mathbb{Z},\ n = 1,2,\ldots)$ be a sequence of the closed sets in W_T defined by

$$E_{n,k} = \Big\{ w \in W_T;\ \frac{k}{n} \leqq F(w) \leqq \frac{k+1}{n} \Big\}.$$

Then

$$\int_{W_T} e^{\frac{F(w)}{\varepsilon}} \mu_T^\varepsilon(dw) \leqq \sum_k e^{\frac{k+1}{n\varepsilon}} \mu_T^\varepsilon(E_{n,k}).$$

Since F is bounded, the sum on the right hand side is a finite one. Hence, by Schilder's theorem,

$$\begin{aligned}
&\limsup_{\varepsilon\downarrow 0}\Big\{\varepsilon \log\Big(\int_{W_T} e^{\frac{F(w)}{\varepsilon}} \mu_T^\varepsilon(dw)\Big)\Big\} \leqq \max_k\Big(\frac{k+1}{n} - \inf_{w\in E_{n,k}} J(w)\Big) \\
&\leqq \max_k\Big(\sup_{w\in E_{n,k}}\Big(\frac{k}{n} - J(w)\Big)\Big) + \frac{1}{n} \leqq \max_k\Big(\sup_{w\in E_{n,k}}(F(w) - J(w))\Big) + \frac{1}{n} \\
&= \sup_{w\in W_T}(F(w) - J(w)) + \frac{1}{n}.
\end{aligned}$$

Since n is arbitrary, we obtain the following estimate from above:

$$\limsup_{\varepsilon\downarrow 0}\Big\{\varepsilon \log\Big(\int_{W_T} e^{\frac{F(w)}{\varepsilon}} \mu_T^\varepsilon(dw)\Big)\Big\} \leqq \sup_{w\in W_T}(F(w) - J(w)).$$

In order to show the opposite inequality, fix $\delta > 0$ and choose $w_\delta \in W_T$ such that

$$F(w_\delta) - J(w_\delta) \geqq \sup_{w\in W_T}(F(w) - J(w)) - \delta. \tag{1.8.5}$$

Since F is continuous, there exists an open set G of W_T such that $w_\delta \in G$ and $F(w) \geqq F(w_\delta) - \delta$ for all $w \in G$. We now apply Schilder's theorem and get

$$\begin{aligned}
\liminf_{\varepsilon\downarrow 0}\Big\{\varepsilon \log\Big(\int_{W_T} e^{\frac{F(w)}{\varepsilon}} \mu_T^\varepsilon(dw)\Big)\Big\} &\geqq \limsup_{\varepsilon\downarrow 0}\Big\{\varepsilon \log\Big(\int_G e^{\frac{F(w)}{\varepsilon}} \mu_T^\varepsilon(dw)\Big)\Big\} \\
&\geqq F(w_\delta) - \delta - \inf_{w\in G} J(w).
\end{aligned}$$

Hence, by (1.8.5),

$$\begin{aligned}
\liminf_{\varepsilon\downarrow 0}\Big\{\varepsilon \log\Big(\int_{W_T} e^{\frac{F(w)}{\varepsilon}} \mu_T^\varepsilon(dw)\Big)\Big\} &\geqq F(w_\delta) - \delta - J(w_\delta) \\
&\geqq \sup_{w\in W_T}\{F(w) - J(w)\} - 2\delta.
\end{aligned}$$

Letting $\delta \to 0$, we obtain

$$\liminf_{\varepsilon\downarrow 0}\Big\{\varepsilon \log\Big(\int_{W_T} e^{\frac{F(w)}{\varepsilon}} \mu_T^\varepsilon(dw)\Big)\Big\} \geqq \sup_{w\in W_T}\{F(w) - J(w)\}. \qquad \square$$

1.9 Analogy to Path Integrals

In this section, by formal considerations, we show a relationship between the Wiener integral and the Feynman path integral. The one-dimensional Wiener measure on $[0,T]$ is a probability measure on the path space W_T such that

$$\mu_T((w(t_1),\dots,w(t_n))\in A)$$
$$=\int\cdots\int_A \frac{1}{(2\pi)^{\frac{n}{2}}}\frac{\exp\left\{-\frac{1}{2}\sum_{i=1}^n\frac{(x_i-x_{i-1})^2}{t_i-t_{i-1}}\right\}}{\sqrt{t_1(t_2-t_1)\cdots(t_n-t_{n-1})}}\,\mathrm{d}x_1\mathrm{d}x_2\cdots\mathrm{d}x_n \quad (1.9.1)$$

for $0=t_0<t_1<\cdots<t_n<T, A\subset\mathbb{R}^n$.

We consider the limit as $n\to\infty$. Setting $x_i=w(t_i)$ for $w\in W_T$, we conclude

$$\sum_{i=1}^n\frac{(x_i-x_{i-1})^2}{t_i-t_{i-1}}=\sum_{i=1}^n\left(\frac{x_i-x_{i-1}}{t_i-t_{i-1}}\right)^2(t_i-t_{i-1})$$
$$\to\int_0^T\left(\frac{\mathrm{d}w(s)}{\mathrm{d}s}\right)^2\mathrm{d}s=\|w\|_{H_T}^2,$$

where w should be assumed to be in the Cameron–Martin subspace and $\|w\|_{H_T}$ is the norm.

If, as a limit of $\mathrm{d}x_1\cdots\mathrm{d}x_n$, there exists a translation invariant measure $\mathscr{D}(\mathrm{d}w)$ on the path space $\mathbb{R}^{[0,T]}$,

$$\mu(C)=\int_C\frac{1}{Z}\exp\left(-\frac{1}{2}\int_0^T\left(\frac{\mathrm{d}w(s)}{\mathrm{d}s}\right)^2\mathrm{d}s\right)\mathscr{D}(\mathrm{d}w)$$
$$=\int_C\frac{1}{Z}\mathrm{e}^{-\frac{\|w\|_H^2}{2}}\mathscr{D}(\mathrm{d}w) \quad (1.9.2)$$

for a subset C of W_T as a limit of (1.9.1). Here Z is the fictitious normalizing constant so that the total mass of W_T is 1.

However, the path of the Brownian motion is nowhere differentiable almost surely and the expression on the exponent has no meaning. Moreover, there is no translation invariant measure on $\mathbb{R}^{[0,T]}$ like the Lebesgue measure. Nevertheless, the Wiener measure on the left hand side of (1.9.2) exists in the mathematical sense and it is an "identity" which shows in a heuristic way that the Wiener measure is a Gaussian measure and that the integral with respect to it is quite similar to the path integral. Furthermore, the Cameron–Martin theorem and the Schilder theorem may be deduced from (1.9.2) in a formal way.

We formally deduce the Cameron–Martin formula in the following way. When h is fixed, the distribution $\mu_{T,h}$ of $w+h$ under μ_T is characterized by the identity

$$\mathbf{E}^{\mu_{T,h}}[F] = \int_{W_T} F(w)\,\mu_{T,h}(\mathrm{d}w) = \int_{W_T} F(w+h)\,\mu_T(\mathrm{d}w)$$

for any function F on W_T. Hence, noting the translation invariance of the "measure" $\mathscr{D}(\mathrm{d}w)$ and using (1.9.2), we obtain

$$\begin{aligned}
\mathbf{E}^{\mu_{T,h}}[F] &= \frac{1}{Z}\int F(w+h)\exp\Big(-\frac{1}{2}\int_0^T\Big(\frac{\mathrm{d}w}{\mathrm{d}s}\Big)^2\mathrm{d}s\Big)\mathscr{D}(\mathrm{d}w)\\
&= \frac{1}{Z}\int F(w)\exp\Big(-\frac{1}{2}\int_0^T\Big(\frac{\mathrm{d}(w-h)}{\mathrm{d}s}\Big)^2\mathrm{d}s\Big)\mathscr{D}(\mathrm{d}w)\\
&= \frac{1}{Z}\int F(w)\exp\Big(\langle w,h\rangle_{H_T} - \frac{1}{2}\|h\|_{H_T}^2 - \frac{1}{2}\int_0^T\Big(\frac{\mathrm{d}w}{\mathrm{d}s}\Big)^2\mathrm{d}s\Big)\mathscr{D}(\mathrm{d}w)
\end{aligned}$$

and, writing $\langle w,h\rangle_{H_T} = \mathscr{I}(h)(w)$,

$$\mathbf{E}^{\mu_{T,h}}[F] = \mathbf{E}^{\mu_T}[F\mathrm{e}^{\mathscr{I}(h)-\frac{1}{2}\|h\|_{H_T}^2}].$$

This leads us to the following Cameron–Martin formula:

$$\mu_{T,h}(\mathrm{d}w) = \exp\Big(\mathscr{I}(h)(w) - \frac{1}{2}\|h\|_{H_T}^2\Big)\mu_T(\mathrm{d}w).$$

The Cameron–Martin theorem also suggests the integration by parts formula on the path space. We have shown in the proof of the Cameron–Martin theorem that, if $F \in L^p(W_T)$, $p > 1$, $h \in H_T$, then

$$\mathbf{E}^{\mu_T}[F \circ T_h] = E^{\mu_T}[FR_h], \qquad (R_h(w) = \mathrm{e}^{\mathscr{I}(h)-\frac{1}{2}\|h\|_{H_T}^2}),$$

where $T_h : W_T \ni w \mapsto w + h$. Replacing h with sh and differentiating in s, we obtain

$$\frac{\mathrm{d}}{\mathrm{d}s}\mathbf{E}^{\mu_T}[F(\cdot\, + sh)]\Big|_{s=0} = \mathbf{E}^{\mu_T}[F\mathscr{I}(h)].$$

Writing the "derivative" of F in the direction of h by D_hF, we have the integration by parts formula

$$\mathbf{E}^{\mu_T}[\mathrm{D}_hF] = \mathbf{E}^{\mu_T}[F\mathscr{I}(h)].$$

In fact, we formulate such an integration by parts formula in the context of the Malliavin calculus in Chapter 5. The formula will play a fundamental role.

It should be noted that an integration by parts formula on a finite dimensional Euclidean space with respect to the Gaussian measure is easily obtained. For example, for a standard normal random variable G, we have

$$\begin{aligned}
\mathbf{E}[\varphi'(G)] &= \int_{\mathbb{R}} \varphi'(t)\frac{1}{\sqrt{2\pi}}\mathrm{e}^{-\frac{1}{2}t^2}\mathrm{d}t\\
&= \int_{\mathbb{R}} \varphi(t)t\frac{1}{\sqrt{2\pi}}\mathrm{e}^{-\frac{1}{2}t^2}\mathrm{d}t = \mathbf{E}[\varphi(G)G].
\end{aligned}$$

This implies the identity $\mathbf{E}[\varphi'(G)] = \mathbf{E}[\varphi(G)G]$ for any $\varphi \in C_0^1(\mathbb{R})$.

Next we consider the measure μ_T^ε in Schilder's theorem (Theorem 1.8.1). We have

$$\mu_T^\varepsilon(C) = \mu_T(\{w;\ \sqrt{\varepsilon}w \in C\}) = \int_{\sqrt{\varepsilon}^{-1}C} \frac{1}{Z} \mathrm{e}^{-\frac{1}{2}\|w\|_{H_T}^2} \mathscr{D}(\mathrm{d}w)$$

for a subset C of the Wiener space and, changing the variable as in the integrals on the Euclidean spaces,

$$\mu_T^\varepsilon(C) = \int_C \frac{1}{Z} \varepsilon^{-\frac{1}{2}\dim(W_T)} \mathrm{e}^{-\frac{1}{2\varepsilon}\|w\|_{H_T}^2} \mathscr{D}(\mathrm{d}w).$$

If we could apply the Laplace method to the integral on the right hand side, we might say that the main term of the asymptotic behavior of $\mu_T^\varepsilon(C)$ as $\varepsilon \to 0$ would be determined by $\inf_{w\in C}\{\frac{1}{2}\|w\|_{H_T}^2\}$. Regarding $\|w\|_{H_T}^2 = \int_0^T (\frac{\mathrm{d}w}{\mathrm{d}s})^2\,\mathrm{d}s$ as the action integral of the path w, we see that the principle of least action works.

2

Stochastic Integrals and Itô's Formula

In this chapter the stochastic integral, which is fundamental in the theory of stochastic analysis, is defined and Itô's formula, the associated chain rule, is shown. Moreover, their applications and some related topics are presented. In the rest of this book, we sometimes say simply "$X = Y$" for "$X = Y$ almost surely" for random variables X, Y.

2.1 Local Martingale

Itô originated stochastic integrals with respect to Brownian motion ([48, 49]). Doob noticed their close relation to martingales as soon as he learned stochastic integrals ([15, Chapter 6]). Nowadays the stochastic integrals are usually formulated in a sophisticated manner developed by Kunita and Watanabe [65] and, also in this book, we follow this way. To have a wide extent of applications, it is necessary to consider stochastic integrals on the space of local martingales. This section is devoted to the explanation of local martingales. In the rest of this book, we assume that $(\Omega, \mathscr{F}, \mathbf{P}, \{\mathscr{F}_t\})$ is a filtered probability space which satisfies the usual condition, if not otherwise defined.

Definition 2.1.1 (1) A right-continuous $\{\mathscr{F}_t\}$-adapted stochastic process $M = \{M(t)\}_{t \geqq 0}$ defined on $(\Omega, \mathscr{F}, \mathbf{P}, \{\mathscr{F}_t\})$ is called an $\{\mathscr{F}_t\}$-**local martingale** if there exists a sequence $\{\sigma_n\}_{n=1}^{\infty}$ of $\{\mathscr{F}_t\}$-stopping times satisfying

$$\mathbf{P}(\sigma_n \leqq \sigma_{n+1}) = 1 \quad \text{and} \quad \mathbf{P}\Big(\lim_{n\to\infty} \sigma_n = \infty\Big) = 1 \qquad (2.1.1)$$

such that each stochastic process $M^{\sigma_n} = \{M(t \wedge \sigma_n)\}_{t \geqq 0}$ is a martingale.
(2) If a sequence $\{\sigma_n\}_{n=1}^{\infty}$ of stopping times satisfying (2.1.1) can be chosen so that $M^{\sigma_n} \in \mathscr{M}^2$, that is, $\mathbf{E}[M(t \wedge \sigma_n)^2] < \infty$ for any $t \geqq 0$, M is called a

locally square-integrable martingale. The space of locally square-integrable martingales is denoted by $\mathscr{M}^2_{\text{loc}}$ and the set of continuous elements in $\mathscr{M}^2_{\text{loc}}$ is denoted by $\mathscr{M}^2_{\text{c,loc}}$.

When the filtration under consideration is clear, the notation $\{\mathscr{F}_t\}$ is omitted. A martingale is a local martingale, but the converse is not true, which causes problems in various situations.

The following is important in applications.

Proposition 2.1.2 *Any non-negative local martingale $X = \{X(t)\}_{t\geqq 0}$ is a supermartingale. Moreover, if $\mathbf{E}[X(t)] = \mathbf{E}[X(0)]$ for any $t \geqq 0$, then X is a martingale.*

Proof Let $\{\sigma_n\}_{n=1}^{\infty}$ be an increasing sequence of stopping times such that $\{X(t \wedge \sigma_n)\}_{t\geqq 0}$ is a martingale. By Fatou's lemma (Theorem 1.4.1(8)) for conditional expectations, for any $s < t$

$$\mathbf{E}[X(t)|\mathscr{F}_s] = \mathbf{E}\Big[\liminf_{n\to\infty} X(t \wedge \sigma_n)\Big|\mathscr{F}_s\Big] \leqq \liminf_{n\to\infty} \mathbf{E}[X(t \wedge \sigma_n)|\mathscr{F}_s]$$
$$= \liminf_{n\to\infty} X(s \wedge \sigma_n) = X(s).$$

The second assertion follows from the fact that, if $\mathbf{E}[N(t)] = \mathbf{E}[N(0)]$ holds for a supermartingale $\{N(t)\}_{t\geqq 0}$, then it is a martingale. □

Also, for a continuous locally square-integrable martingale M, the quadratic variation process is important. The following stochastic process $\langle M\rangle$, which is denoted by the same notation as for martingales, is called the **quadratic variation process** of M.

Theorem 2.1.3 *Let $M, N \in \mathscr{M}^2_{\text{c,loc}}$.*
(1) There exists a unique continuous increasing process $\langle M\rangle = \{\langle M\rangle(t)\}_{t\geqq 0}$ such that $\{M(t)^2 - \langle M\rangle(t)\}_{t\geqq 0} \in \mathscr{M}^2_{\text{c,loc}}$.
(2) M belongs to $\mathscr{M}^2_{\text{c}}$ if and only if $\mathbf{E}[\langle M\rangle(t)] < \infty$ for any $t \geqq 0$. Moreover, if $M(0) = 0$, then $\mathbf{E}[M(t)^2] = \mathbf{E}[\langle M\rangle(t)]$.
(3) There exists a unique continuous process $\langle M, N\rangle = \{\langle M, N\rangle(t)\}_{t\geqq 0}$, which is the difference of two continuous increasing processes, such that $\{M(t)N(t) - \langle M, N\rangle(t)\}_{t\geqq 0} \in \mathscr{M}^2_{\text{c,loc}}$.

Proof (1) By assumption there exists a sequence $\{\sigma_n\}_{n=1}^{\infty}$ of stopping times such that $\sigma_n \uparrow \infty$ almost surely and $M^{\sigma_n} = \{M(t \wedge \sigma_n)\}_{t\geqq 0}$ is a martingale. Set $\tau_n = \inf\{t; |M(t)| \geqq n\}$. Then $M^{(n)} = M^{\sigma_n \wedge \tau_n}$ is a bounded martingale. Hence, by

Corollary 1.5.27, there exists a unique natural continuous increasing process $\{A^{(n)}(t)\}_{t\geqq 0}$ such that $\{M^{(n)}(t)^2 - A^{(n)}(t)\}_{t\geqq 0}$ is a martingale. Since

$$\begin{aligned} &M^{(n+1)}(t \wedge \sigma_n \wedge \tau_n)^2 - A^{(n+1)}(t \wedge \sigma_n \wedge \tau_n) \\ &\quad = M(t \wedge \sigma_n \wedge \tau_n)^2 - A^{(n+1)}(t \wedge \sigma_n \wedge \tau_n) = M^{(n)}(t)^2 - A^{(n+1)}(t \wedge \sigma_n \wedge \tau_n) \end{aligned}$$

and the left hand side defines a martingale, the uniqueness of the quadratic variation process implies $A^{(n+1)}(t \wedge \sigma_n \wedge \tau_n) = A^{(n)}(t)$. Define the continuous increasing process $\{\langle M\rangle(t)\}$ by $\langle M\rangle(t) = \lim_{n\to\infty} A^{(n)}(t)$. Then the stochastic process given by

$$M^{(n)}(t)^2 - \langle M\rangle(t \wedge \sigma_n \wedge \tau_n) = M(t \wedge \sigma_n \wedge \tau_n)^2 - A^{(n)}(t)$$

is a martingale. This means that $\{M(t)^2 - \langle M\rangle(t)\}_{t\geqq 0}$ is a local martingale.
(2) Assume that $M \in \mathscr{M}_c^2$. As was shown above, $\mathbf{E}[M^{(n)}(t)^2 - A^{(n)}(t)] = 0$ and

$$\begin{aligned} \mathbf{E}[\langle M\rangle(t)] &= \mathbf{E}\Big[\lim_{n\to\infty} A^{(n)}(t)\Big] \\ &= \lim_{n\to\infty} \mathbf{E}[A^{(n)}(t)] = \lim_{n\to\infty} \mathbf{E}[(M^{(n)}(t))^2] \end{aligned}$$

by the monotone convergence theorem. Since $|M^{(n)}(t)| \leqq \sup_{0\leqq s\leqq t} |M(s)|$, by Doob's inequality and the Lebesgue convergence theorem, the limit in the right hand side coincides with $\mathbf{E}[M(t)^2]$.

Conversely, assume that $\mathbf{E}[\langle M\rangle(t)] < \infty$ $(t > 0)$. Then, since $\langle M\rangle(t) \geqq A^{(n)}(t)$, Fatou's lemma implies

$$\begin{aligned} \mathbf{E}[M(t)^2] &= \mathbf{E}\Big[\liminf_{n\to\infty}(M(t \wedge \sigma_n \wedge \tau_n)^2)\Big] \\ &\leqq \liminf_{n\to\infty} \mathbf{E}[A^{(n)}(t)] \leqq \mathbf{E}[\langle M\rangle(t)]. \end{aligned}$$

Hence M is square-integrable.
(3) It suffices to set $\langle M, N\rangle(t) = \frac{1}{4}(\langle M + N\rangle(t) - \langle M - N\rangle(t))$. □

2.2 Stochastic Integrals

Let $M = \{M(t)\}_{t\geqq 0}$ be a continuous square-integrable martingale defined on a probability space $(\Omega, \mathscr{F}, \mathbf{P}, \{\mathscr{F}_t\})$. In this section stochastic integrals with respect to M and a continuous locally square-integrable martingale are described. Since M is not of bounded variation (Theorem 1.5.18), stochastic integrals are not defined as Lebesgue–Stieltjes integrals.

We briefly recall the martingale transform in the discrete time case mentioned in Section 1.4. Let $S = \{S_n\}_{n=0}^{\infty}$ be a martingale defined on a probability space with filtration $\{\mathscr{G}_n\}_{n=0}^\infty$. If a bounded stochastic process $H = \{H_n\}_{n=1}^\infty$ is predictable, the stochastic process $X = \{X_n\}_{n=1}^\infty$ defined by

$$X_n = \sum_{k=1}^{n} H_k(S_k - S_{k-1})$$

is a $\{\mathscr{G}_n\}$-martingale and X is called the martingale transform of H with respect to S. The predictability is a natural notion, for example, in mathematical finance. If we consider S as a stock price process and H_k as a share of the stock at time k, the stock at time k was bought at time $k-1$ and H_k should be determined by the information up to time $k-1$. Then, X_n represents the increment of the asset up to time n and is a martingale.

Also, in the continuous time case, the property of predictability of the integrands of stochastic integrals is necessary.

Definition 2.2.1 Let $\mathscr{S}$ be the smallest σ-field on $[0,\infty)\times\Omega$, under which all left-continuous $\{\mathscr{F}_t\}$-adapted $\mathbb{R}$-valued stochastic processes are measurable. A stochastic process X is called **predictable** if X is $\mathscr{S}$-measurable, that is for any $A \in \mathscr{B}(\mathbb{R})$,

$$X^{-1}(A) := \{(t,\omega); X(t,\omega) \in A\} \in \mathscr{S}.$$

For the smallest σ-field $\mathscr{T}$, under which all right-continuous $\{\mathscr{F}_t\}$-adapted stochastic processes are measurable, an $\mathscr{T}$-measurable stochastic process is called **well measurable**.

It is easily seen that a predictable stochastic process is $\{\mathscr{F}_t\}$-adapted.

Definition 2.2.2 The set of $\mathbb{R}$-valued predictable stochastic processes Φ such that

$$[\Phi]_T := \left\{\mathbf{E}\left[\int_0^T \Phi(s)^2 \mathrm{d}\langle M\rangle(s)\right]\right\}^{\frac{1}{2}} < \infty \qquad (T > 0)$$

is denoted by $\mathscr{L}^2(M)$ or $\mathscr{L}^2$. For $\Phi, \Psi \in \mathscr{L}^2$, the distance $[\Phi - \Psi]$ is defined by

$$[\Phi - \Psi] = \sum_{n=1}^{\infty} \frac{1}{2^n}([\Phi - \Psi]_n \wedge 1).$$

Remark 2.2.3 Two elements Φ, Φ' of $\mathscr{L}^2$ satisfying $[\Phi - \Phi'] = 0$ are identified.

We begin with the stochastic integrals for stochastic processes like the step functions.

Definition 2.2.4 The set of measurable $\{\mathscr{F}_t\}$-adapted stochastic processes $\Phi = \{\Phi(t)\}_{t\geqq 0}$ for which there exists an increasing sequence $0 = t_0 < t_1 < \cdots < t_n < \cdots$ with $t_n \to \infty$ $(n \to \infty)$ and $\mathscr{F}_{t_n}$-measurable random variables ξ_n such that

$$\begin{cases} \Phi(0) = \xi_0 \\ \Phi(t) = \xi_n \quad (t_n < t \leqq t_{n+1},\ n = 0, 1, \ldots) \end{cases} \tag{2.2.1}$$

and

$$\sup_{0\leqq n\leqq N,\, \omega\in\Omega} |\xi_n(\omega)| < \infty \qquad (N > 0)$$

is denoted by $\mathscr{L}^0$.

Definition 2.2.5 The stochastic integral of $\Phi \in \mathscr{L}^0$ given by (2.2.1) with respect to $M = \{M(t)\}_{t\geqq 0} \in \mathscr{M}_c^2$ is defined by

$$\int_0^t \Phi(s)\,\mathrm{d}M(s) = \sum_{k=0}^{n-1} \Phi(t_k)(M(t_{k+1}) - M(t_k)) + \Phi(t_n)(M(t) - M(t_n))$$

for $t_n \leqq t \leqq t_{n+1}$ and is also denoted by $I^M(\Phi) = \{I^M(\Phi)(t)\}_{t\geqq 0}$ or $I(\Phi)$.

Remark 2.2.6 The sequence of the times $\{t_n\}$ which defines $\Phi \in \mathscr{L}^0$ is not unique, but the stochastic integral is independent of the choice of such sequence of times.

Lemma 2.2.7 *Let* $\Phi, \Psi \in \mathscr{L}^0$.
(1) [linearity] *For* $a, b \in \mathbb{R}$, $a\Phi + b\Psi = \{a\Phi(t) + b\Psi(t)\}_{t\geqq 0} \in \mathscr{L}^0$ *and*

$$\int_0^t (a\Phi(s) + b\Psi(s))\,\mathrm{d}M(s) = a\int_0^t \Phi(s)\,\mathrm{d}M(s) + b\int_0^t \Psi(s)\,\mathrm{d}M(s).$$

(2) *The stochastic process* $I(\Phi) = \{\int_0^t \Phi(s)\,\mathrm{d}M(s)\}_{t\geqq 0}$ *is continuous.*
(3) [isometry] *For any* $t \geqq 0$

$$\mathbf{E}\Big[\Big(\int_0^t \Phi(s)\,\mathrm{d}M(s)\Big)^2\Big] = \mathbf{E}\Big[\int_0^t \Phi(s)^2\mathrm{d}\langle M\rangle(s)\Big].$$

(4) $I(\Phi)$ *is a continuous square-integrable martingale and, for any* $T > 0$,

$$\mathbf{E}\Big[\sup_{0\leqq t\leqq T}\Big(\int_0^t \Phi(s)\,\mathrm{d}M(s)\Big)^2\Big] \leqq 4\mathbf{E}\Big[\int_0^T \Phi(s)^2\mathrm{d}\langle M\rangle(s)\Big]. \tag{2.2.2}$$

Proof The assertions (1) and (2) are direct conclusions of the definition of stochastic integrals.

To prove (3), we may assume $t = t_N$ in (2.2.1). Then,

$$\int_0^t \Phi(s)\,\mathrm{d}M(s) = \sum_{k=0}^{N-1} \Phi(t_k)(M(t_{k+1}) - M(t_k)).$$

Since M is a martingale and Φ is adapted, we have

$$\begin{aligned}&\mathbf{E}[\Phi(t_k)(M(t_{k+1}) - M(t_k)) \cdot \Phi(t_\ell)(M(t_{\ell+1}) - M(t_\ell))]\\&= \mathbf{E}[\Phi(t_k)(M(t_{k+1}) - M(t_k))\Phi(t_\ell)\mathbf{E}[M(t_{\ell+1}) - M(t_\ell)|\mathscr{F}_{t_\ell}]] = 0\end{aligned}$$

for $k < \ell$. Since $\mathbf{E}[(M_{t_{k+1}} - M_{t_k})^2|\mathscr{F}_{t_k}] = \mathbf{E}[\langle M\rangle_{t_{k+1}} - \langle M\rangle_{t_k}|\mathscr{F}_{t_k}]$, we obtain

$$\begin{aligned}\mathbf{E}\Big[\Big(\int_0^t \Phi(s)\,\mathrm{d}M(s)\Big)^2\Big] &= \sum_{k=0}^{N-1} \mathbf{E}[\Phi(t_k)^2(M(t_{k+1}) - M(t_k))^2]\\&= \sum_{k=0}^{N-1} \mathbf{E}[\Phi(t_k)^2(\langle M\rangle(t_{k+1}) - \langle M\rangle(t_k))] = \mathbf{E}\Big[\int_0^t \Phi(s)^2\mathrm{d}\langle M\rangle(s)\Big].\end{aligned}$$

Thus the assertion (3) follows.

For a proof of (4), let $s < t$. We may assume $s = t_k < t_\ell = t$ in (2.2.1). Then,

$$\begin{aligned}\mathbf{E}\Big[\int_0^t \Phi(u)\,\mathrm{d}M(u)\Big|\mathscr{F}_s\Big] &= \mathbf{E}\Big[\sum_{j=0}^{\ell-1} \Phi(t_j)(M(t_{j+1}) - M(t_j))\Big|\mathscr{F}_{t_k}\Big]\\&= \sum_{j=0}^{k-1} \Phi(t_j)(M(t_{j+1}) - M(t_j)) + \sum_{j=k}^{\ell-1} \mathbf{E}[\Phi(t_j)\mathbf{E}[M(t_{j+1}) - M(t_j)|\mathscr{F}_{t_k}]]\\&= \int_0^s \Phi(u)\,\mathrm{d}M(u)\end{aligned}$$

by the tower property. Thus $I(\Phi)$ is an $\{\mathscr{F}_t\}$-martingale. Its square-integrability has been shown in (3), and (2.2.2) is obtained by (3) and Doob's inequality (Theorem 1.5.14). □

Next, we define the stochastic integral of $\Phi \in \mathscr{L}^2$ with respect to a martingale. It may also be considered as an element of the space of square-integrable martingales. The next proposition is important.

Proposition 2.2.8 *The set $\mathscr{L}^0$ is dense in $\mathscr{L}^2$ with respect to the distance $[\,\cdot\,]$.*

Proof For $\Phi \in \mathscr{L}^2$ and $K, n \in \mathbf{N}$, define Φ^K and Φ^K_n by

$$\begin{aligned}\Phi^K(t,\omega) &= \Phi(t,\omega)\mathbf{1}_{\{|\Phi(t)|\leqq K\}}(\omega),\\\Phi^K_n(t,\omega) &= \sum_{k=-K2^n}^{K2^n} \frac{k}{2^n}\mathbf{1}_{\{\frac{k}{2^n}\leqq \Phi^K < \frac{k+1}{2^n}\}}(t,\omega).\end{aligned}$$

Then, $\Phi^K \in \mathscr{L}^2, [\Phi^K - \Phi] \to 0$ as $K \to \infty$, $\{\frac{k}{2^n} \leqq \Phi^K < \frac{k+1}{2^n}\} \in \mathscr{S}$, and $\Phi_n^K \to \Phi^K$ pointwise as $n \to \infty$.

Let $\mathbf{\Phi}$ be the set of $\Phi \in \mathscr{L}^2$ for which there exists $K > 0$ and $\{\Phi_n\}_{n=1}^\infty \subset \mathscr{L}^0$ such that $|\Phi(t,\omega)| \leqq K$ for any (t,ω) and $[\Phi_n - \Phi] \to 0$. By virtue of the above observation, the proof is completed once we have shown that $\mathbf{1}_B \in \mathbf{\Phi}$ for any $B \in \mathscr{S}$.

To do this, set

$$\mathscr{S}' = \{B; B \text{ is a subset of } [0,\infty) \times \Omega \text{ such that } \mathbf{1}_B \in \mathbf{\Phi}\}.$$

$\mathscr{S}'$ is a Dynkin class (see Definition A.2.1).

For left-continuous $\{\mathscr{F}_t\}$-adapted stochastic processes $Y_1, \dots, Y_k$ and open sets $E_1, \dots, E_k$ in $\mathbb{R}$, we show $\bigcap_{i=1}^k Y_i^{-1}(E_i) \in \mathscr{S}'$. To do it, note

$$\mathbf{1}_{\bigcap_{i=1}^k Y_i^{-1}(E_i)}(t,\omega) = \prod_{i=1}^k \mathbf{1}_{E_i}(Y_i(t,\omega)).$$

Let $\{\phi_n^i\}$ be an increasing sequence of non-negative bounded continuous functions on $\mathbb{R}$ which converges to $\mathbf{1}_{E_i}$ pointwise. Then, since

$$\prod_{i=1}^k \phi_n^i(Y_i(t,\omega)) \uparrow \prod_{i=1}^k \mathbf{1}_{E_i}(Y_i(t,\omega))$$

and the left hand side is a left-continuous $\{\mathscr{F}_t\}$-adapted stochastic process, $\prod_{i=1}^k \mathbf{1}_{E_i}(Y_i(t,\omega))$ is $\mathscr{S}$-measurable. That is, $\bigcap_{i=1}^k Y_i^{-1}(E_i) \in \mathscr{S}'$.

The totality $\mathscr{U}$ of the subsets of $[0,\infty)\times\Omega$ of the form $\bigcap_{i=1}^k Y_i^{-1}(E_i)$ is closed under the finite number of intersections and $\sigma[\mathscr{U}] = \mathscr{S}$ by the definition. On the other hand, the smallest Dynkin class containing $\mathscr{U}$ is $\sigma[\mathscr{U}]$ by the Dynkin class theorem (see, e.g., Theorem A.2.2 and [126, Chapter A1]). Hence we get $\mathscr{S} \subset \mathscr{S}'$ since $\mathscr{U} \subset \mathscr{S}'$. This means that $B \in \mathscr{S}'$ for $B \in \mathscr{S}$, that is, $\mathbf{1}_B \in \mathbf{\Phi}$. □

We define the stochastic integral for $\Phi \in \mathscr{L}^2$. By Proposition 2.2.8, let $\{\Phi_n\}_{n=1}^\infty \in \mathscr{L}^0$ be a sequence such that $[\Phi_n - \Phi] \to 0$ and consider the stochastic integral of Φ_n with respect to $M \in \mathscr{M}_c^2$:

$$I(\Phi_n)(t) = \int_0^t \Phi_n(s)\,\mathrm{d}M(s).$$

Then, by Lemma 2.2.7 (3),

$$\mathbf{E}[(I(\Phi_n)(T) - I(\Phi_m)(T))^2] = [\Phi_n - \Phi_m]_T^2.$$

Thus $\{I(\Phi_n)\}_{n=1}^\infty$ is a Cauchy sequence in the space $\mathscr{M}^2$ of square-integrable martingales. Since $\mathscr{M}_c^2$ is a closed subspace of $\mathscr{M}^2$, $I(\Phi_n)$ converges to some element in $\mathscr{M}_c^2$. It is clear that the limit X does not depend on the choice of the sequence $\{\Phi_n\}_{n=1}^\infty$ and is determined only by Φ.

Definition 2.2.9 $X \in \mathscr{M}_c^2$ determined by Φ as above is called the **stochastic integral** of $\Phi \in \mathscr{L}^2$ with respect to a martingale $M \in \mathscr{M}_c^2$ and also denoted by $I(\Phi)$, $I^M(\Phi)$ or $\int_0^t \Phi(s)\,\mathrm{d}M(s)$.

Proposition 2.2.10 *Let* $\Phi, \Psi \in \mathscr{L}^2$.
(1) *For* $a, b \in \mathbb{R}$ *and* $t > 0$,

$$\int_0^t (a\Phi(s) + b\Psi(s))\,\mathrm{d}M(s) = a\int_0^t \Phi(s)\,\mathrm{d}M(s) + b\int_0^t \Psi(s)\,\mathrm{d}M(s).$$

(2) $I(\Phi) = \{\int_0^t \Phi(u)\mathrm{d}M(u)\}_{t\geqq 0}$ *is a square-integrable martingale and*

$$\mathbf{E}\Big[\Big(\int_s^t \Phi(u)\,\mathrm{d}M(u)\Big)^2\Big|\mathscr{F}_s\Big] = \mathbf{E}\Big[\int_s^t \Phi(u)^2\mathrm{d}\langle M\rangle(u)\Big|\mathscr{F}_s\Big].$$

(3) *For any* $s < t$

$$\mathbf{E}\Big[\int_s^t \Phi(u)\,\mathrm{d}M(u)\int_s^t \Psi(u)\,\mathrm{d}M(u)\Big|\mathscr{F}_s\Big] = \mathbf{E}\Big[\int_s^t \Phi(u)\Psi(u)\,\mathrm{d}\langle M\rangle(u)\Big|\mathscr{F}_s\Big].$$

For $\Phi, \Psi \in \mathscr{L}^0$, we can prove Proposition 2.2.10 in the same way as Lemma 2.2.7. For general $\Phi \in \mathscr{L}^2$, the assertions are shown by taking a sequence $\{\Phi_n\} \subset \mathscr{L}^0$ used to define $I(\Phi)$ and letting $n \to \infty$.

Stochastic integrals with respect to local martingales are given as limits of stochastic integrals with respect to martingales, which are stopped local martingales by stopping times. We also extend the space $\mathscr{L}^2$ of integrands.

Definition 2.2.11 Let $M \in \mathscr{M}_{c,\mathrm{loc}}^2$. The set of $\mathbb{R}$-valued predictable stochastic processes $\Phi = \{\Phi(t)\}_{t\geqq 0}$ such that

$$\int_0^t \Phi(s)^2\mathrm{d}\langle M\rangle(s) < \infty \qquad (t > 0)$$

is denoted by $\mathscr{L}_{\mathrm{loc}}^2(M)$ or $\mathscr{L}_{\mathrm{loc}}^2$.

For $M \in \mathscr{M}_{c,\mathrm{loc}}^2$ and $\Phi \in \mathscr{L}_{\mathrm{loc}}^2(M)$, we define the stochastic integral of Φ with respect to M. First, consider a sequence $\{\sigma_n\}_{n=1}^\infty$ of $\{\mathscr{F}_t\}$-stopping times satisfying (2.1.1) and $\int_0^t \Phi(s)^2\mathrm{d}\langle M\rangle(s) \leqq n$ $(t \leqq \sigma_n)$ and set

$$M_n = M^{\sigma_n}, \qquad \Phi_n(t) = \Phi(t)\mathbf{1}_{\{\sigma_n \geqq t\}}.$$

Since the stochastic integral $I^{M_n}(\Phi_n)$ of $\Phi_n \in \mathscr{L}^2$ with respect to $M_n \in \mathscr{M}^2$ satisfies

$$I^{M_n}(\Phi_m)(t) = I^{M_n}(\Phi_n)(t \wedge \sigma_m) \qquad (m < n),$$

there exists a unique $I^M(\Phi) = \{I^M(\Phi)(t)\}_{t \geqq 0} \in \mathscr{M}^2_{\mathrm{c,loc}}$ such that

$$I^M(\Phi)(t \wedge \sigma_n) = I^{M_n}(\Phi_n)(t) \qquad (t \geqq 0,\ n = 1, 2, \ldots).$$

Definition 2.2.12 The stochastic process $I^M(\Phi)$ defined above for $M \in \mathscr{M}^2_{\mathrm{c,loc}}$ and $\Phi \in \mathscr{L}^2_{\mathrm{loc}}(M)$ is called the **stochastic integral** of Φ with respect to M, which is also denoted by $\int_0^t \Phi(s)\,\mathrm{d}M(s)$.

It is easy to see that $I^M(\Phi)$ does not depend on the choice of the sequence $\{\sigma_n\}_{n=1}^\infty$ of stopping times. Moreover, Proposition 2.2.10 extends to $M, N \in \mathscr{M}^2_{\mathrm{c,loc}}$ and $\Phi \in \mathscr{L}^2_{\mathrm{loc}}(M)$, $\Psi \in \mathscr{L}^2_{\mathrm{loc}}(N)$. In particular,

$$\langle I^M(\Phi), N\rangle(t) = \int_0^t \Phi(s)\,\mathrm{d}\langle M, N\rangle(s),$$

which characterizes the stochastic integrals as (locally) square-integrable martingales.

Proposition 2.2.13 *Let $M \in \mathscr{M}^2_{\mathrm{c,loc}}$ and $\Phi \in \mathscr{L}^2_{\mathrm{loc}}(M)$. If $X \in \mathscr{M}^2_{\mathrm{c,loc}}$ satisfies*

$$\langle X, N\rangle(t) = \int_0^t \Phi(s)\,\mathrm{d}\langle M, N\rangle(s) \qquad (t \geqq 0) \tag{2.2.3}$$

for any $N \in \mathscr{M}^2_{\mathrm{c,loc}}$, then X coincides with $I^M(\Phi)$.

Proof For any $N \in \mathscr{M}^2_{\mathrm{c,loc}}$, $\langle I^M(\Phi) - X, N\rangle(t) = 0$. In particular, letting $N = I^M(\Phi) - X$, we obtain $\mathbf{E}[\langle I^M(\Phi) - X\rangle(t)] = 0$. □

Proposition 2.2.14 (1) *If $M, N \in \mathscr{M}^2_{\mathrm{c,loc}}$ and $\Phi \in \mathscr{L}^2_{\mathrm{loc}}(M) \cap \mathscr{L}^2_{\mathrm{loc}}(N)$, then $\Phi \in \mathscr{L}^2_{\mathrm{loc}}(M + N)$ and*

$$\int_0^t \Phi(s)\,\mathrm{d}(M+N)(s) = \int_0^t \Phi(s)\,\mathrm{d}M(s) + \int_0^t \Phi(s)\,\mathrm{d}N(s).$$

(2) *If $M \in \mathscr{M}^2_{\mathrm{c,loc}}$ and Φ, $\Psi \in \mathscr{L}^2_{\mathrm{loc}}(M)$, then*

$$\int_0^t (\Phi + \Psi)(s)\,\mathrm{d}M(s) = \int_0^t \Phi(s)\,\mathrm{d}M(s) + \int_0^t \Psi(s)\,\mathrm{d}M(s).$$

(3) Let $M \in \mathscr{M}^2_{\mathrm{c,loc}}$ and $\Phi \in \mathscr{L}^2_{\mathrm{loc}}(M)$. Set $N = I^M(\Phi)$. Then, $\Psi\Phi \in \mathscr{L}^2_{\mathrm{loc}}(M)$ for $\Psi \in \mathscr{L}^2_{\mathrm{loc}}(N)$ and

$$\int_0^t (\Psi\Phi)(s)\,\mathrm{d}M(s) = \int_0^t \Psi(s)\,\mathrm{d}N(s).$$

We omit the proofs.

2.3 Itô's Formula

The main stochastic processes treated hereafter in this book are of the following form.

Definition 2.3.1 Let $X(0)$ be an $\mathscr{F}_0$-measurable random variable, M be an element of $\mathscr{M}^2_{\mathrm{c,loc}}$ with $M(0) = 0$, and $A = \{A(t)\}_{t \geqq 0}$ be a continuous $\{\mathscr{F}_t\}$-adapted stochastic process such that $A(0) = 0$ and $t \mapsto A(t)$ is of bounded variation on each finite interval. Then the continuous $\{\mathscr{F}_t\}$-adapted stochastic process $X = \{X(t)\}_{t \geqq 0}$ defined by

$$X(t) = X(0) + M(t) + A(t) \tag{2.3.1}$$

is called a **continuous semimartingale**. An $\mathbb{R}^N$-valued $\{\mathscr{F}_t\}$-adapted continuous stochastic process $X = \{X(t) = (X^1(t), \dots, X^N(t))\}_{t \geqq 0}$ is an N-dimensional semimartingale if each component $\{X^i(t)\}_{t \geqq 0}$ is a continuous semimartingale.

The following stochastic process, called an Itô process, is a typical example of semimartingales.

Definition 2.3.2 The set of $\mathbb{R}$-valued predictable stochastic processes $\{b(t)\}_{t \geqq 0}$ such that $\int_0^T |b(t)|\,\mathrm{d}t < \infty$ for any $T > 0$ is denoted by $\mathscr{L}^1_{\mathrm{loc}}$.

Definition 2.3.3 Let $B = \{B(t) = (B^1(t), \dots, B^d(t))\}_{t \geqq 0}$ be a d-dimensional $\{\mathscr{F}_t\}$-Brownian motion. The continuous and $\{\mathscr{F}_t\}$-adapted stochastic process $\{X(t)\}_{t \geqq 0}$ given by

$$X(t) = X(0) + \sum_{\alpha=1}^d \int_0^t a_\alpha(s)\,\mathrm{d}B^\alpha(s) + \int_0^t b(s)\,\mathrm{d}s \qquad (t \geqq 0)$$

is called an $\mathbb{R}$-valued **Itô process**, where $\{a_\alpha(t)\}_{t \geqq 0} \in \mathscr{L}^2_{\mathrm{loc}}$ $(\alpha = 1, \dots, d)$ and $\{b(t)\}_{t \geqq 0} \in \mathscr{L}^1_{\mathrm{loc}}$. An N-dimensional stochastic process whose components are $\mathbb{R}$-valued Itô processes is called an $\mathbb{R}^N$-valued Itô process.

Itô's formula is the chain rule for semimartingales and is a fundamental tool in the analysis of functions of sample paths of semimartingales.

Theorem 2.3.4 (Itô's formula) *Let $\{X(t) = (X^1(t), \dots, X^N(t))\}_{t \geqq 0}$ be an N-dimensional semimartingale whose components are decomposed as*

$$X^i(t) = X^i(0) + M^i(t) + A^i(t) \qquad (i = 1, \dots, N)$$

and $f \in C^2(\mathbb{R}^N)$. Then, $\{f(X(t))\}_{t \geqq 0}$ is also a semimartingale and, for any $t > 0$, we have

$$f(X(t)) = f(X(0)) + \sum_{i=1}^N \int_0^t \frac{\partial f}{\partial x^i}(X(s))\,\mathrm{d}M^i(s) + \sum_{i=1}^N \int_0^t \frac{\partial f}{\partial x^i}(X(s))\,\mathrm{d}A^i(s)$$
$$+ \frac{1}{2} \sum_{i,j=1}^N \int_0^t \frac{\partial^2 f}{\partial x^i \partial x^j}(X(s))\,\mathrm{d}\langle M^i, M^j \rangle(s). \tag{2.3.2}$$

Itô's formula is written in the following way for Itô processes.

Theorem 2.3.5 *Let $\{a^i_\alpha(t)\}_{t \geqq 0} \in \mathscr{L}^2_{\mathrm{loc}}$ and $\{b^i(t)\}_{t \geqq 0} \in \mathscr{L}^1_{\mathrm{loc}}$ $(\alpha = 1, \dots, d,\ i = 1, \dots, N)$, and $\{X(t)\}_{t \geqq 0}$ be the $\mathbb{R}^N$-valued Itô process defined by*

$$X^i(t) = X^i(0) + \sum_{\alpha=1}^d \int_0^t a^i_\alpha(t)\,\mathrm{d}B^\alpha_s + \int_0^t b^i(s)\,\mathrm{d}s \qquad (i = 1, 2, \dots, N).$$

Then, for any $f \in C^2(\mathbb{R}^N)$,

$$f(X(t)) = f(X(0)) + \sum_{i=1}^N \sum_{\alpha=1}^d \int_0^t \frac{\partial f}{\partial x^i}(X(s)) a^i_\alpha(s)\,\mathrm{d}B^\alpha_s$$
$$+ \sum_{i=1}^N \int_0^t \frac{\partial f}{\partial x^i}(X(s)) b^i(s)\,\mathrm{d}s$$
$$+ \frac{1}{2} \sum_{i,j=1}^N \sum_{\alpha=1}^d \int_0^t \frac{\partial^2 f}{\partial x^i \partial x^j}(X(s)) a^i_\alpha(s) a^j_\alpha(s)\,\mathrm{d}s \quad (t > 0).$$

Proof of Theorem 2.3.4 To avoid unnecessary complexity, we only show the case where $d = 1$ and $N = 1$. The general case can be shown in the same way. We write the semimartingale under consideration as

$$X(t) = X(0) + M(t) + A(t).$$

Denote the total variation of A on $[0, t]$ by $V^A(t)$ and define a sequence $\{\tau_n\}_{n=1}^\infty$ of stopping times by $\tau_n = 0$ if $|X(0)| > n$ and, if $|X(0)| \leqq n$,

$$\tau_n = \inf\{t;\ \max\{|M(t)|, \langle M \rangle(t), V^A(t)\} > n\}.$$

If Itô's formula is proven for $\{X(t \wedge \tau_n)\}_{t \geqq 0}$, we obtain it for $\{X(t)\}_{t \geqq 0}$ by letting $n \to \infty$. Hence, we may assume that all of $X(0), M, \langle M \rangle, V^A$ are bounded and that the support of $f \in C^2(\mathbb{R})$ is compact.

Fix $t > 0$ and let Δ be the partition $0 = t_0 < t_1 < \cdots < t_n = t$ of $[0, t]$. By Taylor's theorem, there exists a ξ_k between $X(t_{k-1})$ and $X(t_k)$ such that

$$\begin{aligned} f(X(t)) - f(X(0)) &= \sum_{k=1}^{n} \{f(X(t_k)) - f(X(t_{k-1}))\} \\ &= \sum_{k=1}^{n} f'(X(t_{k-1}))(X(t_k) - X(t_{k-1})) + \frac{1}{2} \sum_{k=1}^{n} f''(\xi_k)(X(t_k) - X(t_{k-1}))^2. \end{aligned}$$

The first term of the right hand side converges in probability to

$$\int_0^t f'(X(s))\,\mathrm{d}M(s) + \int_0^t f'(X(s))\,\mathrm{d}A(s)$$

by the definition of the stochastic integral. We divide the second term into the sum of

$$\begin{aligned} I_1 &= \frac{1}{2} \sum_{k=1}^{n} f''(\xi_k)(M(t_k) - M(t_{k-1}))^2, \\ I_2 &= \sum_{k=1}^{n} f''(\xi_k)(M(t_k) - M(t_{k-1}))(A(t_k) - A(t_{k-1})), \\ I_3 &= \frac{1}{2} \sum_{k=1}^{n} f''(\xi_k)(A(t_k) - A(t_{k-1}))^2. \end{aligned}$$

Write A as the difference of two increasing processes $A^+, A^- : A(t) = A^+(t) - A^-(t)$. Setting

$$\begin{aligned} J_2 &= \max_{x \in \mathbb{R}} |f''(x)| \cdot \max_{1 \leqq k \leqq n} |M(t_k) - M(t_{k-1})| \cdot (|A^+(t)| + |A^-(t)|), \\ J_3 &= \max_{x \in \mathbb{R}} |f''(x)| \cdot \max_{1 \leqq k \leqq n} |A(t_k) - A(t_{k-1})| \cdot (|A^+(t)| + |A^-(t)|), \end{aligned}$$

we see that $|I_j| \leqq J_j$ $(j = 2, 3)$ and that $J_2, J_3 \to 0$ as $|\Delta| = \max_{1 \leqq k \leqq n}\{t_k - t_{k-1}\} \to 0$.

Next set

$$\begin{aligned} I_1' &= \frac{1}{2} \sum_{k=1}^{n} f''(X(t_{k-1}))(M(t_k) - M(t_{k-1}))^2, \\ J_1 &= \frac{1}{2} \max_{1 \leqq k \leqq n} \zeta_k^\Delta \sum_{k=1}^{n} (M(t_k) - M(t_{k-1}))^2, \end{aligned}$$

where

$$\zeta_k^\Delta = \max_{X(t_k) \wedge X(t_{k+1}) \leqq \xi \leqq X(t_k) \vee X(t_{k+1})} \{|f''(\xi) - f''(X(t_k))|\}.$$

Then, $|I_1 - I_1'| \leqq J_1$. Since

$$\mathbf{E}[J_1] \leqq \frac{1}{2}\Big\{\mathbf{E}\Big[\max_{1\leqq k\leqq n}(\zeta_k^{\Delta})^2\Big]\Big\}^{\frac{1}{2}}\Big\{\mathbf{E}\Big[\Big(\sum_{k=1}^{n}(M(t_k)-M(t_{k-1}))^2\Big)^2\Big]\Big\}^{\frac{1}{2}}$$

by Schwarz's inequality, the boundedness and the uniform continuity of f'' yields $\mathbf{E}\big[\max_{1\leqq k\leqq n}(\zeta_k^{\Delta})^2\big] \to 0$ ($|\Delta| \to 0$). Therefore, if we use the following lemma, we obtain $\mathbf{E}[J_1] \to 0$.

Lemma 2.3.6 *If* $|M(s)| \leqq C$ $(s \in [0,t])$, *then*

$$\mathbf{E}\Big[\Big(\sum_{k=1}^{n}(M(t_k)-M(t_{k-1}))^2\Big)^2\Big] \leqq 6C^4.$$

We postpone a proof of the lemma and continue the proof of Itô's formula. Since

$$\left|\frac{1}{2}\sum_{k=1}^{n} f''(\xi_k)(X(t_k)-X(t_{k-1}))^2 - I_1'\right| \leqq J_1 + J_2 + J_3,$$

we obtain

$$\frac{1}{2}\sum_{k=1}^{n} f''(\xi_k)(X(t_k)-X(t_{k-1}))^2 - I_1' \to 0$$

as $n \to \infty$ in probability. Moreover, set

$$I_1'' = \frac{1}{2}\sum_{k=1}^{n} f''(X(t_{k-1}))(\langle M\rangle(t_k) - \langle M\rangle(t_{k-1})).$$

By the boundedness of f'' and $\langle M\rangle$, the bounded convergence theorem yields

$$\mathbf{E}\Big[\Big|I_1'' - \frac{1}{2}\int_0^t f''(X(s))\,\mathrm{d}\langle M\rangle(s)\Big|\Big] \to 0,\ |\Delta| \to 0.$$

Hence, if we show

$$\mathbf{E}[|I_1' - I_1''|^2] \to 0,$$

we complete the proof of Itô's formula.

For this purpose we write

$$\begin{aligned}
&\mathbf{E}[|I_1' - I_1''|^2]\\
&= \frac{1}{4}\mathbf{E}\Big[\Big\{\sum_{k=1}^{n} f''(X(t_{k-1}))((M(t_k)-M(t_{k-1}))^2 - (\langle M\rangle(t_k)-\langle M\rangle(t_{k-1})))\Big\}^2\Big]\\
&= \frac{1}{4}\mathbf{E}\Big[\sum_{k=1}^{n} f''(X(t_{k-1}))^2((M(t_k)-M(t_{k-1}))^2 - (\langle M\rangle(t_k)-\langle M\rangle(t_{k-1})))^2\Big]
\end{aligned}$$

$$+\frac{1}{2}\sum_{k<\ell}\mathbf{E}\Big[f''(X(t_{k-1}))((M(t_k)-M(t_{k-1}))^2-(\langle M\rangle(t_k)-\langle M\rangle(t_{k-1})))$$
$$\times f''(X(t_{\ell-1}))\mathbf{E}[(M(t_\ell)-M(t_{\ell-1}))^2-(\langle M\rangle(t_\ell)-\langle M\rangle(t_{\ell-1}))|\mathscr{F}_{t_{\ell-1}}]\Big].$$

The second term is zero because

$$\mathbf{E}[(M(t_\ell)-M(t_{\ell-1}))^2-(\langle M\rangle(t_\ell)-\langle M\rangle(t_{\ell-1}))\big|\mathscr{F}_{t_{\ell-1}}]=0.$$

Hence, setting $K = 2\max|f''(x)|^2$, we obtain, by the elementary inequality $(a+b)^2 \leqq 2(a^2+b^2)$,

$$\begin{aligned}
&\mathbf{E}[|I_1-I_1''|^2]\\
&\leqq K\mathbf{E}\Big[\sum_{k=1}^n(M(t_k)-M(t_{k-1}))^4\Big]+K\mathbf{E}\Big[\sum_{k=1}^n(\langle M\rangle(t_k)-\langle M\rangle(t_{k-1}))^2\Big]\\
&\leqq K\mathbf{E}\Big[\max(M(t_k)-M(t_{k-1}))^2\sum_{k=1}^n(M(t_k)-M(t_{k-1}))^2\Big]\\
&\quad+K\mathbf{E}[\max(\langle M\rangle(t_k)-\langle M\rangle(t_{k-1}))\cdot\langle M\rangle(t)]\\
&\leqq K\{\mathbf{E}[\max(M(t_k)-M(t_{k-1}))^4]\}^{\frac{1}{2}}\Big\{\mathbf{E}\Big[\Big(\sum_{k=1}^n(M(t_k)-M(t_{k-1}))^2\Big)^2\Big]\Big\}^{\frac{1}{2}}\\
&\quad+K\mathbf{E}[\max(\langle M\rangle(t_k)-\langle M\rangle(t_{k-1}))\cdot\langle M\rangle(t)].
\end{aligned}$$

By Lemma 2.3.6 and the boundedness of M and $\langle M\rangle$, applying the Lebesgue convergence theorem to the last term, we obtain $\mathbf{E}[|I_1'-I_1''|^2]\to 0$.

Now we have shown Itô's formula (2.3.2) for each $t>0$. Noting that both sides are continuous in t, we complete the proof. □

Proof of Lemma 2.3.6 Write

$$\begin{aligned}
&\Big\{\sum_{k=1}^n(M(t_k)-M(t_{k-1}))^2\Big\}^2\\
&=\sum_{k=1}^n(M(t_k)-M(t_{k-1}))^4+2\sum_{k=1}^n\Big(\sum_{\ell=k+1}^n(M(t_\ell)-M(t_{\ell-1}))^2\Big)(M(t_k)-M(t_{k-1}))^2
\end{aligned}$$

and note

$$\mathbf{E}[(M(v)-M(u))^2|\mathscr{F}_u]=\mathbf{E}[M(v)^2-M(u)^2|\mathscr{F}_u]\qquad(u<v).$$

Then, for the first term,

$$\begin{aligned}
\mathbf{E}\Big[\sum_{k=1}^n(M(t_k)-M(t_{k-1}))^4\Big]&\leqq\mathbf{E}\Big[\sum_{k=1}^n(2C)^2(M(t_k)-M(t_{k-1}))^2\Big]\\
&\leqq 4C^2\mathbf{E}[M(t)^2-M(0)^2]\leqq 4C^4.
\end{aligned}$$

For the second term,

$$\begin{aligned}
&\mathbf{E}\Big[\sum_{k=1}^{n}\Big(\sum_{\ell=k+1}^{n}(M(t_\ell)-M(t_{\ell-1}))^2\Big)(M(t_k)-M(t_{k-1}))^2\Big] \\
&= \mathbf{E}\Big[\sum_{k=1}^{n}\mathbf{E}\Big[\sum_{\ell=k+1}^{n}(M(t_\ell)-M(t_{\ell-1}))^2\Big|\mathscr{F}_{t_{\ell-1}}\Big](M(t_k)-M(t_{k-1}))^2\Big] \\
&= \mathbf{E}\Big[\sum_{k=1}^{n}(M(t)^2-M(t_k)^2)(M(t_k)-M(t_{k-1}))^2\Big] \\
&\leqq \mathbf{E}\Big[\sum_{k=1}^{n}M(t)^2(M(t_k)-M(t_{k-1}))^2\Big] \\
&\leqq C^2\mathbf{E}[M(t)^2-M(0)^2] \leqq C^4.
\end{aligned}$$

Combining the two estimates, we obtain the conclusion. □

Example 2.3.7 For any $n = 1, 2, \ldots$, set $f(x) = x^n$ and apply Itô's formula. Then

$$B(t)^n = B(0)^n + n\int_0^t B(s)^{n-1}\mathrm{d}B(s) + \frac{n(n-1)}{2}\int_0^t B(s)^{n-2}\mathrm{d}s.$$

The second term of the right hand side is a martingale with mean 0. Hence, assuming $B(0) = 0$ and letting $n = 2m\,(m = 1, 2, \ldots)$, we obtain

$$\mathbf{E}[B(t)^{2m}] = m(2m-1)\int_0^t \mathbf{E}[B(s)^{2m-2}]\,\mathrm{d}s.$$

Since $\mathbf{E}[B(s)^2] = s$, this yields $\mathbf{E}[B(t)^4] = 3t^2$. In general, by induction,

$$\mathbf{E}[B(t)^{2m}] = \frac{(2m)!}{2^m m!}t^m \qquad (m = 1, 2, \ldots).$$

Example 2.3.8 For $\sigma \in \mathbb{R}$,

$$\mathrm{e}^{\sigma B(t)-\frac{1}{2}\sigma^2 t} = \mathrm{e}^{\sigma B(0)} + \sigma\int_0^t \mathrm{e}^{\sigma B(s)-\frac{1}{2}\sigma^2 s}\mathrm{d}B(s).$$

This is obtained by applying Itô's formula to $X(t) = \sigma B(t) - \frac{1}{2}\sigma^2 t$ and $f(x) = \mathrm{e}^x$.

Example 2.3.9 Let $B = \{B(t)\}_{t\geqq 0}$ be a one-dimensional Brownian motion and $\phi : [0, \infty) \to \mathbb{R}$ be a function which is square-integrable on each finite interval. Then, setting

$$X(t) = \exp\Big(\int_0^t \phi(s)\,\mathrm{d}B(s) - \frac{1}{2}\int_0^t \phi(s)^2\,\mathrm{d}s\Big),$$

we have

$$X(t) = 1 + \int_0^t \phi(s)X(s)\,\mathrm{d}B(s).$$

Example 2.3.10 Let $B = \{B(t)\}_{t \geqq 0}$ be a one-dimensional Brownian motion with $B(0) = 0$. For $\gamma \in \mathbb{R}$, the stochastic process $X = \{X(t)\}_{t \geqq 0}$ defined by

$$X(t) = x\mathrm{e}^{-\gamma t} + \mathrm{e}^{-\gamma t} \int_0^t \mathrm{e}^{\gamma s} \mathrm{d}B(s)$$

is called an **Ornstein–Uhlenbeck process**. X satisfies

$$X(t) = x + B(t) - \gamma \int_0^t X(s)\,\mathrm{d}s.$$

Example 2.3.11 Let $\{B(t) = (B^1(t), B^2(t), B^3(t))\}_{t \geqq 0}$ be a three-dimensional Brownian motion with $B(0) \neq 0$ and set $\tau_n = \inf\{t; |B(t)| = \frac{1}{n}\}$. If $|B(0)| > \frac{1}{n}$,

$$\frac{1}{|B(t \wedge \tau_n)|} = \frac{1}{|B(0)|} - \sum_{j=1}^{3} \int_0^{t \wedge \tau_n} \frac{B^j(s)}{|B(s)|^3}\,\mathrm{d}B^j(s).$$

Since $\tau_n \to \infty$ almost surely (see Remark 3.2.4), $\{|B(t)|^{-1}\}_{t \geqq 0}$ is a local martingale. $\{|B(t)|^{-1}\}_{t \geqq 0}$ is an example of a local martingale which is not a martingale because $\mathbf{E}[|B(t)|^{-1}] \to 0$ $(t \to \infty)$.

Definition 2.3.12 (Stochastic differential) (1) When a semimartingale X is decomposed as

$$X(t) = X(0) + M_X(t) + A_X(t),$$

we write

$$\mathrm{d}X(t) = \mathrm{d}M_X(t) + \mathrm{d}A_X(t)$$

and call it a representation of X by **stochastic differential**.
(2) Let Y be another semimartingale expressed as $\mathrm{d}Y(t) = \mathrm{d}M_Y(t) + \mathrm{d}A_Y(t)$. The stochastic differential of $\langle M_X, M_Y \rangle$ is denoted by $\mathrm{d}X \cdot \mathrm{d}Y(t)$ and is called the **product** of the stochastic differentials $\mathrm{d}X(t)$ and $\mathrm{d}Y(t)$.
(3) For a continuous $\{\mathscr{F}_t\}$-measurable stochastic process $\Phi = \{\Phi(t)\}_{t \geqq 0}$ which is bounded on each bounded interval, the stochastic integral $\Phi \cdot X$ of Φ with respect to X is defined as a semimartingale with $(\Phi \cdot X)(0) = 0$ whose stochastic differential is $\Phi\,\mathrm{d}X := \Phi\,\mathrm{d}M_X + \Phi\,\mathrm{d}A_X$:

$$(\Phi \cdot X)(t) = \int_0^t \Phi(s)\,\mathrm{d}M_X(s) + \int_0^t \Phi(s)\,\mathrm{d}A_X(s).$$

Remark 2.3.13 (1) For a d-dimensional $\{\mathscr{F}_t\}$-Brownian motion $B = \{B(t) = (B^1(t), \dots, B^d(t))\}_{t\geqq 0}$, we have

$$\mathrm{d}B^i \cdot \mathrm{d}B^j = \delta_{ij}\mathrm{d}t, \quad \mathrm{d}B^i \cdot \mathrm{d}t = 0, \quad \mathrm{d}t \cdot \mathrm{d}t = 0. \tag{2.3.3}$$

Moreover, if $X = \{X(t)\}_{t\geqq 0}$ is an $\mathbb{R}$-valued Itô process given by

$$X(t) = X(0) + \sum_{\alpha=1}^{d} \int_0^t a_\alpha(s)\,\mathrm{d}B^\alpha(s) + \int_0^t b(s)\,\mathrm{d}s \qquad (t \geqq 0),$$

its stochastic differential is

$$\mathrm{d}X(t) = \sum_{\alpha=1}^{d} a_\alpha(t)\,\mathrm{d}B^\alpha(t) + b(t)\,\mathrm{d}t.$$

If $\{Y(t)\}_{t\geqq 0}$ is another Itô process given by

$$\mathrm{d}Y(t) = \sum_{\alpha=1}^{d} p_\alpha(t)\,\mathrm{d}B^\alpha(t) + q(t)\,\mathrm{d}t,$$

(2.3.3) yields

$$\mathrm{d}X \cdot \mathrm{d}Y(t) = \sum_{\alpha=1}^{d} a_\alpha(t)p_\alpha(t)\,\mathrm{d}t.$$

(2) When X, Y, Z are semimartingales, we have

$$(\mathrm{d}X \cdot \mathrm{d}Y) \cdot \mathrm{d}Z = 0.$$

Using the stochastic differentials makes Itô's formula simple.

Theorem 2.3.14 (Itô's formula) *Let $\{X(t) = (X^1(t), \dots, X^N(t))\}_{t\geqq 0}$ be an N-dimensional semimartingale. Then, for any $f \in C^2(\mathbb{R}^N)$,*

$$\mathrm{d}(f(X))(t) = \sum_{i=1}^{N} \frac{\partial f}{\partial x^i}(X(t))\mathrm{d}X^i(t) + \frac{1}{2}\sum_{i,j=1}^{N} \frac{\partial^2 f}{\partial x^i \partial x^j}(X(t))\mathrm{d}X^i \cdot \mathrm{d}X^j(t).$$

Remark 2.3.15 Let ξ be a semimartingale whose martingale part is zero. Then, for any semimartingale Y, $\mathrm{d}\xi \cdot \mathrm{d}Y = 0$. Hence, by a standard approximation by smooth functions, it is easily deduced from Theorem 2.3.14 that, for $f \in C^{1,2}(\mathbb{R} \times \mathbb{R}^N)$,

$$\begin{aligned}\mathrm{d}(f(\xi, X))(t) = {} & \frac{\partial f}{\partial x^0}(\xi(t), X(t))\,\mathrm{d}\xi(t) + \sum_{i=1}^{N} \frac{\partial f}{\partial x^i}(X(t))\mathrm{d}X^i(t) \\ & + \frac{1}{2}\sum_{i,j=1}^{N} \frac{\partial^2 f}{\partial x^i \partial x^j}(X(t))\mathrm{d}X^i \cdot \mathrm{d}X^j(t).\end{aligned}$$

Example 2.3.16 The stochastic process in Example 2.3.9 satisfies

$$\mathrm{d}X(t) = \phi(t)X(t)\,\mathrm{d}B(t).$$

Example 2.3.17 The Ornstein–Uhlenbeck process defined in Example 2.3.10 satisfies

$$\mathrm{d}X(t) = \mathrm{d}B(t) - \gamma X(t)\,\mathrm{d}t.$$

Next we define the stochastic integral due to Stratonovich. If we use this stochastic integral, Itô's formula is of the same form as the chain rule in the usual calculus. It is useful when we apply stochastic calculus to study problems in geometry.

Definition 2.3.18 Let X and Y be semimartingales. Then, $Y \circ \mathrm{d}X$ denotes $Y\,\mathrm{d}X + \frac{1}{2}\,\mathrm{d}X \cdot \mathrm{d}Y$. The corresponding semimartingale starting from 0 is called the **Stratonovich integral** of Y with respect to X and is denoted by

$$\int_0^t Y(s) \circ \mathrm{d}X(s).$$

Theorem 2.3.19 *Let $X = \{X(t)\}_{t \geqq 0}$ be the same N-dimensional semimartingale as Theorem 2.3.4. Then, for any $f \in C^3(\mathbb{R}^N)$,*

$$f(X(t)) = f(X(0)) + \sum_{i=1}^N \int_0^t \frac{\partial f}{\partial x^i}(X(s)) \circ \mathrm{d}X^i(s).$$

Proof We simply write f_i, f_{ij} and f_{ijp} for $\frac{\partial f}{\partial x^i}$, $\frac{\partial^2 f}{\partial x^i \partial x^j}$ and $\frac{\partial^3 f}{\partial x^i \partial x^j \partial x^p}$, respectively. Then, by the definition,

$$\begin{aligned}
\sum_{i=1}^N f_i(X(t)) \circ \mathrm{d}X^i(t) &= \sum_{i=1}^N f_i(X(t))\mathrm{d}X^i(t) + \frac{1}{2}\sum_{i=1}^N \mathrm{d}(f_i(X))(t) \cdot \mathrm{d}X^i(t) \\
&= \sum_{i=1}^N f_i(X(t))\mathrm{d}X^i(t) + \frac{1}{2}\sum_{i,j=1}^N f_{ij}(X(t))\mathrm{d}X^i \cdot \mathrm{d}X^j(t) \\
&\quad + \frac{1}{2}\sum_{i,j,p=1}^N f_{ijp}(X(t))(\mathrm{d}X^i \cdot \mathrm{d}X^j) \cdot \mathrm{d}X^p(t).
\end{aligned}$$

The right hand side coincides with $\mathrm{d}(f(X(t)))$ because the third term is 0 by Remark 2.3.13 (2). □

2.4 Moment Inequalities for Martingales

In the rest of this chapter, several results which are obtained by applying stochastic integrals and Itô's formula are shown. These will be frequently used in the following chapters.

Theorem 2.4.1 (Burkholder–Davis–Gundy inequality) *For any $p > 0$ there exist positive constants c_p and C_p such that*

$$c_p \mathbf{E}[M^*(t)^{2p}] \leqq \mathbf{E}[\langle M\rangle(t)^p] \leqq C_p \mathbf{E}[M^*(t)^{2p}] \tag{2.4.1}$$

for any continuous local martingale $M = \{M(t)\}_{t\geqq 0}$ with $M(0) = 0$, where

$$M^*(t) = \max_{0\leqq s\leqq t} |M(s)|.$$

Proof Following [29], we give a proof by stochastic analysis. See, e.g., [100], for another proof.

We set

$$\tau_n = \inf\{t;\ |M(t)| \geqq n \text{ or } \langle M\rangle(t) \geqq n\}, \quad n = 1, 2, \ldots$$

Then, $\tau_n \to \infty$ almost surely. If we show (2.4.1) for $M^{\tau_n} = \{M(t\wedge\tau_n)\}_{t\geqq 0}$ (c_p and C_p are independent of n), we see (2.4.1) for M by letting $n \to \infty$. Hence we may assume that $\{M(t)\}_{t\geqq 0}$ and $\{\langle M\rangle(t)\}_{t\geqq 0}$ are bounded.

We recall Doob's inequality (Theorem 1.5.14) : for $q > 1$,

$$\mathbf{E}[M^*(t)^q] \leqq \Big(\frac{q}{q-1}\Big)^q \mathbf{E}[|M(t)|^q]. \tag{2.4.2}$$

If $p = 1$, since $\mathbf{E}[\langle M\rangle(t)] = \mathbf{E}[M(t)^2]$, Doob's inequality for $q = 2$ implies (2.4.1) if we set $c_p = \frac{1}{4}$ and $C_p = 1$.

Next we consider the case when $p > 1$. Since $f(x) = |x|^{2p}$ ($x \in \mathbb{R}$) is of C^2-class,

$$\begin{aligned}|M(t)|^{2p} &= \int_0^t 2p|M(s)|^{2p-1}\mathrm{sgn}(M(s))\,\mathrm{d}M(s)\\ &\quad + p(2p-1)\int_0^t |M(s)|^{2p-2}\mathrm{d}\langle M\rangle(s),\end{aligned}$$

where $\mathrm{sgn}(x)$ is a function such that $\mathrm{sgn}(x) = 1$ for $x > 0$ and $\mathrm{sgn}(x) = -1$ for $x \leqq 0$. By Hölder's inequality,

$$\begin{aligned}\mathbf{E}[|M(t)|^{2p}] &= p(2p-1)\mathbf{E}\Big[\int_0^t |M(s)|^{2p-2}\mathrm{d}\langle M\rangle(s)\Big]\\ &\leqq p(2p-1)\mathbf{E}[|M^*(t)|^{2p-2}\langle M\rangle(t)]\\ &\leqq p(2p-1)\{\mathbf{E}[|M^*(t)|^{2p}]\}^{\frac{p-1}{p}}\{\mathbf{E}[\langle M\rangle(t)^p]\}^{\frac{1}{p}}.\end{aligned}$$

Hence we get

$$\mathbf{E}[M^*(t)^{2p}] \leqq \left(\frac{2p}{2p-1}\right)^{2p} p(2p-1)\{\mathbf{E}[M^*(t)^{2p}]\}^{\frac{p-1}{p}}\{\mathbf{E}[\langle M\rangle(t)^p]\}^{\frac{1}{p}}$$

by Doob's inequality. It is now easy to show the first inequality of (2.4.1).

Set $N(t) = \int_0^t \langle M\rangle(s)^{\frac{p-1}{2}} \mathrm{d}M(s)$. Then it is easy to see

$$\langle N\rangle(t) = \int_0^t \langle M\rangle(s)^{p-1}\mathrm{d}\langle M\rangle(s) = \frac{1}{p}\langle M\rangle(t)^p$$

and

$$\mathbf{E}[\langle M\rangle(t)^p] = p\mathbf{E}[\langle N\rangle(t)] = p\mathbf{E}[N(t)^2]. \tag{2.4.3}$$

On the other hand, Itô's formula yields

$$\begin{aligned} M(t)\langle M\rangle(t)^{\frac{p-1}{2}} &= \int_0^t \langle M\rangle(s)^{\frac{p-1}{2}}\mathrm{d}M(s) + \int_0^t M(s)\,\mathrm{d}[\langle M\rangle(s)^{\frac{p-1}{2}}] \\ &= N(t) + \int_0^t M(s)\,\mathrm{d}[\langle M\rangle(s)^{\frac{p-1}{2}}], \end{aligned}$$

which implies

$$|N(t)| \leqq 2M^*(t)\langle M\rangle(t)^{\frac{p-1}{2}}.$$

Hence, by (2.4.3), we get

$$\mathbf{E}[\langle M\rangle(t)^p] \leqq 4p\mathbf{E}[M^*(t)^2\langle M\rangle(t)^{p-1}] \leqq 4p\{\mathbf{E}[M^*(t)^{2p}]\}^{\frac{1}{p}}\{\mathbf{E}[\langle M\rangle(t)^p]\}^{\frac{p-1}{p}}.$$

Thus, for $p > 1$,

$$\mathbf{E}[\langle M\rangle(t)^p] \leqq (4p)^p\mathbf{E}[M^*(t)^{2p}].$$

Finally we show the case when $0 < p < 1$. Note $\langle M\rangle^{-\frac{1-p}{2}} \in \mathscr{L}^2(M)$ and set

$$N(t) = \int_0^t \langle M\rangle(s)^{-\frac{1-p}{2}}\mathrm{d}M(s).$$

Then $\mathbf{E}[\langle M\rangle(t)^p] = p\mathbf{E}[N(t)^2]$ and $M(t) = \int_0^t \langle M\rangle(s)^{\frac{1-p}{2}}\mathrm{d}N(s)$. Itô's formula yields

$$\begin{aligned} N(t)\langle M\rangle(t)^{\frac{1-p}{2}} &= \int_0^t \langle M\rangle(s)^{\frac{1-p}{2}}\mathrm{d}N(s) + \int_0^t N(s)\,\mathrm{d}[\langle M\rangle(s)^{\frac{1-p}{2}}] \\ &= M(t) + \int_0^t N(s)\,\mathrm{d}[\langle M\rangle(s)^{\frac{1-p}{2}}]. \end{aligned}$$

Hence $|M(t)| \leqq 2N^*(t)\langle M\rangle(t)^{\frac{1-p}{2}}$. Therefore, since

$$M^*(t) \leqq 2N^*(t)\langle M\rangle(t)^{\frac{1-p}{2}},$$

we obtain

$$\begin{aligned}\mathbf{E}[M^*(t)^{2p}] &\leqq 2^{2p}\mathbf{E}[N^*(t)^{2p}\langle M\rangle(t)^{p(1-p)}]\\ &\leqq 2^{2p}\{\mathbf{E}[N^*(t)^2]\}^p\{\mathbf{E}[\langle M\rangle(t)^p]\}^{1-p}.\end{aligned}$$

In conjunction with the result in the case when $p = 1$, we obtain

$$\begin{aligned}\mathbf{E}[M^*(t)^{2p}] &\leqq 2^{2p}4^p\{\mathbf{E}[N(t)^2]\}^p\{\mathbf{E}[\langle M\rangle(t)^p]\}^{1-p}\\ &\leqq \left(\frac{16}{p}\right)^p\{\mathbf{E}[\langle M\rangle(t)^p]\}^p\{\mathbf{E}[\langle M\rangle(t)^p]\}^{1-p} = \left(\frac{16}{p}\right)^p\mathbf{E}[\langle M\rangle(t)^p].\end{aligned}$$

Now we get the first inequality of (2.4.1).

To show the second inequality, let $\alpha > 0$ and write

$$\langle M\rangle(t)^p = \{\langle M\rangle(t)^p(\alpha + M^*(t))^{-2p(1-p)}\}(\alpha + M^*(t))^{2p(1-p)}.$$

Then,

$$\mathbf{E}[\langle M\rangle(t)^p] \leqq \{\mathbf{E}[\langle M\rangle(t)(\alpha + M^*(t))^{-2(1-p)}]\}^p\{\mathbf{E}[(\alpha + M^*(t))^{2p}]\}^{1-p}. \tag{2.4.4}$$

Set $\widetilde{N}(t) = \int_0^t(\alpha + M^*(s))^{-(1-p)}\mathrm{d}M(s)$. Then

$$\begin{aligned}\langle\widetilde{N}\rangle(t) &= \int_0^t (\alpha + M^*(s))^{-2(1-p)}\mathrm{d}\langle M\rangle(s)\\ &\geqq (\alpha + M^*(t))^{-2(1-p)}\langle M\rangle(t).\end{aligned} \tag{2.4.5}$$

Itô's formula yields

$$\begin{aligned}M(t)(\alpha + M^*(t))^{-(1-p)} &= \int_0^t (\alpha + M^*(s))^{-(1-p)}\mathrm{d}M(s)\\ &\quad + \int_0^t M(s)\,\mathrm{d}[(\alpha + M^*(s))^{-(1-p)}]\\ &= \widetilde{N}(t) - (1-p)\int_0^t M(s)(\alpha + M^*(s))^{p-2}\mathrm{d}M^*(s).\end{aligned}$$

Hence we get

$$|\widetilde{N}(t)| \leqq M^*(t)^p + (1-p)\int_0^t M^*(s)^{p-1}\mathrm{d}M^*(s) = \frac{1}{p}M^*(t)^p$$

and $\mathbf{E}[\widetilde{N}(t)^2] \leqq p^{-2}\mathbf{E}[M^*(t)^{2p}]$. By (2.4.4) and (2.4.5),

$$\begin{aligned}\mathbf{E}[\langle M\rangle(t)^p] &\leqq \{\mathbf{E}[\langle\widetilde{N}\rangle(t)]\}^p\{\mathbf{E}[(\alpha + M^*(t))^{2p}]\}^{1-p}\\ &\leqq \frac{1}{p^{2p}}\{\mathbf{E}[M^*(t)^{2p}]\}^p\{\mathbf{E}[(\alpha + M^*(t))^{2p}]\}^{1-p}.\end{aligned}$$

Letting $\alpha \downarrow 0$, we obtain $\mathbf{E}[\langle M\rangle(t)^p] \leqq p^{-2p}\mathbf{E}[M^*(t)^{2p}]$. □

2.5 Martingale Characterization of Brownian Motion

First we show the martingale characterization of Brownian motion due to Lévy.

Theorem 2.5.1 *Let $M = \{M(t) = (M^1(t), M^2(t), \dots, M^N(t))\}_{t \geqq 0}$ be an N-dimensional continuous $\{\mathscr{F}_t\}$-local martingale with $M(0) = 0$.*
(1) *If*

$$\langle M^i, M^j\rangle(t) = \delta_{ij} t \quad (i, j = 1, 2, \dots, N), \tag{2.5.1}$$

then M is an N-dimensional $\{\mathscr{F}_t\}$-Brownian motion.
(2) *Suppose that*

$$\exp\Big(\mathrm{i}\sum_{i=1}^N \int_0^t f_i(s)\,\mathrm{d}M^i(s) + \frac{1}{2}\sum_{i=1}^N \int_0^t f_i(s)^2 \mathrm{d}s\Big)$$

is a martingale for any $f_1, \dots, f_N \in L^2([0,\infty))$. Then, M is an N-dimensional $\{\mathscr{F}_t\}$-Brownian motion.

Proof (1) It suffices to show

$$\mathbf{E}[\mathrm{e}^{\mathrm{i}\langle \xi, M(t)-M(s)\rangle} | \mathscr{F}_s] = \mathrm{e}^{-\frac{|\xi|^2(t-s)}{2}}. \tag{2.5.2}$$

for any $\xi \in \mathbb{R}^N$ and $0 \leqq s < t$. The assumption (2.5.1) implies that M is a square-integrable martingale. Moreover, by Itô's formula, it yields

$$\begin{aligned} \mathrm{e}^{\mathrm{i}\langle \xi, M(t)\rangle} - \mathrm{e}^{\mathrm{i}\langle \xi, M(s)\rangle} &= \sum_{i=1}^N \int_s^t \mathrm{i}\,\xi^i \mathrm{e}^{\mathrm{i}\langle \xi, M(u)\rangle} \mathrm{d}M^i(u) \\ &\quad - \frac{1}{2}\sum_{i=1}^N (\xi^i)^2 \int_s^t \mathrm{e}^{\mathrm{i}\langle \xi, M(u)\rangle} \mathrm{d}u. \end{aligned}$$

Since the first term on the right hand side is a square-integrable martingale, we have

$$\mathbf{E}[\mathrm{e}^{\mathrm{i}\langle \xi, M(t)-M(s)\rangle}\mathbf{1}_A] - \mathbf{P}(A) = -\frac{|\xi|^2}{2}\int_s^t \mathbf{E}[\mathrm{e}^{\mathrm{i}\langle \xi, M(u)-M(s)\rangle}\mathbf{1}_A]\,\mathrm{d}u$$

for any $A \in \mathscr{F}_s$. The unique solution of this integral equation is given by

$$\mathbf{E}[\mathrm{e}^{\mathrm{i}\langle \xi, M(t)-M(s)\rangle}\mathbf{1}_A] = \mathbf{P}(A)\mathrm{e}^{-\frac{|\xi|^2(t-s)}{2}}.$$

This means (2.5.2).
(2) For $\xi = (\xi^1, \xi^2, \dots, \xi^N) \in \mathbb{R}^N$ and $T > 0$, set $f_i(s) = \xi^i \mathbf{1}_{[0,T]}(s)$. By the assumption, $\{\mathrm{e}^{\mathrm{i}\sum_i \xi^i M^i(t) - \frac{|\xi|^2 t}{2}}\}_{0 \leqq t \leqq T}$ is a martingale. This means (2.5.2). □

Corollary 2.5.2 *Let $\{B(t) = (B^1(t), B^2(t), \dots, B^N(t))\}_{t\geqq 0}$ be an N-dimensional $\{\mathscr{F}_t\}$-Brownian motion and σ be an $\{\mathscr{F}_t\}$-stopping time which is finite almost surely. Set $B^*(t) = B(\sigma + t)$ and $\mathscr{F}_t^* = \mathscr{F}_{\sigma+t}$ $(t \geqq 0)$. Then, $B^* = \{B^*(t)\}_{t\geqq 0}$ is an N-dimensional $\{\mathscr{F}_t^*\}$-Brownian motion. In particular, $\tilde{B}^*(t) = B(\sigma + t) - B(\sigma)$ is an N-dimensional Brownian motion which is independent of $\mathscr{F}_0^* = \mathscr{F}_\sigma$.*

Proof By the optional sampling theorem, $M^i(t) = B^i(\sigma + t) - B^i(\sigma)$ and $M^i(t)M^j(t) - \delta_{ij}t$ are $\{\mathscr{F}_t^*\}$-local martingales. Hence, Theorem 2.5.1 implies the assertion because $\langle M^i, M^j\rangle(t) = \delta_{ij}t$. □

Next we show that continuous local martingales are given as time changed Brownian motions.

Theorem 2.5.3 (Dambis–Dubins–Schwarz) *Let M be a continuous local martingale such that $\lim_{t\to\infty}\langle M\rangle(t) = \infty$ almost surely and set $\sigma(s) = \inf\{t > 0; \langle M\rangle(t) > s\}$ and $B(s) = M(\sigma(s))$. Then, $B = \{B(s)\}_{s\geqq 0}$ is an $\{\mathscr{F}_{\sigma(s)}\}$-Brownian motion and $M(t) = B(\langle M\rangle(t))$.*

Proof If we show that B is a continuous local martingale satisfying $\langle B\rangle(s) = s$, we are done by Lévy's theorem (Theorem 2.5.1). Note that $\sigma(s)$ is an $\{\mathscr{F}_t\}$-stopping time which is finite almost surely, because $\{\sigma(s) < t\} = \{s < \langle M\rangle(t)\} \in \mathscr{F}_t$ for any t.

To show the continuity of B, we show, for any $t_1 < t_2$,

$$\{\omega; \langle M\rangle(t_1) = \langle M\rangle(t_2)\} \subset \{\omega; M(t) = M(t_1), t_1 \leqq t \leqq t_2\}. \tag{2.5.3}$$

For this purpose, let η be an $\{\mathscr{F}_t\}$-stopping time given by

$$\eta = \inf\{t > t_1; \langle M\rangle(t) > \langle M\rangle(t_1)\}$$

and set

$$N(t) = M((t_1 + t) \wedge \eta) - M(t_1).$$

$\{N(t)\}_{t\geqq 0}$ is a continuous $\{\mathscr{F}_{(t_1+t)\wedge\eta}\}$-local martingale satisfying

$$\langle N\rangle(t) = \langle M\rangle((t_1 + t) \wedge \eta) - \langle M\rangle(t_1).$$

By the definition of η, $\langle N\rangle(t) = 0$ and hence $N(t) = 0\,(t \geqq 0)$ almost surely. Thus (2.5.3) holds.

Due to the continuity of M and $\langle M\rangle$, (2.5.3) implies

$$\mathbf{P}\begin{pmatrix} \text{for all } t_1 \leqq t_2, \text{ if } \langle M\rangle(t_1) = \langle M\rangle(t_2), \\ \text{then } M \text{ is constant on } [t_1, t_2] \end{pmatrix} = 1,$$

which means the continuity of B.

Next we show that $\{B(s)\}_{s \geqq 0}$ is a locally square-integrable martingale satisfying $\langle B \rangle(s) = s$. Let $0 \leqq s_1 < s_2$ and set $\widetilde{M}(t) = M(t \wedge \sigma(s_2))$. $\{\widetilde{M}(t)\}_{t \geqq 0}$ is an $\{\mathscr{F}_t\}$-martingale and

$$\langle \widetilde{M} \rangle(t) = \langle M \rangle(t \wedge \sigma(s_2)) \leqq \langle M \rangle(\sigma(s_2)) = s_2 \qquad (t \geqq 0).$$

This means that $\{\widetilde{M}(t)\}_{t \geqq 0}$ and $\{\widetilde{M}(t)^2 - \langle \widetilde{M} \rangle(t)\}_{t \geqq 0}$ are uniformly integrable martingales. Therefore, by the optional sampling theorem,

$$\mathbf{E}[B(s_2) - B(s_1)|\mathscr{F}_{\sigma(s_1)}] = \mathbf{E}[\widetilde{M}(\sigma(s_2)) - \widetilde{M}(\sigma(s_1))|\mathscr{F}_{\sigma(s_1)}] = 0$$

and

$$\mathbf{E}[(B(s_2) - B(s_1))^2|\mathscr{F}_{\sigma(s_1)}] = \mathbf{E}[\langle \widetilde{M} \rangle(\sigma(s_2)) - \langle \widetilde{M} \rangle(\sigma(s_1))|\mathscr{F}_{\sigma(s_1)}] = s_2 - s_1. \ \square$$

Theorem 2.5.3 was extended to multi-dimensional stochastic processes in [60].

Theorem 2.5.4 *Let $M^i \in \mathscr{M}^2_{\mathrm{c,loc}}$ $(i = 1, 2, \ldots, N)$ and suppose*

$$\langle M^i, M^j \rangle(t) = 0 \qquad (i \neq j)$$

and

$$\lim_{t \to \infty} \langle M^i \rangle(t) = \infty, \ \mathbf{P}\text{-a.s.} \qquad (i = 1, 2, \ldots, N).$$

Then, the stochastic process $\{B(s) = (B^1(s), B^2(s), \ldots, B^N(s))\}_{s \geqq 0}$ defined by

$$B^i(s) = M^i(\tau^i(s)),$$

where $\tau^i(s) = \inf\{t; \langle M^i \rangle(t) > s\}$, is an N-dimensional Brownian motion.

Proof We give a proof following [9]. Since each $B^i = \{B^i(s)\}_{s \geqq 0}$ is a one-dimensional Brownian motion by Theorem 2.5.3, it suffices to show the independence of $B^1, B^2, \ldots, B^N$. For this purpose, let $f_i \in L^2([0, \infty) \to \mathbb{R})$ and set

$$\Phi_i(t) = \exp\Bigl(\mathrm{i} \int_0^t f_i(s)\, \mathrm{d}B^i(s) + \frac{1}{2} \int_0^t f_i(s)^2 \mathrm{d}s\Bigr).$$

Then, it is easy to show by Itô's formula that $\{\Phi_i(t)\}_{t \geqq 0}$ is a bounded martingale and that $\mathbf{E}[\Phi_i(t)] = 1, \mathbf{E}[\Phi_i(\infty)] = 1$. In particular,

$$\mathbf{E}\Bigl[\mathrm{e}^{\mathrm{i} \int_0^\infty f_i(s) \mathrm{d}B^i(s)}\Bigr] = \mathrm{e}^{-\frac{1}{2} \int_0^\infty f_i(s)^2 \mathrm{d}s}. \tag{2.5.4}$$

On the other hand, set $L(t) = \sum_{i=1}^N \int_0^t f_i(\langle M\rangle(u))\,\mathrm{d}M^i(u)$. Then,

$$\langle L\rangle(t) = \sum_{i=1}^N \int_0^t f_i(\langle M\rangle(u))^2\,\mathrm{d}\langle M^i\rangle(u) \leqq \sum_{i=1}^N \int_0^\infty |f_i(s)|^2\mathrm{d}s$$

and $\{L(t)\}_{t\geqq 0}$ is a martingale by Theorem 2.1.3. By Itô's formula again, the stochastic process $\{\mathrm{e}^{\mathrm{i}L(t)+\frac{\langle L\rangle(t)}{2}}\}_{t\geqq 0}$ is a bounded martingale and

$$\mathbf{E}[\mathrm{e}^{\mathrm{i}L(t)+\frac{\langle L\rangle(t)}{2}}] = 1.$$

Hence, denoting the limit of $L(t)$ and $\langle L\rangle(t)$ as $t \to \infty$ by $L(\infty)$ and $\langle L\rangle(\infty)$, we get

$$\mathbf{E}[\mathrm{e}^{\mathrm{i}L(\infty)+\frac{\langle L\rangle(\infty)}{2}}] = 1.$$

Since $\{\tau^i(t) < s\} = \{t < \langle M^i\rangle(s)\}$,

$$\int_0^\infty \mathbf{1}_{(t_1,t_2]}(\langle M^i\rangle(t))\,\mathrm{d}M^i(t) = B^i(t_2) - B^i(t_1) = \int_0^\infty \mathbf{1}_{(t_1,t_2]}(s)\,\mathrm{d}B^i(s)$$

for any $t_1 < t_2$. Let $\{g_{im}\}_{m=1}^\infty$ be a sequence of step functions which are linear combinations of the indicator functions of intervals of the form $(t_1, t_2]$ such that

$$\lim_{m\to\infty} \int_0^\infty |f_i(s) - g_{im}(s)|^2\mathrm{d}s = 0.$$

Then, by the observation above,

$$\int_0^\infty g_{im}(\langle M^i\rangle(t))\,\mathrm{d}M^i(t) = \int_0^\infty g_{im}(s)\,\mathrm{d}B^i(s). \tag{2.5.5}$$

Moreover, since

$$\int_0^\infty |f_i(\langle M^i\rangle(t)) - g_{im}(\langle M^i\rangle(t))|^2\mathrm{d}\langle M^i\rangle(t)$$
$$= \int_0^\infty |f_i(s) - g_{im}(s)|^2\mathrm{d}s \to 0 \qquad (m \to \infty),$$

we obtain

$$\int_0^\infty f_i(\langle M^i\rangle(t))\,\mathrm{d}M^i(t) = \int_0^\infty f_i(s)\,\mathrm{d}B^i(s)$$

by letting $m \to \infty$ in (2.5.5). Hence we have

$$L(\infty) = \sum_{i=1}^N \int_0^\infty f_i(\langle M^i\rangle(t))\,\mathrm{d}M^i(t) = \sum_{i=1}^N \int_0^\infty f_i(s)\,\mathrm{d}B^i(s)$$

and

$$\langle L\rangle(\infty) = \sum_{i=1}^{N} \int_0^\infty f_i(s)^2 \mathrm{d}s.$$

Therefore, by (2.5.4),

$$\mathbf{E}\Big[\mathrm{e}^{\mathrm{i}\sum_{i=1}^N \int_0^\infty f_i(s)\mathrm{d}B^i(s)}\Big] = \prod_{i=1}^{N} \mathrm{e}^{-\frac{1}{2}\int_0^\infty f_i(s)^2\mathrm{d}s} = \prod_{i=1}^{N} \mathbf{E}\Big[\mathrm{e}^{\mathrm{i}\int_0^\infty f_i(s)\mathrm{d}B^i(s)}\Big].$$

This implies the independence of $B^1, B^2, \ldots, B^N$. □

Theorems 2.5.3 and 2.5.4 are extended by removing the assumptions on the divergence of the quadratic variation processes as $t \to \infty$. For this we need to extend the probability spaces and to consider a Brownian motion independent of M.

Theorem 2.5.5 *Let M be a continuous local martingale defined on a probability space $(\Omega, \mathscr{F}, \mathbf{P}, \{\mathscr{F}_t\})$ and set*

$$\sigma(s) = \begin{cases} \inf\{u; \langle M\rangle(u) > s\}, \\ \infty \qquad (\text{if } s \geqq \langle M\rangle(\infty)), \end{cases} \qquad \widehat{\mathscr{F}}_s = \bigvee\nolimits_{u>0} \mathscr{F}_{\sigma(s)\wedge u}.$$

Moreover, let $\{B'(t)\}_{t\geqq 0}$ be a Brownian motion starting from 0 defined on another probability space $(\Omega', \mathscr{F}', \mathbf{P}', \{\mathscr{F}'_t\})$ and $(\widetilde{\Omega}, \widetilde{\mathscr{F}}, \widetilde{\mathbf{P}}, \{\widetilde{\mathscr{F}}_t\})$ be the direct product probability space defined by

$$\widetilde{\Omega} = \Omega\times\Omega', \quad \widetilde{\mathscr{F}} = \mathscr{F}\otimes\mathscr{F}', \quad \widetilde{\mathbf{P}} = \mathbf{P}\times\mathbf{P}', \quad \widetilde{\mathscr{F}}_t = \widehat{\mathscr{F}}_t\otimes\mathscr{F}'_t. \tag{2.5.6}$$

Then, there exists an $\{\widetilde{\mathscr{F}}_t\}$-Brownian motion $\{B(s)\}_{s\geqq 0}$ which is defined on the product probability space and satisfies

$$B(s) = M(\sigma(s)) \qquad (s \in [0, \langle M\rangle(\infty))),$$

that is, $M(t) = B(\langle M\rangle(t))$ $(t \geqq 0)$.

Proof We follow [45]. By the optional sampling theorem (Theorem 1.5.11), if $s \leqq s', u \leqq u'$, then

$$\mathbf{E}[M(\sigma(s')\wedge u')|\mathscr{F}_{\sigma(s)\wedge u}] = M(\sigma(s)\wedge u)$$

and

$$\begin{aligned}&\mathbf{E}[(M(\sigma(s')\wedge u') - M(\sigma(s)\wedge u))^2|\mathscr{F}_{\sigma(s)\wedge u}] \\ &\qquad = \mathbf{E}[\langle M\rangle(\sigma(s')\wedge u') - \langle M\rangle(\sigma(s)\wedge u)|\mathscr{F}_{\sigma(s)\wedge u}].\end{aligned}$$

In particular, for $A \in \mathscr{F}_{\sigma(s)\wedge u}$ $(u > 0)$,

$$\mathbf{E}[M(\sigma(s') \wedge u)\mathbf{1}_A] = \mathbf{E}[M(\sigma(s) \wedge u)\mathbf{1}_A].$$

By the observation above, $M(\sigma(s) \wedge u)$ converges in L^2 as $u \uparrow \infty$. Hence, denoting the limit by $\widetilde{B}(s)$, we have $\mathbf{E}[\widetilde{B}(s')\mathbf{1}_A] = \mathbf{E}[\widetilde{B}(s)\mathbf{1}_A]$. Therefore,

$$\mathbf{E}[\widetilde{B}(s')|\widehat{\mathscr{F}}_s] = \widetilde{B}(s)$$

and

$$\mathbf{E}[(\widetilde{B}(s') - \widetilde{B}(s))^2|\widehat{\mathscr{F}}_s] = \mathbf{E}[(s' \wedge \langle M\rangle(\infty)) - (s \wedge \langle M\rangle(\infty))|\widehat{\mathscr{F}}_s].$$

Set

$$B(s) = B'(s) - B'(s \wedge \langle M\rangle(\infty)) + \widetilde{B}(s) \quad (s \geqq 0).$$

Then, $B = \{B(s)\}_{s\geqq 0}$ is an $\{\widetilde{\mathscr{F}_t}\}$-martingale and $\langle B\rangle(s) = s$. Hence, B is an $\{\widetilde{\mathscr{F}_t}\}$-Brownian motion. □

The direct product probability space in Theorem 2.5.5 is called an **extension** of $(\Omega, \mathscr{F}, \mathbf{P}, \{\mathscr{F}_t\})$. Also, for a multi-dimensional case, the following can be proven. We omit the proof.

Theorem 2.5.6 *Let $M^i = \{M^i(t)\}_{t\geqq 0}$ $(i = 1, 2, \ldots, d)$ be continuous local martingales defined on a probability space $(\Omega, \mathscr{F}, \mathbf{P}, \{\mathscr{F}_t\})$ and assume that*

$$\langle M^i, M^j\rangle(t) = 0 \qquad (i \neq j).$$

Set

$$\sigma^i(s) = \begin{cases} \inf\{u; \langle M^i\rangle(u) > s\} \\ \infty \qquad (\text{if } s \geqq \langle M^i\rangle(\infty)). \end{cases}$$

Then, there exists an extension of $(\Omega, \mathscr{F}, \mathbf{P}, \{\mathscr{F}_t\})$ and a d-dimensional Brownian motion $\{(B^1(s), B^2(s), \ldots, B^d(s))\}_{s\geqq 0}$ defined on it such that

$$B^i(s) = M^i(\sigma^i(s)) \qquad (s \in [0, \langle M^i\rangle(\infty))\,).$$

Next we show that a d-dimensional local martingale can be expressed as a stochastic integral with respect to a Brownian motion.

Theorem 2.5.7 *For $M^i = \{M^i(t)\}_{t\geqq 0} \in \mathscr{M}^2_{\mathrm{c,loc}}$ $(i = 1, 2, \ldots, d)$ with $M^i(0) = 0$ defined on a probability space $(\Omega, \mathscr{F}, \mathbf{P}, \{\mathscr{F}_t\})$, assume that there exists $\Phi_{ij} = \{\Phi_{ij}(s)\}_{s\geqq 0} \in \mathscr{L}^1_{\mathrm{loc}}$ and $\Psi_{ij} = \{\Psi_{ij}(s)\}_{s\geqq 0} \in \mathscr{L}^2_{\mathrm{loc}}$ $(i, j = 1, 2, \ldots, d)$ such that*

$$\langle M^i, M^j\rangle(t) = \int_0^t \Phi_{ij}(s)\,\mathrm{d}s, \quad \Phi_{ij}(s) = \sum_{p=1}^d \Psi_{ip}(s)\Psi_{jp}(s),$$

and $\det((\Psi_{ij}(s))) \neq 0$ *for any* $s \geqq 0$ *almost surely. Then, there exists a* d*-dimensional* $\{\mathscr{F}_t\}$*-Brownian motion* $\{(B^1(t), B^2(t), \dots, B^d(t))\}_{t\geqq 0}$ *satisfying*

$$M^i(t) = \sum_{j=1}^{d} \int_0^t \Psi_{ij}(s)\,\mathrm{d}B^j(s).$$

Proof By the standard argument using stopping times, we may assume that $M^i \in \mathscr{M}_c^2$, $\Phi_{ij} \in \mathscr{L}^1$ and $\Psi_{ij} \in \mathscr{L}^2$. Denote by $(\Psi^{-1})_{ij}(s)$ the (i, j)-component of the inverse of $\Psi(s) = (\Psi_{ij}(s))$ and set

$$\Theta_{ij}^{(N)}(s) = \begin{cases} (\Psi^{-1})_{ij}(s) & (\text{if } |(\Psi^{-1})_{ij}(s)| \leqq N \ (i, j = 1, 2, \dots, d)) \\ 0 & (\text{otherwise}) \end{cases}$$

for $N > 0$. By the dominated convergence theorem,

$$\int_0^t \mathbf{E}\Big[\Big|\sum_{p,q=1}^{d} \Theta_{ip}^{(N)}(s)\Theta_{jq}^{(N)}(s)\Phi_{pq}(s) - \delta_{ij}\Big|^2\Big]\mathrm{d}s \to 0 \qquad (N \to \infty).$$

Set

$$B_{(N)}^i(t) = \sum_{j=1}^{d} \int_0^t \Theta_{ij}^{(N)}(s)\,\mathrm{d}M^j(s).$$

Then, $\{B_{(N)}^i(t)\}_{t\geqq 0} \in \mathscr{M}_c^2$ and the quadratic variation process is given by

$$\langle B_{(N)}^i, B_{(N)}^j \rangle(t) = \int_0^t \sum_{p,q=1}^{d} \Theta_{ip}^{(N)}(s)\Theta_{jq}^{(N)}(s)\Phi_{pq}(s)\,\mathrm{d}s.$$

Hence, $\{B_{(N)}^i(t)\}_{t\geqq 0}$ converges in $\mathscr{M}_c^2$ as $N \to \infty$ and, for the limit $\{B^i(t)\}_{t\geqq 0}$,

$$\langle B^i, B^j \rangle(t) = \delta_{ij} t.$$

By Theorem 2.5.1, $\{(B^1(t), B^2(t), \dots, B^d(t))\}_{t\geqq 0}$ is a d-dimensional $\{\mathscr{F}_t\}$-Brownian motion.

Moreover, setting

$$I_N(s) = \begin{cases} 1 & (\text{if } |(\Psi^{-1})_{ij}(s)| \leqq N \ (i, j = 1, 2, \dots, d)) \\ 0 & (\text{otherwise}), \end{cases}$$

by the definition of $\{B_{(N)}^i(t)\}_{t\geqq 0}$, we obtain

$$\int_0^t I_N(s)\,\mathrm{d}M^i(s) = \sum_{j=1}^{d} \int_0^t \Psi_{ij}(s)\,\mathrm{d}B_{(N)}^j(s).$$

Both sides converge in $\mathscr{M}_c^2$ as $N \to \infty$ and

$$M^i(t) = \sum_{j=1}^{d} \int_0^t \Psi_{ij}(s)\,\mathrm{d}B^j(s). \qquad \square$$

When Φ is degenerate, we need an extension of a probability space.

Theorem 2.5.8 *For $M^i \in \mathscr{M}_{\mathrm{c,loc}}^2$ $(i = 1, 2, \ldots, d)$ with $M^i(0) = 0$ defined on $(\Omega, \mathscr{F}, \mathbf{P}, \{\mathscr{F}_t\})$, assume that there exist $d \times d$ and $d \times r$ matrix-valued predictable processes $\Phi = (\Phi_{ij})$ and $\Psi = (\Psi_{i\alpha})$ such that*

$$\int_0^t \Phi_{ij}(s)\,\mathrm{d}s < \infty, \qquad \int_0^t \Psi_{j\alpha}(s)^2\,\mathrm{d}s < \infty$$
$$(i, j = 1, 2, \ldots, d,\ \alpha = 1, 2, \ldots, r)$$

and

$$\langle M^i, M^j\rangle(t) = \int_0^t \Phi_{ij}(s)\,\mathrm{d}s, \quad \Phi_{ij}(s) = \sum_{\alpha=1}^{r} \Psi_{i\alpha}(s)\Psi_{j\alpha}(s)$$

for every $t > 0$ almost surely. Then there exists an extension of $(\Omega, \mathscr{F}, \mathbf{P}, \{\mathscr{F}_t\})$ and an r-dimensional Brownian motion $\{(B^1(t), B^2(t), \ldots, B^r(t))\}_{t \geqq 0}$ defined on it satisfying

$$M^i(t) = \sum_{\alpha=1}^{r} \int_0^t \Psi_{i\alpha}(s)\,\mathrm{d}B^\alpha(s).$$

Proof We give a constructive proof, following [114, Theorem 4.5.2].

Setting $\Psi_{i,r+1} = \cdots = \Psi_{i,d} = 0$ $(i = 1, 2, \ldots, d)$ if $r < d$ and $M^{d+1} = \cdots = M^r = 0$ if $r > d$, we may assume $r = d$.

Letting $(W, \mathscr{B}(W), \mu)$ be the d-dimensional Wiener space on $[0, \infty)$, we consider the direct product probability space

$$(\widetilde{\Omega}, \widetilde{\mathscr{F}}, \widetilde{\mathbf{P}}) = (\Omega \times W, \mathscr{F} \times \mathscr{B}(W), \mathbf{P} \times \mu).$$

We extend functions on Ω and W to those on $\widetilde{\Omega}$ in a natural way and shall not explain it.

Define $d \times d$-matrix-valued stochastic processes $\{\Pi(t)\}_{t\geqq 0}$, $\{\widetilde{\Pi}(t)\}_{t\geqq 0}$, $\{\alpha(t)\}_{t\geqq 0}$ and $\{\beta(t)\}_{t\geqq 0}$ by

$$\Pi(t) = \lim_{\varepsilon\downarrow 0} \Phi(t)(\varepsilon I + \Phi(t))^{-1}, \qquad \widetilde{\Pi}(t) = \lim_{\varepsilon\downarrow 0} \Psi^*(t)\Psi(t)(\varepsilon I + \Psi^*(t)\Psi(t))^{-1},$$
$$\alpha(t) = \lim_{\varepsilon\downarrow 0} (\varepsilon I + \Phi(t))^{-1}\Pi(t), \qquad \beta(t) = \Psi(t)^*\alpha(t).$$

The diagonalization of symmetric matrices yields the existence of the limits. $\Pi(t)$ and $\widetilde{\Pi}(t)$ are the orthogonal projections onto the images of $\Phi(t)$ and

$\Psi^*(t)\Psi(t)$, respectively. Moreover, by definition, each matrix-valued stochastic process is predictable.

It holds that

$$\Psi(t)\beta(t) = \Pi(t) \quad \text{and} \quad \beta(t)\Psi(t) = \widetilde{\Pi}(t). \tag{2.5.7}$$

We postpone the proof of (2.5.7) and continue the proof of the theorem.

Define a $d \times (2d)$ matrix $\Sigma(t)$ by

$$\Sigma(t) = (\beta(t), I - \widetilde{\Pi}(t))$$

and set

$$B(t) = \int_0^t \Sigma(s) \begin{pmatrix} \mathrm{d}M(s) \\ \mathrm{d}\theta(s) \end{pmatrix},$$

where $\{\theta(t)\}_{t \geqq 0}$ is the coordinate process of W. Then, by (2.5.7) and the fact that $\widetilde{\Pi}(t)$ is an orthogonal projection,

$$(\langle B^i, B^j \rangle(t))_{1 \leqq i,j \leqq d} = \int_0^t \Sigma(s) \begin{pmatrix} \Phi(s) & 0 \\ 0 & I \end{pmatrix} \Sigma(s)^* \mathrm{d}s = tI.$$

This means that $\{B(t)\}_{t \geqq 0}$ is a d-dimensional Brownian motion.

Since $\widetilde{\Pi}(t)$ is the orthogonal projection onto the image of $\Psi^*(t)\Psi(t)$,

$$(I - \widetilde{\Pi}(t))\Psi(t)^*\Psi(t)(I - \widetilde{\Pi}(t)) = 0.$$

That is, $\Psi(t)(I - \widetilde{\Pi}(t)) = 0$. Hence,

$$\begin{aligned} \Psi(t)\mathrm{d}B(t) &= \Psi(t)\Sigma(t) \begin{pmatrix} \mathrm{d}M(t) \\ \mathrm{d}\theta(t) \end{pmatrix} \\ &= \Psi(t)\beta(t)\, \mathrm{d}M(t) + \Psi(t)(I - \widetilde{\Pi}(t))\, \mathrm{d}\theta(t) \\ &= \Psi(t)\beta(t)\, \mathrm{d}M(t). \end{aligned}$$

Moreover, since $\Pi(t)$ is the orthogonal projection onto the image of $\Phi(t)$,

$$(I - \Pi(t))\Phi(t) = 0.$$

Set

$$N(t) = \int_0^t (I - \Pi(s))\, \mathrm{d}M(s).$$

Then,

$$(\langle N^i, N^j \rangle(t))_{1 \leqq i,j \leqq d} = \int_0^t (I - \Pi(s))\Phi(s)(I - \Pi(s))\, \mathrm{d}s = 0.$$

Combining the above observation with (2.5.7), we obtain

$$\Psi(t)\, \mathrm{d}B(t) = \Pi(t)\, \mathrm{d}M(t) = \mathrm{d}M(t). \qquad \square$$

Proof of (2.5.7) The first identity is obtained from

$$\Psi(t)\beta(t) = \Phi(t)\alpha(t) = \lim_{\varepsilon\downarrow 0} \Phi(t)(\varepsilon I + \Phi(t))^{-1}\Pi(t) = \Pi(t)^2 = \Pi(t).$$

To show the second identity, let λ_i be the non-zero eigenvalues of $\Psi^*(t)\Psi(t)$ and $\boldsymbol{f}^i$ $(i = 1, \ldots, m)$ be the corresponding eigenvectors. Moreover, let $\{\boldsymbol{f}^1, \ldots, \boldsymbol{f}^d\}$ be the orthonormal basis of $\mathbb{R}^d$ obtained by extending $\boldsymbol{f}^1, \ldots, \boldsymbol{f}^m$. Then, since $\Psi(t)\boldsymbol{f}^i$ is an eigenvector of $\Phi(t)$ corresponding to the eigenvalue λ_i,

$$\beta(t)\Psi(t)\Big(\sum_{i=1}^d \xi_i \boldsymbol{f}^i\Big) = \Psi^*(t)\alpha(t)\Big(\sum_{i=1}^d \xi_i \Psi(t)\boldsymbol{f}^i\Big) = \Psi^*(t)\Big(\sum_{i=1}^m \frac{\xi_i}{\lambda_i}\Psi(t)\boldsymbol{f}^i\Big)$$

for any $(\xi_1, \ldots, \xi_d) \in \mathbb{R}^d$. Hence we obtain

$$\beta(t)\Psi(t)\Big(\sum_{i=1}^d \xi_i \boldsymbol{f}^i\Big) = \sum_{i=1}^m \xi_i \boldsymbol{f}^i.$$

Thus $\beta(t)\Psi(t)$ is the orthogonal projection onto the image of $\Psi^*(t)\Psi(t)$. □

2.6 Martingales with respect to Brownian Motions

We show that martingales with respect to the filtration generated by a Brownian motion are represented as the stochastic integrals with respect to it. This is given in Theorem 2.6.2 and called **Itô's representation theorem**.

Throughout this section we let $B = \{B(t) = (B^1(t), B^2(t), \ldots, B^d(t))\}_{t\geqq 0}$ be a d-dimensional Brownian motion defined on a complete probability space $(\Omega, \mathscr{F}, \mathbf{P})$ and $\{\mathscr{F}_t^B\}$ be the filtration obtained by the completion of the σ-field generated by sample paths:

$$\mathscr{F}_t^B = \sigma\{B(s); s \leqq t\} \vee \mathscr{N},$$

where $\mathscr{N}$ is the totality of the $\mathbf{P}$-null sets.

Lemma 2.6.1 $\mathscr{F}_{t+0}^B = \mathscr{F}_t^B$ *for any* $t \geqq 0$.

Proof It suffices to show that

$$\varphi(t) := \mathbf{E}[f_1(B(t_1))f_2(B(t_2)) \cdots f_n(B(t_n))\big|\mathscr{F}_t^B]$$

is right-continuous in t for any $0 \leqq t_1 < t_2 < \cdots < t_n$ and $f_1, f_2, \ldots, f_n \in C_\infty(\mathbb{R}^d)$, where $C_\infty(\mathbb{R}^d)$ is the set of continuous functions on $\mathbb{R}^d$ which tend to 0 at infinity and is regarded as a Banach space with the topology of uniform convergence on compacta. If $t_{k-1} \leqq t < t_k$, then

$$\varphi(t) = f_1(B(t_1)) \cdots f_{k-1}(B(t_{k-1}))\mathbf{E}[f_k(B(t_k)) \cdots f_n(B(t_n))|\mathscr{F}_t^B].$$

Let $g(t, x, y)$ be the transition density of a d-dimensional Brownian motion and set

$$(P_t f)(x) = \int_{\mathbb{R}^d} g(t, x, y) f(y)\, \mathrm{d}y \qquad (f \in C_\infty(\mathbb{R}^d)),$$

where $C_\infty(\mathbb{R}^d)$ is the space of continuous functions on $\mathbb{R}^d$ tending to 0 at infinity. Then, $\{P_t\}_{t\geqq 0}$ is a strongly continuous semigroup of linear operators on $C_\infty(\mathbb{R}^d)$. Moreover, define the functions $H_m(t_1, t_2, \ldots, t_m; f_1, f_2, \ldots, f_m) \in C_\infty(\mathbb{R}^d)$ inductively by $H_1(t; f) = P_t f$ and

$$\begin{aligned} &H_m(t_1, t_2, \ldots, t_m; f_1, f_2, \ldots, f_m) \\ &\quad = H_{m-1}(t_1, t_2, \ldots, t_{m-1}; f_1, f_2, \ldots, f_{m-1} P_{t_m - t_{m-1}} f_m) \end{aligned}$$

for $m \geqq 2$. Since

$$\begin{aligned} &\mathbf{E}[f_1(B(t_1)) f_2(B(t_2)) \cdots f_m(B(t_m))] \\ &\quad = \int_{\mathbb{R}^d} H_m(t_1, t_2, \ldots, t_m; f_1, f_2, \ldots, f_m)(x) \nu(\mathrm{d}x), \end{aligned}$$

ν being the probability distribution of $B(0)$, φ is written as

$$\varphi(t) = \prod_{j=1}^{k-1} f_j(B(t_j)) \cdot H_{n-k+1}(t_k - t, t_{k+1} - t, \ldots, t_n - t; f_k, f_{k+1}, \ldots, f_n)(B(t)).$$

It is easy to see that the right hand side is a right-continuous function in t. □

We denote by $\mathscr{M}^2(\{\mathscr{F}_t^B\})$ the set of square-integrable $\{\mathscr{F}_t^B\}$-martingales and by $\mathscr{L}^2(B)$ the set of $\mathbb{R}$-valued predictable stochastic processes Φ satisfying

$$\mathbf{E}\Big[\int_0^T \Phi(s)^2 \mathrm{d}s\Big] < \infty \quad (T > 0).$$

The following theorem shows that every square integrable $\{\mathscr{F}_t^B\}$-martingale is expressed as a stochastic integral with respect to the original Brownian motion.

Theorem 2.6.2 *For any* $M = \{M(t)\}_{t\geqq 0} \in \mathscr{M}^2(\{\mathscr{F}_t^B\})$, *there exist* $\Phi_i \in \mathscr{L}^2(B)$ $(i = 1, 2, \ldots, d)$ *such that*

$$M(t) = M(0) + \sum_{i=1}^{d} \int_0^t \Phi_i(s)\, \mathrm{d}B^i(s). \tag{2.6.1}$$

Remark 2.6.3 Combining a similar argument in the proof below with the localization argument via $\{\mathscr{F}_t^B\}$-stopping times, we can show that for any $M \in \mathscr{M}_{\mathrm{c,loc}}^2(\{\mathscr{F}_t^B\})$ there exist $\Phi_i \in \mathscr{L}_{\mathrm{loc}}^2$ $(i = 1, 2, \ldots, d)$ satisfying (2.6.1).

To avoid unnecessary complexity, we give a proof when $d = 1$. Moreover, it suffices to consider the case when the time interval is $[0, T]$ $(T > 0)$ and $M(0) = 0$. For the proof, setting

$$\mathscr{I}^2(B) = \left\{M \in \mathscr{M}^2(\{\mathscr{F}_t^B\})\,;\; M(t) = \int_0^t \Phi(s)\,\mathrm{d}B(s),\; \Phi \in \mathscr{L}^2(B)\right\},$$

we use the following lemma.

Lemma 2.6.4 *For any $M \in \mathscr{M}^2(\{\mathscr{F}_t^B\})$, there exist unique $M_1 \in \mathscr{I}^2(B)$ and $M_2 \in \mathscr{M}^2(\{\mathscr{F}_t^B\})$ such that $M = M_1 + M_2$ and $\langle M_2, N\rangle = 0$ for all $N \in \mathscr{I}^2(B)$.*

Proof At first we show the uniqueness. Let $M = M_1 + M_2 = M_1' + M_2'$ be the desired decompositions. Then, since $M_2' - M_2 = M_1 - M_1' \in \mathscr{I}^2(B)$, $M_2(M_2' - M_2)$ and $M_2'(M_2' - M_2)$ are martingales. Hence, $(M_2 - M_2')^2$ is also a martingale. This means $\langle M_2 - M_2'\rangle = 0$ and, therefore, $M_2 = M_2'$.

Next let $\mathscr{H}$ be a closed subspace of the L^2-space on $(\Omega, \mathscr{F}_T^B, \mathbf{P})$ defined by

$$\mathscr{H} = \{M_1(T); M \in \mathscr{I}^2(B)\}$$

and $\mathscr{H}^\perp$ be the orthogonal complement of $\mathscr{H}$. For $M \in \mathscr{M}^2(\{\mathscr{F}_t^B\})$ let

$$M(T) = H + K \qquad (H \in \mathscr{H},\ K \in \mathscr{H}^\perp). \tag{2.6.2}$$

be the corresponding decomposition of $M(T)$.

Then, there exists a $\Phi \in \mathscr{L}^2(B)$ such that $H = \int_0^T \Phi(s)\,\mathrm{d}B(s)$. By the right-continuity of $\{\mathscr{F}_t^B\}$, denoting the right-continuous modification of $\mathbf{E}[K|\mathscr{F}_t^B]$ by $\{M_2(t)\}_{0\leqq t\leqq T}$, we have by (2.6.2)

$$M(t) = \int_0^t \Phi(s)\,\mathrm{d}B(s) + M_2(t).$$

Hence, it suffices to show that $\langle M_2, N\rangle(t) = 0$ for any $N \in \mathscr{I}^2(B)$, or that $\{M_2(t)N(t)\}_{0\leqq t\leqq T}$ is an $\{\mathscr{F}_t^B\}$-martingale. To see this, we show $\mathbf{E}[M_2(\sigma)N(\sigma)] = 0$ for any $\{\mathscr{F}_t^B\}$-stopping time σ satisfying $\sigma \leqq T$. This is shown in the following way. Since $N \in \mathscr{I}^2(B)$, there exists a $\Psi \in \mathscr{L}^2(B)$ such that $N(t) = \int_0^t \Psi(s)\,\mathrm{d}B(s)$. Hence we have

$$N(t \wedge \sigma) = \int_0^t \Psi(s)\mathbf{1}_{\{s\leqq\sigma\}}\mathrm{d}B(s) \in \mathscr{I}^2(B).$$

Since $N(\sigma) \in \mathscr{H}$, the optional sampling theorem yields

$$\mathbf{E}[M_2(\sigma)N(\sigma)] = \mathbf{E}[\mathbf{E}[M_2(T)|\mathscr{F}_\sigma^B]N(\sigma)] = \mathbf{E}[KN(\sigma)] = 0. \qquad \square$$

Lemma 2.6.5 *If $M \in \mathscr{M}^2(\{\mathscr{F}_t^B\})$ is bounded, $M(0) = 0$, and $\langle M, N\rangle = 0$ for any $N \in \mathscr{I}^2(B)$, then $M = 0$.*

Proof Let C be a constant satisfying $\mathbf{P}(|M(t)| \leqq C,\ 0 \leqq t \leqq T) = 1$ and set

$$D = 1 + \frac{1}{2C}M(T).$$

Then $D \geqq \frac{1}{2}$ and $\mathbf{E}[D] = 1$. Set $\widetilde{\mathbf{P}}(A) = \mathbf{E}[D\mathbf{1}_A]$ for $A \in \mathscr{F}_T^B$. Then, $\widetilde{\mathbf{P}}$ is a probability measure on $(\Omega, \mathscr{F}_T^B)$. Denote the expectation with respect to $\widetilde{\mathbf{P}}$ by $\widetilde{\mathbf{E}}$. Then, since $\langle M, B\rangle = 0$,

$$\begin{aligned}\widetilde{\mathbf{E}}[B(\sigma)] &= \mathbf{E}[DB(\sigma)] = \mathbf{E}[\mathbf{E}[D|\mathscr{F}_\sigma^B]B(\sigma)] \\ &= \mathbf{E}[B(\sigma)] + \frac{1}{2C}\mathbf{E}[M(\sigma)B(\sigma)] = 0\end{aligned}$$

for any stopping time σ with $\sigma \leqq T$. Hence, $\{B(t)\}_{t\geqq 0}$ is an $\{\mathscr{F}_t^B\}$-martingale under $\widetilde{\mathbf{P}}$.

Similarly, since $B(t)^2 - t = 2\int_0^t B(s)\,\mathrm{d}B(s) \in \mathscr{I}^2(B)$, $\widetilde{\mathbf{E}}[B(\sigma)^2 - \sigma] = 0$ and $\{B(t)^2 - t\}_{t\geqq 0}$ is also an $\{\mathscr{F}_t^B\}$-martingale under $\widetilde{\mathbf{P}}$.

Hence, by Lévy's theorem (Theorem 2.5.1), $\{B(t)\}$ is an $\{\mathscr{F}_t^B\}$-Brownian motion under $\widetilde{\mathbf{P}}$. Since $\widetilde{\mathbf{P}}$ and $\mathbf{P}$ coincide on $\mathscr{F}_T^B$, we obtain $D = 1$ and $M(T) = 0$. $\square$

We are now in a position to give a proof of Theorem 2.6.2.

Proof of Theorem 2.6.2 Following Lemma 2.6.4, write $M(t) = M_1(t) + M_2(t)$. For any fixed $K > 0$, set $\tau_K = \inf\{t; |M_2(t)| \geqq K\}$ and $M_2^K = \{M_2(t \wedge \tau_K)\}_{0\leqq t\leqq T}$. If we have shown $\langle M_2^K, N\rangle = 0$ for any $N \in \mathscr{I}^2(B)$, we obtain $M_2^K = 0$ and, therefore, $M_2 = 0$ by Lemma 2.6.5. This implies the assertion.

We should show the martingale property of $\{M_2^K(t)N(t)\}$, that is,

$$\mathbf{E}[M_2(t \wedge \tau_K)N(t)|\mathscr{F}_s^B] = M_2(s \wedge \tau_K)N(s) \tag{2.6.3}$$

for $s < t$ almost surely.

Let $A \in \mathscr{F}_s^B$. The martingale property of N implies

$$\begin{aligned}&\mathbf{E}[\mathbf{E}[M_2(t \wedge \tau_K)N(t)|\mathscr{F}_s^B]\mathbf{1}_{A\cap\{\tau_K\leqq s\}}] = \mathbf{E}[M_2(t \wedge \tau_K)N(t)\mathbf{1}_{A\cap\{\tau_K\leqq s\}}] \\ &\quad = \mathbf{E}[M_2(s \wedge \tau_K)\mathbf{E}[N(t)|\mathscr{F}_s^B]\mathbf{1}_{A\cap\{\tau_K\leqq s\}}] = \mathbf{E}[M_2(s \wedge \tau_K)N(s)\mathbf{1}_{A\cap\{\tau_K\leqq s\}}].\end{aligned}$$

On the other hand, write

$$\begin{aligned}\mathbf{E}[\mathbf{E}[M_2(t\wedge\tau_K)N(t)|\mathscr{F}_s^B]\mathbf{1}_{A\cap\{\tau_K>s\}}] &= \mathbf{E}[M_2(t\wedge\tau_K)N(t)\mathbf{1}_{A\cap\{\tau_K>s\}}]\\ &= \mathbf{E}[M_2(t\wedge\tau_K)(N(t)-N(t\wedge\tau_K))\mathbf{1}_{A\cap\{\tau_K>s\}}]\\ &\quad+\mathbf{E}[M_2(t\wedge\tau_K)N(t\wedge\tau_K)\mathbf{1}_{A\cap\{\tau_K>s\}}].\end{aligned}$$

Then, since $A\cap\{\tau_K>s\}\in\mathscr{F}_{s\wedge\tau_K}\subset\mathscr{F}_{t\wedge\tau_K}$, the first term is 0. By Lemma 2.6.5, the stochastic process $\{M_2(t)N(t)\}_{t\geqq 0}$ is a martingale and hence

$$\begin{aligned}\mathbf{E}[M_2(t\wedge\tau_K)N(t\wedge\tau_K)\mathbf{1}_{A\cap\{\tau_K>s\}}] &= \mathbf{E}[M_2(s\wedge\tau_K)N(s\wedge\tau_K)\mathbf{1}_{A\cap\{\tau_K>s\}}]\\ &= \mathbf{E}[M_2(s\wedge\tau_K)N(s)\mathbf{1}_{A\cap\{\tau_K>s\}}].\end{aligned}$$

Thus we obtain (2.6.3). □

The following important results are obtained from Itô's representation theorem.

Corollary 2.6.6 *$\mathscr{M}^2(\{\mathscr{F}_t^B\})=\mathscr{M}_c^2(\{\mathscr{F}_t^B\})$, that is, every square-integrable $\{\mathscr{F}_t^B\}$-martingale is continuous.*

Corollary 2.6.7 *Let F be an $\mathscr{F}_T^B$-measurable square-integrable random variable for $T>0$. Then there exist predictable stochastic processes $\{\Phi^i(t)\}_{0\leqq t\leqq T}$ $(i=1,2,\ldots,d)$ with $\mathbf{E}[\int_0^T\Phi_i(s)^2\mathrm{d}s]<\infty$ such that*

$$F=\mathbf{E}[F|\mathscr{F}_0^B]+\sum_{i=1}^d\int_0^T\Phi_i(t)\,\mathrm{d}B^i(t).$$

Proof Set $M(t)=\mathbf{E}[F|\mathscr{F}_t^B]-\mathbf{E}[F|\mathscr{F}_0^B]$. Since $\{M(t)\}_{0\leqq t\leqq T}\in\mathscr{M}^2(\{\mathscr{F}_t^B\})$, Theorem 2.6.2 implies the assertion. □

Remark 2.6.8 (1) The Clark–Ocone formula (Theorem 5.3.5) shows that Φ is given as an expectation of the derivative of F in the sense of the Malliavin calculus.
(2) Corollary 2.6.7 holds when $T=\infty$, that is, for a square-integrable random variable which is measurable under $\bigvee_t\mathscr{F}_t^B=\mathscr{F}_\infty^B$.
(3) If $B(0)$ is non-random, we can first prove Corollary 2.6.7 and, using it conversely, prove the representation theorem (Theorem 2.6.2) ([93]).

2.7 Local Time, Itô–Tanaka Formula

Let $X = \{X(t)\}_{t \geqq 0}$ be a one-dimensional semimartingale, M be its martingale part and $\langle M \rangle$ be the quadratic variation process of M. Then there exists a two-parameter family of random variables (random field) $L = \{L(t,x)\}_{t \geqq 0,\, x \in \mathbb{R}}$, called the **local time** of X, such that

$$\int_0^t f(X(s))\,\mathrm{d}\langle M \rangle(s) = \int_{\mathbb{R}} f(x) L(t,x)\,\mathrm{d}x$$

for a function f on $\mathbb{R}$ which satisfies adequate conditions. If X is a Brownian motion and $f(x) = \mathbf{1}_A(x), A \in \mathscr{B}(\mathbb{R})$, then the left hand side is equal to the total time when X has stayed in A up to time t.

The Lebesgue measure of the set of times when a Brownian motion has stayed at a fixed point is 0.[1] Lévy introduced the notion of local time in order to study the properties of this set in detail. The purpose of this section is to show the existence of the local times of semimartingales and to show an extension of Itô's formula to convex functions, which are not necessarily of C^2-class, by using local times.

Let $X = \{X(t)\}_{t \geqq 0}$ be a continuous semimartingale defined on a probability space $(\Omega, \mathscr{F}, \mathbf{P}, \{\mathscr{F}_t\})$ which has a decomposition (2.3.1). Define the function $\mathrm{sgn} : \mathbb{R} \to \{-1, 1\}$ by

$$\mathrm{sgn}(x) = \begin{cases} -1 & (x \leqq 0) \\ 1 & (x > 0). \end{cases}$$

Theorem 2.7.1 *For each $a \in \mathbb{R}$, there exists a continuous increasing process $L_a = \{L(t,a)\}_{t \geqq 0}$ such that*

$$|X(t) - a| - |X(0) - a| = \int_0^t \mathrm{sgn}(X(s) - a)\,\mathrm{d}X(s) + L(t,a). \tag{2.7.1}$$

Moreover,

$$\int_0^t \mathbf{1}_{\{X(s) \neq a\}} \mathrm{d}L(s,a) = 0, \tag{2.7.2}$$

that is, L_a increases only at times s with $X(s) = a$.

Remark 2.7.2 (1) (2.7.1) is called **Tanaka's formula**.
(2) Combining the trivial identity $X(t) - X(0) = \int_0^t \mathrm{d}X(s)$ with (2.7.1) and setting $x^+ = \max(x, 0)$, we have

[1] For a Brownian motion $B = \{B(t)\}_{t \geqq 0}$, let $\mathscr{Z}$ be the set of zeros of the mapping $t \mapsto B(t)$ and $|\mathscr{Z}|$ be its Lebesgue measure. Then $\mathbf{E}[|\mathscr{Z}|] = \int_0^1 \mathbf{P}(B(s) = 0)\,\mathrm{d}s = 0$.

$$(X(t) - a)^+ - (X(0) - a)^+ = \int_0^t \mathbf{1}_{\{X(s)>a\}} \mathrm{d}X(s) + \frac{1}{2}L(t, a).$$

(3) The second derivative of $\phi(x) = |x - a|$ in the sense of distribution is $2\delta_a(x)$, where δ_a is the Dirac measure concentrated at a. Hence, applying Itô's formula to $|X(t) - a|$ in a formal manner, we get

$$L(t, a) = \int_0^t \delta_a(X(s))\, \mathrm{d}s.$$

It is not easy to give a meaning to the right hand side, but it gives an intuitive understanding to the local time.

Proof It suffices to show the case when $a = 0$. Moreover, we may assume that X, the martingale part M of X, the quadratic variation process $\langle M \rangle$ of M and the increasing process A are all bounded, because, once this is done, the general case can be proven by the localization argument via stopping times.

To approximate $f(x) = |x|$ by C^2-class functions, let $\varphi \in C^\infty(\mathbb{R})$ be a monotone increasing function such that

$$\varphi(x) = -1 \ (x \leqq 0), \qquad \varphi(x) = 1 \ (x \geqq 1)$$

and define a sequence of functions f_n $(n = 1, 2, \dots)$ by

$$f_n(0) = 0, \quad f_n'(x) = \varphi(nx) \qquad (x \in \mathbb{R}).$$

f_n converges to f uniformly and f_n' does to the function sgn pointwise as $n \to \infty$. Moreover, by Itô's formula,

$$f_n(X(t)) - f_n(X(0)) = \int_0^t f_n'(X(s))\, \mathrm{d}X(s) + C_n(t), \tag{2.7.3}$$

where

$$C_n(t) = \frac{1}{2} \int_0^t f_n''(X(s))\, \mathrm{d}\langle M \rangle(s).$$

Since $f_n'' \geqq 0$, $\{C_n(t)\}_{t \geqq 0}$ is increasing and, if $m \leqq n$,

$$\int_0^t \mathbf{1}_{\{|X(s)| > \frac{1}{m}\}}(X(s))\, \mathrm{d}C_n(s) = 0. \tag{2.7.4}$$

Fix $T > 0$. Then,

$$\mathbf{E}\Big[\Big| \int_0^T (\mathrm{sgn}(X(s)) - f_n'(X(s)))\, \mathrm{d}M(s) \Big|^2\Big]$$
$$= \mathbf{E}\Big[\int_0^T (\mathrm{sgn}(X(s)) - f_n'(X(s)))^2 \mathrm{d}\langle M \rangle(s)\Big]$$

converges to 0 by the bounded convergence theorem. Hence, by Doob's inequality (Theorem 1.5.13),

$$\sup_{0\leqq t\leqq T}\left|\int_0^T (\mathrm{sgn}(X(s)) - f_n'(X(s)))\,\mathrm{d}M(s)\right|$$

converges to 0 in L^2 and almost surely if we take a subsequence. On the other hand, it is easy to see that $\int_0^t f_n'(X(s))\,\mathrm{d}A(s)$ converges to $\int_0^t \mathrm{sgn}(X(s))\,\mathrm{d}A(s)$ uniformly in t as $n\to\infty$. Hence, the first term on the right hand side of (2.7.3) converges to $\int_0^t \mathrm{sgn}(X(s))\,\mathrm{d}X(s)$ uniformly on $[0,T]$ almost surely.

The uniform convergence of f_n to f and (2.7.3) yield the uniform convergence of $C_n(t)$. Denote the limit by $L_a = \{L(t,a)\}_{t\geqq 0}$. Then L_a is continuous and increasing in t, and it satisfies (2.7.1). Moreover, letting $n\to\infty$ in (2.7.4), we have

$$\int_0^t \mathbf{1}_{\{|X(s)|>\frac{1}{m}\}}\mathrm{d}L(s,0) = 0$$

and obtain (2.7.2) by letting m tend to ∞. □

On the continuity of $L(t,a)$ as a function of two variables, the following is known ([131]).

Theorem 2.7.3 *The local time $\{L(t,a)\}_{t\geqq 0,\,a\in\mathbb{R}}$ of a continuous semimartingale X has a modification which is continuous in t and right-continuous with the left limits in a.*

Proof As in the proof of Theorem 2.7.1, we may assume that $X, M, \langle M\rangle$ and A are bounded. Set

$$\xi_1(t,a) = \int_0^t \mathrm{sgn}(X(s)-a)\,\mathrm{d}M(s), \quad \xi_2(t,a) = \int_0^t \mathrm{sgn}(X(s)-a)\,\mathrm{d}A(s),$$
$$\xi_3(t,a) = |X(t)-a| - |X(0)-a|.$$

Obviously $\xi_3(t,a)$ is continuous in (t,a). For $\xi_2(t,a)$, the continuity in t follows from that of A and the right-continuity in a follows from the left-continuity of the function $\mathrm{sgn}(x)$.

For $\xi_1(t,a)$, we shall show that, for any $T>0$ and $p>0$, there exists a constant K_p such that

$$\mathbf{E}\Big[\sup_{0\leqq t\leqq T}|\xi_1(t,a)-\xi_1(t,b)|^p\Big] \leqq K_p|a-b|^{\frac{p}{2}}.$$

Once this is done, applying Kolmogorov's continuity theorem (Theorem A.5.1) to a family of random variables $\{L(\cdot,a)\}_{a\in\mathbb{R}}$ with values in the path space, we obtain the continuity of $\xi_1(t,a)$ in two variables (t,a).

If $a < b$, the Burkholder–Davis–Gundy inequality (2.4.1) yields

$$\begin{aligned}\mathbf{E}\Big[\sup_{0\leqq t\leqq T}|\xi_1(t,a)-\xi_1(t,b)|^p\Big] &= 2^p\mathbf{E}\Big[\sup_{0\leqq t\leqq T}\Big|\int_0^t \mathbf{1}_{(a,b]}(X(s))\,\mathrm{d}M(s)\Big|^p\Big] \\ &\leqq \frac{2^p}{c_p}\mathbf{E}\Big[\Big|\int_0^T \mathbf{1}_{(a,b]}(X(s))\,\mathrm{d}\langle M\rangle(s)\Big|^{\frac{p}{2}}\Big].\end{aligned}$$

Let $g \in C^2(\mathbb{R})$ satisfy

$$\begin{aligned}&0 \leqq g'' \leqq 1, \quad \operatorname{supp}[g''] \subseteq [-1,2], \quad g''(x) = 1\ (0 \leqq x \leqq 1), \\ &g'(x) = g(x) = 0\ (x \leqq -1)\end{aligned}$$

and set

$$\varphi(x) = g\Big(\frac{x-a}{b-a}\Big).$$

Since $0 \leqq g' \leqq 3$, we have

$$\begin{aligned}0 &\leqq \int_0^t \mathbf{1}_{(a,b]}(X(s))\,\mathrm{d}\langle M\rangle(s) \leqq (b-a)^2\int_0^t \varphi''(X(s))\,\mathrm{d}\langle M\rangle(s) \\ &= (b-a)^2\Big\{\varphi(X(t)) - \varphi(X(0)) - \int_0^t \varphi'(X(s))\,\mathrm{d}X(s)\Big\} \\ &\leqq (b-a)^2\Big\{|\varphi(X(t)) - \varphi(X(0))| + \Big|\int_0^t \varphi'(X(s))\,\mathrm{d}M(s)\Big| \\ &\qquad\qquad + \int_0^t |\varphi'(X(s))|\,|\mathrm{d}A(s)|\Big\} \\ &\leqq |b-a|\Big\{3|X(t)-X(0)| + \sup_{0\leqq t\leqq T}\Big|\int_0^t g'\Big(\frac{X(s)-a}{b-a}\Big)\mathrm{d}M(s)\Big| + 3\int_0^t |\mathrm{d}A(s)|\Big\},\end{aligned}$$

where $|dA(s)|$ denotes the integral with respect to the total variation of A.

A repeated use of the Burkholder–Davis–Gundy inequality implies

$$\begin{aligned}&\mathbf{E}\Big[\sup_{0\leqq t\leqq T}\Big|\int_0^t g'\Big(\frac{X(s)-a}{b-a}\Big)\mathrm{d}M(s)\Big|^{\frac{p}{2}}\Big] \\ &\quad\leqq C_p\mathbf{E}\Big[\Big(\int_0^T \Big|g'\Big(\frac{X(s)-a}{b-a}\Big)\Big|^2\mathrm{d}\langle M\rangle(s)\Big)^{\frac{p}{4}}\Big] \leqq C_p 3^{\frac{p}{2}}\mathbf{E}[\langle M\rangle(T)^{\frac{p}{4}}].\end{aligned}$$

By the boundedness of X, $\langle M\rangle$ and A, there exists a constant K_p such that

$$\mathbf{E}\Big[\sup_{0\leqq t\leqq T}|\xi_1(t,a)-\xi_1(t,b)|^p\Big] \leqq K_p|b-a|^{\frac{p}{2}}. \qquad \square$$

Remark 2.7.4 From the proof we see that, if $A = 0$, that is, if X is a continuous local martingale, then the local time has a modification which is continuous as a function of the two variables (t, a). This was first proven by Trotter ([119]) when X is a Brownian motion.

Using the local time, we can extend Itô's formula to convex functions or functions given as the difference of two convex functions. The extended formula is called the **Itô–Tanaka formula**.

Let f be a convex function. Since

$$\frac{f(z) - f(x)}{z - x} \leqq \frac{f(z) - f(y)}{z - y}$$

for $x < y < z$, f has a left-derivative $\mathrm{D}_- f$ and a right-derivative $\mathrm{D}_+ f$:

$$(\mathrm{D}_\pm f)(x) = \lim_{h \to \pm 0} \frac{f(x+h) - f(x)}{h}.$$

Note that the left-(right-)derivative is a left-(right-)continuous monotone increasing function and satisfies

$$(\mathrm{D}_+ f)(x) \leqq (\mathrm{D}_- f)(y) \leqq (\mathrm{D}_+ f)(y) \qquad (x < y).$$

Denote by ν_f the measure determined by the monotone increasing function $D_- f$:

$$\nu_f([a, b)) = (\mathrm{D}_- f)(b) - (\mathrm{D}_- f)(a) \qquad (a < b).$$

Theorem 2.7.5 *For a convex function* f,

$$f(X(t)) = f(X(0)) + \int_0^t (\mathrm{D}_- f)(X(s))\,\mathrm{d}X(s) + \frac{1}{2}\int_{\mathbb{R}} L(t, a)\nu_f(\mathrm{d}a).$$

Proof We may assume that all of X, M, $\langle M \rangle$ and A are bounded and ν_f has a compact support by the localization argument. Then, there exist constants α and β such that

$$\begin{aligned} f(x) &= \alpha x + \beta + \frac{1}{2}\int_{\mathbb{R}} |x - a|\nu_f(\mathrm{d}a), \\ (D_- f)(x) &= \alpha + \frac{1}{2}\int_{\mathbb{R}} \operatorname{sgn}(x - a)\nu_f(\mathrm{d}a) \end{aligned} \tag{2.7.5}$$

(see [98, Appendix §3]). Hence, by Tanaka's formula (2.7.1), we have

$$\begin{aligned} f(X(t)) &= \alpha X(t) + \beta + \frac{1}{2}\int_{\mathbb{R}} |X(t) - a|\nu_f(\mathrm{d}a) \\ &= \alpha X(t) + \beta \\ &\quad + \frac{1}{2}\int_{\mathbb{R}}\Big(|X(0) - a| + \int_0^t \operatorname{sgn}(X(s) - a)\,\mathrm{d}X(s) + L(t, a)\Big)\nu_f(\mathrm{d}a) \end{aligned}$$

$$= \alpha(X(t) - X(0)) + f(X(0)) + \frac{1}{2}\int_{\mathbb{R}}\Big(\int_0^t \operatorname{sgn}(X(s)-a)\,\mathrm{d}X(s) + L(t,a)\Big)\nu_f(\mathrm{d}a).$$

Using Fubini's theorem for stochastic integrals (Theorem 2.7.6 below) and (2.7.5), we obtain

$$\begin{aligned}\frac{1}{2}\int_{\mathbb{R}}\Big(\int_0^t \operatorname{sgn}(X(s)-a)\,\mathrm{d}X(s)\Big)\nu_f(\mathrm{d}a) &= \frac{1}{2}\int_0^t\Big(\int_{\mathbb{R}} \operatorname{sgn}(X(s)-a)\nu_f(\mathrm{d}a)\Big)\mathrm{d}X(s)\\ &= \int_0^t (\mathrm{D}_-f)(X(s))\,\mathrm{d}X(s) - \alpha(X(t)-X(0)).\end{aligned}$$

Substituting this into the above identity, we arrive at the conclusion. □

Theorem 2.7.6 *Let ν be a σ-finite measure on $(\mathbb{R}, \mathscr{B}(\mathbb{R}))$ and h be a continuous function on $\mathbb{R}$ with a compact support. Then,*

$$\int_{\mathbb{R}} h(a)\Big(\int_0^t \operatorname{sgn}(X(s)-a)\,\mathrm{d}M(s)\Big)\nu(\mathrm{d}a) = \int_0^t\Big(\int_{\mathbb{R}} h(a)\operatorname{sgn}(X(s)-a)\nu(\mathrm{d}a)\Big)\mathrm{d}M(s). \tag{2.7.6}$$

Proof Suppose that $\operatorname{supp}[h] \subseteq [c,d]$ and set $\xi_k = c + 2^{-n}(d-c)k$ $(k = 0, 1, \ldots, 2^n)$. Define a function F_n by

$$F_n(x) = \sum_{k=0}^{2^n-1} h(\xi_k)\operatorname{sgn}(x-\xi_k)\mu([\xi_k, \xi_{k+1})).$$

Then,

$$\sum_{k=0}^{2^n-1} h(\xi_k)\int_0^t \operatorname{sgn}(X(s)-\xi_k)\,\mathrm{d}M(s)\cdot\mu([\xi_k,\xi_{k+1})) = \int_0^t F_n(X(s))\,\mathrm{d}M(s). \tag{2.7.7}$$

In the proof of Theorem 2.7.3, it was shown that $I(t,a) := \int_0^t \operatorname{sgn}(X(s) - a)\,\mathrm{d}M(s)$ has a modification which is continuous in (t,a). Hence, the left hand side of (2.7.7) converges to that of (2.7.6) almost surely. On the other hand, since $F_n(x)$ converges to $\int_c^d h(a)\operatorname{sgn}(x-a)\nu(\mathrm{d}a)$ uniformly, the right hand side of (2.7.7) converges to that of (2.7.6) in L^2. From these, we obtain the assertion of the theorem. □

Suppose that f is of C^2-class. Then, comparing the development of $f(X(t))$ by Itô's formula and setting $\varphi = f''$, we obtain

$$\int_0^t \varphi(X(s))\,\mathrm{d}\langle M\rangle(s) = \int_{\mathbb{R}} \varphi(a)L(t,a)\,\mathrm{d}a. \tag{2.7.8}$$

By the monotone class theorem (Theorem A.2.5), this result, called the **occupation time formula**, holds for any bounded Borel measurable function φ.

Theorem 2.7.7 *Formula* (2.7.8) *holds for any bounded Borel measurable function φ on $\mathbb{R}$.*

2.8 Reflecting Brownian Motion and Skorohod Equation

Let $B = \{B(t)\}_{t\geqq 0}$ be a one-dimensional Brownian motion starting from $x > 0$ defined on a probability space $(\Omega, \mathscr{F}, \mathbf{P})$ and define $B_+ = \{B_+(t)\}_{t\geqq 0}$ by $B_+(t) = |B(t)|$. Then, for all $0 = t_0 < t_1 < t_2 < \cdots < t_n$ and $A_i \in \mathscr{B}(\mathbb{R}_+)$, we have

$$\begin{aligned}&\mathbf{P}(B_+(t_1) \in A_1, B_+(t_2) \in A_2, \ldots, B_+(t_n) \in A_n)\\ &\quad = \int_{A_1} \mathrm{d}x_1 \int_{A_2} \mathrm{d}x_2 \cdots \int_{A_n} \mathrm{d}x_n \prod_{j=1}^n g_+(t_j - t_{j-1}, x_{j-1}, x_j),\end{aligned}$$

where $x_0 = x$ and

$$g_+(t,x,y) = \frac{1}{\sqrt{2\pi t}}\left(\mathrm{e}^{-\frac{(y-x)^2}{2t}} + \mathrm{e}^{-\frac{(y+x)^2}{2t}}\right) \qquad (x, y \geqq 0,\ t > 0).$$

The stochastic process B_+ on $[0,\infty)$ is called a **reflecting Brownian motion**.

The aim of this section is to present some results related to a reflecting Brownian motion.

Lemma 2.8.1 *Let $x \geqq 0$ and ϕ be an $\mathbb{R}$-valued continuous function on $[0,\infty)$ with $\phi(0) = 0$. Then, there exists a unique continuous function $k : [0,\infty) \to \mathbb{R}$ which satisfies the following three conditions:*

(i) $x(t) := x + \phi(t) + k(t) \geqq 0$ $(t \geqq 0)$,
(ii) $k(0) = 0$ *and k is increasing,*
(iii) $\int_0^t \mathbf{1}_{\{0\}}(x(s))\,\mathrm{d}k(s) = k(t)$, *that is, k increases only at s with $x(s) = 0$.*

Proof First we show the uniqueness. Suppose that a continuous function $\widetilde{k}$ also satisfies the conditions (i), (ii), (iii) and set $\widetilde{x}(t) = x + \phi(t) + \widetilde{k}(t)$. Assume that there exists t_1 such that $x(t_1) > \widetilde{x}(t_1)$ and set

$$t_2 = \max\{t < t_1; x(t) = \widetilde{x}(t)\}.$$

Then, for $t \in (t_2, t_1]$, $x(t) > \widetilde{x}(t) \geqq 0$. By the condition (iii), k is constant on $(t_2, t_1]$ and $k(t_2) = k(t_1)$. Hence,

$$0 < x(t_1) - \widetilde{x}(t_1) = k(t_1) - \widetilde{k}(t_1) = k(t_2) - \widetilde{k}(t_1) \leqq k(t_2) - \widetilde{k}(t_2) = 0,$$

which is a contradiction. Therefore, $x(t) \leqq \widetilde{x}(t)$ $(t \geqq 0)$, which means $k \leqq \widetilde{k}$. The same argument shows $k \geqq \widetilde{k}$ and we obtain $k = \widetilde{k}$.

In order to show the existence, set

$$k(t) = \max\Big[0, \max_{0 \leqq s \leqq t}\{-(x + \phi(s))\}\Big].$$

k is increasing and $k(0) = 0$. Moreover,

$$x + \phi(t) + k(t) \geqq k(t) - \max_{0 \leqq s \leqq t}\{-(x + \phi(s))\} \geqq 0.$$

This means that k satisfies the condition (i).

To show (iii), take arbitrary $\varepsilon > 0$ and let (t_1, t_2) be an interval contained in the open set $\mathscr{O}_\varepsilon = \{s \geqq 0; x(s) > \varepsilon\}$. It suffices to show $k(t_1) = k(t_2)$. For this purpose note

$$-(x + \phi(s)) = k(s) - x(s) \leqq k(t_2) - \varepsilon \qquad (t_1 \leqq s \leqq t_2).$$

Then, we have

$$k(t_2) = \max\Big[k(t_1), \max_{t_1 \leqq s \leqq t_2}\{-(x + \phi(s))\}\Big] \leqq \max\{k(t_1), k(t_2) - \varepsilon\}$$

and obtain $k(t_2) = k(t_1)$. □

Theorem 2.8.2 *Let $x \geqq 0$ and $B = \{B(t)\}_{t \geqq 0}$ be a one-dimensional Brownian motion with $B(0) = 0$. Assume that there exists a continuous stochastic process $\ell = \{\ell(t)\}_{t \geqq 0}$ satisfying the following three conditions:*

(i) $X(t) := x + B(t) + \ell(t) \geqq 0$ $(t \geqq 0)$,
(ii) $\ell(0) = 0$ *and* ℓ *is an increasing process,*
(iii) $\int_0^t \mathbf{1}_{\{0\}}(X(s))\,\mathrm{d}\ell(s) = \ell(t)$.

Then, $X = \{X(t)\}_{t \geqq 0}$ is a reflecting Brownian motion on $[0, \infty)$.

The system of equations satisfying (i)–(iii) is called a **Skorohod equation,** named after Skorohod, who was the first to solve a stochastic differential equation with a reflecting boundary condition.

Proof X and ℓ are uniquely determined from x and the Brownian motion B by Lemma 2.8.1. Hence it is sufficient to show that, letting $W = \{W(t)\}_{t \geqq 0}$ be a one-dimensional Brownian motion and setting $X(t) = |W(t)|$, there exists a Brownian motion $\widetilde{B} = \{\widetilde{B}(t)\}_{t \geqq 0}$ and a continuous increasing process $\widetilde{\ell} = \{\widetilde{\ell}(t)\}_{t \geqq 0}$ which satisfy the conditions (i) and (ii).

Let $\{L(t,a)\}_{t \geqq 0}$ be the local time of W at a and set $L(t) = L(t,0)$. By Tanaka's formula,

$$|W(t)| = \int_0^t \operatorname{sgn}(W(s))\,\mathrm{d}W(s) + L(t).$$

Set

$$\widetilde{B}(t) = \int_0^t \operatorname{sgn}(W(s))\,\mathrm{d}W(s) \quad \text{and} \quad \widetilde{\ell}(t) = L(t).$$

Then, $\langle \widetilde{B} \rangle(t) = t$ and $\widetilde{B}$ is a $\{\mathscr{F}_t^W\}$-Brownian motion by Lévy's theorem (Theorem 2.5.1), where $\mathscr{F}_t^W = \sigma\{W(s); s \leqq t\}$. By Theorem 2.7.7,

$$\int_0^t \mathbf{1}_{(-\varepsilon,\varepsilon)}(W(s))\,\mathrm{d}s = \int_0^t \mathbf{1}_{[0,\varepsilon)}(X(s))\,\mathrm{d}s = \int_{-\varepsilon}^{\varepsilon} L(t,a)\,\mathrm{d}a$$

and

$$\widetilde{\ell}(t) = \lim_{\varepsilon \downarrow 0} \frac{1}{2\varepsilon} \int_0^t \mathbf{1}_{[0,\varepsilon)}(X(s))\,\mathrm{d}s.$$

Hence $\int_0^t \mathbf{1}_{\{0\}}(X(s))\,\mathrm{d}\widetilde{\ell}(s) = \widetilde{\ell}(t)$. Thus the triplet $\{X, \widetilde{B}, \widetilde{\ell}\}$ satisfies the conditions of the theorem. □

Theorem 2.8.3 *Let $B = \{B(t)\}_{t \geqq 0}$ be a one-dimensional Brownian motion starting from 0 and set $m(t) = \min\{B(s); s \leqq t\}$. Moreover, let $\{L(t)\}_{t \geqq 0}$ be the local time at 0 of B. Then, the two-dimensional continuous stochastic processes $\{(B(t) - m(t), -m(t))\}_{t \geqq 0}$ and $\{(|B(t)|, L(t))\}_{t \geqq 0}$ have the same probability law.*

Proof By Tanaka's formula and Theorem 2.5.1, the stochastic process $\{\beta(t)\}$ defined by

$$\beta(t) = |B(t)| - L(t) = \int_0^t \operatorname{sgn}(B(s))\,\mathrm{d}B(s)$$

is a Brownian motion. Consider the decompositions $|B(t)| = \beta(t) + L(t)$ and $B(t) - m(t) = B(t) + (-m(t))$ of Lemma 2.8.1(i) for $\{\beta(t)\}$ and $\{B(t)\}$. $\{(|B(t)|, L(t))\}$ and $\{(B(t) - m(t), -m(t))\}$ are obtained from the Brownian motions $\{\beta(t)\}$ and $\{B(t)\}$, respectively, through the same deterministic procedure. Hence their distributions coincide. □

Corollary 2.8.4 (1) *Let* $M(t) = \max\{B(s); s \leqq t\}$. *Then,* $\{(M(t)-B(t), M(t))\}_{t\geqq 0}$ *and* $\{(|B(t)|, L(t))\}_{t\geqq 0}$ *have the same probability law.*
(2) $\lim_{t\to\infty} L(t) = \infty$ *almost surely.*

Proof (1) Apply the theorem to $\{-B(t)\}_{t\geqq 0}$.
(2) The result follows from the fact that $\sup_{t\geqq 0} B(t) = \infty$ almost surely. □

Remark 2.8.5 (1) Let $\{L(t, a)\}$ be the local time of B at a and set $\tau_a = \inf\{t > 0; B(t) = a\}$. Since $\{B_{\tau_a+t} - a\}_{t\geqq 0}$ is a Brownian motion starting from 0, $\lim_{t\to\infty} L(t, a) = \infty$ almost surely.
(2) In connection with Corollary 2.8.4, an important result is **Pitman's theorem**: $\{2M(t) - B(t)\}_{t\geqq 0}$ has the same probability law as a three dimensional Bessel process $\{\rho(t)\}_{t\geqq 0}$ starting from 0 (Theorem 4.8.7). Moreover, set $J(t) = \inf_{s\geqq t} \rho(s)$. Then, $\{(2M(t) - B(t), M(t))\}_{t\geqq 0}$ has the same probability law as $\{(\rho(t), J(t))\}_{t\geqq 0}$. For the details and related topics, see [45, 83, 98].

2.9 Conformal Martingales

When stochastic analysis is applied to complex analysis, one of the starting points is a complex-valued stochastic process whose real and imaginary parts are both local martingales. The purpose of this section is to present some fundamental properties of such stochastic processes. We set $\mathrm{i} = \sqrt{-1}$.

A complex-valued stochastic process $Z = \{Z(t)\}_{t\geqq 0}$ is a complex-valued continuous locally square-integrable martingale if both its real and imaginary parts are elements of $\mathscr{M}^2_{\mathrm{c,loc}}$, that is, Z is represented as

$$Z(t) = X(t) + \mathrm{i}Y(t)$$

by $X = \{X(t)\}_{t\geqq 0}, Y = \{Y(t)\}_{t\geqq 0} \in \mathscr{M}^2_{\mathrm{c,loc}}$. Such a stochastic process Z is also denoted by $Z = X + \mathrm{i}Y$. The set of complex-valued continuous locally square-integrable martingales is denoted by $\mathscr{M}^2_{\mathrm{c,loc}}(\mathbb{C})$.

For $Z_j = X_j + \mathrm{i}Y_j \in \mathscr{M}^2_{\mathrm{c,loc}}(\mathbb{C})$ $(j = 1, 2)$ we define $\langle Z_1, Z_2\rangle = \{\langle Z_1, Z_2\rangle(t)\}_{t\geqq 0}$ by

$$\langle Z_1, Z_2\rangle(t) = \langle X_1, X_2\rangle(t) - \langle Y_1, Y_2\rangle(t) + \mathrm{i}\{\langle X_2, Y_1\rangle(t) + \langle X_1, Y_2\rangle(t)\}.$$

Note that $\langle\ ,\ \rangle$ is complex bi-linear in two stochastic processes. For $Z \in \mathscr{M}^2_{\mathrm{c,loc}}(\mathbb{C})$, $Z^2 - \langle Z, Z\rangle = \{Z(t)^2 - \langle Z, Z\rangle(t)\}_{t\geqq 0} \in \mathscr{M}^2_{\mathrm{c,loc}}(\mathbb{C})$. Moreover, defining $\overline{Z} \in \mathscr{M}^2_{\mathrm{c,loc}}(\mathbb{C})$ by $\overline{Z} = X - \mathrm{i}Y$ for $Z = X + \mathrm{i}Y \in \mathscr{M}^2_{\mathrm{c,loc}}(\mathbb{C})$, we have

$$\langle Z, \overline{Z}\rangle = \langle X\rangle + \langle Y\rangle. \tag{2.9.1}$$

Definition 2.9.1 $Z = X + \mathrm{i}Y \in \mathscr{M}^2_{\mathrm{c,loc}}(\mathbb{C})$ is called a **conformal martingale** if $\langle X\rangle = \langle Y\rangle$, $\langle X, Y\rangle = 0$. Moreover, when $\mathbf{P}(Z(0) = a) = 1$ for $a \in \mathbb{C}$, it is called a conformal martingale starting from a.

$Z = X + \mathrm{i}Y \in \mathscr{M}^2_{\mathrm{c,loc}}(\mathbb{C})$ is a conformal martingale if and only if $\langle Z, Z\rangle = 0$. Representing Z^2 as

$$Z(t)^2 = X(t)^2 - \langle X\rangle(t) - (Y(t)^2 - \langle Y\rangle(t)) + \langle X\rangle(t) - \langle Y\rangle(t) + 2\mathrm{i}(X(t)Y(t) - \langle X, Y\rangle(t)) + 2\mathrm{i}\langle X, Y\rangle(t),$$

we easily see that Z is a conformal martingale if and only if $Z^2 = \{Z(t)^2\}_{t\geqq 0} \in \mathscr{M}^2_{\mathrm{c,loc}}(\mathbb{C})$. Moreover, by the formula (2.9.1), if Z is a conformal martingale, then $\langle Z, \overline{Z}\rangle = 2\langle X\rangle$.

Let $\{(B_1(t), B_2(t))\}_{t\geqq 0}$ be a two-dimensional Brownian motion. Then,

$$\beta = \{\beta(t) = B_1(t) + \mathrm{i}B_2(t)\}_{t\geqq 0}$$

is a conformal martingale. This is called a **complex Brownian motion**.

Conformal martingales are not closed under summation. For example, let $\beta = \{\beta(t) = B_1(t) + \mathrm{i}B_2(t)\}_{t\geqq 0}$ be a complex Brownian motion starting from $a \in \mathbb{C}$. Then, $\overline{\beta}$ is a conformal martingale starting from $\overline{a} \in \mathbb{C}$. Since $\beta + \overline{\beta} = 2B_1$ and $\langle B_1\rangle(t) = t$, $\beta + \overline{\beta}$ is not a conformal martingale. The sum of two conformal martingales Z and W is a conformal martingale if and only if $\langle Z, W\rangle = 0$.

The next theorem shows that conformal martingales are closed under the transforms defined by holomorphic functions and that they are time changed complex Brownian motions.

Theorem 2.9.2 (1) $Z \in \mathscr{M}^2_{\mathrm{c,loc}}(\mathbb{C})$ *is a conformal martingale if and only if* $f(Z) = \{f(Z(t))\}_{t\geqq 0} \in \mathscr{M}^2_{\mathrm{c,loc}}(\mathbb{C})$ *for any holomorphic function* f.
(2) $Z \in \mathscr{M}^2_{\mathrm{c,loc}}(\mathbb{C})$ *is a conformal martingale if and only if* $f(Z)$ *is a conformal martingale for any holomorphic function* $f : \mathbb{C} \to \mathbb{C}$. *Moreover,*

$$\langle f(Z), \overline{f(Z)}\rangle(t) = \int_0^t |f'(Z(s))|^2 \mathrm{d}\langle Z, \overline{Z}\rangle(s),$$

where f' *is the complex derivative of* f.
(3) *If* $Z = X + \mathrm{i}Y = \{Z(t)\}_{t\geqq 0}$ *is a conformal martingale, then there exists a complex Brownian motion* β *such that* $Z(t) = \beta(\langle X\rangle(t))$.

Proof (1) As was shown above, the sufficiency follows if we show $Z^2 = \{Z(t)^2\}_{t\geqq 0} \in \mathscr{M}^2_{\mathrm{c,loc}}(\mathbb{C})$. This is checked by taking $f(z) = z^2$.

Conversely, assume that $Z = X + \mathrm{i}Y \in \mathscr{M}^2_{\mathrm{c,loc}}(\mathbb{C})$ is a conformal martingale. Let f be a holomorphic function and write $f(x, y) = \varphi(x, y) + \mathrm{i}\psi(x, y)$ with

$\mathbb{R}$-valued functions φ and ψ, where $z = x + \mathrm{i}y$. Since φ and ψ are harmonic functions, Itô's formula yields

$$\begin{aligned} \mathrm{d}(\varphi(X(t), Y(t))) &= \frac{\partial \varphi}{\partial x}(X(t), Y(t))\mathrm{d}X(t) + \frac{\partial \varphi}{\partial y}(X(t), Y(t))\mathrm{d}Y(t), \\ \mathrm{d}(\psi(X(t), Y(t))) &= \frac{\partial \psi}{\partial x}(X(t), Y(t))\mathrm{d}X(t) + \frac{\partial \psi}{\partial y}(X(t), Y(t))\mathrm{d}Y(t). \end{aligned} \tag{2.9.2}$$

Since $X, Y \in \mathscr{M}^2_{\mathrm{c,loc}}$, $f(Z) \in \mathscr{M}^2_{\mathrm{c,loc}}(\mathbb{C})$.

(2) Although the first assertion immediately follows from (1) since the compositions of holomorphic functions are holomorphic, we give an alternative proof.

Write $Z = X + \mathrm{i}Y$ and $f = \varphi + \mathrm{i}\psi$ as above. By (2.9.2) and the Cauchy–Riemann relation, we have

$$\begin{aligned} \mathrm{d}\langle \varphi(X, Y)\rangle(t) &= \left\{\left(\frac{\partial \varphi}{\partial x}(X(t), Y(t))\right)^2 + \left(\frac{\partial \varphi}{\partial y}(X(t), Y(t))\right)^2\right\}\mathrm{d}\langle X\rangle(t) \\ &= \left\{\left(\frac{\partial \psi}{\partial x}(X(t), Y(t))\right)^2 + \left(\frac{\partial \psi}{\partial y}(X(t), Y(t))\right)^2\right\}\mathrm{d}\langle X\rangle(t) \\ &= \mathrm{d}\langle \psi(X, Y)\rangle(t) \end{aligned}$$

and

$$\begin{aligned} \mathrm{d}\langle \varphi(X, Y), \psi(X, Y)\rangle(t) &= \left\{\frac{\partial \varphi}{\partial x}(X(t), Y(t))\frac{\partial \psi}{\partial x}(X(t), Y(t)) \right. \\ &\qquad \left. + \frac{\partial \varphi}{\partial y}(X(t), Y(t))\frac{\partial \psi}{\partial y}(X(t), Y(t))\right\}\mathrm{d}\langle X\rangle(t) \\ &= 0. \end{aligned}$$

Hence, $f(Z) = \varphi(X, Y) + \mathrm{i}\psi(X, Y)$ is a conformal martingale. Moreover, since

$$|f'|^2 = \left(\frac{\partial \varphi}{\partial x}\right)^2 + \left(\frac{\partial \varphi}{\partial y}\right)^2 = \left(\frac{\partial \psi}{\partial x}\right)^2 + \left(\frac{\partial \psi}{\partial y}\right)^2,$$

we have

$$\mathrm{d}\langle \varphi(X, Y)\rangle(t) = \mathrm{d}\langle \psi(X, Y)\rangle(t) = |f'(Z(t))|^2\mathrm{d}\langle X\rangle(t)$$

and

$$\langle f(Z), \overline{f(Z)}\rangle = \langle \varphi(X, Y)\rangle(t) + \langle \psi(X, Y)\rangle(t), \quad \langle Z, \overline{Z}\rangle = 2\langle X\rangle.$$

The assertion follows from these identities.

(3) By Theorem 2.5.6, there exists a two-dimensional Brownian motion $\{(B_1(t), B_2(t))\}_{t \geqq 0}$ such that

$$X(t) = B_1(\langle X\rangle(t)), \quad Y(t) = B_2(\langle Y\rangle(t)).$$

Define a complex Brownian motion β by $\beta = B_1 + \mathrm{i}B_2$. Then, since $\langle X\rangle = \langle Y\rangle$, $Z(t) = \beta(\langle X\rangle(t))$. □

As seen in the next chapter, the probability that a two-dimensional Brownian motion hits a fixed point is 0. We can prove this property for conformal martingales by using the representation via complex Brownian motions.

Theorem 2.9.3 *Let $Z = \{Z(t)\}_{t\geqq 0}$ be a conformal martingale starting from $a \in \mathbb{C}$. Then, for any $b \in \mathbb{C}$, we have*

$$\mathbf{P}(Z(t) \in \mathbb{C} \setminus \{b\}, t \in (0,\infty)) = 1.$$

Proof By Theorem 2.9.2(3), it suffices to show

$$\mathbf{P}(\zeta(t) \in \mathbb{C} \setminus \{b\}, t \in (0,\infty)) = 1$$

for a complex Brownian motion $\zeta = \{\zeta(t)\}_{t\geqq 0}$ with $\mathbf{P}(\zeta(0) = a) = 1$. While this immediately follows from the same property of two-dimensional Brownian motion, we give another proof via complex analysis in the case when $a \neq b$.

Let $\{(B_1(t), B_2(t))\}_{t\geqq 0}$ be a two-dimensional Brownian motion starting from 0. Set $\beta = \{\beta(t) = B_1(t) + \mathrm{i}B_2(t)\}_{t\geqq 0}$ and

$$W(t) = (a-b)\mathrm{e}^{\beta(t)} + b.$$

Then it is easy to show

$$\mathbf{P}(W(t) \neq b,\ t \in (0,\infty)) = 1. \tag{2.9.3}$$

Since $W = \{W(t)\}_{t\geqq 0}$ is a conformal martingale by Theorem 2.9.2 and $\langle \beta, \overline{\beta}\rangle(t) = 2t$, we have

$$\langle W, \overline{W}\rangle(t) = 2\int_0^t |b-a|^2 \mathrm{e}^{2B_1(s)} \mathrm{d}s.$$

For each $n > 0$, $\int_0^n (\max\{B_1(s), 0\})^2 \mathrm{d}s$ has the same probability law as $n^2 \int_0^1 (\max\{B_1(s), 0\})^2 \mathrm{d}s$ (Theorem 1.2.8). Since $\int_0^1 (\max\{B_1(s), 0\})^2 \mathrm{d}s > 0$, $\mathbf{P}$-a.s., by Fatou's lemma and the bounded convergence theorem,

$$\begin{aligned}
&\mathbf{E}\Big[\Big(1 + \int_0^\infty (\max\{B_1(s), 0\})^2 \mathrm{d}s\Big)^{-1}\Big] \\
&\leqq \liminf_{n\to\infty} \mathbf{E}\Big[\Big(1 + \int_0^n (\max\{B_1(s), 0\})^2 \mathrm{d}s\Big)^{-1}\Big] \\
&= \liminf_{n\to\infty} \mathbf{E}\Big[\Big(1 + n^2\int_0^1 (\max\{B_1(s), 0\})^2 \mathrm{d}s\Big)^{-1}\Big] = 0.
\end{aligned}$$

Therefore,

$$\mathbf{P}\Big(\int_0^\infty (\max\{B_1(s), 0\})^2 \mathrm{d}s = \infty\Big) = 1.$$

Moreover, using the elementary inequality $\mathrm{e}^{2x} \geqq (\max\{x, 0\})^2$, we obtain

$$\lim_{t\to\infty} \langle W, \overline{W}\rangle(t) = \lim_{t\to\infty} 2|b-a|^2 \int_0^t \mathrm{e}^{2B_1(s)} \mathrm{d}s = \infty, \quad \mathbf{P}\text{-a.s.} \tag{2.9.4}$$

Using Theorem 2.9.2 again, we see that there exists a complex Brownian motion $\zeta = \{\zeta(t)\}_{t\geqq 0}$ starting from a such that

$$W(t) = \zeta\Big(\int_0^t |b-a|^2 \mathrm{e}^{2B_1(s)} \mathrm{d}s\Big).$$

Hence (2.9.3) and (2.9.4) imply the desired assertion. □

Definition 2.9.4 Let $Z = X + \mathrm{i}Y \in \mathscr{M}^2_{\mathrm{c,loc}}(\mathbb{C})$.
(1) Denote by $\mathscr{L}^2_{\mathrm{loc}}(Z)$ the set of $\mathbb{C}$-valued predictable processes $\Phi = \xi + \mathrm{i}\eta = \{\Phi(t) = \xi(t) + \mathrm{i}\eta(t)\}_{t\geqq 0}$ such that

$$\mathbf{P}\Big(\int_0^t |\Phi(s)|^2 \mathrm{d}\langle Z, \overline{Z}\rangle(s) < \infty\Big) = 1 \quad (t \geqq 0).$$

(2) The complex stochastic integral of $\Phi = \xi + \mathrm{i}\eta \in \mathscr{L}^2_{\mathrm{loc}}(Z)$ with respect to Z,

$$I^Z(\Phi) = \Big\{\int_0^t \Phi(s)\,\mathrm{d}Z(s)\Big\}_{t\geqq 0} \in \mathscr{M}^2_{\mathrm{c,loc}}(\mathbb{C}),$$

is defined by

$$\begin{aligned}\int_0^t \Phi(s)\,\mathrm{d}Z(s) = &\int_0^t \{\xi(s)\,\mathrm{d}X(s) - \eta(s)\,\mathrm{d}Y(s)\} \\ &+ \mathrm{i}\int_0^t \{\eta(s)\,\mathrm{d}X(s) + \xi(s)\,\mathrm{d}Y(s)\}.\end{aligned}$$

Since $\xi, \eta \in \mathscr{L}^2_{\mathrm{loc}}(X) \cap \mathscr{L}^2_{\mathrm{loc}}(Y)$ by (2.9.1), the complex stochastic integral is well defined. For $Z, \widehat{Z} \in \mathscr{M}^2_{\mathrm{c,loc}}(\mathbb{C})$, $\Phi \in \mathscr{L}^2_{\mathrm{loc}}(Z)$ and $\Psi \in \mathscr{L}^2_{\mathrm{loc}}(\widehat{Z})$,

$$\langle I^Z(\Phi), I^{\widehat{Z}}(\Psi)\rangle = \int_0^t \Phi(s)\Psi(s)\,\mathrm{d}\langle Z, \widehat{Z}\rangle(s), \tag{2.9.5}$$

where, for increasing processes φ and ψ, $\mathrm{d}(\varphi + \mathrm{i}\psi) = \mathrm{d}\varphi + \mathrm{i}\mathrm{d}\psi$. Therefore, if Z is a conformal martingale, then this is also the case for $\{\int_0^t \Phi(s)\,\mathrm{d}Z(s)\}_{t\geqq 0}$.

Let $Z_1 = \{Z_1(t)\}_{t\geqq 0}, \ldots, Z_n = \{Z_n(t)\}_{t\geqq 0}$ be a conformal martingale. Itô's formula for a holomorphic function $f : \mathbb{C}^n \to \mathbb{C}$ is given by

$$
\begin{aligned}
&f(Z_1(t), \ldots, Z_n(t)) \\
&\quad = f(Z_1(0), \ldots, Z_n(0)) + \sum_{j=1}^{n} \int_0^t \frac{\partial f}{\partial z_j}(Z_1(s), \ldots, Z_n(s))\, \mathrm{d}Z_j(s) \\
&\qquad + \frac{1}{2} \sum_{1 \leqq j \neq k \leqq n} \int_0^t \frac{\partial^2 f}{\partial z_j \partial z_k}(Z_1(s), \ldots, Z_n(s))\, \mathrm{d}\langle Z_j, Z_k\rangle(s). \qquad (2.9.6)
\end{aligned}
$$

In particular, if $\langle Z_j, Z_k\rangle = 0$ $(j \neq k)$, then $\{f(Z_1(t), \ldots, Z_n(t))\}_{t \geqq 0}$ is a conformal martingale.

By Itô's formula, the logarithm and the p-th root of a conformal martingale are defined.

Theorem 2.9.5 *Let $Z = \{Z(t)\}_{t \geqq 0}$ be a conformal martingale starting from $a \in \mathbb{C} \setminus \{0\}$ and $p \in \mathbb{N}$. Then, there exist conformal martingales $W = \{W(t)\}_{t \geqq 0}$ and $W_p = \{W_p(t)\}_{t \geqq 0}$ such that*

$$
Z(t) = \mathrm{e}^{W(t)} \quad \textit{and} \quad Z(t) = (W_p(t))^p.
$$

Proof It suffices to show the first identity. In fact, setting $W_p(t) = \mathrm{e}^{\frac{W(t)}{p}}$, we obtain $Z(t) = (W_p(t))^p$ from it.

By Theorem 2.9.3,

$$
\mathbf{P}(Z(t) \neq 0, t \in [0, \infty)) = 1.
$$

Combining this identity with the continuity of $\{Z(t)\}_{t \geqq 0}$, we obtain

$$
\left\{\Phi(t) = \frac{1}{Z(t)}\right\}_{t \geqq 0} \in \mathscr{L}^2_{\mathrm{loc}}(Z).
$$

Take $\alpha \in \mathbb{C}$ such that $a = \mathrm{e}^{\alpha}$ and set

$$
W(t) = \alpha + \int_0^t \frac{1}{Z(s)}\, \mathrm{d}Z(s).
$$

Then $W = \{W(t)\}_{t \geqq 0}$ is a conformal martingale.

$\langle Z, W\rangle = 0$ by (2.9.5). Applying Itô's formula (2.9.6) to $f(z, w) = z\mathrm{e}^{-w}$, we see $Z(t)\mathrm{e}^{-W(t)} = 1$. □

An interesting application of conformal martingales is to show the little Picard theorem, which asserts that the range of an entire function on $\mathbb{C}$ is either $\mathbb{C}$ or $\mathbb{C} - a$ for some $a \in \mathbb{C}$. For details, see [12].

3

Brownian Motion and the Laplacian

We prove the Markov and the strong Markov properties of Brownian motion. We also mention its recurrence and transience. The transition density of Brownian motion is given by the fundamental solution to the heat equation for the Laplacian, and Brownian motions are closely related to various differential equations. In this chapter, we show that the solutions of the heat equation and the Dirichlet problem for the Laplacian are represented as expectations with respect to Brownian motions. We also introduce the Feynman–Kac formula, which is associated with the Laplacian and a scalar potential, and its applications.

3.1 Markov and Strong Markov Properties

Suppose that the behavior of a stochastic process $X = \{X(t)\}_{t \geqq 0}$ up to time s is given. If the probability law of the behavior of X after s is determined only by the position $X(s)$ at s, X is said to have the Markov property. Moreover, if X has this property not only for fixed times but also for stopping times, X is said to have the strong Markov property. The purpose of this section is to show that Brownian motions have both of these properties and give applications.

Let $\mathbf{W}$ be the path space given by

$$\mathbf{W} = \{w : [0, \infty) \to \mathbb{R}^d;\ w \text{ is continuous}\}.$$

We denote the coordinate process by $B(t) : \mathbf{W} \ni w \mapsto w(t) \in \mathbb{R}^d$ $(t \geqq 0)$ and let $\mathscr{W}$ be the smallest σ-field under which the coordinate process is measurable. For each $x \in \mathbb{R}^d$, denote by $\mathbf{P}_x$ the probability measure on $\mathbf{W}$ under which the coordinate process $\{B(t)\}_{t \geqq 0}$ is a Brownian motion starting from x. Set $\mathscr{F}_t^0 = \sigma\{B(u); u \leqq t\}$ and let $\{\mathscr{F}_t\}$ be the right-continuous filtration given by

$$\mathscr{F}_t = \bigcap_{s>t} \mathscr{F}_s^0.$$

In this chapter, these $\{B(t)\}_{t \geqq 0}$ and $\mathbf{P}_x$ are dealt with.

Define the **shift** θ_s ($s \geqq 0$) on the path space $\mathbf{W}$ by

$$(\theta_s w)(t) = w(s+t) \qquad (t \geqq 0).$$

If $F : \mathbf{W} \to \mathbb{R}$ is a $\mathscr{W}$-measurable function, $F \circ \theta_s$ is determined by the path of the Brownian motion after s. For example, if F is given by

$$F(w) = \prod_{i=1}^{n} f_i(w(t_i)) \tag{3.1.1}$$

for bounded measurable functions $f_1, f_2, \ldots, f_n$ on $\mathbb{R}^d$ and $0 < t_1 < t_2 < \cdots < t_n$, we have

$$(F \circ \theta_s)(w) = \prod_{i=1}^{n} f_i(w(s+t_i)).$$

Theorem 3.1.1 (Markov property) *Let $s \geqq 0$ and F be a bounded $\mathscr{W}$-measurable function (random variable). Denote the expectation with respect to $\mathbf{P}_x$ by $\mathbf{E}_x$. Then, for any $x \in \mathbb{R}^d$,*

$$\mathbf{E}_x[F \circ \theta_s | \mathscr{F}_s] = \mathbf{E}_{B(s)}[F], \quad \mathbf{P}_x\text{-a.s.}, \tag{3.1.2}$$

where, on the right hand side, $y = B(s)$ is substituted into the function $\varphi(y) = \mathbf{E}_y[F]$ on $\mathbb{R}^d$.

The measurability of the mapping $y \mapsto \mathbf{E}_y[F]$ can be shown by the monotone class theorem. The proof is left to the reader.

Proof What should be shown is the identity

$$\mathbf{E}_x[(F \circ \theta_s)\mathbf{1}_A] = \mathbf{E}_x[\mathbf{E}_{B(s)}[F]\mathbf{1}_A] \tag{3.1.3}$$

for any $A \in \mathscr{F}_s$. First, suppose that F is defined by (3.1.1) and that A is given by

$$A = \{w;\ w(s_1) \in A_1,\ w(s_2) \in A_2, \ldots, w(s_m) \in A_m\} \tag{3.1.4}$$

for $0 < h < t_1$, $0 < s_1 < s_2 < \cdots < s_m \leqq s+h$ and $A_1, A_2, \ldots, A_m \in \mathscr{B}(\mathbb{R}^d)$. Then, letting $g(t,x,y) = (2\pi t)^{-\frac{d}{2}} \mathrm{e}^{-\frac{|y-x|^2}{2t}}$ be the transition density of a Brownian motion, we have

$$\begin{aligned}\varphi(y) = \mathbf{E}_y[F] = &\int_{\mathbb{R}^d} g(t_1, y, y_1) f_1(y_1)\,\mathrm{d}y_1 \int_{\mathbb{R}^d} g(t_2 - t_1, y_1, y_2) f_2(y_2)\,\mathrm{d}y_2 \\ &\times \cdots \times \int_{\mathbb{R}^d} g(t_n - t_{n-1}, y_{n-1}, y_n) f_n(y_n)\,\mathrm{d}y_n.\end{aligned}$$

Set

$$\widetilde{\varphi}(y,h) = \int_{\mathbb{R}^d} g(t_1-h,y,y_1)f_1(y_1)\,\mathrm{d}y_1 \int_{\mathbb{R}^d} g(t_2-t_1,y_1,y_2)f_2(y_2)\,\mathrm{d}y_2$$
$$\times\cdots\times \int_{\mathbb{R}^d} g(t_n-t_{n-1},y_{n-1},y_n)f_n(y_n)\,\mathrm{d}y_n.$$

Then,

$$\mathbf{E}_x[(F\circ\theta_s)\mathbf{1}_A] = \int_{A_1} g(s_1,x,x_1)\,\mathrm{d}x_1 \int_{A_2} g(s_2-s_1,x_1,x_2)\,\mathrm{d}x_2$$
$$\times\cdots\times \int_{A_m} g(s_m-s_{m-1},x_{m-1},x_m)\,\mathrm{d}x_m \int_{\mathbb{R}^d} g(s+h-s_m,x_m,y)\widetilde{\varphi}(y,h)\,\mathrm{d}y$$

and

$$\mathbf{E}_x[(F\circ\theta_s)\mathbf{1}_A] = \mathbf{E}_x[\widetilde{\varphi}(B(s+h),h)\mathbf{1}_A]. \tag{3.1.5}$$

Denote by $\mathscr{G}$ the totality of $A\in\mathscr{F}^0_{s+h}$ which satisfies (3.1.5). Then, by the Dynkin class theorem, (3.1.5) holds for all $A\in\mathscr{F}^0_{s+h}$ and, therefore, for all $A\in\mathscr{F}_s$.

Let $\{y_h\}\subset\mathbb{R}^d$ be a sequence with $y_h\to y$ $(h\to 0)$. Then, by the Lebesgue convergence theorem, $\widetilde{\varphi}(y_h,h)\to\varphi(y)$. Hence, by the bounded convergence theorem,

$$\mathbf{E}_x[(F\circ\theta_s)\mathbf{1}_A] = \mathbf{E}_x[\varphi(B(s))\mathbf{1}_A] = \mathbf{E}_x[\mathbf{E}_{B(s)}[F]\mathbf{1}_A]$$

for any $A\in\mathscr{F}_s$ and F of the form (3.1.1).

Fix $A\in\mathscr{F}_s$ and denote by $\mathscr{H}$ the set of functions F on $\mathbf{W}$ which satisfies (3.1.3). Letting $\mathscr{A}$ be the totality of the subsets of $\mathbf{W}$ given as $\{w;\,w(t_k)\in A_k\,(k=1,2,\ldots,n)\}, A_k\in\mathscr{B}(\mathbb{R}^d)$, we have shown that the indicator functions of the sets in $\mathscr{A}$ belong to $\mathscr{H}$. Therefore, by Theorem A.2.6, (3.1.3) holds for any $\sigma(\mathscr{A})$-measurable, that is, $\mathscr{W}$-measurable bounded function F. □

Theorem 3.1.1 shows that $\{B(s+t)-B(s)\}_{t\geqq 0}$ is a Brownian motion independent of $\mathscr{F}_s$ for any $s\geqq 0$. Moreover, by Theorem 3.1.1, $\mathbf{E}_x[F\circ\theta_s|\mathscr{F}_s]$ is $\mathscr{F}^0_s$-measurable. Hence, by the tower property of the conditional expectation,

$$\mathbf{E}_x[F\circ\theta_s|\mathscr{F}_s] = \mathbf{E}_x[F\circ\theta_s|\mathscr{F}^0_s]. \tag{3.1.6}$$

Put

$$\Phi = \prod_{j=1}^m \varphi_j(w(t_j))$$

for $t_1<t_2<\cdots<t_m$ and bounded measurable functions $\varphi_1,\varphi_2,\ldots,\varphi_m$ on $\mathbb{R}^d$. Then

$$\mathbf{E}_x[\Phi|\mathscr{F}_s] = \mathbf{E}_x[\Phi|\mathscr{F}^0_s]. \tag{3.1.7}$$

In fact, if $t_k \leqq s < t_{k+1}$, decomposing Φ by

$$\Phi = I_1 \cdot I_2, \quad I_1 = \prod_{j=1}^{k} \varphi_j(w(t_j)),\ I_2 = \prod_{k+1}^{m} \varphi_j(w(t_j)),$$

we obtain (3.1.7) by (3.1.6) since I_1 is $\mathscr{F}_s^0$-measurable.

The monotone class theorem yields the following.

Theorem 3.1.2 *Let Φ be a bounded $\mathscr{W}$-measurable random variable. Then, for any $s \geqq 0$ and $x \in \mathbb{R}^d$,*

$$\mathbf{E}_x[\Phi|\mathscr{F}_s] = \mathbf{E}_x[\Phi|\mathscr{F}_s^0], \quad \mathbf{P}_x\text{-a.s.} \tag{3.1.8}$$

From Theorem 3.1.2, the following important result is deduced.

Theorem 3.1.3 (Blumenthal 0-1 law) *For any $x \in \mathbb{R}^d$ and $A \in \mathscr{F}_0$, $\mathbf{P}_x(A)$ is 0 or 1.*

Proof By Theorem 3.1.2,

$$\mathbf{1}_A = \mathbf{E}_x[\mathbf{1}_A|\mathscr{F}_0] = \mathbf{E}_x[\mathbf{1}_A|\mathscr{F}_0^0], \quad \mathbf{P}_x\text{-a.s.}$$

Since $\mathbf{P}_x(B(0) = x) = 1$ and $\mathbf{P}_x(C)$ is 0 or 1 for $C \in \mathscr{F}_0^0$, the conditional expectation in the third term coincides with the usual expectation and $\mathbf{1}_A = \mathbf{P}_x(A)$. Therefore, $\mathbf{P}_x(A)$ is 0 or 1. □

While several properties of Brownian motions are deduced from the Blumenthal 0-1 law, we only show the following. For more results, see, for example, [20, 56].

Theorem 3.1.4 *Set $\tau_+ = \inf\{t \geqq 0; B(t) > 0\}$. Then $\mathbf{P}_0(\tau_+ = 0) = 1$.*

Proof Since $\mathbf{P}_0(\tau_+ \leqq t) \geqq \mathbf{P}_0(B(t) > 0) = \frac{1}{2}$ for $t > 0$,

$$\mathbf{P}_0(\tau_+ = 0) = \lim_{t \downarrow 0} \mathbf{P}_0(\tau_+ \leqq t) \geqq \frac{1}{2}.$$

Hence, by the Blumenthal 0-1 law, $\mathbf{P}_0(\tau_+ = 0) = 1$. □

Also, for $\tau_- = \inf\{t \geqq 0; B(t) < 0\}$, $\mathbf{P}_0(\tau_- = 0) = 1$. Hence, by the continuity of the path of Brownian motions, the following holds.

Corollary 3.1.5 *Set $\tau_0 = \inf\{t > 0; B(t) = 0\}$. Then $\mathbf{P}_0(\tau_0 = 0) = 1$.*

Next we show the strong Markov property of Brownian motions. Let σ be an $\{\mathscr{F}_t\}$-stopping time. The information up to time σ is given by the σ-field

$$\mathscr{F}_\sigma = \{A; A \cap \{\sigma \leqq t\} \in \mathscr{F}_t \text{ for all } t \geqq 0\}.$$

Theorem 3.1.6 (strong Markov property) *Let* $F : [0,\infty)\times\mathbf{W} \mapsto F(s,w) \in \mathbb{R}$ *be bounded and* $\mathscr{B}([0,\infty)) \times \mathscr{W}$*-measurable and* σ *be an* $\{\mathscr{F}_t\}$*-stopping time. Then, for all* $x \in \mathbb{R}^d$,

$$\mathbf{E}_x[F_\sigma \circ \theta_\sigma | \mathscr{F}_\sigma] = \mathbf{E}_y[F_t]\Big|_{y=B(\sigma),t=\sigma}$$

holds $\mathbf{P}_x$-a.s. *on* $\{\sigma < \infty\}$, *where* $(F_\sigma \circ \theta_\sigma)(w) = F(\sigma(w), \theta_{\sigma(w)}w)$ *and* $F_t(w) = F(t,w)$.

Proof For $A \in \mathscr{F}_\sigma$, we shall show

$$\mathbf{E}_x[(F_\sigma \circ \theta_\sigma)\mathbf{1}_{A\cap\{\sigma<\infty\}}] = \mathbf{E}_x[\phi(B(\sigma),\sigma)\mathbf{1}_{A\cap\{\sigma<\infty\}}],$$

where $\phi(y,t) = \mathbf{E}_y[F_t]$.

First consider the case where σ takes only countable values $\{t_n\}_{n=1}^\infty$. Then, since $A \cap \{\sigma = t_n\} \in \mathscr{F}_{t_n}$, the Markov property of Brownian motion (Theorem 3.1.1) implies

$$\begin{aligned}\mathbf{E}_x[(F_\sigma \circ \theta_\sigma)\mathbf{1}_{A\cap\{\sigma<\infty\}}] &= \sum_{n=1}^\infty \mathbf{E}_x[(F_{t_n} \circ \theta_{t_n})\mathbf{1}_{A\cap\{\sigma=t_n\}}] \\ &= \sum_{n=1}^\infty \mathbf{E}_x[\mathbf{E}_{B(t_n)}[F_{t_n}]\mathbf{1}_{A\cap\{\sigma=t_n\}}] = \mathbf{E}_x[\phi(B(\sigma),\sigma)\mathbf{1}_{A\cap\{\sigma<\infty\}}].\end{aligned}$$

For a general stopping time, define σ_n $(n = 1,2,\ldots)$ by $\sigma_n = \frac{k+1}{2^n}$ when $\frac{k}{2^n} \leqq \sigma < \frac{k+1}{2^n}$. Then, σ_n is also an $\{\mathscr{F}_t\}$-stopping time. Suppose that the function F is given by

$$F(s,w) = f_0(s)\prod_{j=1}^n f_j(w(t_j))$$

for $0 < t_1 < t_2 < \cdots < t_n$ and $f_0 \in C_b([0,\infty))$, $f_1,\ldots,f_n \in C_b(\mathbb{R}^d)$. Then

$$\begin{aligned}\phi(y,t) = f_0(t)\int_{\mathbb{R}^d} g(t_1,y,y_1)f_1(y_1)\,\mathrm{d}y_1 \int_{\mathbb{R}^d} g(t_2-t_1,y_1,y_2)f_2(y_2)\,\mathrm{d}y_2 \\ \times\cdots\times \int_{\mathbb{R}^d} g(t_n-t_{n-1},y_{n-1},y_n)f_n(y_n)\,\mathrm{d}y_n.\end{aligned}$$

Thus ϕ is continuous in (y, t). Moreover, since $\sigma \leqq \sigma_n$, $A \in \mathscr{F}_\sigma \subset \mathscr{F}_{\sigma_n}$ and $\{\sigma_n < \infty\} = \{\sigma < \infty\}$, by the above observation,

$$\mathbf{E}_x[(F_{\sigma_n} \circ \theta_{\sigma_n})\mathbf{1}_{A\cap\{\sigma<\infty\}}] = \mathbf{E}_x[\phi(B(\sigma_n), \sigma_n)\mathbf{1}_{A\cap\{\sigma<\infty\}}].$$

Hence, letting $n \to \infty$, we obtain

$$\mathbf{E}_x[(F_\sigma \circ \theta_\sigma)\mathbf{1}_{A\cap\{\sigma<\infty\}}] = \mathbf{E}_x[\phi(B(\sigma), \sigma)\mathbf{1}_{A\cap\{\sigma<\infty\}}].$$

A similar argument to the proof of Theorem 3.1.1 via the monotone class theorem yields the assertion for a general function F. The details are left to the reader. □

We apply Theorem 3.1.6 to show Lévy's formula for the density of the joint distribution of the maximum $M(t) = \max_{0\leqq s\leqq t} B(s)$ and the position $B(t)$ of a one-dimensional Brownian motion for fixed $t > 0$.

Theorem 3.1.7 *Let $d = 1$. Then, for fixed $t > 0$,*

$$\mathbf{P}_0(M(t) \in \mathrm{d}a, B(t) \in \mathrm{d}x) = \frac{2(2a - x)}{\sqrt{2\pi t^3}}\mathrm{e}^{-\frac{(2a-x)^2}{2t}}\,\mathrm{d}a\mathrm{d}x \qquad (a > 0,\ x < a).$$

Proof Set $\tau_a = \inf\{s \geqq 0; B(s) = a\}$, and fix $t > 0$ and $x < a$. Define a function G on $[0, \infty) \times \mathbf{W}$ by

$$G(s, w) = \begin{cases} 1 & (\text{if } s \leqq t \text{ and } w(t - s) \leqq x) \\ 0 & (\text{otherwise}). \end{cases}$$

If $\tau_a(w) \leqq t$,

$$G(\tau_a(w), \theta_{\tau_a(w)}w) = \mathbf{1}_{\{B(t)\leqq x\}}(w). \tag{3.1.9}$$

On the other hand, setting $\psi(z, s) = E_z[G(s, \cdot)]$, we have

$$\psi(z, s) = \mathbf{E}_z[\mathbf{1}_{\{B(t-s)\leqq x\}}] = \mathbf{E}_z[\mathbf{1}_{\{B(t-s)\geqq 2z-x\}}] \quad \text{for} \quad s \leqq t$$

by the symmetry of the Gaussian distribution. Hence, Theorem 3.1.6 implies

$$\mathbf{E}_0[G_{\tau_a} \circ \theta_{\tau_a}|\mathscr{F}_{\tau_a}] = \psi(B(\tau_a), \tau_a) = \mathbf{E}_z[\mathbf{1}_{\{B(t-s)\geqq 2z-x\}}]\Big|_{z=B(\tau_a),\, s=\tau_a}$$

on $\{\tau_a \leqq t\}$. Note that $B(\tau_a) = a$. Then, by multiplying both sides by $\mathbf{1}_{\{\tau_a\leqq t\}}$ and taking the expectation, we obtain

$$\begin{aligned}\mathbf{P}_0(\tau_a \leqq t, B(t) \leqq x) &= \mathbf{E}_0[\mathbf{E}_0[\mathbf{1}_{\{\tau_a\leqq t\}}\mathbf{1}_{\{B((\tau_a+t)-\tau_a)\geqq 2a-x\}}|\mathscr{F}_{\tau_a}]] \\ &= \mathbf{P}_0(\tau_a \leqq t, B(t) \geqq 2a - x) \\ &= \mathbf{P}_0(B(t) \geqq 2a - x)\end{aligned}$$

by (3.1.9) and Theorem 3.1.6. Thus, since $\{\tau_a \leqq t\} = \{M(t) \geqq a\}$,

$$\mathbf{P}_0(M(t) \geqq a, B(t) \leqq x) = \int_{2a-x}^{\infty} \frac{1}{\sqrt{2\pi t}} e^{-\frac{z^2}{2t}} dz.$$

Differentiating with respect to a and x, we arrive at the assertion. □

Corollary 3.1.8 *For any $t > 0$ and $a > 0$,*

$$\mathbf{P}_0(M(t) \in da) = \frac{2}{\sqrt{2\pi t}} e^{-\frac{a^2}{2t}} da, \quad \mathbf{P}_0(\tau_a \in dt) = \frac{a}{\sqrt{2\pi t^3}} e^{-\frac{a^2}{2t}} dt,$$

$$\mathbf{P}_0(M(t) \in da \mid B(t) = 0) = \frac{4a}{t} e^{-\frac{2a^2}{t}} da.$$

Remark 3.1.9 In what follows, we use the (strong) Markov property together with stochastic integrals. However, $\{\mathscr{F}_t\}$ does not satisfy the usual conditions, which we assumed to define stochastic integrals. Therefore, we deal with a special realization of $\mathbf{P}_x$s. Specifically, let $W = \{w \in \mathbf{W}\ ;\ w(0) = 0\}$ and μ be the Wiener measure on it. Through a natural embedding $W \subset \mathbf{W}$, we identify $\mathbf{P}_0$ with μ. Then, $\mathbf{P}_x$ is realized as $\mathbf{P}_0 \circ (x + B(\cdot))^{-1}$, the distribution of $x + B(\cdot)$ under $\mathbf{P}_0$. Thus, we define stochastic integrals only under $\mathbf{P}_0$. In this case, the filtration satisfying the usual condition is given by $\mathscr{F}_t^B = \mathscr{F}_t^0 \vee \mathscr{N}$, where $\mathscr{N}$ is the totality of all null subsets of W with respect to the $\mathbf{P}_0$-outer measure (see Lemma 2.6.1). Moreover, since $\mathbf{P}_x(W) = 0$ for $x \neq 0$, by Theorem 3.1.2, $\mathbf{E}_x[\Phi|\mathscr{F}_s] = \mathbf{E}_x[\Phi|\mathscr{F}_s^0] = \mathbf{E}_x[\Phi|\mathscr{F}_s^B]$, $\mathbf{P}_x$-a.s. for any $x \in \mathbb{R}^d$.

3.2 Recurrence and Transience of Brownian Motions

The purpose of this section is to show that a d-dimensional Brownian motion is recurrent if $d = 1$ or $d = 2$ and is transient if $d \geqq 3$. In this section, to emphasize dimensions, we write $\mathbf{P}_x^{(d)}$ and $\mathbf{E}_x^{(d)}$ for $\mathbf{P}_x$ and $\mathbf{E}_x$, respectively.

Let τ_a be the first hitting time to $a \in \mathbb{R}^d$,

$$\tau_a = \inf\{t > 0; B(t) = a\}.$$

Theorem 3.2.1 *Let $d = 1$.*
(1) *For all $x, y \in \mathbb{R}$, $\mathbf{P}_x^{(1)}(\tau_y < \infty) = 1$.*
(2) *For all $x, y \in \mathbb{R}$ and $s \geqq 0$,*

$$\mathbf{P}_x^{(1)}(\textit{there exists } t \geqq s \textit{ such that } B(t) = y) = 1.$$

(3) *For all $x, y \in \mathbb{R}$,*

$$\mathbf{P}_x^{(1)}\begin{pmatrix}\textit{there exists an increasing sequence } \{t_n\}_{n=1}^{\infty} \\ \textit{with } \lim_{n\to\infty} t_n = \infty \textit{ such that } B(t_n) = y\end{pmatrix} = 1.$$

Proof (1) We may assume that $x = 0$. Then, the assertion follows from the formula (see Theorem 1.5.20)

$$\mathbf{P}_0^{(1)}(\tau_{-a} < \tau_b) = \frac{b}{a+b} \qquad (a, b > 0).$$

(2) By the Markov property of Brownian motion, the probability in question coincides with $\mathbf{E}_x^{(1)}[\mathbf{P}_{B(s)}^{(1)}(\tau_y < \infty)]$, which is 1 by (1).
(3) Take $s = 1, 2, \ldots$ in (2). Then, almost surely, there exists $t_n \geqq n$ such that $B(t_n) = y$ for any $n = 1, 2, \ldots$. □

Set $\phi(x) = \log|x|$ if $d = 2$ and $\phi(x) = |x|^{2-d}$ if $d \geqq 3$. Then

$$\Delta\phi(x) = \sum_{i=1}^{d} \Big(\frac{\partial}{\partial x^i}\Big)^2 \phi(x) = 0 \qquad (x \neq 0),$$

that is, ϕ is a **harmonic function**.

Suppose $r < |x| < R$ in the rest of this section. For a Brownian motion $B = \{B(t)\}_{t \geqq 0}$ starting from x, Itô's formula yields

$$\phi(B(t)) = \phi(B(0)) + \sum_{i=1}^{d} \int_0^t \frac{\partial\phi(x)}{\partial x^i}(B(s))\,\mathrm{d}B^i(s) \qquad (t < \sigma_r \wedge \sigma_R),$$

where $\sigma_r = \inf\{t > 0; |B(t)| = r\}$. Since $\frac{\partial\phi(x)}{\partial x^j}$ is bounded and continuous on $\{y; r < |y| < R\}$, the optional stopping theorem implies

$$\phi(x) = \mathbf{E}_x^{(d)}[\phi(B(t \wedge \sigma_r \wedge \sigma_R))].$$

Letting $t \to \infty$ by the bounded convergence theorem, we obtain

$$\begin{aligned}\phi(x) &= \mathbf{E}_x^{(d)}[\phi(B(\sigma_r \wedge \sigma_R))] \\ &= \phi_0(r)\mathbf{P}_x^{(d)}(\sigma_r < \sigma_R) + \phi_0(R)\mathbf{P}_x^{(d)}(\sigma_R < \sigma_r),\end{aligned}$$

where we have set $\phi_0(|x|) = \phi(x)$. Since $\mathbf{P}_x^{(d)}(\sigma_r < \sigma_R) + \mathbf{P}_x^{(d)}(\sigma_R < \sigma_r) = 1$,

$$\mathbf{P}_x^{(d)}(\sigma_r < \sigma_R) = \frac{\phi_0(R) - \phi(x)}{\phi_0(R) - \phi_0(r)}. \tag{3.2.1}$$

Theorem 3.2.2 *Let G be an open subset of $\mathbb{R}^2$. Then, for any $x \in \mathbb{R}^2$, there exists an increasing sequence $\{t_n\}_{n=1}^\infty$ satisfying $B(t_n) \in G$ and $t_n \to \infty$ almost surely under $\mathbf{P}_x^{(2)}$.*

Proof It suffices to show the case when $G = \{y \in \mathbb{R}^2; |y| < r\},\ r > 0$. Letting $R \to \infty$ in (3.2.1), we have $\mathbf{P}_x^{(2)}(\sigma_r < \infty) = 1$ for all $x \in \mathbb{R}^2$. Hence a similar argument to the proof of Theorem 3.2.1 yields the conclusion. □

Theorem 3.2.3 *If $d \geqq 2$, then $\mathbf{P}_x^{(d)}(\tau_{\{0\}} = \infty) = 1$ for all $x \in \mathbb{R}^2$.*

Proof It is sufficient to show the case when $d = 2$. While we showed the main part of the assertion in Theorem 2.9.3, we here give a proof by using (3.2.1). Let $x \neq 0$. For $R > 0$,

$$\mathbf{P}_x^{(2)}(\sigma_0 < \sigma_R) \leqq \lim_{r\to 0} \mathbf{P}_x^{(2)}(\sigma_r < \sigma_R) = 0.$$

Hence, letting $R \to \infty$, we obtain the assertion when $x \neq 0$.

Suppose that $x = 0$. By the Markov property of Brownian motion and the conclusion when $x \neq 0$,

$$\mathbf{P}_0^{(2)}(\text{there exists a } t \geqq \varepsilon \text{ such that } B(t) = 0) = \mathbf{E}_0^{(2)}[\mathbf{P}_{B(\varepsilon)}^{(2)}(\sigma_0 < \infty)] = 0$$

for any $\varepsilon > 0$, since $\mathbf{P}_0^{(2)}(B(\varepsilon) \neq 0) = 1$. Letting $\varepsilon \to 0$ we obtain the assertion. □

Remark 3.2.4 If $d \geqq 2$ and $x \neq 0$, then $\mathbf{P}_x^{(d)}\big(\lim_{r\downarrow 0} \sigma_r = \infty\big) = 1$.

Theorem 3.2.5 *Let* $d \geqq 3$.
(1) *If* $r < |x|$, *then*

$$\mathbf{P}_x^{(d)}(\sigma_r < \infty) = \left(\frac{r}{|x|}\right)^{d-2}.$$

(2) *For all* $x \in \mathbb{R}^d$, $\mathbf{P}_x^{(d)}\big(\lim_{t\to\infty} |B(t)| = \infty\big) = 1$.

Proof (1) Let $R \to \infty$ in (3.2.1). Then we immediately obtain the assertion.
(2) By Theorem 3.2.1 (1), $\mathbf{P}_0^{(d)}(\sigma_R < \infty) = 1$ for all $R > 0$. Denote by A_n the event that $|B(t)| > \sqrt{n}$ for all $t \geqq \sigma_n$. Then, by the strong Markov property of Brownian motion and (3.2.1), we have

$$\mathbf{P}_x^{(d)}(A_n^c) = \mathbf{E}_x^{(d)}[\mathbf{P}_{B(\sigma_n)}^{(d)}(\sigma_{\sqrt{n}} < \infty)] = \left(\frac{1}{\sqrt{n}}\right)^{d-2}.$$

By the monotonicity of probability measure,

$$\mathbf{P}_x^{(d)}\Big(\limsup_{n\to\infty} A_n\Big) \geqq \limsup_{n\to\infty} \mathbf{P}_x^{(d)}(A_n) = 1.$$

Hence, almost surely, there exists an increasing sequence $\{n_i\}$ with $n_i \to \infty$ such that $|B(t)| > \sqrt{n_i}$ for all $t > \sigma_{n_i}$. This means that $\lim_{t\to\infty} |B(t)| = \infty$ almost surely. □

For a one-dimensional Brownian motion, the expectation of the exit time from an interval was computed in Theorem 1.5.20. Also, for the multi-dimensional case, we have the following.

Theorem 3.2.6 *Let* $d \geqq 2$ *and* $|x| < R$. *Then* $\mathbf{E}_x^{(d)}[\sigma_R] = \frac{R^2-|x|^2}{d}$.

Proof Set $M(t) = |B(t)|^2 - dt$. Then $\{M(t)\}_{t\geqq 0}$ is a martingale and, by the optional stopping theorem,

$$\mathbf{E}_x^{(d)}[|B(\sigma_R \wedge t)|^2 - d(\sigma_R \wedge t)] = |x|^2$$

and

$$\mathbf{E}_x^{(d)}[\sigma_R \wedge t] = \frac{1}{d}(\mathbf{E}_x^{(d)}[|B(\sigma_R \wedge t)|^2] - |x|^2).$$

Letting $t \to \infty$, we obtain the assertion. □

3.3 Heat Equations

Consider the initial value problem for the **heat equation** on $\mathbb{R}^d$. It is a problem to find a function $u(t,x)$ on $[0,\infty) \times \mathbb{R}^d$ which is of C^1-class in t, is of C^2-class in x ($C^{1,2}$-class for short) and satisfies

$$\frac{\partial u}{\partial t}(t,x) = \frac{1}{2}\Delta u(t,x) \qquad (t > 0,\ x \in \mathbb{R}^d), \tag{3.3.1}$$

$$u(0,x) = f(x). \tag{3.3.2}$$

The transition density of a d-dimensional Brownian motion

$$g(t,x,y) = (2\pi t)^{-\frac{d}{2}} \mathrm{e}^{-\frac{|y-x|^2}{2t}} \qquad (t > 0,\ x, y \in \mathbb{R}^d)$$

satisfies $\frac{\partial g}{\partial t} = \frac{1}{2}\Delta_x g$, where Δ_x is the Laplacian acting on functions in x. The next theorem can be proven by justifying the change of order of differentiation and integration. The function $g(t,x,y)$ is called the **fundamental solution** of the heat equation (3.3.1) or the **heat kernel** on $\mathbb{R}^d$.

Theorem 3.3.1 *Let $f : \mathbb{R}^d \to \mathbb{R}$ be continuous and assume that*

$$\frac{1}{|x|^2} \max\{1, \log|f(x)|\} \to 0 \qquad (|x| \to \infty).$$

Then, the function $v(t,x)$ defined by

$$v(t,x) = \int_{\mathbb{R}^d} g(t,x,y) f(y)\,\mathrm{d}y \tag{3.3.3}$$

is of C^∞-class in (t,x) and solves the heat equation (3.3.1). *Moreover,*

$$\lim_{t \downarrow 0} v(t,x) = f(x) \qquad (x \in \mathbb{R}^d).$$

The function $v(t,x)$ defined by (3.3.3) is represented as the expectation,

$$v(t,x) = \mathbf{E}_x[f(B(t))].$$

Hence, if the uniqueness of the solution is proven, this function is the unique solution for the heat equation.

Theorem 3.3.2 *If a bounded function $u(t,x)$ $(t > 0,\ x \in \mathbb{R}^d)$ is a solution to the initial value problem of the heat equation* (3.3.1) *and* (3.3.2)*, then $u(t,x) = v(t,x)$.*

Proof Fix $t > 0$ and set $M(s) = u(t-s, B(s))$. Since u satisfies (3.3.1), by Itô's formula applied to functions of $C^{1,2}$-class (see Remark 2.3.15),

$$M(s) - u(t, B(0)) = \sum_{i=1}^{d} \int_0^s \frac{\partial u}{\partial x^i}(t-r, B(r))\, \mathrm{d}B^i(r).$$

Hence, $\{M(s)\}_{s \geqq 0}$ is a bounded local martingale and, in fact, a martingale. Since $M(s)$ converges to $M(t) = f(B(t))$ as $s \uparrow t$, we obtain

$$u(t,x) = M(0) = \mathbf{E}_x[M(t)] = v(t,x). \qquad \square$$

3.4 Non-Homogeneous Equation

Given $h : (0,\infty) \times \mathbb{R}^d \to \mathbb{R}$ and $f : \mathbb{R}^d \to \mathbb{R}$, consider the problem of finding a function on $[0,\infty) \times \mathbb{R}^d$ which is of $C^{1,2}$-class and satisfies

$$\frac{\partial u}{\partial t} = \frac{1}{2}\Delta u(t,x) + h(t,x) \qquad ((t,x) \in (0,\infty) \times \mathbb{R}^d), \tag{3.4.1}$$

$$u(0,x) = f(x) \quad (x \in \mathbb{R}^d). \tag{3.4.2}$$

If u_1 solves the equation

$$\frac{\partial u_1}{\partial t} = \frac{1}{2}\Delta u_1, \quad u_1(0,x) = f(x),$$

which was considered in the previous section, and if u_2 solves

$$\frac{\partial u_2}{\partial t} = \frac{1}{2}\Delta u_2 + h, \quad u_2(0,x) = 0,$$

then $u = u_1 + u_2$ is a solution of (3.4.1) and (3.4.2). Hence, we concentrate on the case where $f = 0$.

By Itô's formula, we obtain the following.

Proposition 3.4.1 *If u satisfies* (3.4.1), *then the stochastic process* $M = \{M(s)\}_{0\leqq s<t}$ *defined by*

$$M(s) = u(t-s, B(s)) + \int_0^s h(t-r, B(r))\,\mathrm{d}r \tag{3.4.3}$$

is a local martingale.

Let v be the function on $[0,\infty)\times\mathbb{R}^d$ defined by

$$v(t,x) = \mathbf{E}_x\Big[\int_0^t h(t-r, B(r))\,\mathrm{d}r\Big].$$

Proposition 3.4.2 *Assume that the function h is bounded. If the initial value problem* (3.4.1) *and* (3.4.2) *has a bounded solution on* $[0,t]\times\mathbb{R}^d$, *then it coincides with v.*

Proof Let u be a bounded solution for (3.4.1) and (3.4.2). The stochastic process M defined by (3.4.3) is a bounded martingale. Hence, by the martingale convergence theorem (Theorem 1.5.16), there exists the limit $\lim_{s\uparrow t} M(s)$ and it coincides with $\int_0^t h(t-r, B(r))\,\mathrm{d}r$. Therefore, we have

$$u(t,x) = M(0) = \mathbf{E}_x[M(t)] = v(t,x). \qquad \square$$

Proposition 3.4.3 *Assume that h is bounded. If v is of $C^{1,2}$-class, then v satisfies* (3.4.1).

Proof By the Markov property of Brownian motion,

$$\begin{aligned}
&\mathbf{E}_x\Big[\int_0^t h(t-r, B(r))\,\mathrm{d}r\,\Big|\,\mathscr{F}_s^B\Big]\\
&= \int_0^s h(t-r, B(r))\,\mathrm{d}r + \mathbf{E}_x\Big[\int_0^{t-s} h(t-s-r, B(s+r))\,\mathrm{d}r\,\Big|\,\mathscr{F}_s^B\Big]\\
&= \int_0^s h(t-r, B(r))\,\mathrm{d}r + v(t-s, B(s)).
\end{aligned}$$

By the assumption and Itô's formula, the right hand side is equal to

$$v(t,x) + \sum_{i=1}^d \int_0^s \frac{\partial v}{\partial x^i}(t-r, B(r))\,\mathrm{d}B^i(r) + \int_0^s \Big\{-\frac{\partial v}{\partial t} + \frac{1}{2}\Delta v + h\Big\}(t-r, B(r))\,\mathrm{d}r.$$

Since this is a local martingale, the third term is 0. $\square$

Since $v(t,x)\to 0$ as $t\downarrow 0$, it is important to study when $v(t,x)$ is smooth. In fact, if the function h is bounded, then v is continuous and has derivatives of first order in the space variable.

Proposition 3.4.4 *If h is bounded, $v(t,x)$ is continuous on* $(0,\infty)\times\mathbb{R}^d$.

Proof Let $g(s,x,y)=(2\pi t)^{-\frac{d}{2}}\mathrm{e}^{-\frac{|y-x|^2}{2t}}$ and define a function φ by

$$\varphi(s,x)=\mathbf{E}_x[h(t-s,B(s))]=\int_{\mathbb{R}^d} g(s,x,y)h(t-s,y)\,\mathrm{d}y.$$

By the bounded convergence theorem, it is easy to show that $\varphi(s,x)$ is bounded and continuous in x. Hence, $v(t,x)=\int_0^t\varphi(s,x)\,\mathrm{d}s$ is continuous. □

Proposition 3.4.5 *Assume that h is bounded. Then, $\frac{\partial v}{\partial x^i}$ exists and is given by*

$$\frac{\partial v}{\partial x^i}(t,x)=\int_0^t\int_{\mathbb{R}^d}\frac{\partial}{\partial x^i}g(s,x,y)h(t-s,y)\,\mathrm{d}s\mathrm{d}y. \tag{3.4.4}$$

Moreover, $\frac{\partial v}{\partial x^i}$ is continuous on $(0,\infty)\times\mathbb{R}^d$.

Proof Let $e_i=(\overbrace{0,\dots,0}^{i-1},1,0,\dots,0)\in\mathbb{R}^d$. For $\varepsilon\neq0$, by the observation in the Proof of Proposition 3.4.4,

$$\frac{1}{\varepsilon}(v(t,x+\varepsilon e_i)-v(t,x))=\int_0^t\varphi_{i,\varepsilon}(s,x)\,\mathrm{d}s, \tag{3.4.5}$$

where

$$\varphi_{i,\varepsilon}(s,x)=\int_{\mathbb{R}^d}\frac{1}{\varepsilon}\int_0^\varepsilon\frac{\partial}{\partial x^i}g(s,x+\xi e_i,y)h(t-s,y)\,\mathrm{d}\xi\mathrm{d}y.$$

By the Lebesgue convergence theorem, $\varphi_{i,\varepsilon}(s,x)$ converges as $\varepsilon\to0$. Moreover, since

$$\frac{\partial}{\partial x^i}g(s,x,y)=\frac{y^i-x^i}{s}g(s,x,y),$$

$\varphi_{i,\varepsilon}$ is represented as

$$\varphi_{i,\varepsilon}(s,x)=\frac{1}{\varepsilon}\int_0^\varepsilon\mathbf{E}_{x+\xi e_i}\Big[\frac{B^i(s)-x^i-\xi}{s}h(t-s,B(s))\Big]\mathrm{d}\xi.$$

Note that

$$|\varphi_i,\varepsilon(s)|\leqq\sup_{(t,x)\in(0,\infty)\times\mathbb{R}^d}|h(t,x)|\Big(\int_{\mathbb{R}}\frac{1}{\sqrt{2\pi}}|z|\mathrm{e}^{-\frac{|z|^2}{2}}\mathrm{d}z\Big)s^{-\frac{1}{2}}.$$

Hence, by the Lebesgue convergence theorem, the right hand side of (3.4.5) converges to $\int_0^t\mathbf{E}_x[\frac{B^i(s)-x^i}{s}h(t-s,B(s))]\mathrm{d}s$, which is the right hand side of (3.4.4), as $\varepsilon\to0$. □

To show that $v(t,x)$ has second derivatives in the space variable x, we assume the local Hölder continuity of $h(t,x)$ in x.

Proposition 3.4.6 *Let $h : (0,\infty) \times \mathbb{R}^d \to \mathbb{R}$ be a bounded and continuous function and assume that, for any $N > 0$, there exist positive constants α and C such that*

$$|h(t,x) - h(t,y)| \leqq C|x-y|^\alpha \qquad (|x|,|y|,t \leqq N). \tag{3.4.6}$$

(1) *There exists $\frac{\partial^2 v}{\partial x^i \partial x^j}$ which is continuous in (t,x) and given by*

$$\frac{\partial^2 v}{\partial x^i \partial x^j}(t,x) = \int_0^t \int_{\mathbb{R}^d} \frac{\partial^2}{\partial x^i \partial x^j} g(s,x,y) h(t-s,y)\, \mathrm{d}s\mathrm{d}y.$$

(2) *There exists $\frac{\partial v}{\partial t}$ and it satisfies*

$$\frac{\partial v}{\partial t}(t,x) = h(t,x) + \int_0^t \int_{\mathbb{R}^d} \frac{\partial}{\partial t} g(t-r,x,y) h(r,y)\, \mathrm{d}r\mathrm{d}y.$$

Proof (1) Write v_i and g_{ij} for $\frac{\partial v}{\partial x^i}$ and $\frac{\partial^2 g}{\partial x^i \partial x^j}$, respectively. Then we have

$$\frac{1}{\varepsilon}(v_i(t,x+\varepsilon e_j) - v_i(t,x)) = \int_0^t \varphi_{ij,\varepsilon}(s,x)\, \mathrm{d}s, \tag{3.4.7}$$

where

$$\varphi_{ij,\varepsilon}(s,x) = \int_{\mathbb{R}^d} \frac{1}{\varepsilon} \int_0^\varepsilon g_{ij}(s,x+\xi e_j,y) h(t-s,y)\, \mathrm{d}\xi \mathrm{d}y.$$

$\varphi_{ij,\varepsilon}(s,x)$ converges as $\varepsilon \to 0$ by the bounded convergence theorem. Hence, once we show that there exists an integrable function ϕ which is independent of ε and satisfies $|\varphi_{ij,\varepsilon}(s,x)| \leqq \phi(s)$, we obtain the assertion.

Noting that

$$\begin{aligned} 0 = \int_{\mathbb{R}^d} g_{ij}(s,x,y)\, \mathrm{d}y &= \int_{\mathbb{R}^d} \Big(\frac{(y^i - x^i)(y^j - x^j)}{s^2} - \frac{\delta_{ij}}{s} \Big) g(s,x,y)\, \mathrm{d}y \\ &= \mathbf{E}_x \Big[\frac{(B^i(s) - x^i)(B^j(s) - x^j)}{s^2} - \frac{\delta_{ij}}{s} \Big], \end{aligned}$$

we have

$$\begin{aligned} \varphi_{ij,\varepsilon}(s,x) = \frac{1}{\varepsilon} \int_0^\varepsilon \mathbf{E}_{x+\xi e_j} \Big[\Big(&\frac{(B^i(s) - x^i)(B^j(s) - x^j - \xi)}{s^2} - \frac{\delta_{ij}}{s} \Big) \\ &\times (h(t-s,B(s)) - h(t-s,x+\xi e_j)) \Big] \mathrm{d}\xi. \end{aligned}$$

To estimate $\varphi_{ij,\varepsilon}(s,x)$, we apply Schwarz's inequality to the expectation on the right hand side. First, by the scaling property of Brownian motion, observe that

$$\mathbf{E}_{x+\xi e_j} \Big[\Big(\frac{(B^i(s) - x^i)(B^j(s) - x^j - \xi)}{s^2} - \frac{\delta_{ij}}{s} \Big)^2 \Big] = (1 + \delta_{ij}) s^{-2}.$$

Next, by the assumption, there exist positive constants C_1, C_2, and C_3 such that

$$\begin{aligned}
&\mathbf{E}_{x+\xi e_j}[|h(t-s,B(s)) - h(t-s,x+\xi e_j)|^2] \\
&\leqq C_1\mathbf{E}_{x+\xi e_j}[|B(s)-x-\xi e_j|^{2\alpha}\mathbf{1}_{\{|B(s)-x-\xi e_j|\leqq N\}}] \\
&\qquad + C_2\mathbf{P}_{x+\xi e_j}(|B(s)-x-\xi e_j| \geqq N) \\
&= C_1\mathbf{E}_0[|B(s)|^{2\alpha}\mathbf{1}_{\{|B(s)|\leqq N\}}] + C_2\mathbf{P}_0(|B(s)| \geqq N) \\
&\leqq C_3 s^\alpha + C_2\mathbf{P}_0(|B(s)| \geqq N)
\end{aligned}$$

for $|x| \leqq N$. Using the elementary inequality

$$\int_\eta^\infty \mathrm{e}^{-\frac{\xi^2}{2}}\,\mathrm{d}\xi \leqq \frac{1}{\eta}\mathrm{e}^{-\frac{\eta^2}{2}} \qquad (\eta > 0),$$

we obtain

$$\begin{aligned}
\mathbf{P}_0(|B(s)| \geqq N) &\leqq \sum_{i=1}^d \mathbf{P}_0\Big(|B^i(s)| \geqq \frac{N}{\sqrt{d}}\Big) \\
&= 2d\int_{\frac{N}{\sqrt{d}}}^\infty \frac{1}{\sqrt{2\pi s}}\mathrm{e}^{-\frac{\eta^2}{2s}}\,\mathrm{d}\eta \leqq \frac{\sqrt{2d^3 s}}{\sqrt{\pi}N}\mathrm{e}^{-\frac{N^2}{2ds}}.
\end{aligned}$$

Thus there exists a constant C_4, independent of ε, such that

$$|\varphi_{ij,\varepsilon}(s,x)| \leqq C_4(s^{-1+\frac{\alpha}{2}} + s^{-\frac{3}{4}}\mathrm{e}^{-\frac{N^2}{4ds}})$$

for $s > 0$ and $|x| \leqq N$.

(2) The strategy of the proof is similar to that for (1). We start from

$$\frac{1}{\varepsilon}(v(t+\varepsilon,x) - v(t,x)) = \frac{1}{\varepsilon}\mathbf{E}_x\Big[\int_t^{t+\varepsilon} h(r,B(t+\varepsilon-r))\,\mathrm{d}r\Big] + I_\varepsilon, \tag{3.4.8}$$

where

$$I_\varepsilon = \int_0^t \frac{1}{\varepsilon}\mathbf{E}_x[h(r,B(t+\varepsilon-r)) - h(r,B(t-r))]\,\mathrm{d}r.$$

The first term of (3.4.8) converges to $h(t,x)$ by the bounded convergence theorem.

Rewrite I_ε as

$$\begin{aligned}
I_\varepsilon &= \int_0^t \mathrm{d}r\int_{\mathbb{R}^d} \frac{1}{\varepsilon}(g(t+\varepsilon-r,x,y) - g(t-r,x,y))h(r,y)\,\mathrm{d}y \\
&= \int_0^t \mathrm{d}r\int_{\mathbb{R}^d}\frac{1}{\varepsilon}\int_0^\varepsilon \frac{\partial}{\partial t}g(t-r+\xi,x,y)h(r,y)\,\mathrm{d}\xi\mathrm{d}y.
\end{aligned} \tag{3.4.9}$$

Since

$$\int_{\mathbb{R}^d} \frac{\partial}{\partial t} g(t,x,y)\,\mathrm{d}y = \int_{\mathbb{R}^d} \Big(\frac{|y-x|^2}{2t^2} - \frac{d}{2t} \Big) g(t,x,y)\,\mathrm{d}y$$
$$= \mathbf{E}_x\Big[\frac{|B(t)-x|^2 - dt}{2t^2} \Big] = 0,$$

we change the order of the integrations to obtain

$$I_\varepsilon = \int_0^t \mathrm{d}r \,\frac{1}{\varepsilon} \int_0^\varepsilon \mathrm{d}u \int_{\mathbb{R}^d} \frac{\partial}{\partial t} g(t-r+u,x,y)(h(r,y)-h(r,x))\,\mathrm{d}y$$
$$= \int_0^t \mathrm{d}r\,\frac{1}{\varepsilon} \int_0^\varepsilon \mathbf{E}_x\Big[\frac{|B(t-r+u)-x|^2 - d(t-r+u)}{2(t-r+u)^2} \times (h(r,B(t-r+u)) - h(r,x)) \Big] \mathrm{d}u.$$

By the local Hölder continuity of h, there exists an integrable function ϕ on $[0,t)$, independent of ε, such that

$$\Big| \frac{1}{\varepsilon} \int_0^\varepsilon \mathbf{E}_x\Big[\frac{|B(t-r+u)-x|^2 - d(t-r+u)}{2(t-r+u)^2} \times (h(r,B(t-r+u)) - h(r,x)) \Big] \mathrm{d}u \Big| \leqq \phi(r) \quad (r \in [0,t)).$$

Then apply the Lebesgue convergence theorem to the right hand side of (3.4.9) to obtain the assertion. The details are left to the reader. □

3.5 The Feynman–Kac Formula

Given $V, f : \mathbb{R}^d \to \mathbb{R}$, consider the problem of finding a function $u(t,x)$ of $C^{1,2}$-class on $[0,\infty) \times \mathbb{R}^d$ such that

$$\frac{\partial u}{\partial t}(t,x) = \frac{1}{2}\Delta u(t,x) - V(x)u(t,x) \quad ((t,x) \in (0,\infty) \times \mathbb{R}^d), \tag{3.5.1}$$

$$u(0,x) = f(x). \tag{3.5.2}$$

A Feynman path integral presents a representation for a solution to the Schrödinger equation $\frac{1}{\mathrm{i}}\frac{\partial \psi}{\partial t} = -\frac{1}{2}\Delta\psi + V\psi$ via a formal integral of paths. The representation for a solution to (3.5.1) and (3.5.2) via a Brownian motion is called the **Feynman–Kac formula**.

The approach taken in this section is quite similar to those in the previous two sections. Let $B = \{B(s)\}_{s \geqq 0}$ be a d-dimensional Brownian motion. The following can be easily shown as an application of Itô's formula.

Proposition 3.5.1 *If* $u : (0, \infty) \times \mathbb{R}^d \to \mathbb{R}$ *satisfies* (3.5.1), *then the stochastic process* $M = \{M(s)\}_{0 \leqq s < t}$ *defined by*

$$M(s) = u(t - s, B(s)) \exp\Big(-\int_0^s V(B(r))\,\mathrm{d}r\Big) \tag{3.5.3}$$

is a local martingale.

Consider the function $v(t, x)$ on $[0, \infty) \times \mathbb{R}^d$ defined by

$$v(t, x) = \mathbf{E}_x\Big[f(B(t)) \exp\Big(-\int_0^t V(B(r))\,\mathrm{d}r\Big)\Big]. \tag{3.5.4}$$

Proposition 3.5.2 *Assume that* f *and* V *are bounded. Then a bounded solution to* (3.5.1) *and* (3.5.2) *coincides with* v.

Proof Let u be a bounded solution to (3.5.1) and (3.5.2). Then the stochastic process M defined by (3.5.3) is a bounded martingale. Hence, by the martingale convergence theorem (Theorem 1.5.16), there exists $M(t) := \lim_{s \uparrow t} M(s)$ and we have

$$M(t) = f(B(t)) \exp\Big(-\int_0^t V(B(r))\,\mathrm{d}r\Big).$$

Hence, $u(t, x) = M(0) = \mathbf{E}_x[M(t)] = v(t, x)$. □

Proposition 3.5.3 *Assume that* f *and* V *are bounded. If* $v(t, x)$ *is of* $C^{1,2}$*-class, then* v *satisfies* (3.5.1) *for almost all* (t, x).

Proof By the Markov property of Brownian motion, we have

$$\begin{aligned}
&\mathbf{E}_x\Big[f(B(t)) \exp\Big(-\int_0^t V(B(r))\,\mathrm{d}r\Big)\Big|\mathscr{F}_s\Big] \\
&= \exp\Big(-\int_0^s V(B(r))\,\mathrm{d}r\Big)\mathbf{E}_x\Big[f(B(t)) \exp\Big(-\int_s^t V(B(r))\,\mathrm{d}r\Big)\Big|\mathscr{F}_s\Big] \\
&= \exp\Big(-\int_0^s V(B(r))\,\mathrm{d}r\Big)v(t - s, B(s)).
\end{aligned}$$

By the assumption and Itô's formula applied to the right hand side,

$$\begin{aligned}
&\mathbf{E}_x\Big[f(B(t)) \exp\Big(-\int_0^t V(B(r))\,\mathrm{d}r\Big)\Big|\mathscr{F}_s\Big] \\
&= v(t, x) + \sum_{i=1}^d \int_0^t \mathrm{e}^{-\int_0^s V(B(r))\mathrm{d}r} \frac{\partial v}{\partial x^i}(s, B(s))\,\mathrm{d}B^i(s) \\
&\quad + \int_0^t \Big\{-\frac{\partial v}{\partial t} + \frac{1}{2}\Delta v - Vv\Big\}(t - s, B(s))\mathrm{e}^{-\int_0^s V(B(r))\mathrm{d}r}\mathrm{d}s.
\end{aligned}$$

Since this is a martingale, the third term of the right hand side is 0. □

The convergence of $v(t, x)$ as $t \to 0$ is obtained by the bounded convergence theorem.

Proposition 3.5.4 *Assume that V and f are bounded and f is continuous. Then $v(t, x) \to f(x)$ as $t \to 0$ for any x.*

Summing up, under the conditions on V and f in Proposition 3.5.4, once it is shown that $v(t, x)$ is of $C^{1,2}$-class, v gives a probabilistic representation for the unique bounded solution to Equations (3.5.1) and (3.5.2).

To show the smoothness of v, note

$$e^{-\int_0^t V(B(r))dr} = 1 - \int_0^t e^{-\int_\rho^t V(B(r))dr} V(B(\rho))\,d\rho. \tag{3.5.5}$$

Multiply both sides by $f(B(t))$ and take the expectation. Then, by the Markov property of Brownian motion, we obtain

$$\begin{aligned} v(t, x) &= \mathbf{E}_x[f(B(t))] - \int_0^t \mathbf{E}_x\Big[\mathbf{E}_x\Big[f(B(t))e^{-\int_\rho^t V(B(r))dr}\Big|\mathscr{F}_\rho\Big]\Big]V(B(\rho))\,d\rho \\ &= \mathbf{E}_x[f(B(t))] - \mathbf{E}_x\Big[\int_0^t v(t-\rho, B(\rho))V(B(\rho))\,d\rho\Big] \end{aligned}$$

by Fubini's theorem.

If f is bounded and continuous, then $\mathbf{E}_x[f(B(t))]$ is of C^∞-class (Theorem 3.3.1). Since v and V are bounded, we can apply Propositions 3.4.4 and 3.4.5 to obtain the continuity and the existence of the first derivatives with respect to x^j for the second term of the right hand side. Therefore, $v(t, x)$ is of C^1-class with respect to the space variable and, in particular, locally Lipschitz continuous. Hence, by Proposition 3.4.6, we get the following.

Theorem 3.5.5 *Assume that V is bounded and locally Hölder continuous and f is bounded and continuous. Then, the function $v(t, x)$ defined by* (3.5.4) *is the unique bounded solution to Equations* (3.5.1) *and* (3.5.2).

Next we consider the time-independent Schrödinger equation. Let $u(t, x)$ be the solution to (3.5.1) and (3.5.2) and set

$$\psi_\alpha(x) = \int_0^\infty e^{-\alpha t} u(t, x)\,dt.$$

A formal computation yields

$$\frac{1}{2}\Delta\psi_\alpha = \frac{1}{2}\int_0^\infty e^{-\alpha t}\Delta u\, dt$$
$$= \int_0^\infty e^{-\alpha t}\Big(\frac{\partial u}{\partial t} + Vu\Big)dt = (\alpha + V)\psi_\alpha - f.$$

Thus ψ_α seems to satisfy

$$\frac{1}{2}\Delta\psi_\alpha = (\alpha + V)\psi_\alpha - f.$$

We investigate this equation when $d = 1$, following [56], and apply it to a proof of the arcsine law for the sojourn time of a Brownian motion. The expression, given in (3.5.7) below, of the solution is called **Kac's formula** [53].

Definition 3.5.6 Let $f : \mathbb{R} \to \mathbb{R}$ be a function which has left and right limits at every point. If there exists an increasing sequence $\{x_n\}_{n=-\infty}^\infty \subset \mathbb{R}$ with $\lim_{n\to-\infty} x_n = -\infty$, $\lim_{n\to\infty} x_n = \infty$ such that the restriction of f to (x_n, x_{n+1}) is continuous, then f is said to be **piecewise continuous**.

Moreover, if f is m-times differentiable and the derivatives of m-th order are piecewise continuous, f is said to be of **piecewise C^m-class**.

Theorem 3.5.7 *Suppose that f and V are piecewise continuous functions on $\mathbb{R}$ and V is non-negative. Moreover, suppose that there exists an $\alpha > 0$ satisfying*

$$\int_{-\infty}^\infty |f(x+y)|e^{-\sqrt{2\alpha}y}\, dy < \infty \tag{3.5.6}$$

for all x. Then, the function F defined by

$$F(x) = \mathbf{E}_x\Big[\int_0^\infty f(B(t))\exp\Big(-\alpha t - \int_0^t V(B(s))\, ds\Big)dt\Big] \tag{3.5.7}$$

is of piecewise C^2-class and, at every continuum x of both f and V,

$$\frac{1}{2}F''(x) = (\alpha + V(x))F(x) - f(x). \tag{3.5.8}$$

Remark 3.5.8 For $\alpha > 0$ and $x, y \in \mathbb{R}$, the following formula holds:

$$\int_0^\infty e^{-\alpha t}\frac{1}{\sqrt{2\pi t}}e^{-\frac{|y-x|^2}{2t}}\, dt = \frac{1}{\sqrt{2\alpha}}e^{-|y-x|\sqrt{2\alpha}}.$$

Hence, the condition (3.5.6) is equivalent to

$$\mathbf{E}_x\Big[\int_0^\infty e^{-\alpha t}|f(B(t))|\, dt\Big] < \infty \qquad (x \in \mathbb{R}).$$

Proof Define the resolvent operator G_α ($\alpha > 0$) by

$$(G_\alpha g)(x) = \mathbf{E}_x\Big[\int_0^\infty e^{-\alpha t} g(B(t))\,dt\Big],$$

which acts on piecewise continuous functions satisfying (3.5.6). By the preceding remark,

$$\begin{aligned}(G_\alpha g)(x) &= \int_0^\infty e^{-\alpha t} dt \int_{-\infty}^\infty g(y)\frac{1}{\sqrt{2\pi t}} e^{-\frac{|y-x|^2}{2t}}\,dy \\ &= \frac{1}{\sqrt{2\alpha}}\int_{-\infty}^\infty e^{-\sqrt{2\alpha}|y-x|} g(y)\,dy \\ &= \frac{1}{\sqrt{2\alpha}} e^{-\sqrt{2\alpha}x}\int_{-\infty}^x e^{\sqrt{2\alpha}y} g(y)\,dy + \frac{1}{\sqrt{2\alpha}} e^{\sqrt{2\alpha}x}\int_x^\infty e^{-\sqrt{2\alpha}y} g(y)\,dy,\end{aligned}$$

which implies the continuity of $(G_\alpha g)(x)$. Moreover, since g is piecewise continuous, $G_\alpha g$ is differentiable in x and

$$(G_\alpha g)'(x) = -e^{-\sqrt{2\alpha}x}\int_{-\infty}^x e^{\sqrt{2\alpha}y} g(y)\,dy + e^{\sqrt{2\alpha}x}\int_x^\infty e^{-\sqrt{2\alpha}y} g(y)\,dy.$$

Hence $(G_\alpha g)'$ is continuous. Similarly $(G_\alpha g)'$ is also differentiable and, at the continuum x of g,

$$(G_\alpha g)''(x) = 2\alpha (G_\alpha g)(x) - 2g(x). \tag{3.5.9}$$

Next we show $G_\alpha(VF) = G_\alpha f - F$. By (3.5.5),

$$\begin{aligned}(G_\alpha f)(x) - F(x) &= \mathbf{E}_x\Big[\int_0^\infty e^{-\alpha t}\big(1 - e^{-\int_0^t V(B(s))ds}\big) f(B(t))\,dt\Big] \\ &= \mathbf{E}_x\Big[\int_0^\infty e^{-\alpha t} f(B(t))\,dt \int_0^t e^{-\int_s^t V(B(r))dr} V(B(s))\,ds\Big].\end{aligned}$$

Since V is non-negative and

$$\Big|\int_0^t e^{-\int_s^t V(B(r))dr} V(B(s))\,ds\Big| = 1 - e^{-\int_0^t V(B(s))ds} < 1,$$

we have

$$\mathbf{E}_x\Big[\int_0^\infty e^{-\alpha t} |f(B(t))|\,dt \Big|\int_0^t e^{-\int_s^t V(B(r))dr} V(B(s))\,ds\Big|\Big] < \infty,$$

as was remarked in Remark 3.5.8. Hence, changing the order of integrations by Fubini's theorem, we obtain

$$(G_\alpha f)(x) - F(x) = \mathbf{E}_x\Big[\int_0^\infty V(B(s))\mathrm{d}s \int_s^\infty f(B(t))\mathrm{e}^{-\alpha t - \int_s^t V(B(r))\mathrm{d}r}\mathrm{d}t\Big]$$
$$= \mathbf{E}_x\Big[\int_0^\infty \mathrm{e}^{-\alpha s} V(B(s))\,\mathrm{d}s \int_0^\infty f(B(s+\rho))\mathrm{e}^{-\alpha\rho - \int_0^\rho V(B(s+r))\mathrm{d}r}\mathrm{d}\rho\Big].$$

The Markov property of Brownian motion implies

$$(G_\alpha f)(x) - F(x) = \mathbf{E}_x\Big[\int_0^\infty \mathrm{e}^{-\alpha s} V(B(s))F(B(s))\,\mathrm{d}s\Big] = G_\alpha(VF)(x).$$

Applying (3.5.9) with $g = f - VF$, we obtain

$$F''(x) = (G_\alpha(f - VF))''(x) = 2\alpha(G_\alpha(f - VF))(x) - 2(f - VF)(x)$$
$$= 2\alpha F(x) - 2f(x) + 2V(x)F(x)$$

if f and V are continuous at x. □

Example 3.5.9 (Arcsine law) Let $B = \{B(t)\}_{t\geqq 0}$ be a Brownian motion and $\Gamma_+(t)$ be the total time when B stays in $(0,\infty)$ up to time t:

$$\Gamma_+(t) = \int_0^t \mathbf{1}_{(0,\infty)}(B(s))\,\mathrm{d}s.$$

Then, for any positive α and $\beta > 0$,

$$\int_0^\infty \mathrm{e}^{-\alpha t}\mathbf{E}_0\big[\mathrm{e}^{-\beta\Gamma_+(t)}\big]\mathrm{d}t = \frac{1}{\sqrt{\alpha(\alpha+\beta)}},$$
$$\mathbf{P}_0(\Gamma_+(t) \leqq \alpha) = \int_0^{\frac{\alpha}{t}} \frac{\mathrm{d}s}{\pi\sqrt{s(1-s)}} = \frac{2}{\pi}\arcsin\sqrt{\frac{\alpha}{t}} \qquad (0 \leqq \alpha \leqq t).$$

Proof Apply Kac's formula for $f = 1$ and $V(x) = \beta\mathbf{1}_{(0,\infty)}(x)$. The equation (3.5.8) to solve is written as

$$F''(x) = \begin{cases} 2(\alpha F(x) - 1) & (x < 0) \\ 2(\alpha F(x) + \beta - 1) & (x \geqq 0). \end{cases}$$

This equation has a unique solution of piecewise C^2-class, which is given by

$$F(x) = \begin{cases} \frac{\sqrt{\alpha}-\sqrt{\alpha+\beta}}{\alpha\sqrt{\alpha+\beta}}\mathrm{e}^{\sqrt{2\alpha}x} + \frac{1}{\alpha} & (x < 0) \\ \frac{\sqrt{\alpha+\beta}-\sqrt{\alpha}}{(\alpha+\beta)\sqrt{\alpha}}\mathrm{e}^{-\sqrt{2(\alpha+\beta)}x} + \frac{1}{\alpha+\beta} & (x \geqq 0), \end{cases}$$

and we obtain

$$\int_0^\infty \mathrm{e}^{-\alpha t}\mathbf{E}_0\big[\mathrm{e}^{-\beta\Gamma_+(t)}\big]\mathrm{d}t = F(0) = \frac{1}{\sqrt{\alpha(\alpha+\beta)}}.$$

On the other hand, by the elementary formula

$$\int_0^\infty \frac{\mathrm{e}^{-ct}}{\sqrt{t}}\,\mathrm{d}t = \frac{2}{\sqrt{c}}\int_0^\infty \mathrm{e}^{-s^2}\mathrm{d}s = \sqrt{\frac{\pi}{c}} \qquad (c > 0),$$

we have

$$\begin{aligned}\frac{1}{\sqrt{\alpha(\alpha+\beta)}} &= \frac{1}{\pi}\int_0^\infty \frac{\mathrm{e}^{-(\alpha+\beta)s}}{\sqrt{s}}\,\mathrm{d}s\int_0^\infty \frac{\mathrm{e}^{-\alpha t}}{\sqrt{t}}\,\mathrm{d}t \\ &= \frac{1}{\pi}\int_0^\infty \frac{\mathrm{e}^{-(\alpha+\beta)s}}{\sqrt{s}}\,\mathrm{d}s\int_s^\infty \frac{\mathrm{e}^{-\alpha(t-s)}}{\sqrt{t-s}}\,\mathrm{d}t = \int_0^\infty \mathrm{e}^{-\alpha t}\mathrm{d}t\int_0^t \frac{\mathrm{e}^{-\beta s}}{\pi\sqrt{s(t-s)}}\,\mathrm{d}s.\end{aligned}$$

Hence, the uniqueness of Laplace transforms implies

$$\mathbf{E}_0[\mathrm{e}^{-\beta\Gamma_+(t)}] = \int_0^t \frac{\mathrm{e}^{-\beta s}}{\pi\sqrt{s(t-s)}}\,\mathrm{d}s \quad \text{and} \quad \mathbf{P}_0(\Gamma_+(t)\in \mathrm{d}s) = \frac{1}{\pi\sqrt{s(t-s)}}\,\mathrm{d}s. \quad \square$$

The **Kato class** is a natural class of functions which gives the scalar potentials of Schrödinger operators. Its characterization by means of Brownian motion is known.

Definition 3.5.10 Define a function ϕ on $\mathbb{R}^d$ by $\phi(z) = 1$ if $d = 1$, $\phi(z) = -\log|z|$ if $d = 2$ and $\phi(z) = |z|^{2-d}$ if $d \geqq 3$. The Kato class K_d is the set of functions f satisfying

$$\int_0^t \mathbf{E}_x[|f(B(s))|]\,\mathrm{d}s = \int_0^t\int_{\mathbb{R}^d} g(s,x,y)|f(y)|\,\mathrm{d}s\mathrm{d}y < \infty \quad (t > 0) \qquad (3.5.10)$$

and

$$\lim_{\alpha\downarrow 0}\sup_{x\in\mathbb{R}^d}\int_{|y-x|<\alpha}\phi(|y-x|)|f(y)|\,\mathrm{d}y = 0. \qquad (3.5.11)$$

Suppose that V is the difference of two non-negative functions V_+ and V_-, $V = V_+ - V_-$. If $V_- \in K_d$ and $\varphi V_+ \in K_d$ for any $\varphi \in C_0^\infty(\mathbb{R}^d)$, the **semigroup**[1] T_t generated by the Schrödinger operator $-\frac{1}{2}\Delta + V$ defines a bounded operator from $L^p(\mathbb{R}^d)$ into $L^q(\mathbb{R}^d)$ for any p, q with $1 \leqq p \leqq q \leqq \infty$. Moreover, under the same conditions, the semigroup T_t has an integral kernel $p_V(t,x,y)$, which is continuous in $(t,x,y) \in (0,\infty)\times\mathbb{R}^d\times\mathbb{R}^d$. For details, see [2, 10, 106, 108].

We end this section by showing the characterization of the functions in the Kato class K_d via Brownian motion.

[1] For Equations (3.5.1) and (3.5.2), denote the map $f \mapsto u$ by $T_t : u(t,x) = (T_t f)(x)$. Then, $T_t \circ T_s = T_{s+t}$ $(s,t > 0)$.

Theorem 3.5.11 *A function f on $\mathbb{R}^d$ satisfying* (3.5.10) *is of the Kato class K_d if and only if*

$$\lim_{t\downarrow 0}\sup_{x\in\mathbb{R}^d}\mathbf{E}_x\Big[\int_0^t |f(B(s))|\,\mathrm{d}s\Big]=0. \tag{3.5.12}$$

Proof We give a proof when $d \geqq 3$. The proof for the cases when $d = 1$ and 2 can be done in a similar way. For details, see [2].

First let $f \in K_d$. Note that

$$\frac{\partial}{\partial s}g(s,x,y)=\frac{|y-x|^2-ds}{2s^2}g(s,x,y).$$

Since $\frac{\partial g}{\partial s}<0$ if $|y-x|\geqq\alpha$ and $s<\frac{\alpha^2}{d}$, we have

$$g(s,x,y)\leqq g(t,x,y)\qquad\Big(0<s\leqq t\leqq\frac{\alpha^2}{d},\ \ |y-x|\geqq\alpha\Big).$$

Hence, if $t\leqq\frac{\alpha^2}{d}$, then

$$\begin{aligned}\int_0^t \mathbf{E}_x[|f(B(s))|\mathbf{1}_{\{|B(s)-x|\geqq\alpha\}}]\,\mathrm{d}s &= \int_0^t \mathrm{d}s\int_{|y-x|\geqq\alpha}|f(y)|g(s,x,y)\,\mathrm{d}y\\ &\leqq t\int_{|y-x|\geqq\alpha}|f(y)|g(t,x,y)\,\mathrm{d}y.\end{aligned}$$

Moreover, by (3.5.11), there exists a constant $\beta>0$ such that

$$C:=\sup_{x\in\mathbb{R}^d}\int_{|y-x|<\beta}|f(y)|\,\mathrm{d}y<\infty.$$

Let $\alpha<\beta$. For $\boldsymbol{n}=(n_1,n_2,\dots,n_d)\in\mathbb{Z}^d$, define

$$I(\boldsymbol{n})=\Big\{z=(z^1,z^2,\dots,z^d)\,;\ n_i\frac{\alpha}{2}\leqq z^i\leqq(n_i+1)\frac{\alpha}{2},\ i=1,2,\dots,d\Big\}$$

and $|\boldsymbol{n}|=\sum_{i=1}^d|n_i|$. Denote by $I(\boldsymbol{n};x)$ the cube which is obtained by shifting $I(\boldsymbol{n})$ by x, and by $p(\boldsymbol{n})$ the vertex of $I(\boldsymbol{n})$ which is nearest to the origin. Then

$$\begin{aligned}\int_0^t \mathbf{E}_x[|f(B(s))|\mathbf{1}_{\{|B(s)-x|\geqq\alpha\}}]\,\mathrm{d}s &\leqq t\sum_{|\boldsymbol{n}|\geqq 2}\int_{\{|y-x|\geqq\alpha\}\cap I(\boldsymbol{n};x)}|f(y)|g(t,x,y)\,\mathrm{d}y\\ &\leqq Ct\sum_{|\boldsymbol{n}|\geqq 2}\max\Big\{g\Big(\frac{\alpha^2}{d},0,z\Big)\,;\ z\in I(\boldsymbol{n})\Big\}\leqq Ct\sum_{|\boldsymbol{n}|\geqq 2}g\Big(\frac{\alpha^2}{d},0,p(\boldsymbol{n})\Big).\end{aligned}$$

Since $\sum_{|\boldsymbol{n}|\geqq 2}g(\frac{\alpha^2}{d},0,p(\boldsymbol{n}))<\infty$, letting $t\to 0$, we obtain

$$\sup_{x\in\mathbb{R}^d}\int_0^t \mathbf{E}_x[|f(B(s))|\mathbf{1}_{\{|B(s)-x|\geqq\alpha\}}]\,\mathrm{d}s\to 0.$$

On the other hand, since

$$\int_0^\infty g(s,x,y)\,\mathrm{d}s = C_d\phi(y-x),$$

C_d being $(2\pi^{\frac{d}{2}})^{-1}\Gamma(\frac{d-2}{2})$, we have

$$\begin{aligned}\int_0^t \mathbf{E}_x[|f(B(s))|\mathbf{1}_{\{|B(s)-x|<\alpha\}}]\,\mathrm{d}s &\leqq \int_0^\infty \mathrm{d}s \int_{|y-x|<\alpha} |f(y)|g(s,x,y)\,\mathrm{d}y \\ &= C_d \int_{|y-x|<\alpha} |f(y)|\phi(y-x)\,\mathrm{d}y. \qquad (3.5.13)\end{aligned}$$

Fix any $\varepsilon > 0$ and choose $\alpha > 0$ so that the right hand side of (3.5.13) is less than ε for any x. Then, letting $t \to 0$, we obtain

$$\limsup_{t\downarrow 0} \sup_{x\in\mathbb{R}^d} \mathbf{E}_x\Big[\int_0^t |f(B(s))|\,\mathrm{d}s\Big] < \varepsilon,$$

which gives (3.5.12).

Conversely, suppose that f satisfies (3.5.12). Note

$$\int_0^t \frac{1}{(2\pi s)^{\frac{d}{2}}}\mathrm{e}^{-\frac{|z|^2}{2s}}\,\mathrm{d}s = \frac{1}{|z|^{d-2}}\int_0^{\frac{t}{|z|^2}} \frac{1}{(2\pi r)^{\frac{d}{2}}}\mathrm{e}^{-\frac{1}{2r}}\,\mathrm{d}r.$$

Then, it is easy to see that

$$C'\phi(z) \leqq \int_0^t \frac{1}{(2\pi s)^{\frac{d}{2}}}\mathrm{e}^{-\frac{|z|^2}{2s}}\,\mathrm{d}s$$

if $|z| \leqq \sqrt{t}$, where $C' = \int_0^1 \frac{1}{(2\pi r)^{\frac{d}{2}}}\mathrm{e}^{-\frac{1}{2r}}\,\mathrm{d}r$. Hence, we obtain

$$\begin{aligned}&\sup_{x\in\mathbb{R}^d}\int_{|y-x|<\alpha} |f(y)|\phi(y-x)\,\mathrm{d}y \\ &\leqq \frac{1}{C'}\sup_{x\in\mathbb{R}^d}\int_{|y-x|<\alpha}|f(y)|\,\mathrm{d}y \int_0^{\alpha^2}\frac{1}{(2\pi s)^{\frac{d}{2}}}\mathrm{e}^{-\frac{|y-x|^2}{2s}}\,\mathrm{d}s \\ &= \frac{1}{C'}\sup_{x\in\mathbb{R}^d}\mathbf{E}_x\Big[\int_0^{\alpha^2}|f(B(s))|\mathbf{1}_{\{|B(s)-x|<\alpha\}}\mathrm{d}s\Big] \leqq \frac{1}{C'}\sup_{x\in\mathbb{R}^d}\mathbf{E}_x\Big[\int_0^{\alpha^2}|f(B(s))|\,\mathrm{d}s\Big].\end{aligned}$$

By the assumptions, the last term converges to 0 as $\alpha \to 0$. □

3.6 The Dirichlet Problem

We used some special harmonic functions to show the criteria for the recurrence of Brownian motions in Section 3.2. In this section we consider the problem (called the **Dirichlet problem**) of finding a function which is harmonic in the interior of an open set in $\mathbb{R}^d$ and coincides with a given function on the boundary. The expression for the solution via a Brownian motion is shown. Such potential theoretic observation was originated by Kakutani and Wiener.

Let D be an open set in $\mathbb{R}^d$. Given a function f on the boundary ∂D, consider the problem of finding $u \in C^2(D) \cap C(\overline{D})$ such that

$$\Delta u(x) = 0 \qquad (x \in D), \tag{3.6.1}$$

$$u(y) = f(y) \qquad (y \in \partial D). \tag{3.6.2}$$

Let τ be the exit time of the Brownian motion B from $D : \tau = \inf\{t > 0; B(t) \notin D\}$. In the rest of this section, we assume that

$$\mathbf{P}_x(\tau < \infty) = 1$$

for all $x \in D$ unless mentioned otherwise. This condition is fulfilled if D is a bounded domain.

First we show **Koebe's theorem**, which says that a function u is harmonic if and only if u has the mean-value property. We set, for $a \in \mathbb{R}^d$ and $r > 0$,

$$B(a,r) := \{x; |x-a| < r\} \quad \text{and} \quad S(a,r) := \{x; |x-a| = r\}.$$

Definition 3.6.1 $u : D \to \mathbb{R}$ is said to have the **mean-value property** if

$$u(a) = \int_{S(a,r)} u(y)\mu_{a,r}(\mathrm{d}y)$$

holds for all $a \in D$ and $r > 0$ with $\overline{B(a,r)} \subset D$, where $\mu_{a,r}$ is the uniform probability measure on the sphere $S(a,r)$.

Remark 3.6.2 u has the mean-value property if and only if

$$u(a) = \frac{1}{V_r}\int_{B(a,r)} u(y)\,\mathrm{d}y$$

holds for all $a \in D$ and $r > 0$ with $\overline{B(a,r)} \subset D$, where $V_r = \frac{2r^d\pi^{\frac{d}{2}}}{d\Gamma(\frac{d}{2})}$ is the volume of $B(a,r)$.

Proposition 3.6.3 *Let $u \in C^2(D) \cap C(\overline{D})$ be harmonic in D. Then u has the mean-value property.*

Proof Let $a \in D$, $\overline{B(a,r)} \subset D$ and $\tau_r = \inf\{t > 0; |B(t) - a| \geqq r\}$. Then, by Itô's formula,

$$u(B(t \wedge \tau_r)) - u(B(0)) = \sum_{i=1}^{d} \int_0^{t\wedge\tau_r} \frac{\partial u}{\partial x^i}(B(s))\,\mathrm{d}B^i(s).$$

Since $\frac{\partial u}{\partial x^i}$ is bounded and continuous on $B(a,r)$, the expectation under $\mathbf{P}_a$ of the right hand side is 0 and $\mathbf{E}_a[u(B(t \wedge \tau_r))] = u(a)$. Letting $t \uparrow \infty$, by the bounded convergence theorem we obtain

$$u(a) = \mathbf{E}_a[u(B(\tau_r))].$$

This shows the mean-value property of u because the distribution of $B(\tau_r)$ under $\mathbf{P}_a$ is the uniform distribution on $S(a,r)$. □

Corollary 3.6.4 *If u is harmonic in D and there exists $a \in D$ such that*

$$u(a) = \sup_{x\in D} u(x),$$

then u is a constant function.

Proof Take $r > 0$ so that $\overline{B(a,r)} \subset D$. Then, by Proposition 3.6.3,

$$u(a) = \int_{S(a,r)} u(y)\mu_{a,r}(\mathrm{d}y).$$

This means $u(y) = u(a)$ for any $y \in S(a,r)$. □

The converse of Proposition 3.6.3 holds.

Proposition 3.6.5 *If an $\mathbb{R}$-valued function on D satisfies the mean-value property, it is of C^∞-class and harmonic.*

We omit the proof. See, for example, [56, Chapter 4].

We show that the solution of the Dirichlet problem (3.6.1) and (3.6.2) is represented as an expectation with respect to Brownian motion. The following is easily shown by using Itô's formula.

Proposition 3.6.6 *Let $u \in C(\overline{D})$ be harmonic in D. If $a \in D$, then $\{u(B(t \wedge \tau))\}_{t\geqq 0}$ is a local martingale under $\mathbf{P}_a$.*

For a bounded function $f : \partial D \to \mathbb{R}$, define a function v on D by

$$v(x) = \mathbf{E}_x[f(B(\tau))].$$

Proposition 3.6.7 *Suppose that f is bounded. If u is a bounded solution to* (3.6.1) *and* (3.6.2), *then $u = v$.*

Proof For $n = 1, 2, \ldots$, set $D_n = \left\{x \in D; \inf_{y \in \partial D} |x - y| > n^{-1}\right\}$ and

$$\tau_n = \inf\{t; B(t) \notin D_n \text{ or } |B(t)| > n\}.$$

Then, by Itô's formula,

$$u(B(t \wedge \tau_n)) = u(B(0)) + \sum_{i=1}^{d} \int_0^{t \wedge \tau_n} \frac{\partial u}{\partial x^i}(B(s))\, \mathrm{d}B^i(s)$$

and

$$u(a) = \mathbf{E}_a[u(B(t \wedge \tau_n))].$$

Letting $t \to \infty$ and $n \to \infty$, by the bounded convergence theorem we obtain

$$u(a) = \mathbf{E}_a[u(B(\tau))] = \mathbf{E}_a[f(B(\tau))] = v(a).$$

□

Proposition 3.6.8 *If f is bounded, then v is of C^∞-class and harmonic.*

Proof For $x \in D$ and $\delta > 0$ so that $\overline{B(x, \delta)} \subset D$, set

$$\tau_\delta = \inf\{t > 0; B(t) \notin B(x, \delta)\}.$$

Then the distribution of $B(\tau_\delta)$ under $\mathbf{P}_x$ is the uniform distribution on $S(x, \delta)$. Hence, by the strong Markov property of Brownian motion,

$$v(x) = \mathbf{E}_x[\mathbf{E}_{B(\tau_\delta)}[f(B(\tau))]] = \mathbf{E}_x[v(B(\tau_\delta))] = \int_{S(x,\delta)} v(y) \mu_{x,\delta}(\mathrm{d}y).$$

This means that v has the mean-value property and Proposition 3.6.5 implies the assertion. □

For v to satisfy the boundary condition (3.6.2), some condition on ∂D is necessary.

Definition 3.6.9 Let D be an open set in $\mathbb{R}^d$ and τ be the same exit time as above. $y \in \partial D$ is called a **regular point**, if $\mathbf{P}_y(\tau = 0) = 1$.

Since $\{\tau = 0\} \in \mathscr{F}_0$, its probability is 0 or 1 by the Blumenthal 0-1 law.

Proposition 3.6.10 *Assume that f is bounded and continuous and $y \in \partial D$ is a regular point. Then, for any sequence $\{x_n\}_{n=1}^\infty \subset D$ with $x_n \to y \in \partial D$ $(n \to \infty)$, $v(x_n) \to f(y)$ $(n \to \infty)$.*

For the proof, we need a lemma.

Lemma 3.6.11 *Fix $t > 0$. Then, the function $D \ni x \mapsto \mathbf{P}_x(\tau \leqq t)$ is lower-semicontinuous, that is, if $x_n \to y \in \overline{D}$,*

$$\liminf_{n\to\infty} \mathbf{P}_{x_n}(\tau \leqq t) \geqq \mathbf{P}_y(\tau \leqq t).$$

Proof For any $\varepsilon > 0$, by the Markov property

$$\begin{aligned}&\mathbf{P}_x(\text{there exists an } s \in (\varepsilon, t] \text{ such that } B(s) \notin D)\\ &\qquad = \int_{\mathbb{R}^d} g(\varepsilon, x, z)\mathbf{P}_z(\tau \leqq t - \varepsilon)\,\mathrm{d}z. \qquad (3.6.3)\end{aligned}$$

It is easy to see that the right hand side is continuous in x by the Lebesgue convergence theorem. Hence, the left hand side is also continuous in x.

Since

$$\mathbf{P}_{x_n}(\text{there exists an } s \in (\varepsilon, t] \text{ such that } B(s) \notin D) \leqq \mathbf{P}_{x_n}(\tau \leqq t),$$

by the continuity observed above,

$$\mathbf{P}_y(\text{there exists an } s \in (\varepsilon, t] \text{ such that } B(s) \notin D) \leqq \liminf_{n\to\infty} \mathbf{P}_{x_n}(\tau \leqq t).$$

We obtain the assertion by letting $\varepsilon \downarrow 0$. □

Proof of Proposition 3.6.10. Since f is bounded and continuous, it suffices to show that, for any sequence $\{x_n\}_{n=1}^{\infty} \subset D$ with $x_n \to y \in \partial D$ $(n \to \infty)$,

$$\lim_{n\to\infty} \mathbf{P}_{x_n}(B(\tau) \in B(y, \delta)) = 1 \qquad (3.6.4)$$

holds for any $\delta > 0$.

For this purpose, fix arbitrary $\varepsilon > 0$ and take $t > 0$ such that

$$\mathbf{P}_0\Big(\max_{0\leqq s\leqq t} |B(s)| > \frac{\delta}{2}\Big) < \varepsilon.$$

Then, for $x_n \in B(y, \frac{\delta}{2})$, we have

$$\begin{aligned}\mathbf{P}_{x_n}(B(\tau) \in B(y, \delta)) &\geqq \mathbf{P}_{x_n}\Big(\tau \leqq t, \max_{0\leqq s\leqq t} |B(s) - x_n| \leqq \frac{\delta}{2}\Big)\\ &\geqq \mathbf{P}_{x_n}(\tau \leqq t) - \mathbf{P}_{x_n}\Big(\max_{0\leqq s\leqq t} |B(s) - x_n| > \frac{\delta}{2}\Big)\\ &> \mathbf{P}_{x_n}(\tau \leqq t) - \varepsilon.\end{aligned}$$

By Lemma 3.6.11 and the assumption,

$$\liminf_{n\to\infty} \mathbf{P}_{x_n}(\tau \leqq t) \geqq \mathbf{P}_y(\tau \leqq t) = 1.$$

Hence $\mathbf{P}_{x_n}(\tau \leqq t) \to 1$ and

$$\liminf_{n\to\infty} \mathbf{P}_{x_n}(B(\tau) \in B(y, \delta)) \geqq 1 - \varepsilon.$$

Since ε is arbitrary, we get (3.6.4). □

As a sufficient condition for a boundary point to be regular, the **Poincaré cone condition** is well known. For $a \in \mathbb{R}^d$, $\theta \in (0, \pi)$ and $b \in \mathbb{R}^d$, the set

$$V_{a,b,\theta} = \{x \in \mathbb{R}^d;\ \langle x - a, b\rangle \geqq |x - a| \cdot |b| \cos\theta\}$$

is called a **cone** with vertex a, direction b and aperture θ.

Proposition 3.6.12 *$y \in \partial D$ is a regular point if there exist a cone V with vertex y and a positive constant r such that $V \cap B(y, r) \subset D^c$.*

Proof The scaling property (Theorem 1.2.8) of the probability law of Brownian motion implies that $\mathbf{P}_y(B(t) \in V)$ is independent of t. We denote it by γ. If t is sufficiently small, we have

$$\mathbf{P}_y\Big(B(t) \in V, \max_{0 \leqq s \leqq t} |B(s) - y| < r\Big) > \frac{\gamma}{2}.$$

Hence, by the assumption,

$$\liminf_{t \downarrow 0} \mathbf{P}_y(B(t) \notin D) \geqq \frac{\gamma}{2}.$$

Noting that $\mathbf{P}_y(\tau \leqq t) \geqq \mathbf{P}_y(B(t) \notin D)$, we obtain

$$\mathbf{P}_y(\tau = 0) = \liminf_{t \downarrow 0} \mathbf{P}_y(\tau \leqq t) \geqq \liminf_{t \downarrow 0} \mathbf{P}_y(B(t) \notin D) \geqq \frac{\gamma}{2}.$$

Hence, by the Blumenthal 0-1 law, $\mathbf{P}_y(\tau = 0) = 1$. □

As an example of non-regular points, "Lebesgue's thorn" is well known (see, for example, [56, Section 4.2]). Moreover, the criterion by Wiener via the Newtonian capacity is widely known. For this, we refer the reader to [50, Section 7.14] and [94, Section 4.2].

What we have shown in this section is the following.

Theorem 3.6.13 *Let D be an open set in $\mathbb{R}^d$. Assume that $\mathbf{P}_x(\tau < \infty) = 1$ for any $x \in D$ and each point on ∂D is regular. Then, for a bounded and continuous function f on ∂D, $v(x) = \mathbf{E}_x[f(B(\tau))]$ is the unique solution to the Dirichlet problem* (3.6.1) *and* (3.6.2).

When there exists an x such that $\mathbf{P}_x(\tau < \infty) < 1$, a solution to the Dirichlet problem is not unique.

Proposition 3.6.14 *Assume that each point on ∂D is regular and f is bounded and continuous. If there exists an $a \in D$ such that $\mathbf{P}_a(\tau < \infty) < 1$, then the solution to the Dirichlet problem to* (3.6.1) *and* (3.6.2) *is not unique.*

Proof Set $h(x) = \mathbf{P}_x(\tau = \infty)$ and let τ_r be the exit time from $B(x, r)$ of a Brownian motion B. If $\overline{B(x, r)} \subset D$, then the strong Markov property of Brownian motion implies

$$h(x) = \mathbf{E}_x[\mathbf{P}_{B(\tau_r)}(\tau = \infty)] = \mathbf{E}_x[h(B(\tau_r))].$$

Hence, h has the mean-value property and harmonic in D. Moreover, as was shown in the proof of Proposition 3.6.10, $\lim_{D\ni x\to y} \mathbf{P}_x(\tau \leqq t) = 1$ for any $t > 0$ and $\mathbf{P}_y(\tau = \infty) = 0$ for any $y \in \partial D$. $h \not\equiv 0$ by the assumption and h is a solution to the problem (3.6.1) and (3.6.2) for $f \equiv 0$. However, h is different from the obvious solution $v \equiv 0$. □

The non-uniqueness of the solution is caused only by h ([94, Section 2]).

Theorem 3.6.15 *Assume that f is bounded and continuous. If u is a bounded solution to the Dirichlet problem* (3.6.1) *and* (3.6.2)*, then there exists a constant C such that*

$$u(x) = \mathbf{E}_x[f(B(\tau))] + C \cdot \mathbf{P}_x(\tau = \infty).$$

We omit the proof.

Example 3.6.16 (Poisson integral formula on the upper half space) Let $d \geqq 2$ and D be the upper half space of $\mathbb{R}^d : D = \{(x^1, \ldots, x^{d-1}, x^d); x^d > 0\}$. If f is a bounded and continuous function on ∂D, then the solution to the Dirichlet problem is given by

$$u(x) = \frac{\Gamma(\frac{d}{2})}{\pi^{\frac{d}{2}}} \int_{\partial D} \frac{x^d}{|y - x|^d} f(y)\, dy.$$

Proof Let $B = \{B(t)\}_{t\geqq 0}$ be a Brownian motion and set $\tau = \inf\{t > 0; B(t) \notin D\}$. Then, since $\tau = \inf\{t > 0; B^d(t) = 0\}$, we have

$$\mathbf{P}_x(\tau \in dt) = \frac{x^d}{\sqrt{2\pi t^3}} e^{-\frac{(x^d)^2}{2t}}\, dt$$

by Corollary 3.1.8. Since $\{(B^1(t), \ldots, B^{d-1}(t))\}_{t\geqq 0}$ and $\{B^d(t)\}_{t\geqq 0}$ are independent, setting $x = (x', x^d)$ and identifying ∂D with $\mathbb{R}^{d-1}$, we obtain

$$\begin{aligned} u(x) = \mathbf{E}_x[f(B(\tau))] &= \int_{\mathbb{R}^{d-1}} \int_0^\infty \frac{1}{(2\pi t)^{\frac{d-1}{2}}} e^{-\frac{|\xi - x'|^2}{2t}} \frac{x^d}{\sqrt{2\pi t^3}} e^{-\frac{(x^d)^2}{2t}} f(\xi)\, d\xi dt \\ &= \int_{\partial D} \int_0^\infty \frac{x^d}{(2\pi)^{\frac{d}{2}} t^{\frac{d+2}{2}}} e^{-\frac{|y-x|^2}{2t}} f(y)\, dy dt. \end{aligned}$$

Carrying out the integration in t yields the conclusion. □

The Dirichlet problem inside a ball is solved by using the result in Example 3.6.16. For details, see [56, 94].

Example 3.6.17 (Poisson integral formula for a ball) Let $d \geqq 2$ and set $D = B(0, r) = \{x \in \mathbb{R}^d; |x| < r\}$. Then, for a bounded and continuous function f on $\partial B(0, r)$, the unique solution to the Dirichlet problem is given by

$$u(x) = r^{d-2}(r^2 - |x|^2) \int_{\partial B(0,r)} \frac{1}{|y - x|^d} f(y) \mu_{0,r}(\mathrm{d}y) \qquad (x \in B(0, r)).$$

4

Stochastic Differential Equations

We showed in Section 3.1 that Brownian motions have the Markov and strong Markov properties. A continuous stochastic process with the strong Markov property is called a diffusion process and is one of the main objects in probability theory. Kolmogorov constructed diffusion processes with the help of the theory of partial differential equations. On the other hand, Itô originated the theory of stochastic differential equations in order to construct diffusion processes in a pathwise manner. The aim of this chapter is to develop the theory of stochastic differential equations.

4.1 Introduction: Diffusion Processes

In this section we will briefly discuss Markov processes, strong Markov processes, and diffusion processes. We restrict ourselves to time-homogeneous cases. For details, see [25, 45, 100] and references therein.

Let E be a locally compact separable metric space. Typical examples are $\mathbb{R}^d$ and Riemannian manifolds. When E is not compact, we add an extra point $\triangle$ (infinity) to E, and set $E_\triangle = E \cup \{\triangle\}$. A neighborhood of $\triangle$ is the complement of a compact set in E. $E_\triangle$ is called a one-point compactification of E.

Let $\overline{\mathbf{W}}(E)$ be the set of $E_\triangle$-valued functions $w : [0,\infty) \ni t \mapsto w(t) \in E_\triangle$ which satisfies the following condition: for each w, there exists a $\zeta(w) \in [0,\infty]$ such that

(i) $w(t) \in E$ for $t \in [0,\zeta(w))$ and w is continuous on $[0,\zeta(w))$,
(ii) $w(t) = \triangle$ for $t \geqq \zeta(w)$.

The above $\zeta(w)$ is called the **life time** of $w \in \overline{\mathbf{W}}(E)$.

When Markov processes are studied, it is frequently assumed that the sample paths are right-continuous with left limits. The purpose of this book being to study diffusion processes, we have assumed that the sample paths are continuous up to the life time.

As in the case of Wiener space (Section 1.2), a Borel cylinder set of $\overline{\mathbf{W}}(E)$ is defined in the following way. For $0 \leqq t_1 < t_2 < \cdots < t_n$, define the projection $\pi_{t_1,t_2,\dots,t_n}$ by

$$\pi_{t_1,t_2,\dots,t_n} : \overline{\mathbf{W}}(E) \ni w \mapsto (w(t_1), w(t_2), \dots, w(t_n)) \in (E_\triangle)^n.$$

For a Borel set F of $(E_\triangle)^n$, the subset $\pi^{-1}_{t_1,t_2,\dots,t_n}(F)$ of $\overline{\mathbf{W}}(E)$ is called a **Borel cylinder set**.

The σ-field on $\overline{\mathbf{W}}(E)$ generated by the Borel cylinder sets is denoted by $\mathscr{B}(\overline{\mathbf{W}}(E))$ and that generated by the Borel cylinder sets with $t_n \leqq t$ is denoted by $\mathscr{B}_t(\overline{\mathbf{W}}(E))$.

Definition 4.1.1 A family of probability measures $\{P_x\}_{x\in E_\triangle}$ on $\overline{\mathbf{W}}(E)$ satisfying the following conditions is called a (time-homogeneous) **Markov family**:

(i) for all $x \in E_\triangle$, $P_x(w(0) = x) = 1$,
(ii) for all $A \in \mathscr{B}(\overline{\mathbf{W}}(E))$, $E \ni x \mapsto P_x(A)$ is Borel measurable,
(iii) for $0 \leqq s < t$, $A \in \mathscr{B}_s(\overline{\mathbf{W}}(E))$, $F \in \mathscr{B}(E_\triangle)$ and $x \in E_\triangle$,

$$P_x(A \cap \{w(t) \in F\}) = \int_A P_{w'(s)}(w(t-s) \in F)P_x(\mathrm{d}w'). \qquad (4.1.1)$$

Definition 4.1.2 Let m be a probability measure on $(E_\triangle, \mathscr{B}(E_\triangle))$. An $E_\triangle$-valued $\{\mathscr{F}_t\}$-adapted stochastic process $X = \{X(t)\}_{t\geqq 0}$ defined on a filtered probability space $(\Omega, \mathscr{F}, \mathbf{P}, \{\mathscr{F}_t\})$ is called a continuous **Markov process** with initial distribution m if

(i) $X \in \overline{\mathbf{W}}(E)$, $\mathbf{P}$-a.s.,
(ii) for any $F \in \mathscr{B}(E_\triangle)$, $\mathbf{P}(X(0) \in F) = m(F)$,
(iii) for $0 \leqq s < t$ and $F \in \mathscr{B}(E_\triangle)$,

$$\mathbf{P}(X(t) \in F|\mathscr{F}_s) = \mathbf{P}(X(t) \in F|X(s)), \quad \mathbf{P}\text{-a.s.},$$

where $\mathbf{P}(\,\cdot\,|X(s))$ is the conditional probability given the σ-field $\sigma(X(s))$ generated by $X(s)$.

A Markov process corresponds to a Markov family if we set $\Omega = \overline{\mathbf{W}}(E)$ and define $X(t,\omega) = \omega(t)$ for $\omega \in \Omega$.

Definition 4.1.3 Let $\{P_x\}_{x \in E_\triangle}$ be a Markov family. For $t \geqq 0$, $x \in E_\triangle$ and $F \in \mathscr{B}(E_\triangle)$, the probability $P_x(w(t) \in F)$ is denoted by $P(t, x, F)$,

$$P(t, x, F) = P_x(w(t) \in F).$$

The family $\{P(t, x, F)\}$ is called the **transition probability** of a Markov family $\{P_x\}_{x \in E_\triangle}$.

For a cylinder set of $\overline{\mathbf{W}}(E)$, (4.1.1) implies

$$\begin{aligned} &P_x(w(t_1) \in F_1, w(t_2) \in F_2, \dots, w(t_n) \in F_n) \\ &\quad = \int_{F_1} P(t_1, x, \mathrm{d}x_1) \int_{F_2} P(t_2 - t_1, x_1, \mathrm{d}x_2) \cdots \int_{F_n} P(t_n - t_{n-1}, x_{n-1}, \mathrm{d}x_n). \end{aligned}$$

For a Markov family $\{P_x\}_{x \in E_\triangle}$, we define the filtration $\{\mathscr{G}_t = \mathscr{G}_t(\overline{\mathbf{W}}(E))\}$ on the path space $\overline{\mathbf{W}}(E)$ by

$$\mathscr{G}_t = \bigcap_{\varepsilon > 0} \bigcap_{x \in E_\triangle} \overline{\mathscr{B}_{t+\varepsilon}(\overline{\mathbf{W}}(E))}^{P_x} \qquad (t > 0).$$[1]

We set $\mathscr{G}_\infty = \bigvee_{t>0} \mathscr{G}_t$.

Definition 4.1.4 A Markov family $\{P_x\}_{x \in E_\triangle}$ on E is called a **strong Markov family** if, for any $t \geqq 0$, a $\{\mathscr{G}_t\}$-stopping time τ, $A \in \mathscr{G}_\tau$ and $F \in \mathscr{B}(E_\triangle)$,

$$P_x(A \cap \{w(t + \tau(w)) \in F\}) = \int_A P_{w'(\tau(w'))}(w(t) \in F) P_x(\mathrm{d}w')$$

holds for all $x \in E_\triangle$.

Definition 4.1.5 Let $X = \{X(t)\}_{t \geqq 0}$ be an $E_\triangle$-valued continuous Markov process with initial distribution m defined on a probability space $(\Omega, \mathscr{F}, \mathbf{P}, \{\mathscr{F}_t\})$. X is called a **diffusion process** on E if there exists a strong Markov family $\{P_x\}_{x \in E_\triangle}$ on E such that the probability law of X coincides with $P_m(\cdot) = \int_{E_\triangle} P_x(\cdot) m(\mathrm{d}x)$,

$$\mathbf{P}(X \in A) = \int_{E_\triangle} P_x(A) m(\mathrm{d}x) \qquad (A \in \mathscr{B}(\overline{\mathbf{W}}(E))).$$

[1] For a probability space $(\Omega, \mathscr{F}, \mathbf{P})$, the completion of $\mathscr{F}$ by $\mathbf{P}$ is denoted by $\overline{\mathscr{F}}^{\mathbf{P}}$:

$$\overline{\mathscr{F}}^{\mathbf{P}} = \{A \subset \Omega;\ \text{there exist } B \text{ and } B' \in \mathscr{F} \text{ such that } B \subset A \subset B' \text{ and } \mathbf{P}(B) = \mathbf{P}(B')\}.$$

$\mathbf{P}$ is naturally extended to the probability measure $\overline{\mathbf{P}}$ on $(\Omega, \overline{\mathscr{F}}^{\mathbf{P}})$ and $(\Omega, \overline{\mathscr{F}}^{\mathbf{P}}, \overline{\mathbf{P}})$ is a complete probability space.

A diffusion process $X = \{X(t)\}_{t \geqq 0}$ on E defined on $(\Omega, \mathscr{F}, \mathbf{P}, \{\mathscr{F}_t\})$ is called **conservative** if the probability that the life time $\zeta(\omega) = \inf\{t; X(t)(\omega) = \triangle\}$ is finite is 0.

In this book we consider diffusion processes which correspond to second order differential operators. Let $C_{\mathrm{b}}(E_\triangle)$ be the set of real-valued bounded continuous functions on $E_\triangle$ and $L : C_{\mathrm{b}}(E_\triangle) \to C_{\mathrm{b}}(E_\triangle)$ be a linear operator with domain $\mathscr{D}(L)$. We shall mainly deal with the case when L is the Laplacian on $\mathbb{R}^d$ or, more generally, the second order differential operator given by

$$L = \frac{1}{2} \sum_{i,j=1}^{d} a^{ij}(x) \frac{\partial^2}{\partial x^i \partial x^j} + \sum_{i=1}^{d} b^i(x) \frac{\partial}{\partial x^i}, \tag{4.1.2}$$

where a^{ij} and b^i are real functions on $\mathbb{R}^d$ and the matrix (a^{ij}) is symmetric and non-negative definite.

Definition 4.1.6 Let L be a linear operator with domain $\mathscr{D}(L)$ and $\{P_x\}_{x \in E_\triangle}$ be a family of probability measures on $(\overline{\mathbf{W}}(E), \mathscr{B}(\overline{\mathbf{W}}(E)))$ such that the function $x \mapsto P_x(A)$ is Borel measurable for any $A \in \mathscr{B}(\overline{\mathbf{W}}(E))$. If

(i) $P_x(w(0) = x) = 1$ for all $x \in E_\triangle$,
(ii) the stochastic process $\{M_f(t)\}_{t \geqq 0}$ $(f \in \mathscr{D}(L))$ defined by

$$M_f(t) = f(w(t)) - f(w(0)) - \int_0^t (Lf)(w(s))\,\mathrm{d}s$$

is a $\{\mathscr{B}_t(\overline{\mathbf{W}}(E))\}$-martingale under P_x for all $x \in E_\triangle$,

then $\{P_x\}_{x \in E_\triangle}$ is called a family of **diffusion measures** generated by L.

In the rest of this section, let E be $\mathbb{R}^d$ or its subset and L be the second order differential operator given by (4.1.2).

The above construction of a family of diffusion measures generated by L is called a martingale problem by Stroock and Varadhan ([114]). For the martingale problem, see Section 4.5. One of the advantages of the martingale problem is that, if the family of diffusion measures $\{P_x\}_{x \in E_\triangle}$ is unique, then $\{P_x\}_{x \in E_\triangle}$ is a strong Markov family on E.

Definition 4.1.7 If the martingale problem for the differential operator L on E of the form (4.1.2) has a unique solution $\{P_x\}_{x \in E_\triangle}$, the stochastic process whose probability distribution is P_x is called the **diffusion process** starting from $x \in E_\triangle$ generated by L. It is also simply called an L-diffusion or (a,b)-diffusion. L is called the **generator**.

Assume that there exists a diffusion process generated by L. Then for any $f \in \mathscr{D}(L)$ we have

$$\lim_{t \downarrow 0} \frac{P_x[f(X(t))] - f(x)}{t} = (Lf)(x).$$

Moreover, letting f be a function which is equal to x^i or $x^i x^j$ on a bounded domain, we have

$$\lim_{t \downarrow 0} \frac{1}{t} E_x[X^i(t) - x^i] = b^i(x),$$

$$\lim_{t \downarrow 0} \frac{1}{t} E_x[(X^i(t) - x^i)(X^j(t) - x^j)] = a^{ij}(x).$$

Roughly speaking, $b(x) = (b^i(x))$ represents the mean velocity of X and $a(x) = (a^{ij}(x))$ represents the covariance in an infinitesimal sense. b is called a **drift** and a is called a **diffusion matrix**. In the case of Brownian motion, b is 0 and $a^{ij} = \delta_{ij}$.

Example 4.1.8 Let $E = \mathbb{R}^d$. Set

$$\mathscr{D}(L) = \{f \in C_{\mathrm{b}}(E_\triangle); f|_{\mathbb{R}^d} \in C_{\mathrm{b}}^2(\mathbb{R}^d)\}$$

and define Lf for $f \in \mathscr{D}(L)$ by

$$(Lf)(x) = \begin{cases} \frac{1}{2}(\Delta f)(x) & (x \in \mathbb{R}^d) \\ 0 & (x = \triangle). \end{cases}$$

Then there exists a diffusion process on $\mathbb{R}^d$ generated by L. This is nothing but a d-dimensional Brownian motion.

Changing domains of generators, we obtain diffusion processes on subsets.

Example 4.1.9 (Absorbing Brownian motion) Let D be a bounded domain in $\mathbb{R}^d$ with the smooth boundary ∂D and $D_\triangle = D \cup \{\triangle\}$ be its one-point compactification. Denote by $\mathscr{D}(L)$ the space of C^2-functions on D satisfying $f(x) \to 0$ as $D \ni x \to y \in \partial D$ and define L by

$$(Lf)(x) = \begin{cases} \frac{1}{2}(\Delta f)(x) & (x \in D) \\ 0 & (x = \triangle). \end{cases}$$

Then there exists a diffusion process on D generated by L. It is called an **absorbing Brownian motion** on D.

As in the example above, we need some boundary conditions in order to consider diffusion processes on subsets. In this book we do not consider the boundaries except in the case where $d = 1$. For the diffusion processes on subsets with boundaries, see, for example, [45, 114].

Kolmogorov constructed diffusion processes by showing the existence of transition densities with the help of the theory of partial differential equations. Also, in the detailed studies on one-dimensional diffusion processes by Feller, Itô and McKean and others, a lot of important parts of observations rely on the results and arguments associated with differential equations.

On the other hand, the theory of stochastic differential equations gives us a direct method to construct diffusion processes from Brownian motions. In the following sections, this theory will be stated in detail.

For stochastic differential equations, the existence and uniqueness of solutions are equivalent to those of some Markovian family. As was mentioned above, Stroock and Varadhan formulated the latter problem as a martingale problem and solved it in a general framework, where deep results on partial differential equations have again played important roles.

4.2 Stochastic Differential Equations

While a construction of diffusion processes is an important application of stochastic differential equations, there are many other areas where stochastic differential equations play key roles. In these applications, stochastic differential equations of more general form than the ones for diffusion processes appear. Hence we start with a general framework.

Let $\mathbf{W}^d$ be the set of $\mathbb{R}^d$-valued continuous functions on $[0, \infty)$. Set

$$\rho(w_1, w_2) = \sum_{m=1}^{\infty} 2^{-m}\Big(\max_{0 \leqq t \leqq m} |w_1(t) - w_2(t)| \wedge 1\Big) \qquad (w_1, w_2 \in \mathbf{W}^d).$$

Then ρ is a distance function on $\mathbf{W}^d$ and $(\mathbf{W}^d, \rho)$ is a complete separable metric space. Let $\mathscr{B}(\mathbf{W}^d)$ be the topological σ-field on $\mathbf{W}^d$ and $\mathscr{B}_t(\mathbf{W}^d)$ be the sub-σ-field generated by $w(s)$ $(0 \leqq s \leqq t)$. As in Chapter 1, denote the subset of $\mathbf{W}^d$ consisting of the elements with $w(0) = 0$, the d-dimensional Wiener space, by W^d:

$$W^d = \{w \in \mathbf{W}^d; w(0) = 0\}.$$

We define a class of functions which give coefficients of stochastic differential equations. Let $\mathbb{R}^d \otimes \mathbb{R}^r$ be the set of $d \times r$ real matrices.

Definition 4.2.1 Denote by $\mathscr{A}^{d,r}$ the set of $\mathbb{R}^d \otimes \mathbb{R}^r$-valued functions $\alpha(t, w)$ on $[0, \infty) \times \mathbf{W}^d$ such that $\mathbf{W}^d \ni w \mapsto \alpha(t, w) \in \mathbb{R}^d \otimes \mathbb{R}^r$ is $\mathscr{B}_t(\mathbf{W}^d)$-measurable for each $t \geqq 0$.

Let $\alpha \in \mathscr{A}^{d,r}$ and $\beta \in \mathscr{A}^{d,1}$. For an r-dimensional Brownian motion $B = \{B(t)\}_{t \geqq 0}$, we consider the stochastic differential equation

$$\mathrm{d}X^i(t) = \sum_{k=1}^{r} \alpha^i_k(t, X)\,\mathrm{d}B^k(t) + \beta^i(t, X)\,\mathrm{d}t, \tag{4.2.1}$$

where $\alpha(t, w) = (\alpha^i_k(t, w))_{i=1,2,\ldots,d,\,k=1,2,\ldots,r}$, or in matrix notation[2]

$$\mathrm{d}X(t) = \alpha(t, X)\,\mathrm{d}B(t) + \beta(t, X)\,\mathrm{d}t. \tag{4.2.2}$$

First of all we give the definition of solutions.

Definition 4.2.2 Let $(\Omega, \mathscr{F}, \mathbf{P}, \{\mathscr{F}_t\})$ be a filtered probability space. An $\mathbb{R}^d$-valued continuous stochastic process $X = \{X(t)\}_{t \geqq 0}$ is a solution of the stochastic differential equation (4.2.1) if

(i) there exists an r-dimensional $\{\mathscr{F}_t\}$-Brownian motion $B = \{B(t)\}_{t \geqq 0}$ with $B(0) = 0$,
(ii) X is $\{\mathscr{F}_t\}$-adapted,
(iii) $\{\alpha^i_k(t, X)\}_{t \geqq 0} \in \mathscr{L}^2_{\mathrm{loc}}$, $\{\beta^i(t, X)\}_{t \geqq 0} \in \mathscr{L}^1_{\mathrm{loc}}$ $(i = 1, 2, \ldots, d,\ k = 1, 2, \ldots, r)$[3],
(iv) almost surely,

$$X^i(t) = X^i(0) + \sum_{k=1}^{r} \int_0^t \alpha^i_k(s, X)\,\mathrm{d}B^k(s) + \int_0^t b^i(s, X)\,\mathrm{d}s \qquad (i = 1, 2, \ldots, d). \tag{4.2.3}$$

X is called a solution driven by a Brownian motion B. We also say that the pair (X, B) is a solution of (4.2.1). The probability distribution of $X(0)$ is called the initial distribution of X.

The next Markovian type stochastic differential equation is one of the most important objects in this book.

Definition 4.2.3 Let $\sigma(t, x)$ and $b(t, x)$ be Borel measurable functions on $[0, \infty) \times \mathbb{R}^d$ with values in $\mathbb{R}^d \otimes \mathbb{R}^r$ and $\mathbb{R}^d$, respectively. The stochastic differential equation of the form

[2] Every element of $\mathbf{R}^d$ is thought of as a column vector.
[3] For $\mathscr{L}^2_{\mathrm{loc}}$ and $\mathscr{L}^1_{\mathrm{loc}}$, see Definitions 2.2.11 and 2.3.2, respectively.

$$\mathrm{d}X(t) = \sigma(t, X(t))\,\mathrm{d}B(t) + b(t, X(t))\,\mathrm{d}t \tag{4.2.4}$$

is said to be of **Markovian type.**[4] Moreover, if both $\sigma(t,x)$ and $b(t,x)$ depend only on x, the Markovian type equation is called **time-homogeneous**.

If $\sigma \equiv 0$, then the Markovian type stochastic differential equation (4.2.4) is nothing but the ordinary differential equation

$$\frac{\mathrm{d}X(t)}{\mathrm{d}t} = b(t, X(t)).$$

Hence we may regard a Markovian type stochastic differential equation as an ordinary differential equation perturbed by a noise generated by a Brownian motion.

The function $b(t,\cdot) : \mathbb{R}^d \to \mathbb{R}^d$ is identified with a vector field $\sum_{i=1}^d b^i(t,x)\frac{\partial}{\partial x^i}$ on $\mathbb{R}^d$, and $\sigma(t,\cdot) : \mathbb{R}^d \to \mathbb{R}^d \otimes \mathbb{R}^r$ is identified with a set $(V_1, \dots, V_r)$ of vector fields given by

$$V_k = \sum_{i=1}^d \sigma_k^i(t,x)\frac{\partial}{\partial x^i} \qquad (k = 1, 2, \dots, r).$$

Under these identifications we write Equation (4.2.4) as

$$\mathrm{d}X(t) = \sum_{k=1}^r V_k(t, X(t))\,\mathrm{d}B^k(t) + b(t, X(t))\,\mathrm{d}t.$$

It is important to consider solutions which diverge to infinity in finite time (said to **explode**). But, in order to avoid complexity, we do not discuss the explosion problem here, but we do so in Section 4.3, restricting ourselves to the Markovian case.

We present examples of the stochastic differential equations whose solutions are explicitly given. All results are checked by applying Itô's formula.

Example 4.2.4 (Geometric Brownian motion) Let $d = r = 1$. For $\sigma, \rho \in \mathbb{R}$, set

$$X(t) = x\exp\Bigl(\sigma B(t) + \Bigl(\rho - \frac{\sigma^2}{2}\Bigr)t\Bigr).$$

Then $X = \{X(t)\}_{t \geqq 0}$ is a solution of the stochastic differential equation

$$\mathrm{d}X(t) = \sigma X(t)\,\mathrm{d}B(t) + \rho X(t)\,\mathrm{d}t$$

satisfying $X(0) = x$.

[4] $(\alpha_k^i(t,X)) = (\sigma_k^i(t, X(t))) \in \mathscr{A}^{d,r}$, $(\beta^i(t,X)) = (b^i(t, X(t))) \in \mathscr{A}^{d,1}$.

Example 4.2.5 (Ornstein–Uhlenbeck process) For $\alpha \in \mathbb{R}$ and $\beta > 0$, set

$$X(t) = X(0)e^{-\beta t} + \alpha e^{-\beta t} \int_0^t e^{\beta s} dB(s).$$

Then $X = \{X(t)\}_{t \geqq 0}$ satisfies

$$dX(t) = \alpha\, dB(t) - \beta X(t)\, dt.$$

Assume that the initial distribution is Gaussian with mean 0 and variance $\frac{\alpha^2}{2\beta}$. Then, since $X(0)$ and B are independent, $\{X(t)\}_{t \geqq 0}$ is a stationary Gaussian process with covariance $\mathrm{cov}[X(s), X(t)] = \frac{\alpha^2}{2\beta} e^{-\beta|t-s|}$.

Example 4.2.6 Let $d = r = 2$ and set

$$X^1(t) = x^1 + \int_0^t x^2 e^{B^2(s) - \frac{s}{2}} dB^1(s), \qquad X^2(t) = x^2 e^{B^2(t) - \frac{t}{2}}.$$

Then $\{(X^1(t), X^2(t))\}_{t \geqq 0}$ is a solution of

$$dX^1(t) = X^2(t)\, dB^1(t), \qquad dX^2(t) = X^2(t)\, dB^2(t)$$

satisfying $(X^1(0), X^2(0)) = (x^1, x^2)$.

Example 4.2.7 For $\gamma > 0$ and $\sigma \in \mathbb{R}$, set $A = \begin{pmatrix} 0 & \gamma^2 \\ 1 & 0 \end{pmatrix}$. Define an $\mathbb{R}^2$-valued stochastic process $\left\{ Z(t) = \begin{pmatrix} X(t) \\ Y(t) \end{pmatrix} \right\}_{t \geqq 0}$ by

$$Z(t) = e^{tA} \left\{ \begin{pmatrix} x \\ 0 \end{pmatrix} + \int_0^t e^{-sA} \begin{pmatrix} \sigma \\ 0 \end{pmatrix} dB(s) \right\},$$

where e^{tA} is the exponential of the matrix tA, $e^{tA} = \sum_{n=0}^{\infty} \frac{t^n}{n!} A^n$. By Itô's formula,

$$dZ(t) = AZ(t)dt + \begin{pmatrix} \sigma \\ 0 \end{pmatrix} dB(t). \tag{4.2.5}$$

Since $e^{tA} = \begin{pmatrix} \cosh(\gamma t) & \gamma \sinh(\gamma t) \\ \frac{1}{\gamma} \sinh(\gamma t) & \cosh(\gamma t) \end{pmatrix}$, the definition of $\{Z(t)\}_{t \geqq 0}$ implies

$$X(t) = x \cosh(\gamma t) + \sigma \int_0^t \cosh(\gamma(t-s))\, dB(s).$$

Then, it follows from the second line of (4.2.5) that $Y(t) = \int_0^t X(s)\, ds$. Moreover, the first line of (4.2.5) yields that $X = \{X(t)\}_{t \geqq 0}$ obeys the stochastic differential equation

$$\mathrm{d}X(t) = \gamma^2\Big(\int_0^t X(s)\mathrm{d}s\Big)\mathrm{d}t + \sigma\,\mathrm{d}B(t), \quad X(0) = x. \tag{4.2.6}$$

Example 4.2.8 (Doss [16], Sussmann [117]) Let $d = r = 1$ and $\sigma, b \in C_{\mathrm{b}}^2(\mathbb{R})$. Then a solution of the time-homogeneous Markovian type stochastic differential equation

$$\mathrm{d}X(t) = \sigma(X(t))\,\mathrm{d}B(t) + b(X(t))\,\mathrm{d}t$$

is constructed as follows.

For $y \in \mathbb{R}$, denote by $\varphi(x, y)$ the solution of the ordinary differential equation

$$\frac{\mathrm{d}\varphi}{\mathrm{d}x} = \sigma(\varphi), \quad \varphi(0) = y.$$

Then $\varphi(x, y)$ is differentiable in y. Moreover, since

$$\frac{\partial}{\partial x}\Big(\frac{\partial\varphi(x, y)}{\partial y}\Big) = \sigma'(\varphi(x, y))\frac{\partial\varphi(x, y)}{\partial y},$$

we have

$$\frac{\partial\varphi(x, y)}{\partial y} = \exp\Big(\int_0^x \sigma'(\varphi(u, y))\,\mathrm{d}u\Big).$$

Set

$$\widetilde{b}(x) = b(x) - \frac{1}{2}\sigma(x)\sigma'(x) \quad \text{and} \quad f(x, y) = \widetilde{b}(\varphi(x, y))\Big(\frac{\partial\varphi(x, y)}{\partial y}\Big)^{-1}$$

and let $Y(t)$ be the solution of the ordinary differential equation

$$\frac{\mathrm{d}Y}{\mathrm{d}t}(t) = f(B(t), Y(t)).$$

Then $\{X(t) = \varphi(B(t), Y(t))\}_{t \geqq 0}$ is a solution driven by B.[5]

Next we give the definitions of uniqueness of solutions.

Definition 4.2.9 It is said that the **uniqueness in law** of solutions for the stochastic differential equation (4.2.1) holds if the probability laws of X and X' coincide for any solutions X and X' of (4.2.1) with the same initial distribution.

Definition 4.2.10 It is said that the **pathwise uniqueness** of solutions for the stochastic differential equation (4.2.1) holds if $X(t) = X'(t)$ for all $t \geqq 0$ almost surely for any solutions X and X' of (4.2.1) which are defined on the same filtered probability space, driven by the same Brownian motion, and satisfy $X(0) = X'(0)$ almost surely.

[5] Example 4.2.8 is one of the starting points of the representation of solutions of stochastic differential equations (Kunita [64],Yamato [129]).

Definition 4.2.11 $\Phi(x,w) : \mathbb{R}^d \times W^r \to \mathbf{W}^d$ is said to be $\widehat{\mathscr{B}}(\mathbb{R}^d \times W^r)$-measurable if, for any probability measure m on $\mathbb{R}^d$, there exists a function $\widetilde{\Phi}_m : \mathbb{R}^d \times W^r \to \mathbf{W}^d$ such that

(i) for the Wiener measure μ on W^r, $\widetilde{\Phi}_m$ is $\overline{\mathscr{B}(\mathbb{R}^d \times W^r)}^{m\times\mu}$-measurable,
(ii) $\Phi(x,w) = \widetilde{\Phi}_m(x,w)$ for m-a.e. x.

Definition 4.2.12 A solution X of (4.2.1) is called a **strong solution** with a Brownian motion B if there exists a $\widehat{\mathscr{B}}(\mathbb{R}^d \times W^r)$-measurable function $F(x,w)$ such that $w|_{[0,t]} \to F(x,w) \in \mathbf{W}_t^d \equiv C([0,t] \to \mathbb{R}^d)$ is $\overline{\mathscr{B}_t(W^r)}^{\mu}$-measurable for each $x \in \mathbb{R}^d$ and $t \geqq 0$ and $X = F(X(0), B)$ holds almost surely.

Definition 4.2.13 The stochastic differential equation (4.2.1) is said to have a unique strong solution if there exists a function $F(x,w) : \mathbb{R}^d \times W^r \to \mathbf{W}^d$ satisfying the conditions in Definition 4.2.12 such that

(i) for an r-dimensional Brownian motion $B = \{B(t)\}_{t\geqq 0}$ and an $\mathscr{F}_0$-measurable random variable ξ defined on a filtered probability space $(\Omega, \mathscr{F}, \mathbf{P}, \{\mathscr{F}_t\})$, $F(\xi, B)$ is a solution of (4.2.1) with $X(0) = \xi$ almost surely,
(ii) for any solution (X,B) of (4.2.1), $X = F(X(0), B)$ almost surely.

Following [45, 128], we show the fundamental theorem for the uniqueness of the solutions of stochastic differential equations.

Theorem 4.2.14 *Let $\alpha \in \mathscr{A}^{d,r}$ and $\beta \in \mathscr{A}^{d,1}$. Then* (4.2.1) *has a unique strong solution if and only if, for any Borel probability measure m on $\mathbb{R}^d$, there exists a solution with initial distribution m and the pathwise uniqueness holds.*

Proof Assume that (4.2.1) has a unique strong solution, that is, there exists a unique function $F : \mathbb{R}^d \times W^r \to \mathbf{W}^d$ which satisfies the conditions (i) and (ii) in Definition 4.2.13. Take an r-dimensional $\{\mathscr{F}_t\}$-Brownian motion B on a filtered probability space $(\Omega, \mathscr{F}, \mathbf{P}, \{\mathscr{F}_t\})$ and an $\mathscr{F}_0$-measurable random variable ξ whose probability distribution is m. Then $X = F(\xi, B)$ is a solution of (4.2.1) with initial distribution m.

Moreover, if (X,B) and (X',B') are two solutions of (4.2.1) on a same probability space such that $B(t) = B'(t)$ $(t \geqq 0)$ and $X(0) = X'(0)$ almost surely, then we have

$$X = F(X(0), B) = F(X'(0), B') = X'$$

almost surely. Hence the pathwise uniqueness holds for (4.2.1).

We give a sketch of the proof of the converse. Suppose that, for any Borel probability measure m, there exists a solution of (4.2.1) with initial distribution m and the pathwise uniqueness holds. We consider the case when $m = \delta_x$ for a fixed $x \in \mathbb{R}^d$. Let (X, B) and (X', B') be solutions of (4.2.1) starting from x and denote by P_x and P'_x their probability distributions on $\mathbf{W}^d \times W^r$, respectively. Moreover, let $\pi : \mathbf{W}^d \times W^r \ni (w_1, w_2) \mapsto w_2 \in W^r$ be the projection. Then, the distribution of w_2 under P_x or P'_x is the Wiener measure μ on W^r.

Let $Q_{x,w_2}(\mathrm{d}w_1)$ be the regular conditional probability distribution of w_1 under P_x given w_2. Q_{x,w_2} is a probability measure on $\mathbf{W}^d$ which satisfies

(i) for each $w_2 \in W^r$, $Q_{x,w_2}(\mathrm{d}w_1)$ is a probability measure on $(\mathbf{W}^d, \mathscr{B}(\mathbf{W}^d))$,
(ii) for any $A \in \mathscr{B}(\mathbf{W}^d)$, $w_2 \mapsto Q_{x,w_2}(A)$ is $\overline{\mathscr{B}(W^r)}^{\mu}$-measurable,
(iii) for any $A_1 \in \mathscr{B}(\mathbf{W}^d)$ and $A_2 \in \mathscr{B}(W^r)$,

$$P_x(A_1 \times A_2) = \int_{A_2} Q_{x,w_2}(A_1)\mu(\mathrm{d}w_1).$$

We define $Q'_{x,w_2}(\mathrm{d}w_1)$ in the same way.

Set $\Omega = \mathbf{W}^d \times \mathbf{W}^d \times W^r$ and define a probability measure Q_x on Ω by

$$Q_x(\mathrm{d}w_1\mathrm{d}w_2\mathrm{d}w_3) = Q_{x,w_3}(\mathrm{d}w_1)Q'_{x,w_3}(\mathrm{d}w_2)\mu(\mathrm{d}w_3).$$

Let $\mathscr{B}(\Omega)$ be the topological σ-field on Ω and set $\mathscr{F} = \overline{\mathscr{B}(\Omega)}^{Q_x}$. Denoting by $\mathscr{B}_t$ the direct product σ-field $\mathscr{B}_t(\mathbf{W}^d) \otimes \mathscr{B}_t(\mathbf{W}^d) \otimes \mathscr{B}_t(W^r)$ and by $\mathscr{N}$ the totality of the Q_x-null sets, we set $\mathscr{F}_t = \bigcap_{\varepsilon>0}(\mathscr{B}_{t+\varepsilon} \vee \mathscr{N})$. Then, the distributions of (w_1, w_3) and (w_2, w_3) under Q_x are those of (X, B) and (X', B'), respectively.

Hence, both (w_1, w_3) and (w_2, w_3) are solutions of (4.2.1) defined on $(\Omega, \mathscr{F}, \mathbf{P}, \{\mathscr{F}_t\})$, and $Q_x(w_1 = w_2) = 1$ by the assumption, which implies

$$(Q_{x,w} \times Q'_{x,w})(w_1 = w_2) = 1, \quad \mu\text{-a.s.}$$

This means that, for almost all $w \in W^r$, there exists a unique $F_x(w) \in \mathbf{W}^d$ such that $Q_{x,w}$ and $Q'_{x,w}$ are the Dirac measure concentrated on $F_x(w)$. $F_x(w)$ is the desired function on $\mathbb{R}^d \times W^r$ and we can prove that (4.2.1) has a unique strong solution. For details, see [45, p.163]. □

Corollary 4.2.15 *If pathwise uniqueness holds for the stochastic differential equation* (4.2.1), *then uniqueness in law holds.*

4.3 Existence of Solutions

Let $\alpha \in \mathscr{A}^{d,r}$ and $\beta \in \mathscr{A}^{d,1}$, and consider the stochastic differential equation

$$\mathrm{d}X(t) = \alpha(t, X)\,\mathrm{d}B(t) + \beta(t, X)\,\mathrm{d}t. \tag{4.3.1}$$

Define $a = (a^{ij}) : [0, \infty) \times \mathbf{W}^d \to \mathbb{R}^d \otimes \mathbb{R}^d$ by $a = \alpha\alpha^*$, that is,

$$a^{ij}(t, w) = \sum_{k=1}^{r} \alpha^i_k(t, w)\alpha^j_k(t, w),$$

and set

$$(Af)(t, w) = \frac{1}{2}\sum_{i,j=1}^{d} a^{ij}(t, w)\frac{\partial^2 f}{\partial x^i \partial x^j}(w(t)) + \sum_{i=1}^{d} \beta^i(t, w)\frac{\partial f}{\partial x^i}(w(t))$$

for $f \in C^2_b(\mathbb{R}^d)$.

Let (X, B) be a solution of (4.3.1) defined on a filtered probability space $(\Omega, \mathscr{F}, \mathbf{P}, \{\mathscr{F}_t\})$ and set

$$M^X_f(t) = f(X(t)) - f(X(0)) - \int_0^t (Af)(s, X)\,\mathrm{d}s.$$

By Itô's formula

$$M^X_f(t) = \sum_{i=1}^{d}\sum_{k=1}^{r}\int_0^t \alpha^i_k(s, X)\frac{\partial f}{\partial x^i}(X(s))\,\mathrm{d}B^k(s)$$

and $\{M^X_f(t)\}_{t\geqq 0} \in \mathscr{M}^2_{\mathrm{c,loc}}$.

The converse also holds and, as seen below, this is equivalent to the existence of the solution for (4.3.1).

Theorem 4.3.1 *The stochastic differential equation* (4.3.1) *has a solution if and only if there exists a d-dimensional continuous stochastic process* $X = \{X(t)\}_{t\geqq 0}$ *such that* $\{M^X_f(t)\}_{t\geqq 0} \in \mathscr{M}^2_{\mathrm{c,loc}}$ *for any* $f \in C^2_{\mathrm{b}}(\mathbb{R}^d)$.

Proof Assume the existence of X satisfying the condition. Let $R > 0$ and take $f_i \in C^2_{\mathrm{b}}(\mathbb{R})$ such that $f_i(x) = x^i$ for $|x| < R$. Set $\sigma_R = \inf\{t > 0; |X(t)| \geqq R\}$,

$$M^{(R),i}(t) = X^i(t \wedge \sigma_R) - X^i(0) - \int_0^{t\wedge\sigma_R} \beta^i(s, X)\,\mathrm{d}s,$$

and

$$M^i(t) = X^i(t) - X^i(0) - \int_0^t \beta^i(s, X)\,\mathrm{d}s.$$

Then, for $i = 1, 2, \ldots, d$, $\{M^{(R),i}(t)\}_{t\geqq 0} \in \mathscr{M}_c^2$ and it converges to $\{M^i(t)\}_{t\geqq 0}$ as $R \to \infty$. Hence $\{M^i(t)\}_{t\geqq 0} \in \mathscr{M}_{\mathrm{c,loc}}^2$. Moreover, let $f_{ij} \in C_{\mathrm{b}}^2(\mathbb{R})$ satisfy $f_{ij}(x) = x^i x^j$ for $|x| < R$ and set

$$\begin{aligned} M^{ij}(t) &= X^i(t)X^j(t) - X^i(0)X^j(0) \\ &\quad - \int_0^t (a^{ij}(s,X) + \beta^i(s,X)X^j(s) + \beta^j(s,X)X^i(s))\,\mathrm{d}s. \end{aligned}$$

Then, we get, in a similar way, $\{M^{ij}(t)\}_{t\geqq 0} \in \mathscr{M}_{\mathrm{c,loc}}^2$. By Itô's formula

$$\begin{aligned} &M^i(t)M^j(t) - \int_0^t a^{ij}(s,X)\,\mathrm{d}s \\ &\quad = M^{ij}(t) - X^i(0)M^j(t) - X^j(0)M^i(t) \\ &\quad\quad - \int_0^t \Big(\int_0^s \beta^i(u,X)\,\mathrm{d}u\Big)\mathrm{d}M^j(s) - \int_0^t \Big(\int_0^s \beta^j(u,X)\,\mathrm{d}u\Big)\mathrm{d}M^i(s) \end{aligned}$$

and

$$\langle M^i, M^j \rangle(t) = \int_0^t a^{ij}(s,X)\,\mathrm{d}s.$$

Hence, by Theorem 2.5.8, there exists an r-dimensional Brownian motion B such that

$$M^i(t) = \sum_{k=1}^r \int_0^t \alpha_k^i(s,X)\,\mathrm{d}B^k(s)$$

and (X, B) is a solution of (4.3.1). □

Remark 4.3.2 The condition of the theorem is equivalent to the existence of the probability measure on $(\mathbf{W}, \mathscr{B}(\mathbf{W}))$ under which

$$f(w(t)) - f(w(0)) - \int_0^t (Af)(s,w)\,\mathrm{d}s$$

is an $\{\mathscr{F}_t\}$-locally square integrable martingale (see Section 2.1) for any $f \in C_{\mathrm{b}}^2(\mathbb{R}^d)$. In fact, the probability law of X in Theorem 4.3.1 plays the desired role.

The next is a fundamental theorem on the existence of solutions.

Theorem 4.3.3 *Assume that* $\alpha \in \mathscr{A}^{d,r}$ *and* $\beta \in \mathscr{A}^{d,1}$ *are bounded and continuous in* $(t,w) \in [0,\infty) \times \mathbf{W}^d$. *Then, for any* $x_0 \in \mathbb{R}^d$, *there exists a solution* $(X^{(x_0)}, B)$ *of* (4.3.1) *satisfying* $P(X^{(x_0)}(0) = x_0) = 1$.

Proof We give an outline of a proof. For details, see [45].

It suffices to show that there exists a stochastic process $X = \{X(t)\}_{t\geqq 0}$ defined on some filtered probability space $(\Omega, \mathscr{F}, \mathbf{P}, \{\mathscr{F}_t\})$ such that $\mathbf{P}(X(0) = x_0) = 1$ and, for any $f \in C_{\mathrm{b}}^2(\mathbb{R}^d)$,

$$M_f(t) := f(X(t)) - f(X(0)) - \int_0^t (Af)(s, X)\,\mathrm{d}s \in \mathscr{M}_{\mathrm{c,loc}}^2.$$

Take an r-dimensional $\{\mathscr{F}_t\}_{t\geqq 0}$-Brownian motion $B = \{B(t)\}_{t\geqq 0}$ on a filtered probability space $(\Omega, \mathscr{F}, \mathbf{P}, \{\mathscr{F}_t\})$ and define a sequence of d-dimensional stochastic processes $X_n = \{X_n(t)\}_{t\geqq 0}$ $(n = 1, 2, \ldots)$ by $X_n(0) = x_0$ for $t = 0$ and, for t with $\frac{m}{2^n} \leqq t \leqq \frac{m+1}{2^n}$,

$$X_n(t) = X_n\Big(\frac{m}{2^n}\Big) + \alpha\Big(\frac{m}{2^n}, X_{n,m}\Big)\Big(B(t) - B\Big(\frac{m}{2^n}\Big)\Big) + \beta\Big(\frac{m}{2^n}, X_{n,m}\Big)\Big(t - \frac{m}{2^n}\Big),$$

where $\{X_{n,m}(t)\}_{t\geqq 0}$ is given by $X_{n,m}(t) = X_n(t \wedge \frac{m}{2^n})$. Moreover, set $\phi_n(t) = \frac{m}{2^n}$ $(\frac{m}{2^n} \leqq t < \frac{m+1}{2^n})$ and define $\alpha_n \in \mathscr{A}^{d,r}$ and $\beta_n \in \mathscr{A}^{d,1}$ by $\alpha_n(t, w) = \alpha(\phi_n(t), w)$ and $\beta_n(t, w) = \beta(\phi_n(t), w)$. Then, $\{X_n(t)\}_{t\geqq 0}$ is a unique solution of

$$\mathrm{d}X(t) = \alpha_n(t, X)\,\mathrm{d}B(t) + \beta_n(t, X)\,\mathrm{d}t, \qquad X(0) = x_0.$$

By the boundedness and continuity of α and β, we can prove that there exists a subsequence of $\{X_n(t)\}_{t\geqq 0}$ which converges uniformly on compact sets almost surely. The limit is the desired solution of the stochastic differential equation. □

Let P_x be the probability law of the solution X of (4.3.1) with $X(0) = x$. Denote by $\mathscr{P}(\mathbf{W}^d)$ the set of probability measures on $\mathbf{W}^d$ endowed with the topology of weak convergence and by $\mathscr{B}(\mathscr{P}(\mathbf{W}^d))$ its topological σ-field. Then, the mapping $x \mapsto P_x$ is $\mathscr{B}(\mathbb{R}^d)/\mathscr{B}(\mathscr{P}(\mathbf{W}^d))$-measurable ([45, p.171], [114, p.28]). For a probability measure m on $\mathbb{R}^d$, define a probability measure P on $(\mathbf{W}^d, \mathscr{B}(\mathbf{W}^d))$ by

$$P(A) = \int_{\mathbb{R}^d} P_x(A) m(\mathrm{d}x) \qquad (A \in \mathscr{B}(\mathbb{R}^d)).$$

Then, for any $f \in C_{\mathrm{b}}^2(\mathbb{R}^d)$, $\{M_f^X(t)\}_{t\geqq 0}$ is a local martingale under P. Hence there exists a solution of (4.3.1) with initial distribution m.

We devote the rest of this section to time-homogeneous Markovian stochastic differential equations, which are important ingredients in the future chapters. In what follows, we consider solutions which may explode, i.e., they may arrive at infinity in finite time.

For measurable functions $\sigma(x) = (\sigma^i_j(x)) : \mathbb{R}^d \to \mathbb{R}^d \otimes \mathbb{R}^r$ and $b(x) = (b^i(x)) : \mathbb{R}^d \to \mathbb{R}^d$, consider the stochastic differential equation

$$\mathrm{d}X(t) = \sigma(X(t))\,\mathrm{d}B(t) + b(X(t))\,\mathrm{d}t, \tag{4.3.2}$$

or

$$\mathrm{d}X^i(t) = \sum_{k=1}^{r} \sigma^i_k(X(t))\,\mathrm{d}B^k(t) + b^i(X(t))\,\mathrm{d}t \qquad (i = 1, 2, \ldots, d).$$

If σ and b are bounded and continuous, then there exists a solution by Theorem 4.3.3. If they are not bounded, an explosion occurs in general.

Example 4.3.4 Let $d = r = 1$. $X(t) = (1 - B(t))^{-1}$ $(t < \inf\{s; B(s) = 1\})$ is a solution of

$$\mathrm{d}X(t) = X(t)^2 \mathrm{d}B(t) + X(t)^3 \mathrm{d}t, \qquad X(0) = 1.$$

For this stochastic differential equation, the uniqueness of solution holds by Theorem 4.4.5 below.

Let $\widehat{\mathbb{R}^d} = \mathbb{R}^d \cup \{\Delta\}$ be the one-point compactification of $\mathbb{R}^d$ and consider the path space on $\widehat{\mathbb{R}^d}$ given by

$$\widehat{\mathbf{W}}^d = \left\{ w : [0,\infty) \to \widehat{\mathbb{R}^d}; \begin{array}{l} w \text{ is continuous.} \\ \text{If } w(t) = \Delta, \text{ then } w(t') = \Delta\ (t' \geqq t). \end{array} \right\}.$$

Let $\mathscr{B}(\widehat{\mathbf{W}}^d)$ be the σ-field generated by the Borel cylinder sets and $\zeta(w)$ be the life time of $w \in \widehat{\mathbf{W}}^d$,

$$\zeta(w) = \inf\{t; w(t) = \Delta\}.$$

Since we admit explosions, we again give the definition of solutions of stochastic differential equations.

Definition 4.3.5 A $\widehat{\mathbf{W}}^d$-valued random variable $X = \{X(t)\}_{t \geqq 0}$ defined on a filtered probability space $(\Omega, \mathscr{F}, \mathbf{P}, \{\mathscr{F}_t\})$ is called a solution of the stochastic differential equation (4.3.2) if

(i) there exists an r-dimensional $\{\mathscr{F}_t\}$-Brownian motion $B = \{B(t)\}_{t \geqq 0}$ with $B(0) = 0$,
(ii) X is $\{\mathscr{F}_t\}$-adapted, that is, for each $t \geqq 0$, $X(t) \in \widehat{\mathbb{R}^d}$ is $\mathscr{F}_t$-measurable,
(iii) almost surely, for $t < \zeta^X = \inf\{t; X(t) = \Delta\}$,

$$X^i(t) = X^i(0) + \sum_{k=1}^{r} \int_0^t \sigma^i_k(X(s))\,\mathrm{d}B^k(s) + \int_0^t b^i(X(s))\,\mathrm{d}s \quad (i = 1, 2, \ldots, d).$$

Remark 4.3.6 Replacing $\mathbf{W}^d$ with $\widehat{\mathbf{W}}^d$, we can define the uniqueness of solutions and the pathwise uniqueness in the same way.

Remark 4.3.7 If $\zeta^X < \infty$, then $\lim_{t\uparrow\zeta^X} X(t) = \triangle$ almost surely ([45, p.174]).

Theorem 4.3.8 *If $\sigma(x) = (\sigma^i_k(x))$ and $b(x) = (b^i(x))$ are continuous, then, for any $x \in \mathbb{R}^d$, there exists a solution X of (4.3.2) such that $X(0) = x$.*

Proof It is sufficient to show that there exists a $\widehat{\mathbf{W}}^d$-valued random variable $X = \{X(t)\}_{t\geqq 0}$ defined on some filtered probability space $(\Omega, \mathscr{F}, \mathbf{P}, \{\mathscr{F}_t\})$ such that $\mathbf{P}(X(0) = x_0) = 1$ and the stochastic process $\{M_f(t)\}_{t\geqq 0}$ defined by

$$M_f(t) := f(X(t \wedge \tau_n)) - f(X(0)) - \int_0^{t\wedge\tau_n} (Lf)(X(s))\,\mathrm{d}s$$

is an $\{\mathscr{F}_t\}$-martingale for each $f \in C^2_{\mathrm{b}}(\mathbb{R}^d)$, where $\tau_n = \inf\{t; |X(t)| \geqq n\}$ (see Theorem 4.3.1). L is given by

$$(Lf)(x) = \frac{1}{2}\sum_{i,j=1}^d a^{ij}(x)\frac{\partial^2 f}{\partial x^i \partial x^j} + \sum_{i=1}^d b^i(x)\frac{\partial f}{\partial x^i}(x),$$

where

$$a^{ij}(x) = \sum_{k=1}^d \sigma^i_k(x)\sigma^j_k(x).$$

Let $\rho \in C(\mathbb{R}^d \to \mathbb{R})$ be a function with values in $(0, 1]$ such that all of $\rho(x)a^{ij}(x)$ and $\rho(x)b^i(x)$ are bounded. Set

$$(\widetilde{L}f)(x) = \rho(x)(Lf)(x).$$

By Theorem 4.3.3, there exists an $\mathbb{R}^d$-valued stochastic process $\widetilde{X} = \{\widetilde{X}(t)\}_{t\geqq 0}$ starting from x_0 defined on a filtered probability space $(\Omega, \widetilde{\mathscr{F}}, \mathbf{P}, \{\widetilde{\mathscr{F}}_t\})$ such that

$$\widetilde{M}_f(t) := f(\widetilde{X}(t)) - f(\widetilde{X}(0)) - \int_0^t (\widetilde{L}f)(\widetilde{X}(s))\,\mathrm{d}s$$

is an $\{\widetilde{\mathscr{F}}_t\}$-martingale for any $f \in C^2_{\mathrm{b}}(\mathbb{R}^d)$. Define $\{A(t)\}_{t\geqq 0}$ by

$$A(t) = \int_0^t \rho(\widetilde{X}(s))\,\mathrm{d}s$$

and set $e = \lim_{t\to\infty} A(t) \in (0, \infty]$. Since $\rho > 0$, $t \mapsto A(t)$ has an inverse function $\alpha(t)$ $(0 \leqq t < e)$. For fixed $t > 0$, $\alpha(t)$ is an $\{\widetilde{\mathscr{F}}_t\}$-stopping time and $\alpha(t) \to \infty$ as $t \to e$. We now set

$$\mathscr{F}_t = \widetilde{\mathscr{F}}_{\alpha(t)} \quad \text{and} \quad X(t) = \begin{cases} \widetilde{X}(\alpha(t)) & (t < e), \\ \triangle & (t \geqq e) \end{cases}$$

and show that the $\widehat{\mathbf{W}}^d$-valued random variable $X = \{X(t)\}_{t\geqq 0}$ defined on the probability space $(\Omega, \mathscr{F}, \mathbf{P}, \{\mathscr{F}_t\})$ is the desired stochastic process.

Set $\widetilde{\tau}_n = \inf\{t; |\widetilde{X}(t)| \geqq n\}$. By the optional stopping theorem $\{\widetilde{M}_f(t\wedge\widetilde{\tau}_n)\}_{t\geqq 0}$ is an $\{\widetilde{\mathscr{F}}_t\}$-martingale and $\{\widetilde{M}_f(\alpha(t)\wedge\widetilde{\tau}_n)\}_{t\geqq 0}$ is an $\{\mathscr{F}_t\}$-martingale. By definition, $\alpha(t) < \widetilde{\tau}_n$ is equivalent to $t < \tau_n$, and $\alpha(t)\wedge\widetilde{\tau}_n = \alpha(t\wedge\tau_n)$, $\widetilde{X}(\alpha(t)\wedge\widetilde{\tau}_n) = X(t\wedge\tau_n)$.

Moreover, since $t = \int_0^t (\rho(\widetilde{X}(s)))^{-1}\mathrm{d}A(s)$ and

$$\alpha(t) = \int_0^{\alpha(t)} \frac{1}{\rho(\widetilde{X}(s))}\,\mathrm{d}A(s) = \int_0^t \frac{1}{\rho(X(s))}\,\mathrm{d}s,$$

we have

$$\begin{aligned}
\widetilde{M}_f(\alpha(t)\wedge\widetilde{\tau}_n) &= f(\widetilde{X}(\alpha(t)\wedge\widetilde{\tau}_n)) - f(x_0) - \int_0^{\alpha(t\wedge\tau_n)} (\rho Lf)(\widetilde{X}(s))\,\mathrm{d}s \\
&= f(\widetilde{X}(\alpha(t\wedge\tau_n))) - f(x_0) - \int_0^{t\wedge\tau_n} \rho(\widetilde{X}(\alpha(u)))(Lf)(\widetilde{X}(\alpha(u)))\,\mathrm{d}\alpha(u) \\
&= f(X(t\wedge\tau_n)) - f(x_0) - \int_0^{t\wedge\tau_n} (Lf)(X(u))\,\mathrm{d}u.
\end{aligned}$$

Hence, $\{M_f(t)\}_{t\geqq 0}$ is an $\{\mathscr{F}_t\}$-martingale. □

We show a sufficient condition for solutions not to explode.

Theorem 4.3.9 *Assume that $\sigma(x)$ and $b(x)$ are continuous and there exists a positive constant K such that*

$$\|\sigma(x)\|^2 + |b(x)|^2 \leqq K(1+|x|^2) \qquad (x\in\mathbb{R}^d),$$

where

$$\|\sigma(x)\|^2 = \sum_{i=1}^d\sum_{k=1}^r (\sigma_k^i(x))^2, \quad |b(x)|^2 = \sum_{i=1}^d (b^i(x))^2.$$

If $\mathbf{E}[|X(0)|^p] < \infty$ for some $p \geqq 2$, then $\mathbf{E}\Big[\sup_{0\leqq t\leqq T}|X(t)|^p\Big] < \infty$ for any $T > 0$. In particular, no explosion occurs almost surely.

Proof Set $\sigma_n = \inf\{t; |X(t)| \geqq n\}$ and $X_n(t) = X(t\wedge\sigma_n)$. By the Burkholder–Davis–Gundy inequality (Theorem 2.4.1), there exists a constant C_1 such that

$$\mathbf{E}\bigg[\sup_{0\leqq s\leqq t}\bigg|\sum_{k=1}^r\int_0^s \sigma_k^i(X_n(u))\,\mathrm{d}B^k(u)\bigg|^p\bigg] \leqq C_1\mathbf{E}\bigg[\bigg\{\int_0^t |\sigma^i(X_n(u))|^2\mathrm{d}u\bigg\}^{\frac{p}{2}}\bigg]$$

for every $i = 1,2,\ldots,d$. Moreover, by Hölder's inequality,

$$\mathbf{E}\bigg[\bigg\{\int_0^t |\sigma^i(X_n(u))|^2\mathrm{d}u\bigg\}^{\frac{p}{2}}\bigg] \leqq \mathbf{E}\bigg[\int_0^t |\sigma^i(X_n(u))|^p\mathrm{d}u\bigg]t^{\frac{p-2}{2}}$$

and

$$\left|\int_0^t b^i(X_n(u))\,\mathrm{d}u\right|^p \leqq t^{\frac{p-1}{p}} \int_0^t |b^i(X_n(u))|^p \mathrm{d}u.$$

By the assumption, there exists a constant $C_i = C_i(K, p, T)$, independent of n, such that

$$\mathbf{E}\Big[\sup_{0 \leqq s \leqq t} |X_n(s)|^p\Big] \leqq C_2 + C_3 \int_0^t \mathbf{E}\Big[\sup_{0 \leqq u \leqq s} |X_n(u)|^p\Big] \mathrm{d}s$$

for any $t \in [0, T]$. Hence, by Gronwall's inequality (Theorem A.1.1), there exist constants C_4 and C_5 such that

$$\mathbf{E}\Big[\sup_{0 \leqq t \leqq T} |X_n(t)|^p\Big] \leqq C_4 \mathrm{e}^{C_5 T}.$$

We obtain the conclusion by letting $n \to \infty$. □

The explosion problem is studied in [84, Section 4.4] and [114, Chapter 10] from an analytic point of view. In the one-dimensional case, it is completely clarified and the result will be presented in Section 4.8. In that section, as an application of one-dimensional achievements, a sufficient condition for non-explosion of general dimensional diffusions due to Khasminskii is also shown.

4.4 Pathwise Uniqueness

The stochastic differential equations with Lipschitz continuous coefficients have nice properties, as in the case of ordinary differential equations.

Definition 4.4.1 $\gamma \in \mathscr{A}^{1,1}$ is called Lipschitz continuous if there exists a positive constant K such that

$$|\gamma(t, w) - \gamma(t, w')| \leqq K(w - w')^*(t) \qquad (t \geqq 0,\ w, w' \in \mathbf{W}^d),$$

where w^* is given by

$$w^*(t) = \max_{0 \leqq s \leqq t} |w(s)|.$$

The constant K is called a Lipschitz coefficient.

Theorem 4.4.2 *Suppose that each component of $\alpha \in \mathscr{A}^{d,r}$ and $\beta \in \mathscr{A}^{d,1}$ is Lipschitz continuous and, for any $T > 0$, there exists a positive constant C_T such that*

$$\|\alpha(t, 0)\|, |b(t, 0)| \leqq C_T \qquad (0 \leqq t \leqq T).$$

Then the stochastic differential equation (4.3.1) *has a solution and the pathwise uniqueness holds.*

Proof It suffices to consider a solution with $X(0) = x$ for a fixed $x \in \mathbb{R}^d$. We construct it by Picard's method of **successive approximation**.

Let (W^r, μ) be the r-dimensional Wiener space and let $\mathscr{F}_t$ be the σ-field obtained by the completion of $\sigma(\{w(s); s \leqq t\})$ by μ. Set $\mathscr{F}_\infty = \bigvee_{t \geqq 0} \mathscr{F}_t$. Then $(W^r, \mathscr{F}_\infty, \mu, \{\mathscr{F}_t\})$ is a probability space which satisfies the usual condition. Denote by $\{B(t)\}_{t \geqq 0}$ the coordinate process. It is an r-dimensional $\{\mathscr{F}_t\}$-Brownian motion under μ.

Fix $T > 0$ and define a sequence of $\{\mathscr{F}_t\}$-adapted stochastic processes $\{X^{(n)}(t)\}_{t \geqq 0}$ $(n = 0, 1, 2, \ldots)$ by

$$X^{(0)}(t) = x,$$
$$X^{(n+1)}(t) = x + \int_0^t \alpha(s, X^{(n)})\, \mathrm{d}B(s) + \int_0^t \beta(s, X^{(n)})\, \mathrm{d}s.$$

We can easily show by induction that $\mathbf{E}\Big[\sup_{0 \leqq t \leqq T} |X^{(n)}(t)|^2\Big] < \infty$ for each n, where $\mathbf{E}$ denotes the expectation with respect to μ.

To give an estimate for $\mathbf{E}\Big[\sup_{0 \leqq t \leqq T} |X^{(n+1)}(t) - X^{(n)}(t)|^2\Big]$, we show the following.

Lemma 4.4.3 *Let X and Y be $\{\mathscr{F}_t\}$-adapted stochastic processes satisfying*

$$\mathbf{E}\Big[\sup_{0 \leqq t \leqq T} |X(t)|^2\Big] < \infty, \qquad \mathbf{E}\Big[\sup_{0 \leqq t \leqq T} |Y(t)|^2\Big] < \infty$$

for any $T > 0$, and ξ and η be $\mathscr{F}_0$-measurable random variables. Define $\widetilde{X}$ and $\widetilde{Y}$ by

$$\widetilde{X}(t) = \xi + \int_0^t \sigma(s, X)\, \mathrm{d}B(s) + \int_0^t \beta(s, X)\, \mathrm{d}s,$$
$$\widetilde{Y}(t) = \eta + \int_0^t \sigma(s, Y)\, \mathrm{d}B(s) + \int_0^t \beta(s, Y)\, \mathrm{d}s.$$

Then there exists a constant C, dependent only on $T > 0$ and the Lipschitz coefficients of σ and β, such that

$$\mathbf{E}[(\widetilde{X} - \widetilde{Y})^*(t)^2] \leqq C\Big(\mathbf{E}[|\xi - \eta|^2] + \mathbf{E}\Big[\int_0^t (X - Y)^*(s)^2 \mathrm{d}s\Big]\Big)$$

for any $t \in [0, T]$.

Proof By definition

$$|\widetilde{X}(t) - \widetilde{Y}(t)| \leqq |\xi - \eta| + \left| \int_0^t (\sigma(s, X) - \sigma(s, Y))\, \mathrm{d}B(s) \right| + \left| \int_0^t (\beta(s, X) - \beta(s, Y))\, \mathrm{d}s \right|.$$

Applying Doob's inequality to the second term of the right hand side, we obtain

$$\mathbf{E}\left[\sup_{0 \leqq s \leqq t} \left| \int_0^s (\sigma(u, X) - \sigma(u, Y))\, \mathrm{d}B(u) \right|^2 \right]$$
$$\leqq 4\,\mathbf{E}\left[\left| \int_0^t (\sigma(u, X) - \sigma(u, Y))\, \mathrm{d}B(u) \right|^2 \right] = 4\,\mathbf{E}\left[\int_0^t \|\sigma(u, X) - \sigma(u, Y)\|^2 \mathrm{d}u \right].$$

Then the assumption implies

$$\mathbf{E}\left[\sup_{0 \leqq s \leqq t} \left| \int_0^s (\sigma(u, X) - \sigma(u, Y))\, \mathrm{d}B(u) \right|^2 \right] \leqq 4K^2 \mathbf{E}\left[\int_0^t (X - Y)^*(u)^2 \mathrm{d}u \right].$$

Moreover, the assumption and Schwarz's inequality yield

$$\mathbf{E}\left[\sup_{0 \leqq s \leqq t} \left| \int_0^s (\beta(u, X) - \beta(u, Y))\, \mathrm{d}u \right|^2 \right] \leqq t\,\mathbf{E}\left[\sup_{0 \leqq s \leqq t} \int_0^s (X - Y)^*(u)^2 \mathrm{d}u \right]$$
$$= t\,\mathbf{E}\left[\int_0^t (X - Y)^*(u)^2 \mathrm{d}u \right].$$

Combining the above estimates, we obtain the conclusion. □

Proof of Theorem 4.4.2 (continued) Set $\Delta^{(n)}(t) = \mathbf{E}[(X^{(n+1)} - X^{(n)})^*(t)^2]$. By the above lemma,

$$\Delta^{(n)}(t) \leqq C \int_0^t \Delta^{(n-1)}(s)\, \mathrm{d}s.$$

Set $A = \Delta^{(0)}(T)$. Then, by induction,

$$\Delta^{(n)}(T) \leqq A \frac{C^n T^n}{n!}.$$

Hence, Chebyshev's inequality implies

$$\mu\left(\sup_{0 \leqq t \leqq T} |X^{(n+1)}(t) - X^{(n)}(t)| > \frac{1}{2^n} \right) \leqq 4^n \mathbf{E}\left[\sup_{0 \leqq t \leqq T} |X^{(n+1)}(t) - X^{(n)}(t)|^2 \right]$$
$$\leqq A \frac{(4CT)^n}{n!}.$$

Thus

$$\sum_{n=1}^{\infty} \mu\left(\sup_{0 \leqq t \leqq T} |X^{(n+1)}(t) - X^{(n)}(t)| > \frac{1}{2^n} \right) < \infty.$$

By the Borel–Cantelli lemma, for almost all $w \in W^r$, there exists $N(w) \in \mathbb{N}$ such that

$$\sup_{0 \leqq t \leqq T} |X^{(n+1)}(t,w) - X^{(n)}(t,w)| \leqq \frac{1}{2^n}$$

for all $n \geqq N(w)$. This means that

$$X^{(n)}(t) = X^{(0)} + \sum_{j=1}^{n} (X^{(j)}(t) - X^{(j-1)}(t))$$

converges uniformly as $n \to \infty$ almost surely.

Denote the limit by X. Then, by Schwarz's inequality and the estimate for $\Delta^{(n)}(T)$,

$$\mathbf{E}[(X - X^{(n)})^*(T)^2] \leqq \left(\sum_{j=n+1}^{\infty} \sqrt{A \frac{(CT)^j}{j!}} \right)^2 \to 0 \qquad (n \to \infty).$$

On the other hand, setting

$$\overline{X}(t) = x + \int_0^t \sigma(s,X)\,\mathrm{d}B(s) + \int_0^t \beta(s,X)\,\mathrm{d}s,$$

we obtain by the above lemma

$$\mathbf{E}[(\overline{X} - X^{(n)})^*(T)^2] \leqq C \int_0^T \mathbf{E}[(X - X^{(n-1)})^*(t)^2]\,\mathrm{d}t \to 0 \qquad (n \to \infty).$$

Hence, $X = \overline{X}$ and X satisfies the stochastic differential equation (4.3.1).

To show the pathwise uniqueness, let X and X' be the solutions defined on the same probability space. By the lemma, there exists a constant C_T such that

$$\mathbf{E}[(X - X')^*(t)^2] \leqq C_T \int_0^t \mathbf{E}[(X - X')^*(s)^2]\,\mathrm{d}s \quad (t \in [0,T]).$$

This implies that $\mathbf{E}[(X - X')^*(T)^2] = 0$ and $X(t) = X'(t)$ $(0 \leqq t \leqq T)$ almost surely. □

Remark 4.4.4 Suppose that α and β are locally Lipschitz continuous, that is, for any $N > 0$ there exists a constant K_N such that

$$\|\alpha(t,w) - \alpha(t,w')\| + |\beta(t,w) - \beta(t,w')| \leqq K_N (w - w')^*(t)$$

for $w, w' \in \mathbf{W}^d$ and t with $w^*(t), (w')^*(t) \leqq N$ and $0 \leqq t \leqq N$. If we assume that there is no explosion, then the pathwise uniqueness holds. For details, see for example [100, p.132]. A sufficient condition for no explosion is that

$$\|\alpha(t,w)\| + |\beta(t,w)| \leqq K'_N (w)^*(t) \qquad (0 \leqq t \leqq N).$$

For Markovian stochastic differential equations with locally Lipschitz continuous coefficients, the pathwise uniqueness holds. We omit the proof and refer the reader to [45, p.178].

Theorem 4.4.5 *Suppose that $\sigma : \mathbb{R} \to \mathbb{R}^d \otimes \mathbb{R}^r$ and $b : \mathbb{R}^d \to \mathbb{R}^d$ are locally Lipschitz continuous, that is, for any $N > 0$ there exists a positive constant K_N such that*

$$\|\sigma(x) - \sigma(y)\| + |b(x) - b(y)| \leqq K_N|x - y| \quad (|x|, |y| \leqq N).$$

Then the pathwise uniqueness of solutions holds for the stochastic differential equation

$$\mathrm{d}X(t) = \sigma(X(t))\,\mathrm{d}B(t) + b(X(t))\,\mathrm{d}t.$$

Hence this stochastic differential equation has a unique strong solution.

For the one-dimensional stochastic differential equation, deep and interesting results on pathwise uniqueness are known. In the remaining of this section, we list them without proofs. For details, see [69, 89, 128] and also [45, 100].

Theorem 4.4.6 (Yamada [128]) *Let $\sigma, b : \mathbb{R} \to \mathbb{R}$ be bounded functions such that*

(i) *there exists a strictly monotone increasing function ρ on $[0, \infty)$ such that*

$$\rho(0) = 0, \qquad \int_{0+} \rho(u)^{-2}\mathrm{d}u = \infty \quad \text{and} \quad |\sigma(x)-\sigma(y)| \leqq \rho(|x-y|) \ (x, y \in \mathbb{R}),$$

(ii) *there exists a monotone increasing concave function κ on $[0, \infty)$ such that*

$$\kappa(0) = 0, \qquad \int_{0+} \kappa(u)^{-1}\mathrm{d}u = \infty \quad \text{and} \quad |b(x) - b(y)| \leqq \kappa(|x - y|) \ (x, y \in \mathbb{R}).$$

Then, for the stochastic differential equation

$$\mathrm{d}X(t) = \sigma(X(t))\,\mathrm{d}B(t) + b(X(t))\,\mathrm{d}t, \tag{4.4.1}$$

the pathwise uniqueness of solutions holds.

If σ is $\frac{1}{2}$-Hölder continuous and b is Lipschitz continuous, then the conditions of the theorem are fulfilled.

Theorem 4.4.7 (Nakao [89], Le Gall [69]) *Let $\sigma, b : \mathbb{R} \to \mathbb{R}$ be bounded functions. Assume that there exist a positive constant $\varepsilon > 0$ and a bounded and monotone increasing function f such that*

$$\sigma(x) \geqq \varepsilon \quad \text{and} \quad |\sigma(x) - \sigma(y)|^2 \leqq |f(x) - f(y)| \ (x, y \in \mathbb{R}).$$

Then, for the stochastic differential equation (4.4.1), *the pathwise uniqueness of solutions holds.*

Under the assumptions of these theorems, the existence of solutions is seen by Theorem 4.3.8. Hence, by Corollary 4.2.15, the uniqueness in law also holds. The uniqueness in law for the Markovian stochastic differential equations is also discussed in the next section.

4.5 Martingale Problems

Given $a^{ij}, b^i : \mathbb{R}^d \to \mathbb{R}$ $(i, j = 1, 2, \ldots, d)$, assume that (a^{ij}) is symmetric and non-negative definite. We consider the diffusion process corresponding to

$$L = \frac{1}{2}\sum_{i,j=1}^{d} a^{ij}(x)\frac{\partial^2}{\partial x^i \partial x^j} + \sum_{i=1}^{d} b^i(x)\frac{\partial}{\partial x^i}. \tag{4.5.1}$$

While some of the results mentioned below continue to hold in the case where the coefficients a^{ij} and b^i depend on paths, we do not mention them here. For more results, see [100, 114].

Definition 4.5.1 For $y \in \mathbb{R}^d$, a probability measure P^y on the path space $(\mathbf{W}^d, \mathscr{B}(\mathbf{W}^d))$ is called a solution of the **martingale problem** for L starting from y if $P^y(w(0) = y) = 1$ and

$$M_f(t) = f(w(t)) - f(w(0)) - \int_0^t (Lf)(w(s))\,\mathrm{d}s$$

is a $\{\mathscr{B}_t(\mathbf{W}^d)\}$-martingale under P^y for any $f \in C_0^\infty(\mathbb{R}^d)$.

If there exists a unique solution starting from y for any $y \in \mathbb{R}^d$, the martingale problem is called **well posed**.

By Theorems 4.3.1 and 4.3.3, a solution of the stochastic differential equation (4.3.2) exists if and only if so does that for a martingale problem for L. Moreover, the following holds.

Theorem 4.5.2 *Suppose that there exists a bounded continuous function* $\sigma : \mathbb{R}^d \to \mathbb{R} \otimes \mathbb{R}^d$ *such that* $a(x) = \sigma(x)\sigma(x)^*$ *and* $b(x)$ *is also bounded and continuous. Then there exists a solution* P^y *of the martingale problem for* L *starting from any* $y \in \mathbb{R}^d$.

For the uniqueness of solutions, the following conclusive result is known.

Theorem 4.5.3 *Suppose that the diffusion coefficient a and the drift b satisfy*

(i) *a is continuous,*
(ii) *$a(x)$ is strictly positive definite for any x,*
(iii) *there exists a constant K such that $|a^{ij}(x)| \leqq K(1+|x|^2)$, $|b^i(x)| \leqq K(1+|x|)$ $(i,j=1,2,\ldots,d)$.*

Then the martingale problem for L is well posed. In particular, the uniqueness in law holds for the stochastic differential equation (4.3.2).

We omit the proof. See [114, Theorem 7.2.1] and a survey in [100, p.170].

As was mentioned in Section 4.1, one of the advantages of martingale problems is that the strong Markov property of the corresponding diffusion processes is deduced from the well-posedness of them.

Theorem 4.5.4 *Let L be the second order elliptic differential operator defined by* (4.5.1). *Assume that the martingale problem for L is well posed. Then, the coordinate process on $\mathbf{W}^d$ is a strong Markov process under the unique solution P^y starting from y.*

A proof of the theorem is carried out as follows. Let τ be a bounded $\{\mathscr{F}_t\}$-stopping time, where $\{\mathscr{F}_t\}_{t\geqq 0}$ is the filtration described in Section 4.4. Define $\theta_\tau : W^d \to W^d$ by $(\theta_\tau w)(t) = w(\tau(w)+t)$. Moreover, let $(p|\mathscr{F}_\tau)(w,A) : W^d \times \mathscr{F} \to [0,1]$ be the regular conditional probability of P^y given $\mathscr{F}_\tau$. Then, the probability law of θ_τ under $(p|\mathscr{F}_\tau)(w,\cdot)$ is a solution of the martingale problem starting from $X(\tau,w)$. By the uniqueness, it coincides with $P^{X(\tau,w)}$. The details are omitted.

4.6 Exponential Martingales and Transformation of Drift

For $\alpha \in \mathscr{A}^{d,r}$ and $\beta \in \mathscr{A}^{d,1}$, consider the stochastic differential equation (4.2.1) or (4.3.1):

$$\mathrm{d}X(t) = \alpha(t,X)\,\mathrm{d}B(t) + \beta(t,X)\,\mathrm{d}t.$$

The aim of this section is to introduce **transformation of drift**, which guarantees that, if there exists a solution of this equation, then the equation

$$\mathrm{d}X(t) = \alpha(t,X)\,\mathrm{d}B(t) + (\beta(t,X) + \alpha(t,X)\gamma(t,X))\,\mathrm{d}t \tag{4.6.1}$$

also has a solution for suitable $\gamma \in \mathscr{A}^{r,1}$.[6] For this purpose it is important to study whether the local martingale $\{\mathscr{E}_M(t)\}_{t\geqq 0}$ defined by

$$\mathscr{E}_M(t) = \exp\Big(M(t) - \frac{1}{2}\langle M\rangle(t)\Big), \tag{4.6.2}$$

M being a local martingale, is a martingale or not. It should also be noted that transformations of drifts relate changes of probability measures on Wiener spaces.

For an $M = \{M(t)\}_{t\geqq 0} \in \mathscr{L}^2_{\mathrm{c,loc}}$ with $M(0) = 0$ on $(\Omega, \mathscr{F}, \mathbf{P}, \{\mathscr{F}_t\})$, define $\{\mathscr{E}_M(t)\}_{t\geqq 0}$ by (4.6.2). By Itô's formula, we have the following.

Proposition 4.6.1 *$\mathscr{E}_M(t) = 1 + \int_0^t \mathscr{E}_M(s)\,\mathrm{d}M(s)$ and $\mathscr{E}_M = \{\mathscr{E}_M(t)\}_{t\geqq 0}$ is a continuous local martingale.*

Since $\mathscr{E}_M(t) > 0$, $\mathscr{E}_M$ is a supermartingale (Proposition 2.1.2). Hence $\mathscr{E}_M$ is a martingale if and only if

$$\mathbf{E}[\mathscr{E}_M(t)] = 1 \qquad (t \geqq 0).$$

If $\mathscr{E}_M$ is a martingale, a probability measure $\mathbf{P}_M$ on $(\Omega, \mathscr{F}_T)$, which is absolutely continuous with $\mathbf{P}$, is defined by

$$\mathbf{P}_M(A) = \mathbf{E}[\mathscr{E}_M(t)\mathbf{1}_A] \qquad (A \in \mathscr{F}_t,\ 0 \leqq t \leqq T)$$

for any $T > 0$. The following **Girsanov's theorem**, which is often called the Cameron–Martin–Maruyama–Girsanov theorem, is essential to achieve the transformation of drift.

Theorem 4.6.2 *Let $(\Omega, \mathscr{F}, \mathbf{P}, \{\mathscr{F}_t\})$ be a filtered probability space and $B = \{(B^1(t), B^2(t), \dots, B^r(t))\}_{t\geqq 0}$ be an $\{\mathscr{F}_t\}$-Brownian motion. Set $\widehat{B^i}(t) = B^i(t) - \langle B^i, M\rangle(t)$. If $\mathscr{E}_M$ is a martingale, then $\widehat{B} = \{(\widehat{B^1}(t), \widehat{B^2}(t), \dots, \widehat{B^r}(t))\}_{t\geqq 0}$ is an $\{\mathscr{F}_t\}$-Brownian motion on $(\Omega, \mathscr{F}, \mathbf{P}_M, \{\mathscr{F}_t\})$.*

Proof By virtue of the Lévy theorem (Theorem 2.5.1), it suffices to show that $\{\widehat{B^i}(t)\}_{t\geqq 0}$ $(i = 1, 2, \dots, r)$ is a local martingale on $(\Omega, \mathscr{F}, \mathbf{P}_M, \{\mathscr{F}_t\})$.

Since

$$\begin{aligned}
&\widehat{B^i}(t)\mathscr{E}_M(t) - \widehat{B^i}(0) \\
&\quad = \int_0^t \mathscr{E}_M(s)\,\mathrm{d}B^i(s) + \int_0^t B^i(s)\,\mathrm{d}\mathscr{E}_M(s) - \int_0^t \langle B^i, M\rangle(s)\,\mathrm{d}\mathscr{E}_M(s)
\end{aligned}$$

[6] This was first shown by Maruyama [77, 78].

by Itô's formula, $\{\widehat{B^i}(t)\mathscr{E}_M(t)\}_{t\geqq 0}$ is a continuous local martingale under $\mathbf{P}$. Hence, there exists a sequence of stopping times $\{\tau_n\}$ with $\mathbf{P}(\tau_n \to \infty) = 1$ such that, for any $s < t$ and $A \in \mathscr{F}_s$,

$$\mathbf{E}[\widehat{B^i}(t \wedge \tau_n)\mathscr{E}_M(t \wedge \tau_n)\mathbf{1}_A] = \mathbf{E}[\widehat{B^i}(s \wedge \tau_n)\mathscr{E}_M(s \wedge \tau_n)\mathbf{1}_A].$$

Since $A \cap \{\tau_n > s\} \in \mathscr{F}_{s\wedge\tau_n}$, Doob's optional sampling theorem implies

$$\begin{aligned}
\mathbf{E}^{\mathbf{P}_M}[\widehat{B^i}(t \wedge \tau_n)\mathbf{1}_{A\cap\{\tau_n > s\}}] &= \mathbf{E}[\widehat{B^i}(t \wedge \tau_n)\mathscr{E}_M(T)\mathbf{1}_{A\cap\{\tau_n > s\}}] \\
&= \mathbf{E}[\widehat{B^i}(t \wedge \tau_n)\mathscr{E}_M(t \wedge \tau_n)\mathbf{1}_{A\cap\{\tau_n > s\}}] = \mathbf{E}[\widehat{B^i}(s \wedge \tau_n)\mathscr{E}_M(s \wedge \tau_n)\mathbf{1}_{A\cap\{\tau_n > s\}}] \\
&= \mathbf{E}[\widehat{B^i}(s \wedge \tau_n)\mathscr{E}_M(T)\mathbf{1}_{A\cap\{\tau_n > s\}}] = \mathbf{E}^{\mathbf{P}_M}[\widehat{B^i}(s \wedge \tau_n)\mathbf{1}_{A\cap\{\tau_n > s\}}]
\end{aligned}$$

If $\tau_n \leqq s$, then $t \wedge \tau_n = \tau_n$ and

$$\mathbf{E}^{\mathbf{P}_M}[\widehat{B^i}(t \wedge \tau_n)\mathbf{1}_{A\cap\{\tau_n \leqq s\}}] = \mathbf{E}^{\mathbf{P}_M}[\widehat{B^i}(s \wedge \tau_n)\mathbf{1}_{A\cap\{\tau_n \leqq s\}}].$$

Summing up the above, we obtain

$$\mathbf{E}^{\mathbf{P}_M}[\widehat{B^i}(t \wedge \tau_n)\mathbf{1}_A] = \mathbf{E}^{\mathbf{P}_M}[\widehat{B^i}(s \wedge \tau_n)\mathbf{1}_A],$$

which means that each component of $\widehat{B}$ is a local martingale under $\mathbf{P}_M$. □

Remark 4.6.3 The probability measures $\mathbf{P}$ and $\mathbf{P}_M$ are absolutely continuous on $(\Omega, \mathscr{F}_t)$ for any $t > 0$. However, they are not absolutely continuous on $(\Omega, \mathscr{F}_\infty)$ in general, where $\mathscr{F}_\infty = \bigvee_{t\geqq 0} \mathscr{F}_t$.

To see this, let $d = 1$ and set $M(t) = \mu B(t)$ for a non-zero constant μ. Then, $\widehat{B}(t) = B(t) - \mu t$ is a Brownian motion under $\mathbf{P}_M$ and

$$\mathbf{P}_M\Big(\lim_{t\to\infty} t^{-1}B(t) = \mu\Big) = \mathbf{P}_M\Big(\lim_{t\to\infty} t^{-1}\widehat{B}(t) = 0\Big) = 1,$$

but $\mathbf{P}\big(\lim_{t\to\infty} t^{-1}B(t) = \mu\big) = 0$.

If $\mathrm{d}\mathbf{P}_M = \Phi\,\mathrm{d}\mathbf{P}$ on $\mathscr{F}_\infty$, then $\mathscr{E}_M(T) = \mathbf{E}[\Phi|\mathscr{F}_T]$ and $\{\mathscr{E}_M(t)\}_{t\geqq 0}$ is uniformly integrable. Therefore, $\mathbf{P}$ and $\mathbf{P}_M$ are absolutely continuous on $(\Omega, \mathscr{F}_\infty)$ if and only if $\{\mathscr{E}_M(t)\}_{t\geqq 0}$ is uniformly integrable. See Theorem 1.5.16 on the martingale convergence theorem.

Consider a local martingale $\{M(t)\}_{t\geqq 0}$ given by

$$M(t) = \sum_{k=1}^{r} \int_0^t \gamma^k(s, X)\,\mathrm{d}B^k(s), \tag{4.6.3}$$

where $\gamma \in \mathscr{A}^{r,1}$. The following is obtained from Theorem 4.6.2.

Theorem 4.6.4 *Define M by* (4.6.3). *Suppose that $\mathscr{E}_M$ is a martingale. If (X, B) is a solution of the stochastic differential equation* (4.3.1), *then $(X, \widehat{B})$ is a solution of* (4.6.1) *under $\mathbf{P}_M$.*

From the above observations, it is important to have sufficient conditions on M so that $\mathscr{E}_M$ is a martingale. The following two criteria are well known.

Theorem 4.6.5 (Novikov [90, 91]**)** *Let $M \in \mathscr{M}^2_{\mathrm{c,loc}}$. If*

$$\mathbf{E}\Big[\exp\Big(\frac{1}{2}\langle M\rangle(t)\Big)\Big] < \infty \qquad (t \geqq 0), \tag{4.6.4}$$

then $\mathbf{E}[\mathscr{E}_M(t)] = 1$ holds and $\{\mathscr{E}_M(t)\}_{t\geqq 0}$ is a continuous martingale [91, 92].

Theorem 4.6.6 (Kazamaki [59]**)** *Let $M \in \mathscr{M}^2_{\mathrm{c,loc}}$. If*

$$\mathbf{E}\Big[\exp\Big(\frac{1}{2}M(t)\Big)\Big] < \infty \qquad (t \geqq 0), \tag{4.6.5}$$

then $\mathbf{E}[\mathscr{E}_M(t)] = 1$ holds and $\{\mathscr{E}_M(t)\}_{t\geqq 0}$ is a continuous martingale [60].

The **Novikov condition** (4.6.4) implies the **Kazamaki condition** (4.6.5). In fact, for any $\alpha > 0$, Schwarz's inequality implies

$$\begin{aligned}\mathbf{E}[\mathrm{e}^{\alpha M(t)}] &= \mathbf{E}[\mathrm{e}^{\alpha M(t)-\alpha^2\langle M\rangle(t)}\mathrm{e}^{\alpha^2\langle M\rangle(t)}]\\ &\leqq \{\mathbf{E}[\mathrm{e}^{2\alpha M(t)-2\alpha^2\langle M\rangle(t)}]\}^{\frac{1}{2}}\{\mathbf{E}[\mathrm{e}^{2\alpha^2\langle M\rangle(t)}]\}^{\frac{1}{2}}.\end{aligned}$$

Since $\{\exp(2\alpha M(t) - 2\alpha^2\langle M\rangle(t))\}_{t\geqq 0}$ is a supermartingale, we have

$$\mathbf{E}[\mathrm{e}^{\alpha M(t)}] \leqq \{\mathbf{E}[\mathrm{e}^{2\alpha^2\langle M\rangle(t)}]\}^{\frac{1}{2}}.$$

Set $\alpha = \frac{1}{2}$ to see that (4.6.5) is weaker.

Hence, it suffices to show Theorem 4.6.6. However, we give the proofs of both theorems, since they both stem from the following lemma, and have their own interests.

Lemma 4.6.7 *Let $\{B(t)\}_{t\geqq 0}$ be a one-dimensional Brownian motion starting from 0 and set $\sigma_a = \inf\{t > 0; B(t) = t - a\}$ for $a > 0$. Then,*

$$\mathbf{E}[\mathrm{e}^{-\lambda\sigma_a}] = \mathrm{e}^{-(\sqrt{1+2\lambda}-1)a} \qquad (\lambda > 0). \tag{4.6.6}$$

Proof Set $u(t, x) = \exp(-\lambda t - (\sqrt{1 + 2\lambda} - 1)x)$. Then

$$\frac{\partial u}{\partial t} - \frac{\partial u}{\partial x} + \frac{1}{2}\frac{\partial^2 u}{\partial x^2} = 0.$$

Hence, by Itô's formula,

$$u(t, B(t) - t) = 1 + \int_0^t \frac{\partial u}{\partial x}(s, B(s) - s)\,\mathrm{d}B(s).$$

Then, by the optional sampling theorem,

$$\mathbf{E}[u(t \wedge \sigma_a, B(t \wedge \sigma_a) - (t \wedge \sigma_a))] = 1.$$

Since $B(t \wedge \sigma_a) - (t \wedge \sigma_a) \geqq -a$, due to the bounded convergence theorem, we obtain

$$1 = \mathbf{E}[u(\sigma_a, B(\sigma_a) - \sigma_a)] = \mathbf{E}[\mathrm{e}^{-\lambda\sigma_a - (\sqrt{1+2\lambda}-1)(-a)}]. \qquad \square$$

Proofs of Theorems 4.6.5 and 4.6.6 By Theorem 2.5.5, there exists an extension $(\widetilde{\Omega}, \widetilde{\mathscr{F}}, \widetilde{\mathbf{P}}, \{\widetilde{\mathscr{F}}_t\})$ of $(\Omega, \mathscr{F}, \mathbf{P}, \{\mathscr{F}_t\})$ and an $\{\widetilde{\mathscr{F}}_t\}$-Brownian motion B with $B(0) = 0$ such that $M(t) = B(\langle M\rangle(t))$. For each $s \geqq 0$, $\langle M\rangle(s)$ is an $\{\widetilde{\mathscr{F}}_t\}$-stopping time. Setting $\sigma_a = \inf\{t > 0; B(t) = t - a\}$ as in Lemma 4.6.7, we have

$$\mathbf{P}(\sigma_a \in \mathrm{d}t) = \frac{a\mathrm{e}^a}{\sqrt{2\pi t^3}}\mathrm{e}^{-\frac{t}{2}-\frac{a^2}{2t}}\,\mathrm{d}t$$

and

$$\mathbf{E}[\mathrm{e}^{\frac{\sigma_a}{2}}] = \mathrm{e}^a.$$

Set $Y(t) = \exp(B(t \wedge \sigma_a) - \frac{1}{2}(t \wedge \sigma_a))$. By the optional sampling theorem and Fatou's lemma, $\{Y(t)\}_{t\geqq 0}$ is a closable and, hence, uniformly integrable $\{\widetilde{\mathscr{F}}_t\}$-martingale.

First we prove Theorem 4.6.5. By the uniform integrability of $\{Y(t)\}_{t\geqq 0}$, the optional sampling theorem implies

$$\mathbf{E}\Big[\exp\Big(B(\sigma \wedge \sigma_a) - \frac{1}{2}(\sigma \wedge \sigma_a)\Big)\Big] = 1$$

for any $\{\widetilde{\mathscr{F}}_t\}$-stopping time σ. In particular, setting $\sigma = \langle M\rangle(t)$, we have

$$\mathbf{E}[\mathrm{e}^{-a+\frac{\sigma_a}{2}}\mathbf{1}_{\{\sigma_a \leqq \langle M\rangle(t)\}}] + \mathbf{E}[\mathrm{e}^{M(t)-\frac{\langle M\rangle(t)}{2}}\mathbf{1}_{\{\sigma_a > \langle M\rangle(t)\}}] = 1.$$

Since $\mathbf{E}[\mathrm{e}^{-a+\frac{\sigma_a}{2}}\mathbf{1}_{\{\sigma_a \leqq \langle M\rangle(t)\}}] \leqq \mathrm{e}^{-a}\mathbf{E}[\mathrm{e}^{\frac{\langle M\rangle(t)}{2}}]$, by the assumption, the first term of the left hand side tends to 0 as $a \to \infty$. Therefore we obtain $\mathbf{E}[\mathrm{e}^{M(t)-\frac{\langle M\rangle(t)}{2}}] = 1$ and the conclusion of Theorem 4.6.5.

Next we prove Theorem 4.6.6. Since $\{Y(t)\}_{t\geqq 0}$ is uniformly integrable, by the optional sampling theorem, $\mathbf{E}[Y(\langle M\rangle(t))] = 1$. Rewrite as

$$\mathbf{E}[Y(\langle M\rangle(t))\mathbf{1}_{\{\sigma_a > \langle M\rangle(t)\}}] + \mathbf{E}[Y(\langle M\rangle(t))\mathbf{1}_{\{\sigma_a \leqq \langle M\rangle(t)\}}] = 1. \tag{4.6.7}$$

If $\sigma_a > \langle M\rangle(t)$, then $Y(\langle M\rangle(t)) = \mathscr{E}_M(t)$. Hence

$$\mathbf{E}[Y(\langle M\rangle(t))\mathbf{1}_{\{\sigma_a > \langle M\rangle(t)\}}] \leqq \mathbf{E}[\mathscr{E}_M(t)]. \tag{4.6.8}$$

For the second term of (4.6.7), we have

$$\mathbf{E}[Y(\langle M\rangle(t))\mathbf{1}_{\{\sigma_a\leqq\langle M\rangle(t)\}}] \leqq \mathbf{E}[\mathrm{e}^{\frac{1}{4}B(\sigma_a\wedge\langle M\rangle(t))}\mathrm{e}^{\frac{3}{4}B(\sigma_a)-\frac{1}{2}\sigma_a}] \leqq I_1^{\frac{1}{2}}I_2^{\frac{1}{2}},$$

where I_1 and I_2 are given by

$$I_1 = \mathbf{E}[\mathrm{e}^{\frac{1}{2}B(\sigma_a\wedge\langle M\rangle(t))}] \qquad \text{and} \qquad I_2 = \mathbf{E}[\mathrm{e}^{\frac{3}{2}B(\sigma_a)-\sigma_a}].$$

To estimate I_1, we consider an $\{\widetilde{\mathscr{F}}_t\}$-stopping time $T_a = \inf\{t \geqq 0; \langle M\rangle(t) \geqq \sigma_a\}$. Since $\langle M\rangle(t\wedge T_a) = \langle M\rangle(t)\wedge\sigma_a$ and $\{M(t) = B(\langle M\rangle(t))\}_{t\geqq 0}$ is an $\{\widetilde{\mathscr{F}}_{\langle M\rangle(t)}\}$-martingale, by the optional sampling theorem, we have

$$\mathbf{E}[B(\langle M\rangle(t))|\widetilde{\mathscr{F}}_{\langle M\rangle(t\wedge T_a)}] = B(\langle M\rangle(t\wedge T_a)) = B(\langle M\rangle(t)\wedge\sigma_a).$$

Hence, by Jensen's inequality,

$$\begin{aligned} I_1 &= \mathbf{E}\Big[\exp\Big(\frac{1}{2}\mathbf{E}[B(\langle M\rangle(t))|\widetilde{\mathscr{F}}_{\langle M\rangle(t\wedge T_a)}]\Big)\Big] \\ &\leqq \mathbf{E}\Big[\exp\Big(\frac{1}{2}B(\langle M\rangle(t))\Big)\Big] = \mathbf{E}\Big[\exp\Big(\frac{M(t)}{2}\Big)\Big]. \end{aligned}$$

On the other hand, since $B(\sigma_a) - \sigma_a = -a$,

$$I_2 = \mathbf{E}[\mathrm{e}^{\frac{\sigma_a}{2}}]\mathrm{e}^{-\frac{3a}{2}} = \mathrm{e}^{-\frac{a}{2}}.$$

Hence, the second term of the left hand side of (4.6.7) tends to 0 as $a\to\infty$. By (4.6.7) and (4.6.8), we obtain $\mathbf{E}[\mathscr{E}_M(t)] \geqq 1$ for every $t \geqq 0$ and the conclusion. □

Finally we present a necessary and sufficient condition for exponential local martingales to be martingales by an explosion problem of solutions of the corresponding stochastic differential equations.

Let b and $V_k : \mathbb{R}^d \to \mathbb{R}^d$ $(k = 1, 2, \ldots, r)$ be continuous and identify them with vector fields on $\mathbb{R}^d$. Suppose that the stochastic differential equation

$$\mathrm{d}X(t) = \sum_{k=1}^{r} V_k(X(t))\,\mathrm{d}B^k(t) + b(X(t))\,\mathrm{d}t, \quad X(0) = x$$

has a unique solution $X = \{X(t)\}_{t\geqq 0}$ and denote its explosion time by ζ.

Let f_k $(k = 1, 2, \ldots, r)$ be Borel functions on $\mathbb{R}^d$ and set $\widetilde{b} = b - \sum_{k=1}^{r} f_k V_k$. Suppose also that the stochastic differential equation

$$\mathrm{d}Y(t) = \sum_{k=1}^{r} V_k(Y(t))\,\mathrm{d}B^k(t) + \widetilde{b}(Y(t))\,\mathrm{d}t, \quad Y(0) = x$$

has a unique solution $Y = \{Y(t)\}_{t \geqq 0}$ and define a local martingale $M = \{M(t)\}_{t \geqq 0}$ by

$$M(t) = \exp\Big(\sum_{k=1}^{r} \int_0^t f_k(Y(s))\,\mathrm{d}B^k(s) - \frac{1}{2}\sum_{k=1}^{r} \int_0^t f_k(Y(s))^2 \mathrm{d}s\Big).$$

Then, we have the following ([84, Section 3.7]).

Theorem 4.6.8 *Assume that Y does not explode. Then, for any $T > 0$ and Borel subset A in $\mathbf{W}_T^d$,*

$$\mathbf{P}(X \in A, \zeta > T) = \mathbf{E}[M(T)\mathbf{1}_{\{Y \in A\}}].$$

X does not explode if and only if the identity $\mathbf{E}[M(T)] = 1$ holds, which is equivalent to M being a martingale.

Proof Let $\varphi_n \in C^\infty(\mathbb{R}^d)$ be a function such that

$$\varphi_n(x) = \begin{cases} 1 & (|x| \leqq n+1), \\ 0 & (|x| \geqq n+2) \end{cases}$$

and set

$$V_k^{(n)} = \varphi_n V_k, \quad b^{(n)} = \varphi_n b, \quad \widetilde{b}^{(n)} = \varphi_n \widetilde{b}.$$

Denote by $X^{(n)} = \{X^{(n)}(t)\}_{t \geqq 0}$ and $Y^{(n)} = \{Y^{(n)}(t)\}_{t \geqq 0}$ the solutions of the stochastic differential equations

$$\mathrm{d}X(t) = \sum_{k=1}^{r} V_k^{(n)}(X(t))\,\mathrm{d}B^k(t) + b^{(n)}(X(t))\,\mathrm{d}t, \quad X(0) = x,$$

$$\mathrm{d}Y(t) = \sum_{k=1}^{r} V_k^{(n)}(Y(t))\,\mathrm{d}B^k(t) + \widetilde{b}^{(n)}(Y(t))\,\mathrm{d}t, \quad Y(0) = x,$$

respectively. Set

$$\zeta_n = \inf\{t \geqq 0; |X^{(n)}(t)| \geqq n\} \quad \text{and} \quad \zeta_n^Y = \inf\{t \geqq 0; |Y^{(n)}(t)| \geqq n\}.$$

By the assumption, $\lim_{n\to\infty} \zeta_n = \zeta$ and $\lim_{n\to\infty} \zeta_n^Y = \infty$ almost surely.

Now define $M^{(n)} = \{M^{(n)}(t)\}_{t \geqq 0}$ by

$$M^{(n)}(t) = \exp\Big(\sum_{k=1}^{r} \int_0^t \varphi_{n+1}(Y^{(n)}(s)) f_k(Y^{(n)}(s))\,\mathrm{d}B^k(s) - \frac{1}{2}\sum_{k=1}^{r} \int_0^t \varphi_{n+1}(Y^{(n)}(s))^2 f_k(Y^{(n)}(s))^2 \mathrm{d}s\Big).$$

Then $M^{(n)}$ is a martingale and, by Theorem 4.6.4,

$$\mathbf{P}(X^{(n)} \in A') = \mathbf{P}[M^{(n)}(T)\mathbf{1}_{\{Y^{(n)} \in A'\}}]$$

for any Borel subset A' in $\mathbf{W}_T^d$. Hence, setting

$$\tau_n(w) = \inf\{t \geqq 0; |w(t)| \geqq n\} \qquad (w \in \mathbf{W}^d)$$

and putting $A' = A \cap \{\tau_n > T\}$, we obtain

$$\mathbf{P}(X \in A, \zeta_n > T) = \mathbf{E}[M(T)\mathbf{1}_{\{Y \in A, \zeta_n^Y > T\}}].$$

Let $n \to \infty$ to obtain the desired conclusion.

The assertion (2) follows from (1). □

4.7 Solutions by Time Change

Time changes, whose importance was seen in the representation theorems for martingales, are also useful to solve stochastic differential equations. In fact, in the proof of Theorem 4.3.8, we have already used a similar argument as described below.

A non-negative $\{\mathscr{F}_t\}$-adapted stochastic process $\phi = \{\phi(t)\}_{t \geqq 0}$ defined on a probability space $(\Omega, \mathscr{F}, \mathbf{P}, \{\mathscr{F}_t\})$ is called a **time change process** with respect to $\{\mathscr{F}_t\}$ if

(i) $\phi(0) = 0$,
(ii) ϕ is continuous and strictly increasing,
(iii) $\lim_{t \uparrow \infty} \phi(t) = \infty$.

Denote the inverse function of $t \mapsto \phi(t, \omega)$ by ϕ^{-1}. For arbitrarily fixed $u \geqq 0$, $\{\phi^{-1}(u) \leqq t\} = \{\phi(t) \geqq u\} \in \mathscr{F}_t$ for any t. Thus $\phi^{-1}(u)$ is an $\{\mathscr{F}_t\}$-stopping time.

Given an $\{\mathscr{F}_t\}$-adapted stochastic process $X = \{X(t)\}_{t \geqq 0}$, the stochastic process $T^\phi X = \{(T^\phi X)(t)\}_{t \geqq 0}$ defined by $(T^\phi X)(t) = X(\phi^{-1}(t))$ is $\{\mathscr{F}_{\phi^{-1}(t)}\}$-adapted. $T^\phi X$ is called a **time change** of X by the time change process ϕ.

Write simply $\widetilde{\mathscr{F}_t}$ for the σ-field $\mathscr{F}_{\phi^{-1}(t)}$ and denote by $\widetilde{\mathscr{M}^2_{\mathrm{c,loc}}}$ the set of square integrable continuous $\{\widetilde{\mathscr{F}_t}\}$-local martingales.

Doob's optional sampling theorem implies the following.

Lemma 4.7.1 (1) *If* $M \in \mathscr{M}^2_{\mathrm{c,loc}}$, *then* $T^\phi M \in \widetilde{\mathscr{M}^2_{\mathrm{c,loc}}}$.
(2) *If* $M_1, M_2 \in \mathscr{M}^2_{\mathrm{c,loc}}$, *then* $\langle T^\phi M_1, T^\phi M_2 \rangle = T^\phi \langle M_1, M_2 \rangle$.

First we discuss the time change in the framework of martingale problems. Let L be the second order elliptic differential operator on $\mathbb{R}^d$ defined by (4.5.1) and P^x ($x \in \mathbb{R}^d$) be a solution of the martingale problem for L starting from x, using the notation in Section 4.5.

Let $\rho : \mathbb{R}^d \to (0, \infty)$ be a measurable function and set

$$\varphi(t) = \int_0^t \rho(X(s))\,\mathrm{d}s.$$

Suppose that $\lim_{t\uparrow\infty} \varphi(t) = \infty$, P^y-almost surely, and set $\psi = \varphi^{-1}$. Then, $\varphi = \{\varphi(t)\}_{t\geqq 0}$ is a time change process with respect to the natural filtration $\{\mathscr{F}_t\}$. Hence, by the optional sampling theorem,

$$\widetilde{C}_f(t) \equiv f(X(\psi(t))) - f(x) - \int_0^{\psi(t)} (Lf)(X(s))\,\mathrm{d}s$$

is an $\{\mathscr{F}_{\psi(t)}\}$-local martingale for any $f \in C_0^\infty(\mathbb{R}^d)$. Setting $\widetilde{X}(t) = X(\psi(t))$, we have

$$\widetilde{C}_f(t) = f(\widetilde{X}(t)) - f(x) - \int_0^t \rho(\widetilde{X}(u))^{-1}(Lf)(\widetilde{X}(u))\,\mathrm{d}u.$$

This means that the probability law of $\{\widetilde{X}(t)\}_{t\geqq 0}$ is a solution of the martingale problem for $\rho^{-1}L$.

We shall discuss time changes in detail in the case of $d = 1$. Consider the one-dimensional stochastic differential equation without drift

$$\mathrm{d}X(t) = \alpha(t, X)\,\mathrm{d}B(t), \quad X(0) = 0 \tag{4.7.1}$$

for $\alpha \in \mathscr{A}^{1,1}$. Solutions for stochastic differential equations with drifts can be constructed from the solution for this equation by applying the transformation of drifts.

In the rest of this section, we assume that there exist positive constants C_1 and C_2 such that

$$C_1 \leqq \alpha(t, w) \leqq C_2 \qquad (t \geqq 0,\ w \in \mathbf{W}^1).$$

For the methods of time change in a more general setting, see for example [56].

Theorem 4.7.2 (1) *Let* $\theta = \{\theta(t)\}_{t\geqq 0}$ *be a one-dimensional Brownian motion starting from* 0 *defined on a probability space* $(\Omega, \mathscr{F}, \mathbf{P}, \{\mathscr{F}_t\})$. *Set* $\xi(t) = X(0) + \theta(t)$ *and suppose that there exists a time change process* ϕ *such that*

$$\phi(t) = \int_0^t \alpha(\phi(s), T^\phi\xi)^{-2}\mathrm{d}s. \tag{4.7.2}$$

Moreover, set

$$X = T^\phi\xi, \quad \widetilde{\mathscr{F}}_t = \mathscr{F}_{\phi^{-1}(t)}.$$

Then, there exists an $\{\widetilde{\mathscr{F}_t}\}$*-Brownian motion* $B = \{B(t)\}_{t \geqq 0}$ *satisfying* (4.7.1).
(2) *Conversely, let* (X, B) *be a solution of* (4.7.1) *defined on a probability space* $(\Omega, \mathscr{F}, \mathbf{P}, \{\mathscr{F}_t\})$. *Then, there exist a filtration* $\{\mathscr{G}_t\}$, *a* $\{\mathscr{G}_t\}$*-Brownian motion* $\theta = \{\theta(t)\}_{t \geqq 0}$ *starting from* 0, *and a time change process* $\phi = \{\phi(t)\}_{t \geqq 0}$ *with respect to* $\{\mathscr{G}_t\}$ *such that* $\xi = \{\xi(t) = X(0) + \theta(t)\}_{t \geqq 0}$ *satisfies* (4.7.2) *and* $X = T^{\phi}\xi$.
(3) *If there exists a unique time change process* ϕ *satisfying* (4.7.2), $X = T^{\phi}\xi$ *is the unique solution of* (4.7.1).

Proof (1) We have $M = T^{\phi}\theta \in \widetilde{\mathscr{M}^2_{\mathrm{c,loc}}}$, $M(t) = X(t) - X(0)$ and $\langle M\rangle(t) = \phi^{-1}(t)$. Since $t = \int_0^t \alpha(\phi(s), X)^2 \mathrm{d}\phi(s)$ by (4.7.2),

$$\langle M\rangle(t) = \phi^{-1}(t) = \int_0^{\phi^{-1}(t)} \alpha^2(\phi(s), X)^2 \mathrm{d}\phi(s) = \int_0^t \alpha(s, X)^2 \mathrm{d}s. \tag{4.7.3}$$

Set

$$B(t) = \int_0^t \alpha(s, X)^{-1} \mathrm{d}M(s).$$

Then, $B = \{B(t)\}_{t \geqq 0} \in \widetilde{\mathscr{M}^2_{\mathrm{c,loc}}}$ and (4.7.3) implies

$$\langle B\rangle(t) = \int_0^t \alpha(s, X)^{-2} \mathrm{d}\langle M\rangle(s) = t.$$

Hence, B is an $\{\widetilde{\mathscr{F}_t}\}$-Brownian motion and

$$X(t) - X(0) = M(t) = \int_0^t \alpha(s, X) \mathrm{d}B(s).$$

(2) Set $M(t) = X(t) - X(0)$. Then $M \in \mathscr{M}^2_{\mathrm{c,loc}}$ and $\langle M\rangle(t) = \int_0^t \alpha(s, X)^2 \mathrm{d}s$. Moreover, set

$$\psi(t) = \langle M\rangle(t), \quad \phi(t) = \psi^{-1}(t) \quad \text{and} \quad \widetilde{\mathscr{F}_t} = \mathscr{F}_{\psi^{-1}(t)}.$$

Then $\phi = \{\phi(t)\}_{t \geqq 0}$ is a time change process with respect to $\{\widetilde{\mathscr{F}_t}\}$. Hence, by the martingale representation theorem (Theorem 2.5.3), the stochastic process $\theta = \{\theta(t)\}_{t \geqq 0}$ given by $\theta(t) = M(\phi(t))$ is an $\{\widetilde{\mathscr{F}_t}\}$-Brownian motion.

Now set $\xi(t) = X(0) + \theta(t)$. Then, since $T^{\phi}\xi = X$ and $\int_0^t \alpha(s, X)^{-2} \mathrm{d}\psi(s) = t$, we have

$$\phi(t) = \int_0^{\phi(t)} \alpha(s, X)^{-2} \mathrm{d}\psi(s) = \int_0^t \alpha(\phi(u), X)^{-2} \mathrm{d}u = \int_0^t \alpha(\phi(u), T^{\phi}\xi)^{-2} \mathrm{d}u.$$

This means that ϕ satisfies (4.7.2).

The assertion of (3) follows from those of (1) and (2). □

We give some examples. We use the same notations as in Theorem 4.7.2 and its proof.

Example 4.7.3 Let $a : \mathbb{R}^1 \to [0,\infty)$ be a bounded measurable function and assume that there exists a positive constant C such that $a(x) \geqq C$ $(x \in \mathbb{R}^1)$, and let θ be a one-dimensional Brownian motion starting from 0. Set $\xi(t) = X(0) + \theta(t)$ and

$$\phi(t) = \int_0^t a(\xi(s))^{-2}\mathrm{d}s.$$

Then, the time change process satisfying (4.7.2) exists uniquely and $X(t) = \xi(\phi^{-1}(t))$ satisfies the following time-homogeneous Markovian stochastic differential equation:

$$\mathrm{d}X(t) = a(X(t))\,\mathrm{d}B(t).$$

Example 4.7.4 Let $a : [0,\infty)\times\mathbb{R}^1 \to [0,\infty)$ be a bounded measurable function and assume that there exists a positive constant C such that $a(t,x) \geqq C$ $((t,x) \in [0,\infty)\times\mathbb{R}^1)$. In order to solve the stochastic differential equation

$$\mathrm{d}X(t) = a(t,X(t))\,\mathrm{d}B(t)$$

by the method of time change, we need a time change process $\phi = \{\phi(t)\}_{t\geqq 0}$ satisfying

$$\phi(t) = \int_0^t a(\phi(s), T^{\phi}\xi(\phi(s)))^{-2}\mathrm{d}s = \int_0^t a(\phi(s),\xi(s))^{-2}\mathrm{d}s.$$

This problem of finding ϕ is equivalent to solving the ordinary differential equation

$$\frac{\mathrm{d}\phi(t)}{\mathrm{d}t} = a(\phi(t),\xi(t))^{-2}, \quad \phi(0) = 0$$

for each sample path $\{\xi(t)\}_{t\geqq 0}$. Hence, under the assumption on the existence and uniqueness of solutions of this ordinary differential equation (for example, suppose that $a(t,x)$ is Lipschitz continuous in t and the Lipschitz constant is independent of x), the unique solution of the above stochastic differential equation is given by $X(t) = \xi(\phi^{-1}(t))$.

4.8 One-Dimensional Diffusion Process

The purpose of this section is to study the explosion problem for the one-dimensional diffusion processes which are determined by stochastic differential equations. As an application, we present a result on the explosion problem for multi-dimensional diffusions. For one-dimensional diffusion processes, the explosion problem is solved almost completely and the results are shown in detail in Itô and McKean [50]. See also [46, 56, 100].

Let $-\infty \leqq \ell < r \leqq \infty$ and $I = (\ell, r)$ be an open interval. For continuous functions σ and b on I such that $\sigma(x) > 0$ $(x \in I)$, consider the stochastic differential equation

$$\mathrm{d}X(t) = \sigma(X(t))\,\mathrm{d}B(t) + b(X(t))\,\mathrm{d}t, \qquad X(0) = x \in I.$$

Assume that this equation has a unique solution $X^x = \{X^x(t)\}_{t \leqq \zeta}$, ζ being its life time. Let a_n and b_n $(n = 1, 2, \ldots)$ be sequences satisfying $a_n \downarrow \ell$ and $b_n \uparrow r$, respectively, and set

$$\tau_n = \inf\{t > 0; X^x(t) = a_n \text{ or } b_n\}.$$

Then, $\zeta = \lim_{n\to\infty} \tau_n$. Moreover, if $\zeta < \infty$, then there exists $\lim_{t\uparrow\zeta} X^x(t)$ and it is ℓ or r.[7]

Define $X^x(t) = \lim_{s\uparrow\zeta} X^x(s)$ for $t \geqq \zeta$. Then X^x is a random variable with values in

$$\widehat{W}_I = \left\{ w \in \mathbf{W}([\ell, r])\,;\, \begin{array}{l} w(t) \in (\ell, r)\ (0 \leqq t < \zeta(w)) \\ \text{and } w(t) = w(\zeta(w))\ (t \geqq \zeta(w)) \end{array} \right\},$$

where $\zeta(w) = \inf\{t > 0; w(t) = \ell \text{ or } r\}$. Denote by $\mathbf{P}_x$ the probability law of X^x on $\widehat{W}_I$. Then, $\{\mathbf{P}_x\}_{x\in I}$ is a diffusion measure on I with generator

$$L = \frac{1}{2}\sigma(x)^2 \frac{\mathrm{d}^2}{\mathrm{d}x^2} + b(x)\frac{\mathrm{d}}{\mathrm{d}x}.$$

Fix $c \in I$ and set

$$s(x) = \int_c^x \exp\Big(-\int_c^y \frac{2b(z)}{\sigma(z)^2}\,\mathrm{d}z\Big)\mathrm{d}y.$$

s is called a **scale function**. It is strictly increasing on I and satisfies

$$Ls(x) = 0 \qquad (x \in I). \tag{4.8.1}$$

Defining $m'(x)$ by

$$m'(x) = \frac{2}{\sigma(x)^2} \exp\Big(\int_c^x \frac{2b(z)}{\sigma(z)^2}\,\mathrm{d}z\Big),$$

we have

$$L = \frac{1}{m'(x)}\frac{\mathrm{d}}{\mathrm{d}x}\Big(\frac{1}{s'(x)}\frac{\mathrm{d}}{\mathrm{d}x}\Big).$$

[7] We do not need to consider one-point compactification in the one-dimensional case.

The measure m with density $m'(x)$ is called a **speed measure**:

$$m([a,b]) = \int_a^b m'(\xi)\,\mathrm{d}\xi.$$

Proposition 4.8.1 *Let $a, b, x \in I$ satisfy $a < x < b$ and set*

$$\tau_a = \inf\{t > 0; X(t) = a\}, \quad \tau_b = \inf\{t > 0; X(t) = b\}.$$

Then

$$\mathbf{P}_x(\tau_b < \tau_a) = \frac{s(x) - s(a)}{s(b) - s(a)}.$$

Proof Set $\tau = \tau_a \wedge \tau_b$. By Itô's formula and (4.8.1),

$$s(X(t \wedge \tau)) - s(x) = \int_0^{t\wedge\tau} s'(X(u))\sigma(X(u))\,\mathrm{d}B(u)$$

and $\mathbf{E}_x[s(X(t \wedge \tau))] = s(x)$. Hence, by letting $t \uparrow \infty$ and applying the bounded convergence theorem, we obtain

$$s(x) = \mathbf{E}[s(X(\tau))] = s(a)\mathbf{P}_x(\tau_a < \tau_b) + s(b)\mathbf{P}_x(\tau_a > \tau_b).$$

Combining this with $\mathbf{P}_x(\tau_a < \tau_b) + \mathbf{P}_x(\tau_a > \tau_b) = 1$, we obtain the conclusion. □

Example 4.8.2 (Brownian motion with constant drift) Let $\{B(t)\}_{t\geqq 0}$ be a one-dimensional Brownian motion with $B(0) = 0$ and $\mu \in \mathbb{R} \setminus \{0\}$. Set $X(t,x) = x + B(t) + \mu t$. Then, $\{X(t,x)\}_{t\geqq 0}$ determines a diffusion process on $\mathbb{R}$ with generator

$$L = \frac{1}{2}\frac{\mathrm{d}^2}{\mathrm{d}x^2} + \mu\frac{\mathrm{d}}{\mathrm{d}x} = \frac{1}{2}\frac{1}{\mathrm{e}^{2\mu x}}\frac{\mathrm{d}}{\mathrm{d}x}\Big(\frac{1}{\mathrm{e}^{-2\mu x}}\frac{\mathrm{d}}{\mathrm{d}x}\Big).$$

A scale function is given by

$$s(x) = \int_0^x \mathrm{e}^{-2\mu x}\mathrm{d}x = \frac{1 - \mathrm{e}^{-2\mu x}}{2\mu}.$$

Hence, if $a < x < b$, we have

$$\mathbf{P}_x(\tau_b < \tau_a) = \frac{\mathrm{e}^{-2\mu a} - \mathrm{e}^{-2\mu x}}{\mathrm{e}^{-2\mu a} - \mathrm{e}^{-2\mu b}}.$$

The next is a fundamental theorem for the explosion of one-dimensional diffusion processes.

Theorem 4.8.3 *Set $s(\ell+) = \lim_{x\downarrow\ell} s(x)$ and $s(r-) = \lim_{x\uparrow r} s(x)$, and let $x \in I$.*
(1) If $s(\ell+) = -\infty$ and $s(r-) = \infty$, then

$$\mathbf{P}_x(\zeta = \infty) = \mathbf{P}_x\Big(\sup_{0\leqq t<\infty} X(t) = r\Big) = \mathbf{P}_x\Big(\inf_{0\leqq t<\infty} X(t) = \ell\Big) = 1.$$

(2) If $s(\ell+) = -\infty$ and $s(r-) < \infty$, then

$$\mathbf{P}_x\Big(\lim_{t\uparrow\zeta} X(t) = r\Big) = \mathbf{P}_x\Big(\inf_{0\leqq t<\zeta} X(t) > \ell\Big) = 1.$$

(3) If $s(\ell+) > -\infty$ and $s(r-) = \infty$, then

$$\mathbf{P}_x\Big(\lim_{t\uparrow\zeta} X(t) = \ell\Big) = \mathbf{P}_x\Big(\sup_{0\leqq t<\zeta} X(t) < r\Big) = 1.$$

(4) If $s(\ell+) > -\infty$ and $s(r-) < \infty$, then

$$\mathbf{P}_x\Big(\lim_{t\uparrow\zeta} X(t) = \ell\Big) = 1 - \mathbf{P}_x\Big(\lim_{t\uparrow\zeta} X(t) = r\Big) = \frac{s(r-) - s(x)}{s(r-) - s(\ell+)}.$$

Proof (1) By Proposition 4.8.1, if $\ell < a < x < b < r$, then

$$\mathbf{P}_x\Big(\sup_{0\leqq t<\zeta} X(t) \geqq b\Big) \geqq \mathbf{P}_x(\tau_b < \tau_a) = \frac{s(x) - s(a)}{s(b) - s(a)}.$$

Letting $a \downarrow \ell$, we obtain by the assumption

$$\mathbf{P}_x\Big(\sup_{0\leqq t<\zeta} X(t) \geqq b\Big) = 1.$$

Since $b < r$ is arbitrary, $\mathbf{P}_x(\sup_{0\leqq t<\zeta} X(t) = r) = 1$. We have, in the same way, $\mathbf{P}_x(\inf_{0\leqq t<\zeta} X(t) = \ell) = 1$. These two identities show $\mathbf{P}_x(\zeta = \infty) = 1$.
(2) By the assumption $s(\ell+) = -\infty$ and the same argument as in the proof of (1),

$$\mathbf{P}_x\Big(\sup_{0\leqq t<\zeta} X(t) = r\Big) = 1.$$

On the other hand, by Proposition 4.8.1,

$$\mathbf{P}_x(\text{there exists a } t \in [0,\zeta) \text{ such that } X(t) = a) = \frac{s(r-) - s(x)}{s(r-) - s(a)}.$$

Hence, letting $a \downarrow \ell$, we obtain $\mathbf{P}_x\Big(\inf_{0\leqq t<\zeta} X(t) = \ell\Big) = 0$ and

$$\mathbf{P}_x\Big(\sup_{0\leqq t<\zeta} X(t) = r\Big) = \mathbf{P}_x\Big(\inf_{0\leqq t<\zeta} X(t) > \ell\Big) = 1. \tag{4.8.2}$$

Next set $Y(t) = s(r-) - s(X(t \wedge \zeta))$. Since $Ls = 0$, $\{Y(t)\}_{t \geqq 0}$ is a non-negative local martingale, and, therefore, a supermartingale (Proposition 2.1.2). Hence, $Y(t)$ converges as $t \to \infty$ and, by (4.8.2),

$$\mathbf{P}_x\Big(\lim_{t \to \zeta} X(t) = r\Big) = 1.$$

(3) can be proven in the same way as (2).

(4) is obtained by letting $a \downarrow \ell$ and $b \uparrow r$ in Proposition 4.8.1. □

Remark 4.8.4 (1) is the only case where $\{X(t)\}_{t \geqq 0}$ is recurrent in the sense that $\mathbf{P}_x(\tau_y < \infty) = 1$ for any $x, y \in I$.

Fix $c \in (\ell, r)$, and define the functions ν and μ on I by

$$\nu(x) = \int_c^x m([c, \xi]) s'(\xi)\, \mathrm{d}\xi = \int_c^x (s(x) - s(\eta))\, m'(\eta) \mathrm{d}\eta,$$

$$\mu(x) = \int_c^x s(\xi)\, m'(\xi) \mathrm{d}\xi = \int_c^x m([\eta, x]) s'(\eta)\, \mathrm{d}\eta.$$

Definition 4.8.5 (Feller's classification of boundaries) The boundary ℓ is called

(i) **regular** if $\nu(\ell+) < \infty$ and $\mu(\ell+) < \infty$,
(ii) **exit** if $\nu(\ell+) < \infty$ and $\mu(\ell+) = \infty$,
(iii) **entrance** if $\nu(\ell+) = \infty$ and $\mu(\ell+) < \infty$,
(iv) **natural** if $\nu(\ell+) = \infty$ and $\mu(\ell+) = \infty$.

The boundary r is classified in a similar way.

The probability for diffusion processes to reach regular or exit boundaries in finite time is positive. For the other two boundaries the probability is 0. It is possible to construct diffusion processes starting from regular or entrance boundaries. For the other two boundaries it is impossible.

Remark 4.8.6 If $\nu(r-) < \infty$, then $s(r-) < \infty$. If $\nu(\ell+) < \infty$, then $s(\ell+) > -\infty$. In fact, if $\xi \geqq r - \delta$ for $\delta > 0$, then $m([c, \xi]) \geqq C$ for some constant C. Hence, if $x \geqq r - \delta$,

$$\nu(x) = \int_c^x m([c, \xi]) s'(\xi)\, \mathrm{d}\xi \geqq C(s(x) - s(r - \delta)).$$

Let $x \uparrow r$ to see $s(r-) < \infty$. The second assertion on $s(\ell+)$ can be shown in the same way.

Example 4.8.7 (Bessel processes) For $\delta \in \mathbb{R}$ and $x > 0$, consider the stochastic differential equation

$$dX(t) = dB(t) + \frac{\delta - 1}{2X(t)}dt, \quad X(0) = x.$$

Its unique solution defines a diffusion process on $(0, \infty)$ whose generator is given by

$$L_\delta = \frac{1}{2}\frac{d^2}{dx^2} + \frac{\delta - 1}{2x}\frac{d}{dx} = \frac{1}{2x^{\delta-1}}\frac{d}{dx}\left(\frac{1}{x^{1-\delta}}\frac{d}{dx}\right).$$

It is called a **δ-dimensional Bessel process**. The number ν defined by

$$\nu = \frac{\delta - 2}{2}$$

is called the **index** of this Bessel process. If δ is a positive integer, the Bessel process has the same probability law as the radial part of a δ-dimensional Brownian motion. After this fact, δ is called the dimension of the Bessel process. The transition densities are expressed by means of the modified Bessel functions with index ν. For details, see the references cited at the beginning of this section.

If $\delta \neq 2$, we can take $s'(x) = x^{1-\delta}$ and $m'(x) = 2x^{\delta-1}$. Letting the base point be 1, we have

$$\nu(x) = \frac{2}{\delta}\int_x^1 (\xi^{1-\delta} - \xi)\, d\xi \qquad \text{and} \qquad \mu(x) = \frac{2}{2-\delta}\int_x^1 (\xi^{\delta-1} - \xi)\, d\xi.$$

Hence, ∞ is natural for any δ, while 0 is entrance if $\delta > 2$, regular if $0 < \delta < 2$ and exit if $\delta \leqq 0$.

When $\delta = 2$, we can take the scale function as $s(x) = \log x$. The boundary 0 is entrance.

Theorem 4.8.8 (1) *If $\nu(r-) = \nu(\ell+) = \infty$, then $\mathbf{P}_x(\zeta = \infty) = 1$ for any $x \in I$.*
(2) *If either $\nu(r-)$ or $\nu(\ell+)$ is finite, then $\mathbf{P}_x(\zeta < \infty) > 0$ for any $x \in I$.*
(3) *$\mathbf{P}_x(\zeta < \infty) = 1$ for any $x \in I$ if and only if one of the following conditions holds:*

(i) *$\nu(r-) < \infty$ and $\nu(\ell+) < \infty$,*
(ii) *$\nu(r-) < \infty$ and $s(\ell+) = -\infty$,*
(iii) *$\nu(\ell+) < \infty$ and $s(r-) = \infty$.*

We prepare a lemma to prove the theorem.

Lemma 4.8.9 *For $\alpha > 0$ there exist unique solutions $e_1(x;\alpha)$ and $e_2(x;\alpha)$ of $Lu = \alpha u$ such that*

$$\begin{aligned} &e_1(c;\alpha) = 1, \qquad \frac{1}{s'(c)}\frac{\mathrm{d}}{\mathrm{d}x}e_1(c;\alpha) \equiv \frac{\mathrm{d}e_1}{\mathrm{d}s}(c;\alpha) = 0, \\ &e_2(c;\alpha) = 0, \qquad \frac{\mathrm{d}e_2}{\mathrm{d}s}(c;\alpha) = 1. \end{aligned} \tag{4.8.3}$$

Moreover, for $e_1(x;\alpha)$,

$$1 + \alpha v(x) \leqq e_1(x;\alpha) \leqq \exp(\alpha v(x)) \qquad (x \in I). \tag{4.8.4}$$

Proof It suffices to show that the integral equations

$$\begin{aligned} e_1(x) &= 1 + \alpha \int_c^x s'(\xi)\,\mathrm{d}\xi \int_c^\xi e_1(\eta)m'(\eta)\,\mathrm{d}\eta, \\ e_2(x) &= s(x) - s(c) + \alpha \int_c^x s'(\xi)\,\mathrm{d}\xi \int_c^\xi e_2(\eta)m'(\eta)\,\mathrm{d}\eta \end{aligned} \tag{4.8.5}$$

have unique solutions. Since we only need the function $e_1(x;1)$ and the estimate (4.8.4) for it in the proof of Theorem 4.8.8, we only present a construction of $e_1(x;\alpha)$ by a successive approximation and give an estimate for it. For details, see, for example, [46].

Define a sequence of functions $\{u_n\}_{n=0}^\infty$ inductively by $u_0(x) \equiv 1$ and

$$u_{n+1}(x) = \alpha \int_c^x s'(\xi)\,\mathrm{d}\xi \int_c^\xi u_n(\eta)m'(\eta)\,\mathrm{d}\eta \qquad (n \geqq 0). \tag{4.8.6}$$

We have $u_1(x) = \alpha v(x)$ and $u_n(x) \geqq 0$ $(n \geqq 1)$.

We show by induction

$$u_n(x) \leqq \frac{\alpha^n v(x)^n}{n!} \qquad (n = 0, 1, 2, \ldots). \tag{4.8.7}$$

The identity holds for $n = 0$ by the definition. Suppose that the inequality (4.8.7) holds for n. Since v is increasing for $x > c$, we have

$$\begin{aligned} u_{n+1}(x) &\leqq \alpha \int_c^x s'(\xi)\,\mathrm{d}\xi \int_c^\xi \frac{\alpha^n v(\eta)^n}{n!} m'(\eta)\,\mathrm{d}\eta \\ &\leqq \frac{\alpha^{n+1}}{n!} \int_c^x v(\xi)^n s'(\xi) m([c,\xi])\,\mathrm{d}\xi \\ &= \frac{\alpha^{n+1}}{n!} \int_c^x v(\xi)^n v'(\xi)\,\mathrm{d}\xi = \frac{\alpha^{n+1} v(x)^{n+1}}{(n+1)!}. \end{aligned}$$

The same estimate for $x < c$ can be seen in a similar way. Thus (4.8.7) holds.

By (4.8.7), the right hand side of

$$u(x) = \sum_{n=0}^{\infty} u_n(x)$$

converges uniformly on compact sets. Therefore, summing up both sides of (4.8.6) in n, we see that u satisfies (4.8.5). The estimate (4.8.4) follows from (4.8.7). □

Proof of Theorem 4.8.8 Denote by $u(x)$ the solution $e_1(x; 1)$ of the differential equation in Lemma 4.8.9. Let $\ell < a < x < b < r$ and set $\tau = \inf\{t > 0; X(t) \notin (a, b)\}$. Then, by Itô's formula,

$$\mathrm{d}(\mathrm{e}^{-t}u(X(t))) = \mathrm{e}^{-t}u'(X(t))\sigma(X(t))\mathrm{d}B(t)$$

and $\{\mathrm{e}^{-(t\wedge\tau)}u(X(t\wedge\tau))\}_{t\geqq 0}$ is a positive martingale. Hence, by letting $a \downarrow \ell$, $b \uparrow r$ and applying Fatou's lemma, we see that $\{\mathrm{e}^{-(t\wedge\zeta)}u(X(t\wedge\zeta))\}_{t\geqq 0}$ is a non-negative supermartingale.

(1) By (4.8.4), $\lim_{x\downarrow\ell} u(x) = \lim_{x\uparrow r} u(x) = \infty$. Since $t \mapsto \mathrm{e}^{-(t\wedge\zeta)}u(X(t \wedge \zeta))$ is a non-negative supermartingale, it converges as $t \uparrow \infty$ almost surely. By its boundedness, we get $\mathbf{P}_x(\zeta = \infty) = 1$.

(2) We only give a proof under the assumption $\nu(r-) < \infty$. The other case can be proven in the same way. We may assume $c < x$, because c is chosen arbitrarily.

By (4.8.4), we have $u(r-) < \infty$. Set $\tau' = \inf\{t > 0; X(t) = c\}$. Then, since $\{\mathrm{e}^{-(t\wedge\tau')}u(X(t \wedge \tau'))\}_{t\geqq 0}$ is a bounded martingale,

$$\begin{aligned} u(x) &= \mathbf{E}_x[\mathrm{e}^{-(t\wedge\tau')}u(X(t \wedge \tau'))] \\ &= \mathbf{E}_x[\mathrm{e}^{-\zeta}u(r-)\mathbf{1}_{\{\lim_{t\uparrow\zeta\wedge\tau'} X(t)=r\}}] + \mathbf{E}_x[\mathrm{e}^{-\tau'}u(c)\mathbf{1}_{\{\lim_{t\uparrow\zeta\wedge\tau'} X(t)=c\}}]. \end{aligned}$$

Now suppose that the first term of the right hand side is 0. Then,

$$u(x) = u(c)\mathbf{E}_x[\mathrm{e}^{-\tau'}\mathbf{1}_{\{\lim_{t\uparrow\zeta\wedge\tau'} X(t)=c\}}] \leqq u(c) = 1,$$

which contradicts (4.8.4). Hence, we have $\mathbf{E}_x[\mathrm{e}^{-\zeta}\mathbf{1}_{\{\lim_{t\uparrow\zeta\wedge\tau'} X(t)=r\}}] > 0$ and, therefore, $\mathbf{P}_x(\zeta < \infty) > 0$.

(3) We first show the "only if" part. To do this, suppose that $\mathbf{P}_x(\zeta < \infty) = 1$ for any $x \in I$. The proof is divided into two parts according as $\nu(r-)$ is finite or not.

First assume that $\nu(r-) < \infty$. Then, if none of (i), (ii), and (iii) holds, then $\nu(\ell+) = \infty$ and $s(\ell+) > -\infty$. Since $s(r-) < \infty$ by Remark 4.8.6, we have

$$\mathbf{P}_x\Big(\lim_{t\uparrow\zeta} X(t) = \ell\Big) > 0$$

by Theorem 4.8.3. $\nu(\ell+) = \infty$ implies $\lim_{x\downarrow\ell} u(x) = \infty$. Moreover, since $\{e^{-(t\wedge\zeta)}u(X(t\wedge\zeta))\}_{t\geqq 0}$ is a non-negative supermartingale, $\{\lim_{t\uparrow\zeta} X(t) = \ell\} \subset \{\zeta = \infty\}$, $\mathbf{P}_x$-a.s., which implies $\mathbf{P}_x(\zeta = \infty) > 0$. This is a contradiction. Thus one of the conditions (i)–(iii) holds.

Secondly assume that $\nu(r-) = \infty$. If (iii) does not hold, then $\nu(\ell+) = \infty$ or $s(r-) < \infty$. When $\nu(\ell+) = \infty$, $\mathbf{P}_x(\zeta = \infty) = 1$ by (1). This contradicts the assumption. When $s(r-) < \infty$, $\mathbf{P}_x(\lim_{t\uparrow\zeta} X(t) = r) > 0$ by Theorem 4.8.3. Combining this with $u(r-) = \infty$, we have $\mathbf{P}_x(\zeta = \infty) > 0$. This contradicts the assumption. Thus the condition (iii) holds.

We now proceed to the proof of the "if" part.

At first assume (i). Define a function $G(x, y)$ by

$$G(x,y) = \begin{cases} \dfrac{(s(x) - s(\ell+))(s(r-) - s(y))}{s(r-) - s(\ell+)} & (x < y), \\[2mm] \dfrac{(s(y) - s(\ell+))(s(r-) - s(x))}{s(r-) - s(\ell+)} & (y \leqq x). \end{cases}$$

For a bounded continuous function f on I, set

$$v(x) = \int_I G(x,y) f(y)\, m'(y) \mathrm{d}y.$$

Then, under the condition (i), v is a bounded function of C^2-class on I and

$$v(\ell+) = v(r-) = 0, \qquad Lv = -f. \tag{4.8.8}$$

In particular, setting $f = 1$ and defining

$$u_1(x) = \int_I G(x,y)\, m'(y) \mathrm{d}y,$$

we obtain by Itô's formula

$$u_1(X(t\wedge\zeta)) - u_1(x) = \int_0^{t\wedge\zeta} u_1'(X(s))\, \mathrm{d}B(s) - (t\wedge\zeta)$$

and $\mathbf{E}_x[t\wedge\zeta] = u_1(x) - \mathbf{E}_x[u_1(X(t\wedge\zeta))]$. Letting $t \uparrow \infty$, by Theorem 4.8.3(4) and (4.8.8), we get $\mathbf{E}_x[\zeta] = u_1(x)$. Hence $\mathbf{P}_x(\zeta < \infty) = 1$.

Next assume (ii). Set $\tau_n = \inf\{t > 0; X(t) = \ell + n^{-1}\}$ and $\sigma_r = \inf\{t > 0; X(t) = r\}$. Then, $\zeta = \lim_{n\to\infty}(\tau_n \wedge \sigma_r)$ and $\mathbf{P}_x(\tau_n \wedge \sigma_r < \infty) = 1$ for any $x \in I$ by (i). Since $s(r-) < \infty$, $\lim_{n\to\infty} \mathbf{P}_x(\sigma_r < \tau_n) = 1$ by Theorem 4.8.3(2). Hence $\mathbf{P}_x(\sigma_r < \infty) = 1$. Since $\zeta \leqq \sigma_r$, we have $\mathbf{P}_x(\zeta < \infty) = 1$.

When (iii) is assumed, we obtain the conclusion by changing ℓ and r in the argument of the proof of (ii). □

Example 4.8.10 Let C and δ be positive constants and consider a diffusion process $\mathbb{R}$ with generator

$$L_{C,\delta} = \frac{1}{2}\frac{\mathrm{d}}{\mathrm{d}x^2} + C|x|^{\delta}\frac{\mathrm{d}}{\mathrm{d}x}.$$

We take a scale function given by

$$s(x) = \int_0^x \exp\Bigl(-\frac{2C}{\delta+1}\xi^{\delta+1}\Bigr)\mathrm{d}\xi \qquad (x > 0).$$

For $x < 0$, set $s(x) = -s(|x|)$. Define the speed measure by $m'(x) = 2(s'(x))^{-1}$.

By L'Hospital's rule,

$$\lim_{x\to\infty} \frac{m([0,x])}{x^{-\delta}m'(x)} = \frac{1}{2C}.$$

If $\delta > 1$, then

$$v(\infty) = \int_0^{\infty} m([0,\xi])s'(\xi)\,\mathrm{d}\xi < \infty.$$

By symmetry, we also have $v(-\infty) < \infty$. Hence, in this case, the diffusion process corresponding to $L_{C,\delta}$ explodes in finite time almost surely for any $C > 0$. If $0 \leqq \delta \leqq 1$, then we do not have explosion almost surely.

As an application of Theorem 4.8.8, we show a sufficient condition for general dimensional diffusion processes not to explode or to explode almost surely. It is called **Khasminskii's condition**.

For $d \geqq 2$, let $\sigma : \mathbb{R}^d \to \mathbb{R}^d \otimes \mathbb{R}^d$ and $b : \mathbb{R}^d \to \mathbb{R}^d$ be continuous functions. Consider the Markovian stochastic differential equation

$$\mathrm{d}X(t) = \sigma(X(t))\,\mathrm{d}B(t) + b(X(t))\,\mathrm{d}t. \tag{4.8.9}$$

Denote by $\langle\ ,\ \rangle$ the inner product in $\mathbb{R}^d$ and define functions a, a_0, b_0 on $\mathbb{R}^d$ by

$$a(x) = \sigma(x)\sigma(x)^*, \quad a_0(x) = \langle x, a(x)x\rangle,$$

$$b_0(x) = \frac{1}{a_0(x)}\Bigl(\sum_{i=1}^{d} a^{ii}(x) + 2\langle b(x), x\rangle\Bigr).$$

We assume that the diffusion matrix $a(x)$ is non-degenerate for every $x \in \mathbb{R}^d$.

For $r > 0$, set

$$a_+(r) = \max_{|x|=r} a_0(x), \quad b_+(r) = \max_{|x|=r} b_0(x),$$

$$a_-(r) = \min_{|x|=r} a_0(x), \quad b_-(r) = \min_{|x|=r} b_0(x)$$

and define $s'_{\pm}, m'_{\pm} : [0, \infty) \to [0, \infty)$ by

$$s'_{\pm}(r) = \exp\Big(-\int_1^r b_{\pm}(\xi)\xi\,\mathrm{d}\xi\Big), \quad m'_{\pm}(r) = \frac{2}{a_{\pm}(r)s'_{\pm}(r)}.$$

Theorem 4.8.11 *Suppose that the stochastic differential equation* (4.8.9) *has a unique solution* $X = \{X(t)\}_{t\geqq 0}$ *starting from* $x \in \mathbb{R}^d$. *Denote its probability law by* $\mathbf{P}_x$ *and let* ζ *be the life time.*
(1) *If* $\int_1^\infty s'_+(r)r\,\mathrm{d}r \int_1^r m'_+(\xi)\xi\,\mathrm{d}\xi = \infty$, *then* $\mathbf{P}_x(\zeta = \infty) = 1$ *for every* $x \in \mathbb{R}^d$.
(2) *If* $\int_1^\infty s'_-(r)r\,\mathrm{d}r \int_1^r m'_-(\xi)\xi\,\mathrm{d}\xi < \infty$, *then* $\mathbf{P}_x(\zeta < \infty) = 1$ *for every* $x \in \mathbb{R}^d$.

Proof We follow [84, Section 4.5]. See also [100, Section V.52] and [114, Chapter 10].
(1) For $r \geqq 1$, set $r = \sqrt{2\rho}$ and

$$u_0(\rho) = 1, \quad u_n(\rho) = \int_1^{\sqrt{2\rho}} s'_+(\xi)\xi\,\mathrm{d}\xi \int_1^{\xi} u_{n-1}\Big(\frac{\eta^2}{2}\Big)m'_+(\eta)\eta\,\mathrm{d}\eta.$$

Then, the solution for the ordinary differential equation

$$u(\rho) = \frac{1}{2}a_+(\sqrt{2\rho})(u''(\rho) + b_+(\sqrt{2\rho})u'(\rho)) \Big(= \frac{1}{m'_+(\sqrt{2\rho})}\frac{\mathrm{d}}{\mathrm{d}\rho}\Big(\frac{1}{s'_+(\sqrt{2\rho})}\frac{\mathrm{d}u}{\mathrm{d}\rho}(\rho)\Big)\Big),$$

$$u\Big(\frac{1}{2}\Big) = 1, \quad \frac{\mathrm{d}u}{\mathrm{d}\rho}\Big(\frac{1}{2}\Big) = 0$$

is given by $u(\rho) = \sum_{n=0}^\infty u_n(\rho)$ (see Lemma 4.8.9). We extend u for $\rho < \frac{1}{2}$ so that $u \in C^\infty(\mathbb{R})$.

Under the assumption, we have

$$u(\rho) \geqq u_1(\rho) \to \infty \ (\rho \to \infty), \ \ u'(\rho) > 0, \ \ u''(\rho) + b_+(\sqrt{2\rho})u'(\rho) \geqq 0 \ (\rho > \frac{1}{2}).$$

Let L be the generator of $\{X(t)\}_{t\geqq 0}$:

$$L = \frac{1}{2}\sum_{i,j=1}^d a^{ij}(x)\frac{\partial^2}{\partial x^i \partial x^j} + \sum_{j=1}^d b^i(x)\frac{\partial}{\partial x^i}.$$

Set $v(x) = u(\frac{|x|^2}{2})$ for $x \in \mathbb{R}^d$. Then,

$$Lv(x) = \frac{1}{2}a_0(x)\Big\{u''\Big(\frac{|x|^2}{2}\Big) + b_0(x)u'\Big(\frac{|x|^2}{2}\Big)\Big\}.$$

Hence

$$-v(x) + Lv(x) \leqq 0 \quad \text{if } |x| \geqq 1.$$

By Itô's formula and the time change argument, there exists a one-dimensional Brownian motion $\{\beta(t)\}_{t\geqq 0}$ and an increasing process $\{A(t)\}_{t\geqq 0}$ such that

$$e^{-t}v(X(t)) - v(x) = \beta(A(t)) + \int_0^t e^{-s}(-v + Lv)(X(s))ds.$$

Let $\gamma = \max\{t > 0\,;\ |X(t)| = 1\}$ and $\gamma = 0$ if $\{\ldots\} = \emptyset$. Take $\omega \in \{\zeta < \infty\}$. Then $\gamma(\omega) < \zeta(\omega)$ and $|X(s,\omega)| > 1$ for any $s \in (\gamma(\omega), \zeta(\omega))$, and $\lim_{t\to\zeta(\omega)} v(X(t)) = \infty$. Moreover,

$$\begin{aligned} e^{-t}v(X(t,\omega)) - e^{-\gamma(\omega)}u(\tfrac{1}{2}) &= \beta(A(t,\omega),\omega) - \beta(A(\gamma(\omega),\omega),\omega) \\ &\quad + \int_{\gamma(\omega)}^t e^{-s}(-v + Lv)(X(s,\omega))\,ds \\ &\leqq \beta(A(t,\omega),\omega) - \beta(A(\gamma(\omega),\omega),\omega) \end{aligned}$$

for any $t \in (\gamma(\omega), \zeta(\omega))$. This implies $\liminf_{t\to\zeta(\omega)} e^{-t}v(X(t,\omega)) < \infty$, which contradicts $\lim_{t\to\zeta(\omega)} v(X(t)) = \infty$ shown above.

Thus $\mathbf{P}_x(\zeta = \infty) = 1$.

(2) Under the assumption, the solution u of the ordinary differential equation

$$u(\rho) = \frac{1}{2}a_-(\sqrt{2\rho})(u''(\rho) + b_-(\sqrt{2\rho})u'(\rho)) \left(= \frac{1}{m'_-(\sqrt{2\rho})}\frac{d}{d\rho}\Big(\frac{1}{s'_-(\sqrt{2\rho})}\frac{du}{d\rho}(\rho)\Big)\right),$$

$$u(\tfrac{1}{2}) = 1, \quad \frac{du}{d\rho}(\tfrac{1}{2}) = 0$$

satisfies

$$1 < u(\rho) < \exp\Big(\int_1^{\sqrt{2\rho}} s'_-(\xi)\xi\,d\xi \int_1^{\xi} m'_-(\eta)\eta\,d\eta\Big) \qquad (\rho > \tfrac{1}{2})$$

and is bounded.

By the same argument as (1), $v(x) = u(\frac{|x|^2}{2})$ satisfies $-v+Lv \geqq 0$ ($|x| \geqq 1$). Let $T_R = \inf\{t \geqq 0; |X(t)| = R\}$. Then, by Itô's formula and the optional sampling theorem, we obtain for $|x| > 1$

$$\mathbf{E}_x[e^{-T_n}\mathbf{1}_{\{T_n<T_1\}}]u(\frac{n^2}{2}) + \mathbf{E}_x[e^{-T_1}\mathbf{1}_{\{T_n\geqq T_1\}}]u(\frac{1}{2}) \geqq u(\frac{|x|^2}{2}).$$

Letting $n \to \infty$ and setting $u(\infty) = \lim_{\rho\to\infty} u(\rho)$, we have

$$u(\infty)\mathbf{E}_x[e^{-\zeta}\mathbf{1}_{\{\zeta<T_1\}}] + u(\frac{1}{2})\mathbf{E}_x[e^{-T_1}\mathbf{1}_{\{\zeta\geqq T_1\}}] \geqq u(\frac{|x|^2}{2}).$$

Then

$$\liminf_{|x|\to\infty}\Big\{u(\infty)\mathbf{E}_x[e^{-\zeta}\mathbf{1}_{\{\zeta<T_1\}}] + u(\frac{1}{2})\mathbf{E}_x[e^{-T_1}\mathbf{1}_{\{\zeta\geqq T_1\}}]\Big\} \geqq u(\infty).$$

Since

$$u(\infty) > u(\frac{1}{2}) \quad \text{and} \quad \mathbf{E}_x[\mathrm{e}^{-\zeta}\mathbf{1}_{\{\zeta<T_1\}}] + \mathbf{E}_x[\mathrm{e}^{-T_1}\mathbf{1}_{\{\zeta\geqq T_1\}}] \leqq 1,$$

we obtain

$$\liminf_{|x|\to\infty} \mathbf{E}_x[\mathrm{e}^{-\zeta}\mathbf{1}_{\{\zeta<T_1\}}] = 1.$$

Moreover, since $\mathbf{E}_x[\mathrm{e}^{-\zeta}\mathbf{1}_{\{\zeta<T_1\}}] \leqq \mathbf{P}_x(\zeta < \infty)$,

$$\liminf_{|x|\to\infty} \mathbf{P}_x(\zeta < \infty) = 1.$$

Hence, for any $\varepsilon > 0$, there exists $R > 0$ such that $\mathbf{P}_x(\zeta < \infty) > 1 - \varepsilon$ if $|x| > R$.

On the other hand, the non-degeneracy of the diffusion matrix a yields $\mathbf{P}_x(T_R < \infty) = 1$ for $|x| < R$ (see Theorem 4.8.12 below). Therefore, by the strong Markov property of X,

$$\mathbf{P}_x(\zeta < \infty) = \mathbf{E}_x\big[\mathbf{E}_{X(T_R)}[\mathbf{1}_{\{\zeta<\infty\}}]\big] > 1 - \varepsilon$$

holds also for x with $|x| < R$.

Hence, $\mathbf{P}_x(\zeta < \infty) > 1 - \varepsilon$ holds for all $x \in \mathbb{R}^d$. Since $\varepsilon > 0$ is arbitrary, $\mathbf{P}_x(\zeta < \infty) = 1$. □

The fact used in the above proof of Theorem 4.8.11 is important and we give a proof.

Theorem 4.8.12 *Let K be a compact set containing $X(0) = x$ and τ_K be the exit time from K of X. Then, under the same assumption as Theorem 4.8.11, $\mathbf{E}_x[\tau_K] < \infty$.*

Proof We may assume that K is a ball $B(R)$ with center at the origin and radius R. Fix $T > 0$ and set $\tau^T = \tau_{B(R)} \wedge T$. Then, $\phi(x) = -\exp(\alpha x_1)$ satisfies

$$\begin{aligned} L\phi(x) &= \frac{1}{2}\sum_{i,j=1}^{d} a^{ij}(x)\frac{\partial^2\phi(x)}{\partial x^i \partial x^j} + \sum_{i=1}^{d} b^i(x)\frac{\partial\phi(x)}{\partial x^i} \\ &= -\Big(\frac{1}{2}\alpha^2 a^{11}(x) + \alpha b^1(x)\Big)\mathrm{e}^{\alpha x_1} \end{aligned}$$

and $L\phi(x) \leqq -1$ $(x \in B(R))$ for sufficiently large α by the assumption. Then, by Itô's formula, we get $\mathbf{E}_x[\phi(X(\tau^T))] \leqq \phi(x) - \mathbf{E}_x[\tau^T]$ and, letting $T \to \infty$, $\mathbf{E}_x[\tau_{B(R)}] < \infty$. □

The essence of Khasminskii's theorem is a comparison theorem for one-dimensional diffusion processes. In fact, in [45], the comparison theorem is proven first, and, as an application of it, Theorem 4.8.11 is shown. For another approach to the explosion problem, we refer the reader to [39].

4.9 Linear Stochastic Differential Equations

Consider an ordinary differential equation with constant coefficients:

$$\frac{\mathrm{d}x(t)}{\mathrm{d}t} = ax(t) + b.$$

Since $(\mathrm{e}^{-at}x(t))' = \mathrm{e}^{-at}(x'(t) - ax(t)) = b\mathrm{e}^{-at}$, this equation can be solved as

$$x(t) = \mathrm{e}^{at}\Big(x(0) + b\int_0^t \mathrm{e}^{-as}\mathrm{d}s\Big).$$

For a one-dimensional stochastic differential equation with constant coefficients

$$\mathrm{d}X(t) = \sigma\,\mathrm{d}B(t) + (aX(t) + b)\,\mathrm{d}t,$$

the situation is similar and it can be checked by Itô's formula that the solution is given by

$$X(t) = \mathrm{e}^{at}\Big(X(0) + \sigma\int_0^t \mathrm{e}^{-as}\,\mathrm{d}B(s) + b\int_0^t \mathrm{e}^{-as}\mathrm{d}s\Big).$$

In this section we generalize this observation to multi-dimensional cases. As applications, a higher-dimensional Ornstein–Uhlenbeck process and a Brownian bridge will be investigated.

Let $\sigma(t)$, $A(t)$, and $b(t)$ be locally bounded Borel measurable functions on $[0,\infty)$ with values in $\mathbb{R}^d\otimes\mathbb{R}^r$, $\mathbb{R}^d\otimes\mathbb{R}^d$, and $\mathbb{R}^d$, respectively. Let $\{B(t)\}_{t\geqq 0}$ be an r-dimensional $\{\mathscr{F}_t\}$-Brownian motion defined on a filtered probability space $(\Omega,\mathscr{F},\mathbf{P},\{\mathscr{F}_t\})$ and $X(0)$ be an $\mathscr{F}_0$-measurable $\mathbb{R}^d$-valued random variable. Consider the stochastic differential equation

$$\mathrm{d}X(t) = \sigma(t)\,\mathrm{d}B(t) + (A(t)X(t) + b(t))\,\mathrm{d}t, \quad X(0) = X_0. \tag{4.9.1}$$

Denote the unique strong solution by $X = \{X(t)\}_{t\geqq 0}$.

Associated with (4.9.1), we consider the ordinary differential equation

$$\frac{\mathrm{d}x(t)}{\mathrm{d}t} = A(t)x(t) + b(t), \quad x(0) = x_0 \tag{4.9.2}$$

and denote the solution by $x(t)$. Moreover, denote by $\Phi(t)$ the solution of the matrix equation

$$\frac{\mathrm{d}\Phi(t)}{\mathrm{d}t} = A(t)\Phi(t), \quad \Phi(0) = I, \tag{4.9.3}$$

where I is the d-dimensional unit matrix. For the function $\Psi(t)$ given as the solution of the equation

$$\frac{\mathrm{d}\Psi(t)}{\mathrm{d}t} = -\Psi(t)A(t), \quad \Psi(0) = I,$$

we have $\frac{\mathrm{d}}{\mathrm{d}t}(\Psi(t)\Phi(t)) = 0$ and $\Psi(t)\Phi(t) = I$. Hence, $\Phi(t)$ is non-degenerate for any $t \geqq 0$ and

$$x(t) = \Phi(t)\Big(x(0) + \int_0^t \Phi^{-1}(s)b(s)\,\mathrm{d}s\Big).$$

Also, for the stochastic differential equation (4.9.1), the solution is explicitly expressed.

Proposition 4.9.1 (1) *Let* $\Phi(t)$ *be a solution of* (4.9.3). *Then the solution of the stochastic differential equation* (4.9.1) *is given by*

$$X(t) = \Phi(t)\Big\{X_0 + \int_0^t \Phi^{-1}(s)\sigma(s)\,\mathrm{d}B(s) + \int_0^t \Phi^{-1}(s)b(s)\,\mathrm{d}s\Big\}. \tag{4.9.4}$$

(2) $m(t) = \mathbf{E}[X(t)]$ *is a solution of* (4.9.2) *satisfying* $m(0) = \mathbf{E}[X_0]$.
(3) *The covariance matrix* $R(s,t) = \mathbf{E}[(X(s) - m(s))(X(t) - m(t))^*]$ *is given by*

$$R(s,t) = \Phi(s)\Big[R(0,0) + \int_0^{s\wedge t} \Phi^{-1}(u)\sigma(u)\sigma(u)^*\Phi^{-1}(u)^*\mathrm{d}u\Big]\Phi(t)^*,$$

where A^* *is the transpose matrix of* A. *Moreover,* $R(t) = R(t,t)$ *satisfies*

$$\frac{\mathrm{d}R(t)}{\mathrm{d}t} = A(t)R(t) + R(t)A(t)^* + \sigma(t)\sigma(t)^*. \tag{4.9.5}$$

Proof (1) is verified by Itô's formula.
(2) is obtained by taking the expectation of both sides of (4.9.1).
By (1),

$$\begin{aligned}
R(s,t) &= \Phi(s)\mathbf{E}\Big[\Big(\int_0^s \Phi^{-1}(u)\sigma(u)\,\mathrm{d}B(u)\Big)\Big(\int_0^t \Phi^{-1}(u)\sigma(u)\,\mathrm{d}B(u)\Big)^*\Big]\Phi(t)^* \\
&\quad + \Phi(s)\mathbf{E}[(X_0 - m(0))(X_0 - m(0))^*]\Phi(t)^* \\
&= \Phi(s)\Big\{R(0,0) + \mathbf{E}\Big[\Big(\int_0^{s\wedge t} \Phi^{-1}(u)\sigma(u)\,\mathrm{d}B(u)\Big) \\
&\qquad\qquad \times \Big(\int_0^{s\wedge t} \Phi^{-1}(u)\sigma(u)\,\mathrm{d}B(u)\Big)^*\Big]\Big\}\Phi(t)^*.
\end{aligned}$$

Computing the expectation of the right hand sides by using Itô's isometry, we obtain the first assertion of (3).
(4.9.5) follows from (4.9.3). □

Remark 4.9.2 If the initial distribution is d-dimensional Gaussian, then that of $X(t)$ $(t > 0)$ is also Gaussian.

Example 4.9.3 (Multi-dimensional Ornstein–Uhlenbeck process) When $\sigma(t)$ and $A(t)$ are constant matrices, say σ and A, respectively, $\Phi(t) = e^{tA}$, and

$$X(t) = e^{tA}X_0 + \int_0^t e^{(t-s)A}\sigma\,dB(s) + \int_0^t e^{(t-s)A}b(s)\,ds,$$

the covariance matrix is given by

$$R(t) = e^{tA}R(0)e^{tA^*} + \int_0^t e^{(t-u)A}\sigma\sigma^* e^{(t-u)A^*}\,du.$$

Suppose that the real part of each eigenvalue of A is negative. Set

$$R = \int_0^\infty e^{uA}\sigma\sigma^* e^{uA^*}\,du$$

and assume that $R(0) = R$. Then $R(t)$ does not depend on t. In fact, since

$$AR + RA^* = \int_0^\infty \frac{d}{du}\Big(e^{uA}\sigma\sigma^* e^{uA^*}\Big)du = -\sigma\sigma^*,$$

we have

$$\begin{aligned} e^{tA}\Big(\int_0^t e^{-uA}\sigma\sigma^* e^{-uA^*}du\Big)e^{tA^*} &= -e^{tA}\Big(\int_0^t e^{-uA}(AR + RA^*)e^{-uA^*}du\Big)e^{tA^*} \\ &= e^{tA}\Big(\int_0^t \frac{d}{du}\Big(e^{-uA}Re^{-uA^*}\Big)du\Big)e^{tA^*} = R - e^{tA}Re^{tA^*} \end{aligned}$$

and $R(t) = R$ $(t \geqq 0)$. The covariance matrix is given by

$$R(s,t) = \begin{cases} e^{(s-t)A}R & (0 \leqq t \leqq s) \\ Re^{(t-s)A^*} & (0 \leqq s \leqq t). \end{cases}$$

Example 4.9.4 (Brownian bridge) Let $r = d = 1$, and fix $x, y \in \mathbb{R}$ and $T > 0$. The solution $\{X^{x,y}(t)\}_{0 \leqq t < T}$ of

$$dX(t) = dB(t) + \frac{y - X(t)}{T - t}\,dt, \quad X(0) = x$$

is called the **Brownian bridge** connecting x and y, or the **pinned Brownian motion**. By Proposition 4.9.1,

$$X^{x,y}(t) = x\Big(1 - \frac{t}{T}\Big) + y\frac{t}{T} + (T - t)\int_0^t \frac{1}{T - s}\,dB(s).$$

As $t \to T$, $\mathbf{E}[|X^{x,y}(t) - y|^2] \to 0$. Moreover, the third term, say $\{Y(t)\}_{t \geqq 0}$, of the right hand side obeys a Gaussian distribution with mean 0 for each $t \geqq 0$ and the covariance is given by

$$\mathbf{E}[Y(s)Y(t)] = (s \wedge t) - \frac{st}{T} \qquad (0 \leqq s, t \leqq T).$$

A finite dimensional distribution of $\{X^{x,y}(t)\}_{0\leqq t\leqq T}$ is given by

$$\mathbf{P}(X^{x,y}(t_1) \in \mathrm{d}x_1, X^{x,y}(t_2) \in \mathrm{d}x_2, \ldots, X^{x,y}(t_n) \in \mathrm{d}x_n) = \prod_{i=1}^{n} g(t_i - t_{i-1}, x_{i-1}, x_i) \cdot \frac{g(T - t_n, x_n, y)}{g(T, x, y)} \mathrm{d}x_1 \mathrm{d}x_2 \cdots \mathrm{d}x_n$$

for $0 = t_0 < t_1 < t_2 < \cdots < t_n < T$, where $x_0 = x, x_n = y$ and $g(t, z, z') = \frac{1}{\sqrt{2\pi t}} \mathrm{e}^{-\frac{(z-z')^2}{2t}}$.

Furthermore, the probability distribution of $\{X^{x,y}(t)\}_{0\leqq t\leqq T}$ coincides with the conditional probability distribution of a one-dimensional Brownian motion $\{B(t)\}_{0\leqq t\leqq T}$ starting from x, given $B(T) = y$. The stochastic process $\{B^{x,y}(t)\}_{0\leqq t\leqq T}$ given by

$$B^{x,y}(t) = B(t) - \frac{t}{T} B(T) + \frac{t}{T} y$$

also has the same probability distribution.

4.10 Stochastic Flows

For $V : \mathbb{R}^d \to \mathbb{R}^d$, consider the ordinary differential equation

$$\frac{\mathrm{d}\xi(t)}{\mathrm{d}t} = V(\xi(t)), \quad \xi(0) = x.$$

If V is smooth, then the solution $\xi(t, x)$ depends smoothly on the initial value x and the mapping $\xi(t) : x \mapsto \xi(t, x)$ is a diffeomorphism of $\mathbb{R}^d$. Since $\xi(t)\circ\xi(s) = \xi(t + s)$, $\{\xi(t)\}_{t\in\mathbb{R}}$ is called a one-parameter group of diffeomorphisms of $\mathbb{R}^d$.

The aim of this section is to show similar facts for stochastic differential equations, following Kunita [62, 64]. Stochastic processes with values in $C(\mathbb{R}^d \to \mathbb{R}^d)$ or on the space of homeomorphisms of $\mathbb{R}^d$ are called **stochastic flows**. We consider only stochastic flows which are defined via solutions of stochastic differential equations.

Let $V_0, V_1, \ldots, V_r : \mathbb{R}^d \to \mathbb{R}^d$ be bounded and continuous functions. For an r-dimensional Brownian motion $\{(B^1(t), B^2(t), \ldots, B^r(t))\}_{t\geqq 0}$ defined on a probability space $(\Omega, \mathscr{F}, \mathbf{P}, \{\mathscr{F}_t\})$, consider the stochastic differential equation

$$\mathrm{d}X(t) = \sum_{k=1}^{r} V_k(X(t))\,\mathrm{d}B^k(t) + V_0(X(t))\,\mathrm{d}t, \quad X(0) = x. \tag{4.10.1}$$

We assume that each coefficient V_i is Lipschitz continuous, that is, there exists a positive constant K such that

$$|V_k(x) - V_k(y)| \leqq K|x - y| \qquad (x, y \in \mathbb{R}^d,\ k = 0, 1, 2, \ldots, r).$$

Denote the unique strong solution of (4.10.1) by $\{X(t, x)\}_{t \geqq 0}$.

Proposition 4.10.1 *Let $T > 0$ and $p \geqq 2$. Then, there exists a positive constant C such that*

$$\mathbf{E}[|X(t, x) - X(s, y)|^p] \leqq C(|x - y|^p + |t - s|^{\frac{p}{2}})$$

for all $x, y \in \mathbb{R}^d$ and $s, t \in [0, T]$. Moreover, $\{X(t, x)\}_{t \geqq 0, x \in \mathbb{R}^d}$ has a modification which is Hölder continuous in (t, x).

Proof The second assertion is an immediate consequence of the first one and Kolmogorov's continuity theorem (Theorem A.5.1). Thus we give a proof of only the first assertion.

First we consider the case where $x = y$. Let $s < t$. By the elementary inequality

$$|a_1 + a_2 + \cdots + a_{r+1}|^p \leqq (r + 1)^{p-1}(|a_1|^p + |a_2|^p + \cdots + |a_{r+1}|^p),$$

we have

$$\begin{aligned}\mathbf{E}[|X(t, x) - X(s, x)|^p] \leqq (r + 1)^{p-1} \sum_{k=1}^{r} \mathbf{E}\Big[\Big|\int_s^t V_k(X(u, x))\,\mathrm{d}B^k(u)\Big|^p\Big] \\ + (r + 1)^{p-1}\mathbf{E}\Big[\Big|\int_s^t V_0(X(u, x))\,\mathrm{d}u\Big|^p\Big].\end{aligned}$$

Moreover, by the Burkholder–Davis–Gundy inequality (Theorem 2.4.1) and the boundedness of the V_ks, there exist positive constants C_1 and C_2 such that[8]

$$\begin{aligned}\mathbf{E}[|X(t, x) - X(s, x)|^p] &\leqq C_1 \sum_{k=1}^{r} \mathbf{E}\Big[\Big(\int_s^t |V_k(X(u, x))|^2\mathrm{d}u\Big)^{\frac{p}{2}}\Big] + C_1(t - s)^p \\ &\leqq C_2((t - s)^{\frac{p}{2}} + (t - s)^p) \quad (0 \leqq s < t \leqq T, x \in \mathbb{R}^d).\end{aligned}$$

Hence there exists a constant C_3 such that

$$\mathbf{E}[|X(t, x) - X(s, x)|^p] \leqq C_3(t - s)^{\frac{p}{2}} \quad (s, t \in [0, T], x \in \mathbb{R}^d).$$

Next consider the case where $s = t$. Since

[8] Here and in the remainder of this section, constants $C_1, C_2, \ldots$ depend only on d, r, p, T, and the bounds and the Lipschitz constants of the V_ks.

$$X(t,x) - X(t,y) = x - y + \sum_{k=1}^{r} \int_0^t (V_k(X(u,x)) - V_k(X(u,y)))\,\mathrm{d}B^k(u)$$
$$+ \int_0^t (V_0(X(u,x)) - V_0(X(u,y)))\,\mathrm{d}u,$$

by the Burkholder–Davis–Gundy inequality, there exists a constant C_4 such that

$$\begin{aligned}
&\mathbf{E}[|X(t,x) - X(t,y)|^p] \\
&\leqq C_4|x-y|^p + C_4 \sum_{k=1}^{r} E\Big[\Big(\int_0^t |V_k(X(u,x)) - V_k(X(u,y))|^2 \mathrm{d}u\Big)^{\frac{p}{2}}\Big] \\
&\qquad + C_4 E\Big[\Big|\int_0^t (V_0(X(u,x)) - V_0(X(u,y)))\,\mathrm{d}u\Big|^p\Big].
\end{aligned}$$

By the Lipschitz continuity of the V_ks and Hölder's inequality, there exists a constant C_5 such that

$$\mathbf{E}[|X(t,x) - X(t,y)|^p] \leqq C_4|x-y|^p + C_5 \int_0^t E[|X(u,x) - X(u,y)|^p]\,\mathrm{d}u.$$

Hence, by Gronwall's inequality (Theorem A.1.1), there exists a constant C_6 such that

$$\mathbf{E}[|X(t,x) - X(t,y)|^p] \leqq C_6|x-y|^p \qquad (t \in [0,T],\ x,y \in \mathbb{R}^d).$$ □

Next we prove that the continuous mapping $X(t,\cdot)$ is injective. For this purpose we give the following proposition.

Proposition 4.10.2 *For any $T > 0$ and $p \in \mathbb{R}$, there exists a positive constant C such that*

$$\mathbf{E}[|X(t,x) - X(t,y)|^p] \leqq C|x-y|^p \tag{4.10.2}$$

for every $x, y \in \mathbb{R}^d$ $(x \neq y)$ and $t \in [0,T]$.

Proof For $\varepsilon > 0$ and $x, y \in \mathbb{R}^d$ with $|x-y| > \varepsilon$, set $\tau_\varepsilon = \inf\{t; |X(t,x)-X(t,y)| < \varepsilon\}$. Let $f(z) = |z|^p$, $f_i = \frac{\partial f}{\partial z^i}$, $f_{ij} = \frac{\partial^2 f}{\partial z^i \partial z^j}$. By Itô's formula,

$$|X(t,x) - X(t,y)|^p - |x-y|^p = I_1(t) + I_2(t)$$

for $t < \tau_\varepsilon$, where

$$I_1(t) = \sum_{i=1}^{d}\sum_{k=1}^{r}\int_0^t f_i(X(s,x)-X(s,y))(V_k^i(X(s,x)) - V_k^i(X(s,y)))\,\mathrm{d}B^k(s)$$
$$+ \sum_{i=1}^{d}\int_0^t f_i(X(s,x)-X(s,y))(V_0^i(X(s,x)) - V_0^i(X(s,y)))\,\mathrm{d}s,$$
$$I_2(t) = \frac{1}{2}\sum_{i,j=1}^{d}\sum_{k=1}^{r}\int_0^t f_{ij}(X(s,x)-X(s,y))(V_k^i(X(s,x)) - V_k^i(X(s,y)))$$
$$\times (V_k^j(X(s,x)) - V_k^j(X(s,y)))\,\mathrm{d}s.$$

Since $f_i(z) = p|z|^{p-2}z^i$, $f_{ij}(z) = p|z|^{p-2}\delta^{ij} + p(p-2)|z|^{p-4}z^iz^j$, and the V_ks are Lipschitz continuous, there exist constants C_1 and C_2 such that

$$|\mathbf{E}[I_1(t\wedge\tau_\varepsilon)]| \leqq C_1\int_0^t \mathbf{E}[|X(s\wedge\tau_\varepsilon,x) - X(s\wedge\tau_\varepsilon,y)|^p]\,\mathrm{d}s$$

and

$$|\mathbf{E}[I_2(t\wedge\tau_\varepsilon)]| \leqq C_2\int_0^t \mathbf{E}[|X(s\wedge\tau_\varepsilon,x) - X(s\wedge\tau_\varepsilon,y)|^p]\,\mathrm{d}s.$$

Hence, there exists a positive constant C_3 such that

$$\mathbf{E}[|X(t\wedge\tau_\varepsilon,x) - X(t\wedge\tau_\varepsilon,y)|^p]$$
$$\leqq |x-y|^p + C_3\int_0^t \mathbf{E}[|X(s\wedge\tau_\varepsilon,x) - X(s\wedge\tau_\varepsilon,y)|^p]\,\mathrm{d}s$$

and, by Gronwall's inequality (Theorem A.1.1), there exists a positive constant C_4 such that

$$\mathbf{E}[|X(t\wedge\tau_\varepsilon,x) - X(t\wedge\tau_\varepsilon,y)|^p] \leqq C_4|x-y|^p \qquad (t\in[0,T]).$$

Letting $\varepsilon \downarrow 0$ and setting $\tau = \inf\{t; X(t,x) = X(t,y)\}$, we obtain

$$\mathbf{E}[|X(t\wedge\tau,x) - X(t\wedge\tau,y)|^p] \leqq C_4|x-y|^p \qquad (t\in[0,T]).$$

Substituting $p=-1$, we have $\mathbf{P}(\tau<\infty)=0$. Thus (4.10.2) holds. □

By Propositions 4.10.1 and 4.10.2, for $t \geqq 0$ and $x,y\in\mathbb{R}^d$ with $x\neq y$, $X(t,x)\neq X(t,y)$ almost surely. However, the exceptional set $\{X(t,x)\neq X(t,y)\}$ depends on (x,y). Hence these propositions are not enough to see the injectivity of $X(t,\cdot)$.

Proposition 4.10.3 *Set* $D = \{(x,x); x\in\mathbb{R}^d\}$. *Then,* $Z(t,x,y) = |X(t,x) - X(t,y)|^{-1}$ *has a modification which is continuous on* $[0,\infty)\times(\mathbb{R}^d\times\mathbb{R}^d\setminus D)$.

Proof Let $p > 2(2d+1)$. Then,

$$|Z(t,x,y) - Z(t',x',y')|^p \leqq 2^p Z(t,x,y)^p Z(t',x',y')^p \times \{|X(t,x) - X(t',x')|^p + |X(t,y) - X(t',y')|^p\}.$$

By Hölder's inequality,

$$\begin{aligned}&\mathbf{E}[|Z(t,x,y) - Z(t',x',y')|^p] \\ &\leqq 2^p \mathbf{E}[Z(t,x,y)^{4p}]^{\frac{1}{4}} \mathbf{E}[Z(t',x',y')^{4p}]^{\frac{1}{4}} \\ &\qquad \times \{\mathbf{E}[|X(t,x) - X(t',x')|^{2p}]^{\frac{1}{2}} + \mathbf{E}[|X(t,y) - X(t',y')|^{2p}]^{\frac{1}{2}}\}.\end{aligned}$$

Hence, by Propositions 4.10.1 and 4.10.2, there exists a positive constant C_1 such that

$$\mathbf{E}[|Z(t,x,y) - Z(t',x',y')|^p] \leqq C_1 |x-y|^{-p} |x'-y'|^{-p} \times \{|x-x'|^p + |y-y'|^p + |t-t'|^{\frac{p}{2}}\}.$$

For $\delta > 0$, let $D_\delta = \{(x,y)\ x,y \in \mathbb{R}^d, |x-y| > \delta\}$. The above inequality implies

$$\mathbf{E}[|Z(t,x,y) - Z(t',x',y')|^p] \leqq C_1 \delta^{-2p} \{|x-x'|^p + |y-y'|^p + |t-t'|^{\frac{p}{2}}\}$$

for any $t,t' \in [0,T]$ and $(x,y),(x',y') \in D_\delta$. By Kolmogorov's continuity theorem (Theorem A.5.1), $Z(t,x,y)$ has a modification which is continuous on $[0,T] \times D_\delta$. Since $T > 0$ and $\delta > 0$ are arbitrary, we obtain the conclusion. □

We show the surjectivity of $X(t,\cdot)$.

Proposition 4.10.4 *For $T > 0$ and $\in \mathbb{R}$, there exists a positive constant C such that*

$$\mathbf{E}[(1+|X(t,x)|^2)^p] \leqq C(1+|x|^2)^p \qquad (x \in \mathbb{R}^d,\ t \in [0,T]).$$

Set $f(z) = (1+|z|^2)^p$ and apply Itô's formula to $f(X(t,x))$. Then we can show the proposition in the same way as Proposition 4.10.2. We omit the proof.

Proposition 4.10.5 *Let $\widehat{\mathbb{R}}^d = \mathbb{R}^d \cup \{\triangle\}$ be a one-point compactification of $\mathbb{R}^d$ and set*

$$\eta(t,x) = \begin{cases} (1+|X(t,x)|)^{-1} & (x \in \mathbb{R}^d) \\ 0 & (x = \triangle). \end{cases}$$

Then, η is continuous on $[0,\infty) \times \widehat{\mathbb{R}}^d$.

Proof The continuity of η on $[0,\infty)\times\mathbb{R}^d$ follows from Proposition 4.10.1.

Let $p > 2(2d+1)$. By a similar argument to that in the proof of Proposition 4.10.3, we can show that there exists a positive constant C such that

$$\mathbf{E}[|\eta(t,x)-\eta(s,y)|^p] \leqq C(1+|x|)^{-p}(1+|y|)^{-p}(|x-y|^p+|t-s|^{\frac{p}{2}})$$

for all $x,y\in\mathbb{R}^d$ and $s,t\in[0,T]$. For $x=(x^1,x^2,\ldots,x^d)$, set

$$x^{-1}=\Big(\frac{x_1}{|x|^2},\frac{x_2}{|x|^2},\ldots,\frac{x_d}{|x|^2}\Big).$$

Since

$$\frac{|x-y|}{|x|\,|y|}=|x^{-1}-y^{-1}|,$$

we have

$$\mathbf{E}[|\eta(t,x)-\eta(s,y)|^p] \leqq C(|x^{-1}-y^{-1}|^p+|t-s|^{\frac{p}{2}}).$$

Moreover, set

$$\widetilde{\eta}(t,x)=\begin{cases}\eta(t,x^{-1}) & (x\neq 0)\\ 0 & (x=0).\end{cases}$$

Then, since $(x^{-1})^{-1}=x$ $(x\in\mathbb{R}^d\setminus\{0\})$,

$$\mathbf{E}[|\widetilde{\eta}(t,x)-\widetilde{\eta}(s,y)|^p] \leqq C(|x-y|^p+|t-s|^{\frac{p}{2}}) \text{ and } \mathbf{E}[|\widetilde{\eta}(t,x)|^p] \leqq C|x|^p.$$

Hence, by Kolmogorov's continuity theorem (Theorem A.5.1), $\widetilde{\eta}$ is continuous on $[0,\infty)\times\mathbb{R}^d$. This implies that $\eta(t,x)$ is continuous on $[0,\infty)\times(\{|x|>R\}\cup\{\triangle\})$ for any $R>0$. □

Proposition 4.10.6 *The mapping $X(t,\cdot)$ is surjective for any $t>0$ almost surely.*

Proof Define a stochastic process $\{\widehat{X}(t,x)\}_{t\geqq 0}$ $(x\in\widehat{\mathbb{R}}^d)$ on $\widehat{\mathbb{R}}^d$ by

$$\widehat{X}(t,x)=\begin{cases}X(t,x) & (x\in\mathbb{R}^d)\\ \triangle & (x=\triangle).\end{cases}$$

By Proposition 4.10.5, $\widehat{X}(t,x)$ is continuous on $[0,\infty)\times\widehat{\mathbb{R}}^d$ and, for any $t\geqq 0$, the mapping $\widehat{X}(t,\cdot)$ is homotopic with the identity mapping on $\widehat{\mathbb{R}}^d$.[9] On the other hand, since $\widehat{\mathbb{R}}^d$ and a d-dimensional sphere S^d are homeomorphic, $\widehat{X}(t,\cdot)$ induces a continuous and injective mapping on S^d. It is well known in the

[9] Mappings f and g from a topological space M onto M are said to be homotopic if there exists a continuous mapping $\phi(t,x)$, $(t,x)\in[0,1]\times M$ such that $\phi(0,x)=f(x)$ and $\phi(1,x)=g(x)$.

theory of topology that the identity mapping on S^d is not homotopic with the constant mapping.

Suppose now that $\widehat{X}(t,\cdot)$ is not surjective. Then, $\widehat{X}(t,\cdot)$ is homotopic with a continuous mapping from $\widehat{\mathbb{R}}^d$ into $\mathbb{R}^d$ and, therefore, with a constant mapping. This contradicts the fact that $\widehat{X}(t,\cdot)$ is homotopic with the identity mapping on $\widehat{\mathbb{R}}^d$. Hence, the mapping $\widehat{X}(t,\cdot)$ and, by definition, $X(t,\cdot)$ are surjective. □

Theorem 4.10.7 *Let $\{X(t,x)\}_{t\geqq 0}$ be a unique strong solution of the stochastic differential equation* (4.10.1). *Then, the mapping defined by $\mathbb{R}^d \ni x \mapsto X(t,x) \in \mathbb{R}^d$ is a homeomorphism on $\mathbb{R}^d$ for all $t \geqq 0$ almost surely.*

Proof The assertion follows by summing up Propositions 4.10.1, 4.10.3, and 4.10.6. □

Finally, we discuss the differentiability of the mapping $X(t,\cdot)$.

Theorem 4.10.8 *For $m \geqq 1$, assume that $V_0, V_1, \ldots, V_r$ are bounded and of C^m-class with bounded derivatives up to m-th order. Then,*
(1) *the mapping $X(t,\cdot)$ is of C^{m-1}-class;*
(2) *the Jacobian matrix $Y(t,x) = (\frac{\partial X^i(t,x)}{\partial x^j})_{i,j=1,2,\ldots,d}$ satisfies*

$$Y(t,x) = I + \sum_{k=0}^{r} \int_0^t V_k'(X(s,x))Y(s,x)\,\mathrm{d}B^k(s),$$

where I is an identity mapping on $\mathbb{R}^d$, $V_k'(x) = (\frac{\partial V_k^i(x)}{\partial x^j})_{i,j=1,2,\ldots,d}$ and $B^0(s) = s$;
(3) *$Y(t,x)$ is non-degenerate for all $t > 0$ and $x \in \mathbb{R}^d$ almost surely;*
(4) *the mapping $x \mapsto X(t,x)$ is a diffeomorphism of C^{m-1}-class for any t almost surely.*

Proof Let $\varepsilon \in (0,1]$ and $e_j = (\overbrace{0,\ldots,0}^{j-1},1,0,\ldots,0) \in \mathbb{R}^d$ $(j=1,2,\ldots,d)$. Set

$$Y_j^\varepsilon(t,x) = \frac{1}{\varepsilon}(X(t,x+\varepsilon e_j) - X(t,x)).$$

Then we have

$$\begin{aligned} Y_j^\varepsilon(t,x) &= e_j + \sum_{k=0}^{r} \int_0^t \frac{1}{\varepsilon}(V_k(X(s,x+\varepsilon e_j)) - V_k(X(s,x)))\mathrm{d}B^k(s) \\ &= e_j + \sum_{k=0}^{r} \int_0^t \Big\{ \int_0^1 V_k'(X(s,x) + u(X(s,x+\varepsilon e_j) - X(s,x)))\mathrm{d}u \Big\} \\ &\qquad \times Y_j^\varepsilon(s,x)\,\mathrm{d}B^k(s). \end{aligned} \tag{4.10.3}$$

Hence, $\{(X(t, x + \varepsilon e_j), X(t, x), Y_j^\varepsilon(t, x))\}_{t \geqq 0}$ is a solution of a stochastic differential equation with Lipschitz continuous coefficients on $\mathbb{R}^{3d}$. By Proposition 4.10.1, it is continuous in $(t, x, \varepsilon) \in [0, T] \times \mathbb{R}^d \times ([-1, 1] \setminus \{0\})$ and extended to a continuous mapping on $[0, T] \times \mathbb{R}^d \times [-1, 1]$. This means that $\lim_{\varepsilon \to 0} Y_j^\varepsilon(t, x) = \frac{\partial X(t,x)}{\partial x^j}$ exists almost surely and is continuous in (t, x).

(2) is obtained from (4.10.3) by letting $\varepsilon \to 0$.

To show (3), let $Z = \{Z(t, x)\}_{t \geqq 0}$ be a solution of a matrix-valued stochastic differential equation

$$Z(t, x) = I - \sum_{k=0}^{r} \int_0^t Z(s, x)V_k'(X(s, x))\,\mathrm{d}B^k(s) + \sum_{k=1}^{r} \int_0^t Z(s, x)(V_k'(X(s, x)))^2\mathrm{d}s.$$

By Itô's formula, we obtain $\mathrm{d}(Z(t, x)Y(t, x)) = 0$ and $Z(t, x)Y(t, x) = I$ for any $t \geqq 0$. Hence, $Y(t, x)$ is non-degenerate.

By a similar argument,[10] we can show that $\{(X(t, x), Y(t, x))\}_{t \geqq 0}$ is differentiable with respect to x, there exists $\frac{\partial^2 X^i(t,x)}{\partial x_j \partial x_\ell}$ and it is continuous. Moreover, by induction, we can prove that $X(t, x)$ is of C^{m-1}-class in x and obtain (1).

(4) follows from the assertions (1)–(3). □

Also, from a solution of the Stratonovich type stochastic differential equation

$$\mathrm{d}X(t) = \sum_{k=1}^{r} V_k(X(t)) \circ \mathrm{d}B^k(t) + V_0(X(t))\,\mathrm{d}t,$$

a stochastic flow of diffeomorphisms is defined. If the coefficients are of C^m-class, then the mapping $X(t, \cdot)$ is a diffeomorphism of C^{m-2}-class.

4.11 Approximation Theorem

Let $V_0, V_1, \ldots, V_r : \mathbb{R}^d \to \mathbb{R}^d$ be bounded and C^∞ functions with bounded derivatives of all orders. Consider the Stratonovich type stochastic differential equation

$$\begin{cases} \mathrm{d}X(t) = \sum_{k=1}^{r} V_k(X(t)) \circ \mathrm{d}B^k(t) + V_0(X(t))\,\mathrm{d}t, \\ X(0) = x. \end{cases} \tag{4.11.1}$$

Denote the unique strong solution by $\{X(t, x)\}_{t \geqq 0}$.

Consider a sequence of curves which converges to a Brownian motion $B = \{(B^1(t), B^2(t), \ldots, B^r(t))\}_{t \geqq 0}$; for example, curves obtained by piecewise linear approximations or by approximations via mollifiers. A natural question

[10] While we need to show that $X(t, x)$ and $Y(t, x)$ have moments of any order, we omit the proof. See [62, 111] for details.

is whether solutions of ordinary differential equations driven by these curves converge to those of stochastic ones. This problem is important in other fields than probability theory, for example, in numerical analysis for solutions of stochastic differential equations. In this section we consider a piecewise linear approximation of Brownian motion and discuss the corresponding approximation for solutions of stochastic differential equations, originated by Wong and Zakai [127]. For other approximations, see [45, 62, 79].

For $n = 1, 2, \ldots$ and $m = 1, 2, \ldots$, set

$$\Delta_{m,n} = B\Big(\frac{m+1}{2^n}\Big) - B\Big(\frac{m}{2^n}\Big)$$

and define a piecewise linear approximation $\{B_n(t)\}_{t\geqq 0}$ of a Brownian motion B by

$$B_n(t) = B\Big(\frac{m}{2^n}\Big) + 2^n\Big(t - \frac{m}{2^n}\Big)\Delta_{m,n} \qquad \Big(\frac{m}{2^n} \leqq t \leqq \frac{m+1}{2^n}\Big).$$

Let $\{X_n(t,x)\}_{t\geqq 0}$ be the solution of the random ordinary differential equation

$$\begin{cases} \dfrac{\mathrm{d}}{\mathrm{d}t}X_n(t) = \displaystyle\sum_{k=1}^{r} V_k(X_n(t))\dot{B}_n^k(t) + V_0(X_n(t)) \\ X_n(0) = x, \end{cases}$$

where $\dot{B}_n(t) = \frac{\mathrm{d}}{\mathrm{d}t}B_n(t)$ for $t \neq \frac{m}{2^n}$ $(m = 1, 2, \ldots)$ and $\dot{B}_n(t) = 0$ for $t = \frac{m}{2^n}$ $(m = 1, 2, \ldots)$.

The following is the main result in this section.

Theorem 4.11.1 *For any $T > 0$ and $p \geqq 2$,*

$$\lim_{n\to\infty} \sup_{x\in\mathbb{R}^d} \mathbf{E}\Big[\sup_{0\leqq t\leqq T} |X_n(t,x) - X(t,x)|^p\Big] = 0.$$

Proof For simplicity write $B^0(t) = t$, and denote $X(t,x)$ and $X_n(t,x)$ by $X(t)$ and $X_n(t)$, respectively. Let $V_k = (V_k^1, V_k^2, \ldots, V_k^d)$ and define the function $V_k[V_\ell] = ((V_k[V_\ell])^i) : \mathbb{R}^d \to \mathbb{R}^d$ by

$$(V_k[V_\ell])^i = \sum_{j=1}^{d} V_k^j \frac{\partial V_\ell^i}{\partial x^j}.$$

For $\frac{m}{2^n} \leqq t \leqq \frac{m+1}{2^n}$, we have by Taylor's theorem

$$X_n(t) = X_n\Big(\frac{m}{2^n}\Big) + 2^n\Big(t - \frac{m}{2^n}\Big)\sum_{k=0}^{r} V_k\Big(X_n\Big(\frac{m}{2^n}\Big)\Big)\Delta_{m,n}^k$$
$$+ \frac{1}{2}2^{2n}\Big(t - \frac{m}{2^n}\Big)^2 \sum_{k,\ell=1}^{r} V_k[V_\ell]\Big(X_n\Big(\frac{m}{2^n}\Big)\Big)\Delta_{m,n}^k\Delta_{m,n}^\ell + R_n^m(t),$$

where

$$R_n^m(t) = \sum_{k=0}^{r} \int_{m2^{-n}}^{t} \mathrm{d}s \int_{m2^{-n}}^{s} V_k[V_0](X_n(u))\,\mathrm{d}u \times 2^n \Delta_{n,m}^k$$
$$+ 2^{3n} \sum_{k,\ell,p=1}^{r} \int_{m2^{-n}}^{t} \mathrm{d}s \int_{m2^{-n}}^{s} \mathrm{d}u \int_{m2^{-n}}^{u} \mathrm{d}v\, V_k[V_\ell[V_p]](X_n(v))$$
$$\times \Delta_{m,n}^k \Delta_{m,n}^\ell \Delta_{m,n}^p.$$

In particular, we have

$$X_n\Big(\frac{m+1}{2^n}\Big) = X_n\Big(\frac{m}{2^n}\Big) + \sum_{k=1}^{r} V_k\Big(X_n\Big(\frac{m}{2^n}\Big)\Big)\Delta_{m,n}^k + \widetilde{V_0}\Big(X_n\Big(\frac{m}{2^n}\Big)\Big)\Delta_{n,m}^0$$
$$+ \frac{1}{2}\sum_{k=1}^{r} V_k[V_k]\Big(X_n\Big(\frac{m}{2^n}\Big)\Big)\{(\Delta_{m,n}^k)^2 - 2^{-n}\}$$
$$+ \sum_{k\neq\ell} V_k[V_\ell]\Big(X_n\Big(\frac{m}{2^n}\Big)\Big)\Delta_{m,n}^k\Delta_{m,n}^\ell + R_n^m\Big(\frac{m+1}{2^n}\Big),$$

where $\widetilde{V_0}$ is given by

$$\widetilde{V_0} = V_0 + \frac{1}{2}\sum_{k=1}^{r} V_k[V_k].$$

Hence, setting $[s]_n = \frac{[2^n s]}{2^n} \in \{\frac{m}{2^n}\}_{m=0}^{\infty}$, $[u]$ being the largest integer less than or equal to u, we have for $t > 0$

$$X_n(t) = x + \sum_{k=1}^{r} \int_0^{[t]_n} V_k(X_n([s]_n))\,\mathrm{d}B^k(s) + \int_0^{[t]_n} \widetilde{V_0}(X_n([s]_n))\,\mathrm{d}s$$
$$+ I_n^1(t) + I_n^2(t) + \sum_{m=0}^{[2^n t]-1} R_n^m\Big(\frac{m+1}{2^n}\Big) + X_n(t) - X_n([t]_n),$$

where

$$I_n^1(t) = \frac{1}{2}\sum_{m=1}^{[2^n t]}\sum_{k=1}^{r} V_k[V_k]\Big(X_n\Big(\frac{m}{2^n}\Big)\Big)\{(\Delta_{m,n}^k)^2 - 2^{-n}\},$$
$$I_n^2(t) = \frac{1}{2}\sum_{m=1}^{[2^n t]}\sum_{k\neq\ell} V_k[V_\ell]\Big(X_n\Big(\frac{m}{2^n}\Big)\Big)\Delta_{m,n}^k\Delta_{m,n}^\ell.$$

It is easy to show that $\sup_x \mathbf{E}\Big[\sup_{0\leqq t\leqq T} |X_n(t) - X_n([t]_n)|^p\Big] \to 0$ and

$$\sup_x \mathbf{E}\Big[\sup_{0\leqq t\leqq T}\Big|\sum_{m=1}^{[2^n t]} R_n^m\Big(\frac{m+1}{2^n}\Big)\Big|^p\Big] \to 0$$

as $n \to \infty$. By Itô's formula, we have

$$(\Delta^k_{m,n})^2 - \frac{1}{2^n} = 2\int_{m2^{-n}}^{(m+1)2^{-n}} \Big(B^k(s) - B^k\Big(\frac{m}{2^n}\Big)\Big)\mathrm{d}B^k(s),$$

and hence

$$I^1_n(t) = \sum_{k=1}^r \int_0^{[t]_n} V_k[V_k](X_n([s]_n))(B^k(s) - B^k([s]_n))\,\mathrm{d}B^k(s).$$

Thus, by the continuity of paths of Brownian motion, we obtain

$$\lim_{n\to\infty} \sup_{x\in\mathbb{R}^d} \mathbf{E}\Big[\sup_{0\leqq t\leqq T} |I^1_n(t)|^p\Big] = 0.$$

If $k \neq \ell$, then

$$\Delta^k_{m,n}\Delta^\ell_{m,n} = \int_{m2^{-n}}^{(m+1)2^{-n}} \Big(B^k(s) - B^k\Big(\frac{m}{2^n}\Big)\Big)\mathrm{d}B^\ell(s) + \int_{m2^{-n}}^{(m+1)2^{-n}} \Big(B^\ell(s) - B^\ell\Big(\frac{m}{2^n}\Big)\Big)\mathrm{d}B^k(s).$$

Hence we obtain $\sup_x \mathbf{E}\Big[\sup_{0\leqq t\leqq T} |I^2_n(t)|^p\Big] \to 0$ in a similar way to $I^1_n(t)$.

From the above observations, there exists an $I_n(t)$ with

$$\lim_{n\to\infty} \sup_{x\in\mathbb{R}^d} \mathbf{E}\Big[\sup_{0\leqq t\leqq T} |I_n(t)|^p\Big] = 0$$

such that

$$X(t) - X_n(t) = \sum_{k=1}^r \int_0^{[t]_n} (V_k(X(s)) - V_k(X_n(s)))\mathrm{d}B^k(s) + \int_0^{[t]_n} (\widetilde{V}_0(X(s)) - \widetilde{V}_0(X_n(s)))\mathrm{d}s + I_n(t).$$

Hence, in a similar way to the proof of Proposition 4.10.1, we obtain the conclusion by Doob's inequality (Theorem 1.5.13), the Burkholder–Davis–Gundy inequality (Theorem 2.4.1), and Gronwall's inequality. The details are left to the reader. □

The problem of characterizing the subset of path space which consists of the paths of the solutions of a stochastic differential equation of the form (4.11.1) is called a **support problem**, and is closely related to approximation of solutions of stochastic differential equations.

Let $\mathbf{P}_x$ be the probability law of the solution $X = \{X(t)\}_{t\geqq 0}$ of (4.11.1). Let $\mathscr{C}^1$ be the subspace of the Wiener space $W^r = \{w \in \mathbf{W}^r; w(0) = 0\}$ consisting of paths of C^1-class. For $\phi \in \mathscr{C}^1$, let $\xi(x, \phi)$ be the solution of the ordinary differential equation

$$\begin{cases} \mathrm{d}\xi(t) = \sum_{k=1}^{r} V_k(\xi(t))\dot{\phi}^k(t)\,\mathrm{d}t + V_0(\xi(t))\,\mathrm{d}t, \\ \xi(0) = x. \end{cases}$$

Then the following is known ([115]).

Theorem 4.11.2 *For any $x \in \mathbb{R}^d$, the topological support of $\mathbf{P}_x$, that is, the smallest closed subset of $\mathbf{W}^d$ whose $\mathbf{P}_x$-measure is 1, coincides with the closure of $\{\xi(x, \phi); \phi \in \mathscr{C}^1\}$.*

Moreover, if the rank of the Lie algebra generated by $V_1, V_2, \ldots, V_r$ is equal to d at every point in $\mathbb{R}^d$, then the closure of $\{\xi(x, \phi); \phi \in \mathscr{C}^1\}$ is $W_x^d = \{w \in \mathbf{W}^d; w(0) = x\}$ ([63]) and the topological support of $\mathbf{P}_x$ coincides with $\mathbf{W}_x^d$. For details, see the references cited at the beginning of this section and [45, 114].

5

Malliavin Calculus

In 1976, Malliavin ([74, 75]) proposed a new calculus on Wiener spaces and achieved purely probabilistic proofs of results related to diffusion processes, which, before him, were shown based on outcomes in other mathematical fields like partial differential equations. For example, he proved the existence and smoothness of the transition densities of diffusion processes in a purely probabilistic manner. This method has been developed into a theoretical system, which is nowadays called the **Malliavin calculus** [43, 73, 104, 122]. It plays an important role in stochastic analysis together with the Itô calculus, consisting of stochastic integrals, stochastic differential equations, and so on. The aim of this chapter is to introduce the Malliavin calculus. We will use the fundamental terminologies and notions in functional analysis without detailed explanation. For these, consult [1, 19, 58].

5.1 Sobolev Spaces and Differential Operators

Throughout this chapter, let $T > 0$, $d \in \mathbb{N}$ and $(W_T, \mathscr{B}(W_T), \mu_T)$ be the d-dimensional Wiener space on $[0, T]$ (Definition 1.2.2). For $t \in [0, T]$, define $\theta(t) : W_T \to \mathbb{R}^d$ by $\theta(t)(w) = w(t)$ $(w \in W_T)$. Moreover, let H_T be the Cameron–Martin subspace of W_T. Then, identifying H_T with its dual space H_T^* in a natural way, we obtain the relation $W_T^* \subset H_T^* = H_T \subset W_T$ (see Section 1.2).

Let E be a real separable Hilbert space and $L^p(\mu_T; E)$ be the space of E-valued p-th integrable functions with respect to μ_T on W_T. $L^p(\mu_T; \mathbb{R})$ is simply written as $L^p(\mu_T)$. Denote the norm in $L^p(\mu_T; E)$ by $\|\cdot\|_p$ or $\|\cdot\|_{p,E}$ when emphasizing E is necessary.

Let $\mathscr{P}$ be the set of functions $\phi : W_T \to \mathbb{R}$ of the form $\phi = f(\ell_1, \dots, \ell_n)$ for $\ell_1, \dots, \ell_n \in W_T^*$ and a polynomial $f : \mathbb{R}^n \to \mathbb{R}$, that is,

$$\phi(w) = f(\ell_1(w), \dots, \ell_n(w)) \qquad (w \in W_T).$$

Set

$$\mathscr{P}(E) = \Big\{\sum_{j=1}^{m} \phi_j e_j \,;\, \phi_j \in \mathscr{P},\ e_j \in E,\ j = 1, \ldots, m,\ m \in \mathbb{N}\Big\}.$$

For $\phi = f(\ell_1, \ldots, \ell_n) \in \mathscr{P}$, define $\nabla\phi \in \mathscr{P}(H_T)$ by

$$\nabla\phi = \sum_{i=1}^{n} \frac{\partial f}{\partial x^i}(\ell_1, \ldots, \ell_n)\ell_i.$$

Moreover, for $\phi = \sum_{j=1}^{m} \phi_j e_j \in \mathscr{P}(E)$, define $\nabla\phi \in \mathscr{P}(H_T \otimes E)$ by

$$\nabla\phi = \sum_{j=1}^{m} \nabla\phi_j \otimes e_j,$$

where, for real separable Hilbert spaces E_1 and E_2, $E_1 \otimes E_2$ denotes the Hilbert space of Hilbert–Schmidt operators $A : E_1 \to E_2$ and, for $e^{(1)} \in E_1$ and $e^{(2)} \in E_2$, $e^{(1)} \otimes e^{(2)}$ denotes the Hilbert–Schmidt operator such that $E_1 \ni e \mapsto \langle e^{(1)}, e\rangle_{E_1} e^{(2)} \in E_2$. The Hilbert space $E_1 \otimes E_2$ has an inner product given by

$$\langle A, B\rangle_{E_1\otimes E_2} = \sum_{n=1}^{\infty} \langle Ae_n^{(1)}, Be_n^{(1)}\rangle_{E_2} \qquad (A, B \in E_1 \otimes E_2),$$

where $\{e_n^{(1)}\}_{n=1}^{\infty}$ is an orthonormal basis of E_1. It should be noted that the above definition of $\nabla\phi$ does not depend on the expression of $\phi \in \mathscr{P}$, because

$$\langle \nabla\phi(w), h\rangle_{H_T} = \frac{\mathrm{d}}{\mathrm{d}\xi}\Big|_{\xi=0} \phi(w + \xi h) \qquad (w \in W_T,\ h \in H_T). \tag{5.1.1}$$

Example 5.1.1 Let $d = 1$. For $t \in [0, T]$, the coordinate function $\theta(t) : W_T \to \mathbb{R}$ satisfies

$$\langle \nabla\theta(t), h\rangle_{H_T} = h(t) = \int_0^T \mathbf{1}_{[0,t]}(s)\dot{h}(s)\,\mathrm{d}s \qquad (h \in H_T).$$

Hence, defining $\ell_{[0,t]} \in H_T$ by $\dot{\ell}_{[0,t]}(s) = \mathbf{1}_{[0,t]}(s)$ $(s \in [0, t])$, we have $\nabla\theta(t) = \ell_{[0,t]}$.

Lemma 5.1.2 *Let $p > 1$. For $F \in L^p(\mu_T; E)$, $\ell \in W_T^*$, $\phi \in \mathscr{P}$ and $e \in E$, the mapping $\mathbb{R} \ni \xi \mapsto \int_{W_T} \langle F(\cdot + \xi\ell), \phi e\rangle_E \,\mathrm{d}\mu_T$ is differentiable and*

$$\frac{\mathrm{d}}{\mathrm{d}\xi}\Big|_{\xi=0} \int_{W_T} \langle F(\cdot + \xi\ell), \phi e\rangle_E \,\mathrm{d}\mu_T = \int_{W_T} \langle F, e\rangle_E \partial_\ell\phi \,\mathrm{d}\mu_T, \tag{5.1.2}$$

where

$$\partial_\ell\phi(w) = \ell(w)\phi(w) - \langle \nabla\phi(w), \ell\rangle_{H_T}.$$

In particular, the mapping $\nabla : L^p(\mu_T; E) \supset \mathscr{P}(E) \ni \phi \mapsto \nabla\phi \in L^p(\mu_T; H_T \otimes E)$ *is closable, that is,* ∇ *is extended to a unique closed operator whose domain is a dense subset of* $L^p(\mu_T; E)$.

Proof By the Cameron–Martin theorem (Theorem 1.7.2),

$$\int_{W_T} \langle F(w + \xi\ell), \phi(w)e\rangle_E \mu_T(\mathrm{d}w)$$
$$= \int_{W_T} \langle F(w), e\rangle_E \phi(w - \xi\ell) \exp\Bigl(\xi\ell(w) - \frac{\xi^2}{2}\|\ell\|_{H_T}^2\Bigr)\mu_T(\mathrm{d}w).$$

It is easy to see that the right hand side is differentiable in ξ, and, hence, so is the left hand side. Differentiating both sides in $\xi = 0$, we obtain (5.1.2) by (5.1.1).

By (5.1.2), we have, for any $\psi \in \mathscr{P}(E)$, $e \in E$, $\ell \in W_T^*$ and $\phi \in \mathscr{P}$,

$$\int_{W_T} \langle \nabla\psi, \ell \otimes e\rangle_{H_T \otimes E} \phi \, \mathrm{d}\mu_T = \int_{W_T} \langle \psi, e\rangle_E \partial_\ell \phi \, \mathrm{d}\mu_T.$$

Hence, if $\{\psi_n\}_{n=1}^\infty \subset \mathscr{P}(E)$ and $G \in L^p(\mu_T; H_T \otimes E)$ satisfy $\|\psi_n\|_p \to 0$ and $\|\nabla\psi_n - G\|_p \to 0$ $(n \to \infty)$, then

$$\int_{W_T} \langle G, \ell \otimes e\rangle_{H_T \otimes E} \phi \, \mathrm{d}\mu_T = 0.$$

Since this holds for any e, ℓ, and ϕ, we obtain that $G = 0$, μ_T-a.s. and ∇ is closable. □

Remark 5.1.3 (1) By (5.1.1) and (5.1.2),

$$\int_{W_T} \langle \nabla F, \phi\ell \otimes e\rangle_{H_T \otimes E} \, \mathrm{d}\mu_T = \int_{W_T} \langle F, (\partial_\ell \phi)\, e\rangle_E \, \mathrm{d}\mu_T \tag{5.1.3}$$

for any $F \in \mathscr{P}(E)$. Denote the dual operator of ∇ by ∇^*. Then, the left hand side is equal to $\int_{W_T} \langle F, \nabla^*(\phi\ell \otimes e)\rangle_E \, \mathrm{d}\mu_T$. Thus

$$\nabla^*(\phi\ell \otimes e) = (\partial_\ell \phi)\, e \tag{5.1.4}$$

since F is arbitrary. Identity (5.1.3) is a prototype of the integration by parts formula on the Wiener space presented in Section 5.4.
(2) Set $E = \mathbb{R}$ and $\phi = 1$ in (5.1.4). Then

$$(\nabla^*\ell)(w) = \ell(w), \quad \mu_T\text{-a.s.}, \tag{5.1.5}$$

where $\ell \in W_T^*$ is regarded as an H_T-valued constant function on the left hand side and as a random variable $\ell : W_T \to \mathbb{R}$ on the right hand side.

Moreover, if $\|\ell\|_{H_T} = 1$ and $F = \varphi(\ell)$, then (5.1.3) corresponds to the following elementary identity for a standard normal random variable X,

$$\mathbf{E}[\varphi'(X)] = \int_{\mathbb{R}} \varphi'(x) \frac{1}{\sqrt{2\pi}} e^{-\frac{x^2}{2}} dx = \int_{\mathbb{R}} \varphi(x) x \frac{1}{\sqrt{2\pi}} e^{-\frac{x^2}{2}} dx = \mathbf{E}[\varphi(X)X].$$

On account of the closability of ∇, we introduce **Sobolev spaces** over the Wiener space.

Definition 5.1.4 Let $p \geqq 1$ and $k \in \mathbb{N}$. For $\phi \in \mathscr{P}(E)$, set

$$\|\phi\|_{(k,p)} = \sum_{j=0}^{k} \|\nabla^j \phi\|_p$$

and denote the completion of $\mathscr{P}(E)$ with respect to $\|\cdot\|_{(k,p)}$ by $\mathbb{D}^{k,p}(E)$. Simply write $\mathbb{D}^{k,p}$ for $\mathbb{D}^{k,p}(\mathbb{R})$. Denote by the same ∇ the extension of $\nabla : \mathscr{P}(E) \to \mathscr{P}(H_T \otimes E)$ to $\mathbb{D}^{k,p}(E)$ and by ∇^* the adjoint operator of the closed operator $\nabla : L^p(\mu_T; E) \to L^p(\mu_T; H_T \otimes E)$.

If $k \leqq k'$ and $p \leqq p'$, then $\mathbb{D}^{k',p'}(E) \subset \mathbb{D}^{k,p}(E)$. By definition, ∇ is defined consistently on each $\mathbb{D}^{k,p}(E)$. Moreover, by (5.1.4), for $F \in \mathscr{P}(H_T \otimes E)$ of the form

$$F = \sum_{j=1}^{m} \phi_j \ell_j \otimes e_j$$

with $\phi_j \in \mathscr{P}$, $\ell_j \in W_T^*$ and $e_j \in E$ $(j = 1, \dots, m)$, we have

$$\nabla^* F = \sum_{j=1}^{m} (\partial_{\ell_j} \phi_j) e_j.$$

Hence, ∇^* is also defined consistently on each $L^p(\mu_T; E)$. Because of this consistency, we may use the simple notations ∇ and ∇^* without referring to the dependency on k and p.

Example 5.1.5 Let $\ell \in W_T^*$. Set $f(x) = x$ $(x \in \mathbb{R})$ and write $\ell(w) = f(\ell(w))$ $(w \in W_T)$. Then, by definition, the derivative of $\ell : W_T \ni w \to \ell(w) \in \mathbb{R}$ is given by

$$(\nabla \ell)(w) = \ell, \quad \mu_T\text{-a.s. } w \in W_T.$$

Combining this identity with (5.1.5), we obtain

$$\nabla(\nabla^* \ell)(w) = \ell, \quad \mu_T\text{-a.s. } w \in W_T. \tag{5.1.6}$$

We now show that this identity is extended to H_T.

Let $h \in H_T$ and take $\ell_n \in W_T^*$ $(n = 1, 2, \dots)$ so that $\|\ell_n - h\|_{H_T} \to 0$. Setting

$$A_p = \Big(\int_{\mathbb{R}} |x|^p \frac{1}{\sqrt{2\pi}} \mathrm{e}^{-\frac{x^2}{2}} \mathrm{d}x \Big)^{\frac{1}{p}},$$

we have

$$\int_{W_T} \|\ell_n(w) - \ell_m(w)\|^p \mu_T(\mathrm{d}w) = A_p^p \big\|\ell_n - \ell_m\big\|_{H_T}^p.$$

By (5.1.5), this implies

$$\lim_{n,m\to\infty} \|\nabla^* \ell_n - \nabla^* \ell_m\|_p = 0. \tag{5.1.7}$$

Since ∇^* is a closed operator, h belongs to the domain of ∇^* as a constant H_T-valued function and

$$\lim_{n\to\infty} \|\nabla^* \ell_n - \nabla^* h\|_p = 0.$$

Combining this with (5.1.5), we obtain

$$\nabla^* h = \mathscr{I}(h), \tag{5.1.8}$$

where $\mathscr{I}(h)$ is the Wiener integral of $\dot{h}$ (see Section 1.7).

On the other hand, (5.1.6) implies

$$\lim_{n,m\to\infty} \|\nabla(\nabla^* \ell_n) - \nabla(\nabla^* \ell_m)\|_p = \lim_{n,m\to\infty} \|\ell_n - \ell_m\|_{H_T} = 0.$$

By (5.1.7) and the closedness of ∇, $\nabla^* h$ belongs to the domain of ∇ and

$$\nabla(\nabla^* h) = h. \tag{5.1.9}$$

In order to develop the theory of distributions on Wiener spaces, we need to consider the Sobolev spaces $\mathbb{D}^{k,p}(E)$ for $k \in \mathbb{R}$. For this extension, we introduce the Wiener chaos decomposition of $L^2(\mu_T)$, which plays an important role in several areas of stochastic analysis.

Define the **Hermite polynomials** $\{H_n\}_{n=0}^\infty$ by

$$H_n(x) = \frac{(-1)^n}{n!} \mathrm{e}^{\frac{1}{2}x^2} \frac{\mathrm{d}^n}{\mathrm{d}x^n} (\mathrm{e}^{-\frac{1}{2}x^2}) \qquad (x \in \mathbb{R}).$$

We have

$$\mathrm{e}^{-\frac{1}{2}(x-y)^2} = \sum_{n=0}^\infty \frac{1}{n!} \frac{\mathrm{d}^n}{\mathrm{d}x^n} (\mathrm{e}^{-\frac{1}{2}x^2})(-y)^n = \mathrm{e}^{-\frac{1}{2}x^2} \sum_{n=0}^\infty H_n(x) y^n$$

and the generating function for the Hermite polynomials is

$$\sum_{n=0}^\infty H_n(x) y^n = \mathrm{e}^{xy - \frac{1}{2}y^2}. \tag{5.1.10}$$

$\{H_n\}_{n=0}^{\infty}$ forms an orthogonal basis of the L^2-space on $\mathbb{R}$ with respect to the standard normal distribution and

$$\int_{\mathbb{R}} H_i(x)H_j(x)\frac{1}{\sqrt{2\pi}}\mathrm{e}^{-\frac{1}{2}x^2}\mathrm{d}x = \frac{1}{j!}\delta_{ij}.$$

This identity is shown by inserting (5.1.10) into the left hand side of

$$\int_{\mathbb{R}} \mathrm{e}^{sx-\frac{1}{2}s^2}\mathrm{e}^{tx-\frac{1}{2}t^2}\frac{1}{\sqrt{2\pi}}\mathrm{e}^{-\frac{1}{2}x^2}\mathrm{d}x = \mathrm{e}^{st}$$

and comparing the coefficients of $s^i t^j$.

By using the Hermite polynomials, we construct an orthonormal basis of $L^2(\mu_T)$ in the following way. Let $\mathscr{A}$ be the set of sequences of non-negative integers with a finite number of non-zero elements:

$$\mathscr{A} = \Big\{\boldsymbol{\alpha} = \{\alpha_j\}_{j=1}^{\infty}\,;\,\alpha_j \in \mathbb{Z}_+, \sum_{j=1}^{\infty}\alpha_j < \infty\Big\}.$$

For $\boldsymbol{\alpha} \in \mathscr{A}$, define $|\boldsymbol{\alpha}|$ and $\boldsymbol{\alpha}!$ by

$$|\boldsymbol{\alpha}| = \sum_{j=1}^{\infty}\alpha_j \quad \text{and} \quad \boldsymbol{\alpha}! = \prod_{j:\alpha_j\neq 0}\alpha_j!.$$

Fix an orthonormal basis $\{h_n\}_{n=1}^{\infty}$ of the Cameron–Martin subspace H_T and define a family H_α ($\alpha \in \mathscr{A}$) of functions on H by

$$H_\alpha(w) = \prod_{j=1}^{\infty} H_{\alpha_j}(\mathscr{I}(h_j)(w)),$$

where $\mathscr{I}(h)$ is the Wiener integral of $h \in H_T$.

Theorem 5.1.6 *$\{\sqrt{\alpha!}H_\alpha, \alpha \in \mathscr{A}\}$ forms an orthonormal basis of $L^2(\mu_T)$. Moreover, $L^2(\mu_T)$ admits the orthogonal decomposition*

$$L^2(\mu_T) = \bigoplus_{n=0}^{\infty}\mathscr{H}_n, \tag{5.1.11}$$

where $\mathscr{H}_n$ $(n = 0, 1, 2, \dots)$ is the closed subspace of $L^2(\mu_T)$ spanned by $\{H_\alpha;\ |\alpha| = n\}$. $\mathscr{H}_n$ does not depend on the choice of the orthonormal basis of H_T.

Proof $\mathscr{I}(h_j)$ is a standard normal random variable. Hence, the orthonormality of $\{\sqrt{n!}H_n\}$ with respect to the standard Gaussian measure implies that of $\{\sqrt{\alpha!}H_\alpha\}$.

We next show that $\bigoplus_{n=0}^{\infty} \mathscr{H}_n$ is dense. Let $X \in L^2(\mu_T)$ and suppose that $\int_{W_T} XY \,\mathrm{d}\mu_T = 0$ for any $Y \in \bigoplus_{n=0}^{\infty} \mathscr{H}_n$. This implies that X is orthogonal to all polynomials of $\mathscr{I}(h_j)$ and that

$$\int_{W_T} X \exp\Big(\mathrm{i} \sum_{j=1}^{n} a_j \mathscr{I}(h_j)\Big) \mathrm{d}\mu_T = 0$$

for all $n \in \mathbb{N}$ and $a_j \in \mathbb{R}$. Hence,

$$\int_{W_T} X f(\mathscr{I}(h_1), \mathscr{I}(h_2), \ldots, \mathscr{I}(h_n)) \,\mathrm{d}\mu_T = 0$$

for any $f \in C_0^\infty(\mathbb{R}^n)$ $(n \in \mathbb{N})$, which means $X = 0$ because, by the Itô–Nisio theorem (Theorem 1.2.5), $\sum_j \mathscr{I}(h_j)h_j$ converges almost surely and the distribution of the limit is μ_T.

If $\|h_n - h\|_{H_T} \to 0$, then $\mathscr{I}(h_n) \to \mathscr{I}(h)$ in $L^2(\mu_T)$. Hence, $\mathscr{H}_n$ does not depend on the choice of the orthonormal basis of H_T. □

Definition 5.1.7 The orthogonal decomposition (5.1.11) of $L^2(\mu_T)$ is called the **Wiener chaos decomposition** and an element in $\mathscr{H}_n$ is called an n-th **Wiener chaos**.

Let $J_n : L^2(\mu_T) \to L^2(\mu_T)$ be the orthogonal projection onto $\mathscr{H}_n$. Extend J_n to $\mathscr{P}(E)$ so that

$$J_n F = \sum_{j=1}^{m} (J_n F_j) e_j$$

for $F = \sum_{j=1}^{m} F_j e_j$ $(F_j \in \mathscr{P},\ e_j \in E\ (j = 1, \ldots, m))$. If $G \in \mathscr{P}$, then there exist an $N \in \mathbb{N}$ and $c_\alpha \in \mathbb{R}$ $(|\alpha| \leqq N)$ such that

$$G = \sum_{|\alpha| \leqq N} c_\alpha H_\alpha,$$

where H_α is given by $H_\alpha = \prod_{j=1}^{\infty} H_{\alpha_j}(\ell_j)$ with an orthonormal basis $\{\ell_j\}_{j=1}^{\infty}$ such that G is expressed as $G = g(\ell_1, \ldots, \ell_M)$ for some $M \in \mathbb{N}$ and a polynomial $g : \mathbb{R}^M \to \mathbb{R}$. Hence $J_n G \in \mathscr{P}$ and $J_n G = 0$ if $n > N$.

Definition 5.1.8 Let $r \in \mathbb{R}$ and $p > 1$. Define $(I - L)^r : \mathscr{P}(E) \to \mathscr{P}(E)$ by

$$(I - L)^r = \sum_{n=0}^{\infty} (1 + n)^r J_n \tag{5.1.12}$$

and set

$$\|F\|_{r,p} = \|(I - L)^{\frac{r}{2}} F\|_p.$$

Denote the completion of $\mathscr{P}(E)$ with respect to $\|\cdot\|_{r,p}$ by $\mathbb{D}^{r,p}(E)$ and write $\mathbb{D}^{r,p}$ for $\mathbb{D}^{r,p}(\mathbb{R})$.

The infinite sum on the right hand side of (5.1.12) is a finite one for $G \in \mathscr{P}(E)$. $\mathbb{D}^{0,p}(E)$ is equal to $L^p(\mu_T; E)$. It is known as **Meyer's equivalence** that, for $k \in \mathbb{Z}_+$, Definitions 5.1.4 and 5.1.8 are consistent, that is, they define the same space $\mathbb{D}^{k,p}(E)$.

Theorem 5.1.9 ([104, Theorem 4.4]) *For any $k \in \mathbb{Z}_+$ and $p > 1$, there exist $a_{k,p}$ and $A_{k,p} > 0$ such that*

$$a_{k,p}\|\nabla^k F\|_p \leqq \|F\|_{k,p} \leqq A_{k,p} \sum_{j=0}^{k} \|\nabla^j F\|_p \quad (F \in \mathscr{P}(E)).$$

The family of Sobolev spaces $\mathbb{D}^{r,p}(E)$ $(r \in \mathbb{R}, p > 1)$ has the following consistency.

Theorem 5.1.10 (1) *For $r, r' \in \mathbb{R}$ and $p, p' > 1$ with $r \leqq r'$, $p \leqq p'$, the inclusion mapping $\mathbb{D}^{r',p'}(E) \subset \mathbb{D}^{r,p}(E)$ is a continuous embedding.*
(2) *Let $(\mathbb{D}^{r,p}(E))^*$ be the dual space of $\mathbb{D}^{r,p}(E)$. Under the identification of $(L^p(\mu_T; E))^*$ and $L^q(\mu_T; E)$, where $\frac{1}{p} + \frac{1}{q} = 1$,*

$$\mathbb{D}^{-r,q}(E) = (\mathbb{D}^{r,p}(E))^*.$$

For the proof, we prepare some lemmas. Define L and $T_t : \mathscr{P}(E) \to \mathscr{P}(E)$ $(t > 0)$ by

$$LG = \sum_{n=0}^{\infty} (-n) J_n G \quad \text{and} \quad T_t G = \sum_{n=0}^{\infty} e^{-nt} J_n G \quad (G \in \mathscr{P}(E)), \tag{5.1.13}$$

respectively. Since J_n is the orthogonal projection onto $\mathscr{H}_n$, we have

$$\|T_t F\|_2^2 = \sum_{n=0}^{\infty} \mathrm{e}^{-2nt} \|J_n F\|_2^2 \leqq \|F\|_2^2 \qquad (F \in \mathscr{P}(E)).$$

Since $\mathscr{P}(E)$ is dense in $L^2(\mu_T; E)$, T_t is extended to a contraction operator on $L^2(\mu_T; E)$, which is also denoted by T_t. Moreover,

$$T_t(T_s F) = T_{t+s} F \quad (F \in L^2(\mu_T; E))$$

by definition and $\{T_t\}_{t \geqq 0}$ defines a contraction semigroup on $L^2(\mu_T; E)$ satisfying

$$\frac{\mathrm{d}}{\mathrm{d}t} T_t F = L T_t F \qquad (F \in \mathscr{P}(E)). \tag{5.1.14}$$

$\{T_t\}_{t \geqq 0}$ and L are called the **Ornstein–Uhlenbeck semigroup** and the **Ornstein–Uhlenbeck operator**, respectively. Moreover, by definition, for $H_\alpha \in \mathscr{H}_{|\alpha|}$ $(\alpha \in \mathscr{A})$, we have

$$(I - L)^r H_\alpha = (1 + |\alpha|)^r H_\alpha, \quad LH_\alpha = -|\alpha|\, H_\alpha, \quad T_t H_\alpha = \mathrm{e}^{-|\alpha|t} H_\alpha. \tag{5.1.15}$$

Lemma 5.1.11 *Let $p > 1$, $F \in \mathscr{P}(E)$ and $G \in L^p(\mu_T; E)$.*
(1) *For any $t \geqq 0$ and $w \in W_T$,*

$$T_t F(w) = \int_{W_T} F(\mathrm{e}^{-t} w + \sqrt{1 - \mathrm{e}^{-2t}}\, w')\mu_T(\mathrm{d}w'). \tag{5.1.16}$$

(2) $\|T_t F\|_p \leqq \|F\|_p$ *holds. In particular, $T_t : L^p(\mu_T; E) \supset \mathscr{P}(E) \ni F \mapsto T_t F \in \mathscr{P}(E) \subset L^p(\mu_T; E)$ is extended to a bounded linear operator.*
(3) $\lim_{t\to 0} \|T_t G - G\|_p = 0$ *holds.*

Proof (1) It suffices to show the case where $E = \mathbb{R}$. Let $F \in \mathscr{P}$. Then, there exist an $N \in \mathbb{N}$, a polynomial $f : \mathbb{R}^N \to \mathbb{R}$, and an orthonormal system $\ell_1, \ldots, \ell_N \in W_T^*$ of H_T such that $F = f(\ell_1, \ldots, \ell_N)$. Set $\boldsymbol{\ell}(w) = (\ell_1(w), \ldots, \ell_N(w))$ for $w \in W_T$. Then, denoting the right hand side of (5.1.16) by $S_t F(w)$, we have

$$S_t F(w) = \int_{\mathbb{R}^N} f(\mathrm{e}^{-t}\boldsymbol{\ell}(w) + y) g_N(1 - \mathrm{e}^{-2t}, y)\, \mathrm{d}y,$$

where $g_N(s, y) = (2\pi s)^{-\frac{N}{2}} \exp(-\frac{|y|^2}{2s})$. Since

$$\int_{\mathbb{R}^N} g_N(s, z - y) g_N(t, y)\, \mathrm{d}y = g_N(s + t, z) \quad \text{and} \quad \frac{\partial g_N}{\partial s} = \frac{1}{2}\Delta g_N,$$

setting

$$\widetilde{f}(x) = \int_{\mathbb{R}^n} f(\mathrm{e}^{-t} x + y) g_N(1 - \mathrm{e}^{-2t}, y)\, \mathrm{d}y, \qquad \widetilde{F} = \widetilde{f}(\ell_1, \ldots, \ell_N),$$

we have $S_t F = \widetilde{F}$,

$$S_s(S_t F)(w) = S_{s+t} F(w), \tag{5.1.17}$$

and

$$\frac{\mathrm{d}}{\mathrm{d}t}\Big|_{t=0} S_t F(w) = \Delta f(\boldsymbol{\ell}(w)) - \sum_{j=1}^N \ell_j(w) \frac{\partial f}{\partial x^j}(\boldsymbol{\ell}(w)). \tag{5.1.18}$$

Extend $\{\ell_j\}_{j=1}^N$ to an orthonormal basis $\{\ell_j\}_{j=1}^\infty$ of H_T and set

$$H_\alpha = \prod_{j=1}^\infty H_{\alpha_j}(\ell_j) \quad (\boldsymbol{\alpha} \in \mathscr{A}).$$

Since $H_n''(x) - xH_n'(x) = -nH_n(x)$, (5.1.18) yields

$$\frac{\mathrm{d}}{\mathrm{d}t}\Big|_{t=0} S_t H_\alpha(w) = -|\alpha| H_\alpha(w).$$

Combining this with (5.1.17), we have

$$\frac{\mathrm{d}}{\mathrm{d}t} S_t H_\alpha(w) = -|\alpha| S_t H_\alpha(w).$$

Since $S_0 H_\alpha(w) = H_\alpha(w)$, by this ordinary differential equation and (5.1.15), we obtain

$$S_t H_\alpha(w) = \mathrm{e}^{-|\alpha|t} H_\alpha(w) = T_t H_\alpha(w).$$

Thus (5.1.16) holds for H_α. Since $F \in \mathscr{P}$ is written as a linear combination of H_αs, (5.1.16) is satisfied.
(2) For the same $F = f(\ell_1, \dots, \ell_N)$ as above, we have by Hölder's inequality

$$\begin{aligned}\|T_t F\|_p^p &\leqq \int_{W_T}\int_{W_T} |F(\mathrm{e}^{-t}w + \sqrt{1-\mathrm{e}^{-2t}}\,w')|^p \mu_T(\mathrm{d}w)\mu_T(\mathrm{d}w') \\ &= \int_{\mathbb{R}^N}\int_{\mathbb{R}^N} |f(x+y)|^p g_N(\mathrm{e}^{-2t}, x) g_N(1-\mathrm{e}^{-2t}, y)\,\mathrm{d}x\mathrm{d}y \\ &= \int_{\mathbb{R}^N} |f(z)|^p g_N(1,z)\,\mathrm{d}z = \|F\|_p^p.\end{aligned}$$

Hence, $\|T_t F\|_p \leqq \|F\|_p$.
(3) Let $K \in \mathscr{P}(E)$. By (5.1.16), $\lim_{t\to 0} \|T_t K(w) - K(w)\|_E = 0$ $(w \in W_T)$. Since $\|T_t K\|_{2p} \leqq \|K\|_{2p}$ by (2), $\{T_t K\}_{t\in[0,T]}$ is uniformly integrable (Theorem A.3.4). Hence $\lim_{t\to 0} \|T_t K - K\|_p = 0$. Using (2) again, we obtain

$$\|T_t G - G\|_p \leqq \|T_t K - K\|_p + 2\|G - K\|_p.$$

Since $\mathscr{P}(E)$ is dense in $L^p(\mu_T; E)$, this inequality implies the conclusion. □

Lemma 5.1.12 *Let $r, r' \in \mathbb{R}$ and $p, p' > 1$.*
(1) If $r \leqq r'$ and $p \leqq p'$, then $\|F\|_{r,p} \leqq \|F\|_{r',p'}$ $(F \in \mathscr{P}(E))$.
(2) If $F_n \in \mathscr{P}(E)$ satisfies

$$\lim_{n\to\infty} \|F_n\|_{r,p} = 0, \qquad \lim_{n,m\to\infty} \|F_n - F_m\|_{r',p'} = 0,$$

then $\lim_{n\to\infty} \|F_n\|_{r',p'} = 0$.

Proof (1) Let $s \geqq 0$. Then, by Definition 5.1.8 and (5.1.13),

$$(I - L)^{-s} F = \frac{1}{\Gamma(s)} \int_0^\infty t^{s-1} \mathrm{e}^{-t} T_t F\,\mathrm{d}t$$

for $F \in \mathscr{P}(E)$. Since $\|(I-L)^{-s}F\|_p \leqq \|F\|_p$ by Lemma 5.1.11, we obtain

$$\begin{aligned}\|F\|_{r,p} &= \|(I-L)^{-\frac{r'-r}{2}}(I-L)^{\frac{r'}{2}}F\|_p \leqq \|(I-L)^{\frac{r'}{2}}F\|_p \\ &\leqq \|(I-L)^{\frac{r'}{2}}F\|_{p'} = \|F\|_{r',p'}.\end{aligned}$$

(2) Set $G_n = (I-L)^{\frac{r'}{2}}F_n$. Since $\|F_n-F_m\|_{r',p'} \to 0$, $\{G_n\}_{n=1}^\infty$ is a Cauchy sequence in $L^{p'}(\mu_T;E)$. Hence, $\lim_{n\to\infty}\|G_n - G\|_{p'} = 0$ holds for some $G \in L^{p'}(\mu_T;E)$. Since $\|F_n\|_{r,p} \to 0$,

$$\lim_{n\to\infty}\|(I-L)^{\frac{r-r'}{2}}G_n\|_p = 0.$$

Hence, we have, for any $K \in \mathscr{P}(E)$,

$$\begin{aligned}\int_{W_T}\langle G,K\rangle_E\,\mathrm{d}\mu_T &= \lim_{n\to\infty}\int_{W_T}\langle G_n,K\rangle_E\,\mathrm{d}\mu_T \\ &= \lim_{n\to\infty}\int_{W_T}\langle (I-L)^{\frac{r-r'}{2}}G_n,(I-L)^{\frac{r'-r}{2}}K\rangle_E\,\mathrm{d}\mu_T = 0\end{aligned}$$

and $G = 0$. Therefore, $\|F_n\|_{r',p'} = \|G_n\|_{p'} \to 0$. □

Proof of Theorem 5.1.10 (1) The assertion follows from Lemma 5.1.12.
(2) For $p > 1$, $q = \frac{p}{p-1}$ and $G \in \mathscr{P}(E)$, we have

$$\begin{aligned}\|G\|_{-r,q} &= \|(I-L)^{-\frac{r}{2}}G\|_q \\ &= \sup\Big\{\int_{W_T}\langle (I-L)^{-\frac{r}{2}}G,F\rangle_E\,\mathrm{d}\mu_T\,;\,F \in \mathscr{P}(E),\ \|F\|_p \leqq 1\Big\} \\ &= \sup\Big\{\int_{W_T}\langle G,(I-L)^{-\frac{r}{2}}F\rangle_E\,\mathrm{d}\mu_T\,;\,F \in \mathscr{P}(E),\ \|F\|_p \leqq 1\Big\} \\ &= \sup\Big\{\int_{W_T}\langle G,K\rangle_E\,\mathrm{d}\mu_T\,;\,K \in \mathscr{P}(E),\ \|K\|_{r,p} \leqq 1\Big\}.\end{aligned}$$

This implies the assertion (2). □

Definition 5.1.13 (1) Define

$$\mathbb{D}^{r,\infty-}(E) = \bigcap_{p\in(1,\infty)}\mathbb{D}^{r,p}(E), \qquad \mathbb{D}^{\infty,p}(E) = \bigcap_{r\in\mathbb{R}}\mathbb{D}^{r,p}(E),$$

$$\mathbb{D}^{r,1+}(E) = \bigcup_{p\in(1,\infty)}\mathbb{D}^{r,p}(E), \qquad \mathbb{D}^{-\infty,p}(E) = \bigcup_{r\in\mathbb{R}}\mathbb{D}^{r,p}(E),$$

$$\mathbb{D}^{\infty,\infty-}(E) = \bigcap_{r\in\mathbb{R},\,p\in(1,\infty)}\mathbb{D}^{r,p}(E), \quad \mathbb{D}^{-\infty,1+}(E) = \bigcup_{r\in\mathbb{R},\,p\in(1,\infty)}\mathbb{D}^{r,p}(E).$$

(2) An element $\Phi \in \mathbb{D}^{-\infty,1+}(E)$ is called a **generalized Wiener functional**.

$\mathbb{D}^{\infty,\infty-}(E)$ is a Fréchet space and $\mathbb{D}^{-\infty,1+}(E)$ is its dual space. The value $\Phi(F)$ of $\Phi \in \mathbb{D}^{-\infty,1+}(E) = (\mathbb{D}^{\infty,\infty-}(E))^*$ at $F \in \mathbb{D}^{\infty,\infty-}(E)$ is denoted by $\int_{W_T} \langle F, \Phi\rangle_E \, d\mu_T$ or $\mathbf{E}[\langle F, \Phi\rangle_E]$:

$$\Phi(F) = \int_{W_T} \langle F, \Phi\rangle_E \, d\mu_T = \mathbf{E}[\langle \Phi, F\rangle_E].$$

When $E = \mathbb{R}$, we simply write the above as $\int_{W_T} F\Phi \, d\mu_T$ or $\mathbf{E}[F\Phi]$. Moreover, if $F = 1$, it is also written as $\int_{W_T} \Phi \, d\mu_T$ or $\mathbf{E}[\Phi]$. These notations come from the fact that, if $F \in L^q(\mu_T; E)$ and $\Phi \in L^p(\mu_T; E)$, then $\langle F, \Phi\rangle_E \in L^1(\mu_T)$ and $\int_{W_T} \langle F, \Phi\rangle_E d\mu_T$ is a usual integral.

5.2 Continuity of Operators

The aim of this section is to prove the continuity of the operators ∇, ∇^*, and T_t and to present their applications.

Theorem 5.2.1 (1) *For any $r \in \mathbb{R}$ and $p > 1$, $\nabla : \mathscr{P}(E) \to \mathscr{P}(H_T \otimes E)$ is extended to a unique linear operator $\overline{\nabla} : \mathbb{D}^{-\infty,1+}(E) \to \mathbb{D}^{-\infty,1+}(H_T \otimes E)$ whose restriction $\overline{\nabla} : \mathbb{D}^{r+1,p}(E) \to \mathbb{D}^{r,p}(H_T \otimes E)$ is continuous.*
(2) *For any $r \in \mathbb{R}$ and $p > 1$, the adjoint operator ∇^* of ∇ is extended to a unique linear operator $\overline{\nabla}^* : \mathbb{D}^{-\infty,1+}(H_T \otimes E) \to \mathbb{D}^{-\infty,1+}(E)$ whose restriction $\overline{\nabla}^* : \mathbb{D}^{r+1,p}(H_T \otimes E) \to \mathbb{D}^{r,p}(E)$ is continuous.*
(3) *For any $t > 0$ and $p > 1$, $T_t(L^p(\mu_T; E)) \subset \mathbb{D}^{\infty,p}(E)$. In particular, if a measurable function $F : W_T \to E$ is bounded, then $T_t F \in \mathbb{D}^{\infty,\infty-}(E)$.*

In the following, the extensions $\overline{\nabla}$ and $\overline{\nabla}^*$ of ∇ and ∇^* will also be denoted by ∇ and ∇^*.

For a proof of the theorem, we prepare a lemma. For $\phi = \{\phi_n\}_{n=0}^{\infty} \subset \mathbb{R}$, define the mapping $M_\phi : \mathscr{P}(E) \to \mathscr{P}(E)$ by

$$M_\phi F = \sum_{n=0}^{\infty} \phi_n J_n F \qquad (F \in \mathscr{P}(E)). \tag{5.2.1}$$

Lemma 5.2.2 *For $\phi = \{\phi_n\}_{n=0}^{\infty} \subset \mathbb{R}$, set $\phi^+ = \{\phi_{n+1}\}_{n=0}^{\infty}$. Then, for any $F \in \mathscr{P}(E)$,*

$$\nabla M_\phi F = M_{\phi^+} \nabla F.$$

In particular, $\nabla(J_n F) = J_{n-1}(\nabla F)$, $n = 1, 2, \ldots$

Proof We may assume that $E = \mathbb{R}$ and F is a function of the form $F = H_\alpha = \prod_{j=1}^\infty H_{\alpha_j}(\ell_j)$, $\{\ell_j\}_{j=1}^\infty$ being an orthonormal basis of H_T. Then, since

$$\nabla M_\phi H_\alpha = \phi_{|\alpha|}\nabla H_\alpha$$

and $H_n' = H_{n-1}$, we have

$$\nabla H_\alpha = \sum_{j:\alpha_j>0} H_{\alpha_j-1}(\ell_j)\Big(\prod_{i\neq j} H_{\alpha_i}(\ell_i)\Big)\,\ell_j.$$

Since $H_{\alpha_j-1}(\ell_j)\big(\prod_{i\neq j} H_{\alpha_i}(\ell_i)\big) \in \mathscr{H}_{|\alpha|-1}$, we obtain

$$M_{\phi^+}\nabla H_\alpha = \phi_{|\alpha|}\nabla H_\alpha = \nabla M_\phi H_\alpha. \qquad \square$$

Lemma 5.2.3 (Hypercontractivity of $\{T_t\}$) *Let $p > 1$ and $t \geqq 0$, and set $q(t) = \mathrm{e}^{2t}(p-1)+1$. Then, for any $F \in L^p(\mu_T)$,*

$$\|T_tF\|_{q(t)} \leqq \|F\|_p. \tag{5.2.2}$$

The proof is omitted. See [104, Theorem 2.11].

Lemma 5.2.4 *For any $p > 1$ and $n \in \mathbb{Z}_+$, there exists a constant $b_{p,n} > 0$ such that*

$$\|J_nF\|_p \leqq b_{p,n}\|F\|_p \tag{5.2.3}$$

for any $F \in \mathscr{P}$. In particular, J_n defines a bounded operator on $L^p(\mu_T)$.

Proof For $p > 1$, define $c(p)$ by

$$c(p) = \begin{cases} (p-1)^{\frac{1}{2}} & (p \geqq 2) \\ (p-1)^{-\frac{1}{2}} & (1 < p < 2). \end{cases}$$

First we assume $p \geqq 2$. Let $t \geqq 0$ so that $\mathrm{e}^{2t} = p-1$. Then, since $\|T_tF\|_{1+\mathrm{e}^{2t}} \leqq \|F\|_2$ by Lemma 5.2.3, we have $\|T_tF\|_p \leqq \|F\|_2$. Moreover, since J_n is an orthogonal projection on $L^2(\mu_T)$,

$$\|T_tJ_nF\|_p \leqq \|J_nF\|_2 \leqq \|F\|_2 \leqq \|F\|_p.$$

Hence, from the identity $T_tJ_nF = \mathrm{e}^{-nt}J_nF = c(p)^{-n}J_nF$, taking $b_{p,n} = c(p)^n$, we obtain (5.2.3).

Second, we assume $1 < p < 2$. Then, since $\frac{p}{p-1} > 2$ and $c(\frac{p}{p-1}) = c(p)$, we obtain from the above consideration

$$\|J_nF\|_{\frac{p}{p-1}} \leqq c(p)^n\|F\|_{\frac{p}{p-1}}.$$

Due to the duality,

$$\|J_n^* F\|_p \leqq c(p)^n \|F\|_p.$$

Since J_n is an orthogonal projection on $L^2(\mu_T)$, we have $J_n^* F = J_n F$ and obtain (5.2.3), by setting $b_{p,n} = c(p)^n$ again. □

Lemma 5.2.5 *For any $p > 1$ and $n \in \mathbb{Z}_+$, there exists a constant $C_{n,p} > 0$ such that*

$$\|T_t(I - J_0 - \cdots - J_{n-1})F\|_p \leqq C_{n,p} \mathrm{e}^{-nt} \|F\|_p \tag{5.2.4}$$

for any $t > 0$ and $F \in L^p(\mu_T)$.

Proof If $p = 2$, then, by the definition of T_t,

$$\big\|T_t(I - J_0 - \cdots - J_{n-1})F\big\|_2^2 = \sum_{k=n}^{\infty} \mathrm{e}^{-2kt} \big\|J_k F\big\|_2^2 \leqq \mathrm{e}^{-2nt} \big\|F\big\|_2^2$$

and (5.2.4) holds.

Assume that $p > 2$. Set $p = \mathrm{e}^{2t_0} + 1$ for $t_0 > 0$. For $t > t_0$, by Lemma 5.2.3 and the above observation,

$$\begin{aligned}\|T_t(I - J_0 - \cdots - J_{n-1})F\|_p &\leqq \|T_{t-t_0}(I - J_0 - \cdots - J_{n-1})F\|_2 \\ &\leqq \mathrm{e}^{-n(t-t_0)} \|F\|_2 \leqq \mathrm{e}^{nt_0} \mathrm{e}^{-nt} \|F\|_p.\end{aligned}$$

For $t \leqq t_0$, by Lemmas 5.1.11 and 5.2.4,

$$\begin{aligned}&\|T_t(I - J_0 - \cdots - J_{n-1})F\|_p \leqq \|(I - J_0 - \cdots - J_{n-1})F\|_p \\ &\leqq \Big(\sum_{k=0}^{n-1} b_{p,k}\Big)\|F\|_p \leqq \mathrm{e}^{nt_0}\Big(\sum_{k=0}^{n-1} b_{p,k}\Big)\mathrm{e}^{-nt}\|F\|_p.\end{aligned}$$

Hence, we have (5.2.4) also for $p > 2$.

If $p \in (1, 2)$, we can prove the conclusion by the duality between $L^p(\mu_T)$ and $L^{\frac{p}{p-1}}(\mu_T)$ in the same way as in the proof of Lemma 5.2.4. □

Lemma 5.2.6 *Let $\delta > 0$ and $\psi : (-\delta, \delta) \to \mathbb{R}$ be real analytic. Suppose that, for $\alpha \in (0, 1]$, $\phi = \{\phi_n\}_{n=0}^{\infty}$ satisfies $\phi_n = \psi(n^{-\alpha})$ for $n \geqq \delta^{-\frac{1}{\alpha}}$. Then, for each $p > 1$, there exists a constant C_p such that*

$$\|M_\phi F\|_p \leqq C_p \|F\|_p \qquad (F \in \mathscr{P}). \tag{5.2.5}$$

Proof Fix $n \in \mathbb{N}$ so that $\frac{1}{n^\alpha} < \delta$ and set

$$M_\phi^{(1)} = \sum_{k=0}^{n-1} \phi_k J_k \quad \text{and} \quad M_\phi^{(2)} = \sum_{k=n}^{\infty} \phi_k J_k.$$

Since $M_\phi^{(1)}$ is a bounded operator on $L^p(\mu_T)$ by Lemma 5.2.4, it suffices to prove the following inequality:

$$\sup\{\|M_\phi^{(2)}F\|_p\,;\ F\in\mathscr{P},\ \|F\|_p \leqq 1\} < \infty. \tag{5.2.6}$$

First we show (5.2.6) when $\alpha = 1$. Define the operator R by

$$R = \int_0^\infty T_t(I - J_0 - \cdots - J_{n-1})\,\mathrm{d}t.$$

Then, we have

$$R^jF = \int_0^\infty \cdots \int_0^\infty T_{t_1+t_2+\cdots+t_j}(I - J_0 - \cdots - J_{n-1})F\,\mathrm{d}t_1\cdots\mathrm{d}t_j.$$

By Lemma 5.2.5,

$$\|R^jF\|_p \leqq C_{n,p}\frac{1}{n^j}\|F\|_p. \tag{5.2.7}$$

Moreover, by the definition of R,

$$RJ_kF = \frac{1}{k}J_kF, \quad R^jJ_kF = \frac{1}{k^j}J_kF \qquad (k \geqq n).$$

Combining this with the series expansion of ψ,

$$\psi(x) = \sum_{j=0}^\infty a_jx^j \qquad (x\in(-\delta,\delta)),$$

we obtain

$$\phi_kJ_kF = \psi(k^{-1})J_kF = \sum_{j=0}^\infty a_j\frac{1}{k^j}J_kF = \sum_{j=0}^\infty a_jR^jJ_kF.$$

Hence

$$M_\phi^{(2)}F = \sum_{j=0}^\infty a_jR^jF.$$

From this identity and (5.2.7), we obtain

$$\|M_\phi^{(2)}F\|_p \leqq C_{n,p}\sum_{j=0}^\infty |a_j|\Big(\frac{1}{n}\Big)^j\|F\|_p$$

and (5.2.6).

Second, we show (5.2.6) when $\alpha < 1$. For $t \geqq 0$, let ν_t be a probability measure on $[0,\infty)$ such that

$$\int_0^\infty \mathrm{e}^{-\lambda s}\nu_t(\mathrm{d}s) = \mathrm{e}^{-\lambda^\alpha t} \qquad (\lambda > 0).$$

Set

$$Q_t = \int_0^\infty T_s \nu_t(\mathrm{d}s)$$

and define

$$Q = \int_0^\infty Q_t(I - J_0 - \cdots - J_{n-1})\,\mathrm{d}t.$$

By Lemma 5.2.5,

$$\|Q^j F\|_p \leqq C_{n,p}\Big(\frac{1}{n^\alpha}\Big)^j \|F\|_p.$$

Moreover, by definition,

$$Q^j J_k F = \Big(\frac{1}{k^\alpha}\Big)^j J_k F$$

for $k \geqq n$. From these observations we obtain (5.2.6) by a similar argument to the case where $\alpha = 1$. □

By the following lemma, the assertions of Lemmas 5.2.4, 5.2.5, and 5.2.6 also hold if we replace $\mathscr{P}$ and $L^p(\mu_T)$ by $\mathscr{P}(E)$ and $L^p(\mu_T; E)$.

Lemma 5.2.7 *Let K be a real separable Hilbert space and $1 < p \leqq q < \infty$. Suppose that a linear operator $A : \mathscr{P} \to \mathscr{P}(K)$ is extended to a continuous operator $L^p(\mu_T) \to L^q(\mu_T; K)$. Define $A(G\,e) = (AG) \otimes e$ ($G \in \mathscr{P}$, $e \in E$) and extend A to $\mathscr{P}(E)$. Then, $A : \mathscr{P}(E) \to \mathscr{P}(K \otimes E)$ is extended to a continuous linear operator $L^p(\mu_T; E) \to L^q(\mu_T; K \otimes E)$.*

Proof We use the following Khinchin's inequality (see [112]): Let $\{r_n\}_{n=1}^\infty$ be a Bernoulli sequence on a probability space $(\Omega, \mathscr{F}, P)$, that is, $r_1, r_2, \ldots$ are independent and satisfy $P(r_i = 1) = P(r_i = -1) = \frac{1}{2}$ ($i = 1, 2, \ldots$). Then, for any $p > 1$, $N \in \mathbb{N}$, $e_1, \ldots, e_N \in E$,

$$\frac{1}{B_p}\Big(\mathbf{E}^P\Big[\Big\|\sum_{m=1}^N r_m e_m\Big\|_E^p\Big]\Big)^{\frac{1}{p}} \leqq \Big(\sum_{m=1}^N \|e_m\|_E^2\Big)^{\frac{1}{2}} \leqq B_p\Big(\mathbf{E}^P\Big[\Big\|\sum_{m=1}^N r_m e_m\Big\|_E^p\Big]\Big)^{\frac{1}{p}},$$

where $B_p = (p-1) \vee (\frac{1}{p-1})$.

For $F \in \mathscr{P}(E)$, take an orthonormal basis $\{e_n\}_{n=1}^\infty$ of E so that $F = \sum_{n=1}^N F_n e_n$ for some $N \in \mathbb{N}$ and $F_n \in \mathscr{P}$ ($n = 1, \ldots, N$). Denoting by L_A the operator norm of $A : L^p(\mu_T) \to L^q(\mu_T; K)$, by Khinchin's inequality, we obtain

$$\|AF\|_{q,K\otimes E}^q = \int_{W_T} \Big(\sum_{n=1}^N \|AF_n(w)\|_K^2\Big)^{\frac{q}{2}} \mu_T(\mathrm{d}w)$$

$$\leqq B_q^q \int_{W_T} \mathbf{E}^P\Big[\Big\|\sum_{n=1}^N r_n A F_n(w)\Big\|_K^q\Big]\mu_T(\mathrm{d}w)$$

$$\leqq B_q^q L_A^q \Big(\mathbf{E}^P\Big[\int_{W_T}\Big|\sum_{n=1}^N r_n F_n(w)\Big|^p \mu_T(\mathrm{d}w)\Big]\Big)^{\frac{q}{p}}$$

$$\leqq B_q^q L_A^q B_p^q \Big(\int_{W_T}\Big(\sum_{n=1}^N F_n(w)^2\Big)^{\frac{p}{2}}\mu_T(\mathrm{d}w)\Big)^{\frac{q}{p}}$$

$$= B_q^q L_A^p B_p^q \|F\|_{p,E}^q.$$

Hence, $A : \mathscr{P}(E) \to \mathscr{P}(K \otimes E)$ is extended continuously. □

Proof of Theorem 5.2.1. (1) Let $r \in \mathbb{R}$ and $p > 1$. Define $\phi = \{\phi_n\}_{n=0}^\infty$ by

$$\phi_0 = 0, \quad \phi_n = \Big(\frac{n}{1+n}\Big)^{\frac{r}{2}} = \Big(\frac{1}{1+\frac{1}{n}}\Big)^{\frac{r}{2}} \qquad (n \geqq 1).$$

By Lemma 5.2.6, there exists a constant C_p such that

$$\|M_\phi F\|_p \leqq C_p\|F\|_p$$

for any $F \in \mathscr{P}(E)$. By Lemma 5.2.2, we have

$$\nabla M_\phi (I-L)^{\frac{r}{2}} F = (I-L)^{\frac{r}{2}} \nabla F.$$

Moreover, by Theorem 5.1.9, there exists a constant C_p' such that

$$\|\nabla F\|_p \leqq C_p' \|(I-L)^{\frac{1}{2}} F\|_p.$$

Summing up the above observations, we obtain

$$\begin{aligned}
\|(I-L)^{\frac{r}{2}}\nabla F\|_p &= \|\nabla M_\phi (I-L)^{\frac{r}{2}} F\|_p \\
&\leqq C_p'\|(I-L)^{\frac{1}{2}} M_\phi (I-L)^{\frac{r}{2}} F\|_p = C_p'\|M_\phi (I-L)^{\frac{r+1}{2}} F\|_p \\
&\leqq C_p' C_p \|(I-L)^{\frac{r+1}{2}} F\|_p = C_p' C_p \|F\|_{r+1,p}.
\end{aligned}$$

Hence, $\nabla : \mathbb{D}^{r+1,p}(E) \to \mathbb{D}^{r,p}(H_T \otimes E)$ is continuous.
(2) The assertion follows from (1) and the duality.
(3) Let $\{\ell_n\}_{n=1}^\infty \subset W_T^*$ be an orthonormal basis of H_T. By (5.1.5), $\nabla^*\ell_n(w) = \ell_n(w)$, μ_T-a.s. $w \in W_T$. Hence, by Lemma 5.1.11,

$$\begin{aligned}
\langle \nabla T_t F(w), \ell_n\rangle_{H_T} &= \mathrm{e}^{-t}\int_{W_T} \langle (\nabla F)(\mathrm{e}^{-t}w + \sqrt{1-\mathrm{e}^{-2t}}w'), \ell_n\rangle_{H_T}\mu_T(\mathrm{d}w') \\
&= \frac{\mathrm{e}^{-t}}{\sqrt{1-\mathrm{e}^{-2t}}}\int_{W_T} \langle \nabla[F(\mathrm{e}^{-t}w + \sqrt{1-\mathrm{e}^{-2t}}\,\cdot\,)](w'), \ell_n\rangle_{H_T}\mu_T(\mathrm{d}w') \\
&= \frac{\mathrm{e}^{-t}}{\sqrt{1-\mathrm{e}^{-2t}}}\int_{W_T} F(\mathrm{e}^{-t}w + \sqrt{1-\mathrm{e}^{-2t}}w')\ell_n(w')\mu_T(\mathrm{d}w').
\end{aligned}$$

Since $\{\ell_n\}_{n=1}^{\infty} \subset \mathscr{H}_1$ is an orthonormal basis of $\mathscr{H}_1$,

$$\begin{aligned}\left\|\nabla T_t F(w)\right\|_{H_T}^2 &= \sum_{n=1}^{\infty} \frac{\mathrm{e}^{-2t}}{1-\mathrm{e}^{-2t}}\Big(\int_{W_T} F(\mathrm{e}^{-t}w + \sqrt{1-\mathrm{e}^{-2t}}w')\ell_n(w')\mu_T(\mathrm{d}w')\Big)^2 \\ &= \frac{\mathrm{e}^{-2t}}{1-\mathrm{e}^{-2t}}\left\|J_1[F(\mathrm{e}^{-t}w + \sqrt{1-\mathrm{e}^{-2t}}\ \cdot)]\right\|_2^2.\end{aligned}$$

Set

$$A_p = \Big(\int_{\mathbb{R}} \frac{1}{\sqrt{2\pi}}|x|^p \mathrm{e}^{-\frac{x^2}{2}}\mathrm{d}x\Big)^{\frac{1}{p}}.$$

Then, we have

$$\Big(\int_{\mathbb{R}} \frac{1}{\sqrt{2\pi t}}|y|^p \mathrm{e}^{-\frac{y^2}{2t}}\mathrm{d}y\Big)^{\frac{1}{p}} = A_p\sqrt{t}.$$

In particular, since $G \in \mathscr{H}_1$ is a Gaussian random variable with mean 0 and variance $\|G\|_2^2$, $\|G\|_p = A_p\|G\|_2$. Combining this with Lemma 5.2.4, we obtain

$$\begin{aligned}&\int_{W_T} \left\|\nabla T_t F\right\|_{H_T}^p \mathrm{d}\mu_T \\ &= \Big(\frac{\mathrm{e}^{-t}}{\sqrt{1-\mathrm{e}^{-2t}}}\Big)^p \int_{W_T} \left\|J_1[F(\mathrm{e}^{-t}w + \sqrt{1-\mathrm{e}^{-2t}}\ \cdot)]\right\|_2^p \mu_T(\mathrm{d}w) \\ &= A_p^{-p}\Big(\frac{\mathrm{e}^{-t}}{\sqrt{1-\mathrm{e}^{-2t}}}\Big)^p \int_{W_T} \left\|J_1[F(\mathrm{e}^{-t}w + \sqrt{1-\mathrm{e}^{-2t}}\ \cdot)]\right\|_p^p \mu_T(\mathrm{d}w) \\ &\leqq A_p^{-p}b_{p,1}^p\Big(\frac{\mathrm{e}^{-t}}{\sqrt{1-\mathrm{e}^{-2t}}}\Big)^p \int_{W_T} \left\|F(\mathrm{e}^{-t}w + \sqrt{1-\mathrm{e}^{-2t}}\ \cdot)\right\|_p^p \mu_T(\mathrm{d}w) \\ &= A_p^{-p}b_{p,1}^p\Big(\frac{\mathrm{e}^{-t}}{\sqrt{1-\mathrm{e}^{-2t}}}\Big)^p \int_{W_T} T_t|F|^p \mathrm{d}\mu_T = A_p^{-p}b_{p,1}^p\Big(\frac{\mathrm{e}^{-t}}{\sqrt{1-\mathrm{e}^{-2t}}}\Big)^p \left\|F\right\|_p^p,\end{aligned}$$

where the last identity follows from

$$\int_{W_T} G(T_tK)\,\mathrm{d}\mu_T = \int_{W_T} (T_tG)K\,\mathrm{d}\mu_T \quad (G, K \in \mathscr{P}) \quad \text{and} \quad T_t1 = 1.$$

Hence, by Lemma 5.2.7, $\nabla T_t : \mathscr{P}(E) \to \mathscr{P}(H_T \otimes E)$ is extended to a continuous linear operator from $L^p(\mu_T; E)$ into $L^p(\mu_T; H_T \otimes E)$. Therefore,

$$T_t(L^p(\mu_T; E)) \subset \mathbb{D}^{1,p}(E).$$

Repeating the above arguments inductively, we obtain the assertion. □

We end this section by showing the fundamental properties of ∇ and ∇^*.

Theorem 5.2.8 *Let $p, q, r > 1$ be such that $\frac{1}{r} = \frac{1}{p} + \frac{1}{q}$ and E, E_1, E_2 be real separable Hilbert spaces.*

(1) *Let* $F \in \mathbb{D}^{1,p}(E_1)$, $G_1 \in \mathbb{D}^{1,q}(E_2)$, $G_2 \in \mathbb{D}^{1,q}(H_T \otimes E_2)$ *and* $K \in \mathbb{D}^{1,p}$. *Then,* $F \otimes G_1 \in \mathbb{D}^{1,r}(E_1 \otimes E_2)$, $KG_2 \in \mathbb{D}^{1,r}(H_T \otimes E_2)$ *and*

$$\nabla(F \otimes G_1) = F \otimes \nabla G_1 + \nabla F \otimes G_1, \tag{5.2.8}$$

$$\nabla^*(KG_2) = K\nabla^* G_2 - \langle \nabla K, G_2 \rangle_{H_T}, \tag{5.2.9}$$

where $E_1 \otimes H_T \otimes E_2$ *is identified with* $H_T \otimes E_1 \otimes E_2$.
(2) *Let* $k \in \mathbb{Z}_+$. *Both of the following mappings are bounded and bilinear:*

$$\mathbb{D}^{k,p}(E_1) \times \mathbb{D}^{k,q}(E_2) \ni (F, G) \mapsto F \otimes G \in \mathbb{D}^{k,r}(E_1 \otimes E_2),$$
$$\mathbb{D}^{k,p}(E) \times \mathbb{D}^{k,q}(E) \ni (F, G) \mapsto \langle F, G \rangle_E \in \mathbb{D}^{k,r}.$$

In particular, if $F, G \in \mathbb{D}^{\infty,\infty-}$, *then* $FG \in \mathbb{D}^{\infty,\infty-}$.

Proof (1) Let $F \in \mathscr{P}(E_1)$ and $G_1 \in \mathscr{P}(E_2)$. By (5.1.1), the $E_1 \otimes E_2$-valued random variable $\langle \nabla(F \otimes G_1), h \rangle_{H_T}$ is obtained by

$$\langle \nabla(F \otimes G_1)(w), h \rangle_{H_T} = \frac{\mathrm{d}}{\mathrm{d}\xi}\Big|_{\xi=0} (F \otimes G_1)(w + \xi h) \qquad (w \in W_T,\ h \in H_T).$$

Hence, $\nabla(F \otimes G_1) = F \otimes \nabla G_1 + \nabla F \otimes G_1$. By the continuity of ∇, (5.2.8) holds for any $F \in \mathbb{D}^{1,p}(E_1)$ and $G_1 \in \mathbb{D}^{1,q}(E_2)$.

Next, let $G_2 \in \mathscr{P}(H_T \otimes E_2)$, $K \in \mathscr{P}$ and $\psi \in \mathscr{P}(E_2)$. By (5.2.8), we have

$$\begin{aligned}\int_{W_T} \langle KG_2, \nabla\psi \rangle_{H_T \otimes E_2} \mathrm{d}\mu_T &= \int_{W_T} \langle G_2, \nabla(K\psi) - \nabla K \otimes \psi \rangle_{H_T \otimes E_2} \mathrm{d}\mu_T \\ &= \int_{W_T} \langle K\nabla^* G_2 - \langle \nabla K, G_2 \rangle_{H_T}, \psi \rangle_{E_2} \mathrm{d}\mu_T.\end{aligned}$$

By the continuity of ∇ and ∇^*, (5.2.9) holds for any $G_2 \in \mathbb{D}^{1,q}(H_T \otimes E_2)$ and $K \in \mathbb{D}^{1,p}$.
(2) The assertion is trivial by (1) and the definition of the inner product. □

Proposition 5.2.9 *For* $G \in \mathbb{D}^{1,2}(E)$, *if* $\nabla G = 0$, *then there exists an* $e \in E$ *such that* $G = e$, μ_T*-a.s.*

Proof We may assume $E = \mathbb{R}$. Let $\{\ell_j\}_{j=1}^\infty \subset W_T^*$ be an orthonormal basis of H_T. As in Theorem 5.1.6, we set $H_{\boldsymbol{\alpha}} = \prod_{j=0}^\infty H_{\alpha_j}(\ell_j)$ for $\boldsymbol{\alpha} = (\alpha_1, \alpha_2, \dots) \in \mathscr{A}$. Since $H_n'(x) = xH_n(x) - (n+1)H_{n+1}(x)$ and $\nabla^*\ell_j = \ell_j$, by Theorem 5.2.8,

$$\nabla^*(H_{\boldsymbol{\alpha}}\ell_j) = (\alpha_j + 1)H_{\boldsymbol{\alpha}+\boldsymbol{\delta}_j}, \tag{5.2.10}$$

where $\boldsymbol{\delta}_j = (\delta_{ji})_{i \in \mathbb{N}}$.

For $\alpha \in \mathscr{A}$ with $|\alpha| \neq 0$, fix $j \in \mathbb{N}$ satisfying $\alpha_j \neq 0$. Since $H_\alpha = \nabla^*(\alpha_j^{-1} H_{\alpha-\delta_j}\ell_j)$ by (5.2.10), we have

$$\int_{W_T} G H_\alpha \mathrm{d}\mu_T = \int_{W_T} \langle \nabla G, \alpha_j^{-1} H_{\alpha-\delta_j}\ell_j \rangle_{H_T} \mathrm{d}\mu_T = 0.$$

Hence, by Theorem 5.1.6, G is a constant. □

5.3 Characterization of Sobolev Spaces

The aim of this section is to present explicit criteria for generalized Wiener functionals to belong to $\mathbb{D}^{r,p}(E)$, by using the continuity of ∇ and ∇^*.

The following characterization of $\mathbb{D}^{r,p}(E)$ holds as in the theory of distributions on finite dimensional spaces.

Theorem 5.3.1 *Let $r \in \mathbb{R}$, $k \in \mathbb{Z}_+$ and $p > 1$.*
(1) *$\Phi \in \mathbb{D}^{-\infty,1+}(E)$ belongs to $\mathbb{D}^{r,p}(E)$ if and only if*

$$\sup\Big\{\int_{W_T} \langle \Phi, F\rangle_E \,\mathrm{d}\mu_T \,;\, F \in \mathscr{P}(E),\ \|F\|_{-r,q} \leqq 1\Big\} < \infty,$$

where $q = \frac{p}{p-1}$.
(2) *$F \in L^p(\mu_T; E)$ belongs to $\mathbb{D}^{k,p}(E)$ if and only if there exists an $F_k \in L^p(\mu_T; H_T^{\otimes k} \otimes E)$ such that*

$$\int_{W_T} \langle F, (\nabla^*)^k G\rangle_E \,\mathrm{d}\mu_T = \int_{W_T} \langle F_k, G\rangle_{H_T^{\otimes k}\otimes E} \,\mathrm{d}\mu_T$$

for any $G \in \mathscr{P}(H_T^{\otimes k} \otimes E)$, where $H_T^{\otimes k} = \underbrace{H_T \otimes \cdots \otimes H_T}_{k \text{ times}}$. Moreover, in this case, $F_k = \nabla^k F$.

Proof (1) The necessity is trivial by the definition. We show the sufficiency. If $\sup\{\cdots\} < \infty$, then the mapping $\mathbb{D}^{-r,q}(E) \ni F \mapsto \int_{W_T} \langle \Phi, F\rangle_E \mathrm{d}\mu_T$ is extended to a bounded linear operator on $\mathbb{D}^{-r,q}(E)$. Since the dual space of $\mathbb{D}^{-r,q}(E)$ is $\mathbb{D}^{r,p}(E)$, there exists a $G \in \mathbb{D}^{r,p}(E)$ such that

$$\int_{W_T} \langle \Phi, F\rangle_E \,\mathrm{d}\mu_T = \int_{W_T} \langle G, F\rangle_E \,\mathrm{d}\mu_T$$

for all $F \in \mathscr{P}(E)$. Hence, $\Phi = G \in \mathbb{D}^{r,p}(E)$.
(2) The necessity and the identity $F_k = \nabla^k F$ are trivial by definition. We only show the sufficiency. Set

$$R_0 = \sum_{n=1}^{\infty} \frac{1}{n} J_n = \int_0^{\infty} T_t(I - J_0)\,\mathrm{d}t.$$

By Lemma 5.2.5, R_0 is a bounded operator on $L^q(\mu_T; E)$. By Lemma 5.2.4, the operator $R_0(I - L)$,

$$R_0(I - L) = \sum_{n=1}^{\infty} \frac{1 + n}{n} J_n = R_0 + (I - J_0),$$

is also a bounded operator on $L^q(\mu_T; E)$. Hence, we have

$$\begin{aligned} \|R_0 F\|_{s+2,q} &= \|(I - L)^{\frac{s+2}{2}} R_0 F\|_q = \|R_0(I - L)(I - L)^{\frac{s}{2}} F\|_q \\ &\leqq \|R_0(I - L)\|_{q\to q} \|F\|_{s,q} \qquad (F \in \mathscr{P}(E)), \end{aligned}$$

where $\|R_0(I-L)\|_{q\to q}$ is the operator norm of $R_0(I-L) : L^q(\mu_T; E) \to L^q(\mu_T; E)$. Hence, $R_0 : \mathbb{D}^{s,q}(E) \to \mathbb{D}^{s+2,q}(E)$ is a bounded operator. In particular, the power

$$(\nabla R_0)^n : \mathbb{D}^{-n,q}(E) \to L^q(\mu_T; H_T^{\otimes n} \otimes E) \qquad (n \in \mathbb{Z}_+)$$

is a bounded operator. Denote the operator norm of $(\nabla R_0)^n$ by $B_{n,q}$.

Let $\{H_\alpha\}_{\alpha\in\mathscr{A}}$ be the orthonormal basis of $L^2(\mu_T)$ as in the proof of Lemma 5.2.2. By (5.2.9), $\nabla^*\nabla H_\alpha = |\alpha| H_\alpha$. Hence,

$$\nabla^*\nabla = \sum_{n=0}^{\infty} n J_n$$

and

$$\nabla^*\nabla R_0 = I - J_0.$$

Suppose that $(\nabla^*)^n(\nabla R_0)^n = I - J_0 - \cdots - J_{n-1} = \sum_{k=n}^{\infty} J_k$. By the commutativity of R_0 and J_k and Lemma 5.2.2, we have

$$\begin{aligned} (\nabla^*)^{n+1}(\nabla R_0)^{n+1} &= \nabla^* \sum_{k=n}^{\infty} J_k \nabla R_0 = \nabla^*\nabla R_0 \sum_{k=n}^{\infty} J_{k+1} \\ &= (I - J_0)(I - J_0 - \cdots - J_n) = I - J_0 - \cdots - J_n. \end{aligned}$$

Hence, by induction,

$$(\nabla^*)^n(\nabla R_0)^n = I - J_0 - \cdots - J_{n-1} \qquad (n \in \mathbb{Z}_+).$$

Thus, for $G \in \mathscr{P}(E)$, $(I - J_0 - \cdots - J_{k-1})F \in L^p(\mu_T; E) \subset \mathbb{D}^{-\infty,1+}(E)$ satisfies

$$\begin{aligned} &\Big|\int_{W_T} \langle (I - J_0 - \cdots - J_{k-1})F, G\rangle_E \,\mathrm{d}\mu_T\Big| \\ &= \Big|\int_{W_T} \langle F, (\nabla^*)^k(\nabla R_0)^k G\rangle_E \,\mathrm{d}\mu_T\Big| = \Big|\int_{W_T} \langle F_k, (\nabla R_0)^k G\rangle_{H_T^{\otimes k}\otimes E} \,\mathrm{d}\mu_T\Big| \\ &\leqq \|(\nabla R_0)^k G\|_q \|F_k\|_p \leqq B_{k,q} \|F_k\|_p \|G\|_{-k,q}. \end{aligned}$$

By (1), $(I - J_0 - \cdots - J_{k-1})F \in \mathbb{D}^{k,p}(E)$. Since, by Lemma 5.2.4,

$$\|J_nG\|_{k,p} = (1+n)^{\frac{k}{2}}\|J_nG\|_p \leqq (1+n)^{\frac{k}{2}} b_{p,n}\|G\|_p,$$

$J_n(L^p(\mu_T; E)) \subset \mathbb{D}^{k,p}(E)$ holds. Hence, $J_nF \in \mathbb{D}^{k,p}(E)$ $(n = 0, 1, \ldots, k-1)$ and $F \in \mathbb{D}^{k,p}(E)$. □

By Theorem 5.3.1, we can prove the chain rule for the composition of differentiable functions and smooth functionals.

Let $C^k_{\exp}(\mathbb{R}^n)$ be the space of C^k-functions $f : \mathbb{R}^n \to \mathbb{R}$ which and whose derivatives of all orders are of at most exponential growth, that is, for each $i_1, \ldots, i_m \in \{1, \ldots, n\}$, $m \leqq k$, there exist positive constants C_1 and C_2 such that $|\frac{\partial^m f}{\partial x^{i_1} \cdots \partial x^{i_m}}(x)| \leqq C_1 e^{C_2|x|}$ $(x \in \mathbb{R}^n)$. Moreover, let $C^k_{\nearrow}(\mathbb{R}^n)$ be the space of C^k-functions $f : \mathbb{R}^n \to \mathbb{R}$ which and whose derivatives of all orders are of at most polynomial growth, that is, for each $i_1, \ldots, i_m \in \{1, \ldots, n\}$, $m \leqq k$, there exist $C \geqq 0$ and $r \in \mathbb{N}$ such that $|\frac{\partial^m f}{\partial x^{i_1} \cdots \partial x^{i_m}}(x)| \leqq C(1 + |x|)^r$ $(x \in \mathbb{R}^n)$.

Corollary 5.3.2 (1) *Let $f \in C^k_{\exp}(\mathbb{R}^n)$ and $\ell_1, \ldots, \ell_n \in W_T^*$. Then, $F = f(\ell_1, \ldots, \ell_n) \in \mathbb{D}^{k,\infty-}$ and*

$$\nabla^j F = \sum_{i_1,\ldots,i_j=1}^{n} \frac{\partial^j f}{\partial x^{i_1} \cdots \partial x^{i_j}}(\ell_1, \ldots, \ell_n)\ell_{i_1} \otimes \cdots \otimes \ell_{i_j}.$$

(2) *Let $f \in C^k_{\nearrow}(\mathbb{R}^n)$ and $F_1, \ldots, F_n \in \mathbb{D}^{\infty,\infty-}$. Then, $f(F_1, \ldots, F_n) \in \mathbb{D}^{k,\infty-}$ and*

$$\nabla(f(F_1, \ldots, F_n)) = \sum_{i=1}^{n} \frac{\partial f}{\partial x^i}(F_1, \ldots, F_n)\nabla F_i. \tag{5.3.1}$$

Proof (1) We may assume that $\ell_1, \ldots, \ell_n$ are orthonormal. Extend this system to an orthonormal basis $\{\ell_k\}_{k=1}^\infty$ of H_T. Since

$$\int_{W_T} g(\ell_1, \ldots, \ell_m)\, d\mu_T = \int_{\mathbb{R}^m} g(x) \frac{1}{\sqrt{2\pi}^m} e^{-\frac{|x|^2}{2}} dx \qquad (m \in \mathbb{N}),$$

we have

$$\int_{W_T} F\nabla^* G\, d\mu_T = \int_{W_T} \Big\langle \sum_{j=1}^{n} \frac{\partial f}{\partial x^i}(\ell_1, \ldots, \ell_n)\ell_i, G \Big\rangle_{H_T} d\mu_T$$

for any $G \in \mathscr{P}(H_T)$. By Theorem 5.3.1, $F \in \mathbb{D}^{1,\infty-}$ and

$$\nabla F = \sum_{j=1}^{n} \frac{\partial f}{\partial x^i}(\ell_1, \ldots, \ell_n)\ell_i.$$

Applying Theorem 5.2.8, we obtain the conclusion.
(2) If $F_i \in \mathscr{P}$, the conclusion holds by (1). Take $r > 0$ so that

$$\sup_{x\in\mathbb{R}} \frac{|f(x)| + \sum_{i=1}^n \left|\frac{\partial f}{\partial x^i}(x)\right|}{(1+|x|)^r} < \infty$$

and let $p > 2r$. Choose $F_i^m \in \mathscr{P}$ so that $\|F_i^m - F_i\|_{1,p} = 0$ $(m \to \infty)$. Then, for $G \in \mathscr{P}(H_T)$,

$$\begin{aligned}
\int_{W_T} f(F_1,\dots,F_n)\nabla^* G\,\mathrm{d}\mu_T &= \lim_{m\to\infty}\int_{W_T} f(F_1^m,\dots,F_n^m)\nabla^* G\,\mathrm{d}\mu_T \\
&= \lim_{m\to\infty}\int_{W_T}\Big\langle\sum_{i=1}^n \frac{\partial f}{\partial x^i}(F_1^m,\dots,F_n^m)\nabla F_i^m, G\Big\rangle_{H_T}\mathrm{d}\mu_T \\
&= \int_{W_T}\Big\langle\sum_{i=1}^n \frac{\partial f}{\partial x^i}(F_1,\dots,F_n)\nabla F_i, G\Big\rangle_{H_T}\mathrm{d}\mu_T.
\end{aligned}$$

Since p is arbitrary, Theorem 5.3.1 implies $f(F_1,\dots,F_n) \in \mathbb{D}^{1,\infty-}$ and (5.3.1). Repeating this argument, we obtain $f(F_1,\dots,F_n) \in \mathbb{D}^{\infty,\infty-}$. □

The operator ∇^* is a generalization of stochastic integrals.[1]

Theorem 5.3.3 (1) *Let $\mathscr{N} \subset \mathscr{B}(W_T)$ be the totality of sets of zero μ_T-outer measure and $\mathscr{F}_t = \sigma(\mathscr{N} \cup \sigma(\{\theta(u), u \leqq t\}))$. Let $\{u(t) = (u^1(t),\dots,u^d(t))\}_{t\in[0,T]}$ be an $\mathbb{R}^d$-valued $\{\mathscr{F}_t\}$-predictable stochastic process such that*

$$\int_{W_T}\Big(\int_0^T |u(t)|^2\,\mathrm{d}t\Big)\mathrm{d}\mu_T < \infty.$$

Define $\Phi_u : W_T \to H_T$ by

$$\Phi_u(w)(t) = \int_0^t u(s)(w)\,\mathrm{d}s \qquad (t \in [0,T]).$$

Then, $\Phi_u \in L^2(\mu_T; H_T)$ and

$$\nabla^*\Phi_u = \sum_{\alpha=1}^d \int_0^T u^\alpha(t)\,\mathrm{d}\theta^\alpha(t). \tag{5.3.2}$$

(2) *Let $f \in C^\infty_{\exp}(\mathbb{R}^d)$. Then, for any $t \in [0,T]$ and $\alpha = 1,\dots,d$,*

$$\int_0^t f(\theta(s))\,\mathrm{d}\theta^\alpha(s) \in \mathbb{D}^{\infty,\infty-}.$$

[1] ∇^* coincides with the Skorohod integral, which is a generalization of stochastic integrals ([28, 92]).

Remark 5.3.4 In the above assumption on u, μ_T is extended to $\mathscr{F}_T$ naturally, and so is the measurability. Even so, we may think of Φ_u and $\int_0^T u^\alpha(t)\mathrm{d}\theta^\alpha(t)$ as $\mathscr{B}(W_T)$-measurable functions. To see this, let $\mathscr{F}_t^0 = \sigma(\{\theta(s); s \leqq t\})$. Notice that every $\mathscr{F}_t$-measurable F possesses an $\mathscr{F}_t^0$-measurable modification $\widetilde{F}$. Hence every $v = \{v(t)\}_{t\in[0,T]} \in \mathscr{L}^0(\{\mathscr{F}_t\})$, the $\mathscr{L}^0$-space with respect to $\{\mathscr{F}_t\}$ (Definition 2.2.4), admits $\widetilde{v} = \{\widetilde{v(t)}\}_{t\in[0,T]} \in \mathscr{L}^0(\{\mathscr{F}_t^0\})$ such that $\mu_T(v(t) = \widetilde{v}(t)\ (0 \leqq t \leqq T)) = 1$. Therefore, by Proposition 2.2.8, there exists $u_n = \{u_n(t) = (u_n^1(t), \dots, u_n^d(t))\}_{t\in[0,T]} \in \mathscr{L}^0(\{\mathscr{F}_t^0\})$ such that

$$\int_{W_T}\Big(\int_0^T |u_n(t) - u(t)|^2 \mathrm{d}t\Big)\mathrm{d}\mu_T \to 0 \quad (n \to \infty).$$

Then, defining

$$\Phi_u(w)^\alpha(t) = \int_0^t \limsup_{n\to\infty} u_n^\alpha(s)\,\mathrm{d}s$$

and

$$\int_0^T u^\alpha(t)\,\mathrm{d}\theta^\alpha(t) = \limsup_{n\to\infty} \int_0^T u_n^\alpha(t)\,\mathrm{d}\theta^\alpha(t) \quad (\alpha = 1, \dots, d),$$

we obtain the desired $\mathscr{B}(W_T)$-modifications.

Proof (1) Take a sequence $\{\{u_n(t) = (u_n^1(t), \dots, u_n^d(t))\}_{t\in[0,T]}\}_{n=1}^\infty$ of $\mathbb{R}^d$-valued stochastic processes with $\{u_n^\alpha(t)\}_{t\in[0,T]} \in \mathscr{L}^0(\{\mathscr{F}_t^0\})$ $(\alpha = 1, \dots, d)$ (see Definition 2.2.4) such that

$$\lim_{n\to\infty} \int_{W_T}\Big(\int_0^T |u_n(t) - u(t)|^2 \mathrm{d}t\Big)\mathrm{d}\mu_T = 0.$$

By the definition of $\mathscr{L}^0(\{\mathscr{F}_t^0\})$, there exist an increasing sequence $0 = t_0^n < t_1^n < \cdots < t_k^n < \cdots < t_{m_n}^n = T$ and bounded, $\mathscr{F}_{t_k^n}$-measurable $\mathbb{R}^d$-valued random variables $\xi_{n,k} = (\xi_{n,k}^1, \dots, \xi_{n,k}^d)$ such that

$$u_n^\alpha(t) = \xi_{n,k}^\alpha \qquad (t_k^n < t \leqq t_{k+1}^n,\ k = 0, \dots, m_n - 1,\ \alpha = 1, \dots, d).$$

Since $\mathscr{F}_t^0$ is generated by $\theta(s)$ $(s \leqq t)$, we may assume that, taking a subsequence if necessary, there exist $0 < s_1^{k,n} < \cdots < s_{j_{k,n}}^{k,n} \leqq t_k^n$ and $\phi_{n,k}^\alpha \in C_b^\infty(\mathbb{R}^{dj_{k,n}})$ such that

$$\xi_{n,k}^\alpha = \phi_{n,k}^\alpha(\theta(s_1^{k,n}), \dots, \theta(s_{j_{k,n}}^{k,n})). \tag{5.3.3}$$

For $\alpha = 1, \dots, d$, let $e_\alpha = (\overbrace{0, \dots, 0}^{\alpha-1}, 1, 0, \dots, 0) \in \mathbb{R}^d$. For $0 \leqq s < t \leqq T$, define $\ell_{(s,t]}^\alpha \in H_T$ by

$$\dot{\ell}_{(s,t]}^\alpha(v) = \mathbf{1}_{(s,t]}(v)\mathrm{e}^\alpha \qquad (v \in [0, T]),$$

that is, $\langle \ell^\alpha_{(s,t]}, h\rangle_{H_T} = h^\alpha(t) - h^\alpha(s)$ ($h \in H_T$). Then,

$$\Phi_{u_n} = \sum_{k=0}^{m_n-1}\sum_{\alpha=1}^{d} \xi^\alpha_{n,k}\ell^\alpha_{(t^n_k,t^n_{k+1}]}.$$

Hence, by (5.2.9),

$$\nabla^*\Phi_{u_n} = \sum_{k=0}^{m_n-1}\sum_{\alpha=1}^{d}\{\xi^\alpha_{n,k}\nabla^*\ell^\alpha_{(t^n_k,t^n_{k+1}]} - \langle\nabla\xi^\alpha_{n,k}, \ell^\alpha_{(t^n_k,t^n_{k+1}]}\rangle_{H_T}\}.$$

By the expression (5.3.3) and Corollary 5.3.2,

$$\langle\nabla\xi^\alpha_{n,k}, \ell^\alpha_{(t^n_k,t^n_{k+1}]}\rangle_{H_T} = 0 \qquad (k = 0,\dots,m_n - 1).$$

By (5.1.5), $\{u_n(t)\}_{t\in[0,T]}$ satisfies (5.3.2). In particular, for $F \in \mathscr{P}$, we have

$$\int_{W_T}\langle\Phi_{u_n}, \nabla F\rangle_{H_T}\,\mathrm{d}\mu_T = \int_{W_T}\Bigl(\sum_{\alpha=0}^{d}\int_0^T u^\alpha_n(t)\,\mathrm{d}\theta^\alpha(t)\Bigr)F\,\mathrm{d}\mu_T.$$

Letting $n \to \infty$, we arrive at

$$\int_{W_T}\langle\Phi_{u}, \nabla F\rangle_{H_T}\,\mathrm{d}\mu_T = \int_{W_T}\Bigl(\sum_{\alpha=0}^{d}\int_0^T u^\alpha(t)\,\mathrm{d}\theta^\alpha(t)\Bigr)F\,\mathrm{d}\mu_T,$$

which gives (5.3.2).
(2) Define $\{u(s)\}_{s\in[0,T]}$ by $u^\alpha(s) = f(\theta(s))\mathbf{1}_{[0,t]}(s)$ and $u^\beta(s) = 0$ ($\beta \neq \alpha$). Since $\exp\bigl(\max_{0\leqq s\leqq T}|\theta(s)|\bigr) \in \bigcap_{p\in(1,\infty)} L^p(\mu_T)$, Corollary 5.3.2 implies $\Phi_u \in \mathbb{D}^{\infty,\infty-}(H_T)$. (1) and the continuity of ∇^* yields the conclusion. □

By Theorem 5.3.3, we obtain an explicit formula for the integrand in Itô's representation theorem (Theorem 2.6.2) for martingales, as will be seen below. The result is called the **Clark–Ocone formula**, which, for example, plays an important role in the theory of mathematical finance to obtain the hedging strategy for derivatives.

Theorem 5.3.5 *Let $\mathscr{F}_t$ be as in Theorem 5.3.3. For $F \in \mathbb{D}^{1,2}$, set $f^\alpha(t,w) = \mathbf{E}[(\overbrace{(\nabla F)}^{\cdot}(w))^\alpha(t)|\mathscr{F}_t]$, where $(\overbrace{(\nabla F)}^{\cdot}(w))^\alpha(t)$ is the α-th component of the value at time t of the derivative of $(\nabla F)(w) \in H_T$ and $\mathbf{E}[\,\cdot\,|\mathscr{F}_t]$ is the conditional expectation with respect to the natural extension of μ_T to $\mathscr{F}_t$. Then,*

$$F = \mathbf{E}[F] + \sum_{\alpha=1}^{d}\int_0^T f^\alpha(t)\,\mathrm{d}\theta^\alpha(t). \tag{5.3.4}$$

Proof By Itô's representation theorem (Theorem 2.6.2), there exists some $\{g^\alpha(t)\}_{t\in[0,T]} \in \mathscr{L}^2$ $(\alpha = 1, \ldots, d)$ such that

$$F = \mathbf{E}[F] + \sum_{\alpha=1}^{d} \int_0^T g^\alpha(t)\, \mathrm{d}\theta^\alpha(t).$$

What is to be shown is $g^\alpha(t) = f^\alpha(t)$ $(\alpha = 1, \ldots, d)$.

Let $\{u^\alpha(t)\}_{t\in[0,T]}$ be as in Theorem 5.3.3. Since stochastic integrals are isometries (Proposition 2.2.10), Theorem 5.3.3 implies

$$\begin{aligned}
&\int_{W_T} \sum_{\alpha=1}^{d} \Big(\int_0^T u^\alpha(t) g^\alpha(t)\, \mathrm{d}t \Big) \mathrm{d}\mu_T \\
&= \int_{W_T} \Big(\sum_{\alpha=1}^{d} \int_0^T u^\alpha(t)\, \mathrm{d}\theta^\alpha(t) \Big) \Big(\sum_{\alpha=1}^{d} \int_0^T g^\alpha(t)\, \mathrm{d}\theta^\alpha(t) \Big) \mathrm{d}\mu_T \\
&= \int_{W_T} (\nabla^* \Phi_u)(F - \mathbf{E}[F])\, \mathrm{d}\mu_T. \qquad (5.3.5)
\end{aligned}$$

By the definitions of dual operators and the inner product, the last term is rewritten as

$$\int_{W_T} \langle \Phi_u, \nabla F \rangle_{H_T}\, \mathrm{d}\mu_T = \int_{W_T} \sum_{\alpha=1}^{d} \Big(\int_0^T u^\alpha(t) \big(\dot{\widehat{(\nabla F)}} \big)^\alpha(t)\, \mathrm{d}t \Big) \mathrm{d}\mu_T.$$

Moreover, since $\{u^\alpha(t)\}$ is $\{\mathscr{F}_t\}$-adapted, it is equal to

$$\int_{W_T} \sum_{\alpha=1}^{d} \Big(\int_0^T u^\alpha(t) f^\alpha(t)\, \mathrm{d}t \Big) \mathrm{d}\mu_T$$

by Fubini's theorem. Comparing this with (5.3.5), we obtain $g^\alpha(t) = f^\alpha(t)$ $(\alpha = 1, \ldots, d)$ since $\{u(t)\}_{t\in[0,T]}$ is arbitrary. □

Next, we show that the Lipschitz continuity of a Wiener functional implies its differentiability.

Theorem 5.3.6 *Suppose that, for $F \in \bigcap_{p\in(1,\infty)} L^p(\mu_T)$, there exists $\widetilde{F}$ with $\widetilde{F} = F$, μ_T-a.s., and a constant C such that*

$$|\widetilde{F}(w + h) - \widetilde{F}(w)| \leqq C \|h\|_{H_T} \qquad (5.3.6)$$

for any $w \in W_T$ and $h \in H_T$. Then, $F \in \mathbb{D}^{1,\infty-}$ and $\|\nabla F\|_{H_T} \leqq C$, μ_T-a.s.

Proof Let $\ell \in W_T^*$ $(\ell \neq 0)$. Define $\pi_\ell : W_T \to W_T$ by $\pi_\ell(w) = w - \|\ell\|_{H_T}^{-2} \ell(w)\ell$ and decompose W_T into an orthogonal sum

$$W_T = \pi_\ell(W_T) \oplus \mathbb{R}\ell = \{w' + \xi\ell\,;\ w' \in \pi_\ell(W_T),\ \xi \in \mathbb{R}\}.$$

Then, by the Itô–Nisio theorem (Theorem 1.2.5), we have

$$\mu_T = (\mu_T \circ \pi_\ell^{-1}) \otimes \frac{1}{\sqrt{2\pi\|\ell\|_{H_T}^2}} \mathrm{e}^{-\frac{\xi^2}{2\|\ell\|_{H_T}^2}} \mathrm{d}\xi.$$

Let $w' \in \pi_\ell(W_T)$. Since $\mathbb{R} \ni \xi \mapsto \widetilde{F}(w' + \xi\ell)$ is absolutely continuous by the assumption, the set

$$\Big\{\xi \in \mathbb{R}\,;\ \frac{\widetilde{F}(w' + (\xi+\varepsilon)\ell) - \widetilde{F}(w' + \xi\ell)}{\varepsilon} \text{ does not converge as } \varepsilon \to 0\Big\}$$

has Lebesgue measure 0. Hence, setting

$$A(\ell) = \Big\{w \in W_T\,;\ \lim_{\varepsilon\to 0} \frac{\widetilde{F}(w+\varepsilon\ell) - \widetilde{F}(w)}{\varepsilon} \text{ exists}\Big\},$$

we have $\mu_T(A(\ell)) = 1$ by Fubini's theorem. Set

$$G(w,\ell) = \mathbf{1}_{A(\ell)}(w) \lim_{\varepsilon\to 0} \frac{\widetilde{F}(w+\varepsilon\ell) - \widetilde{F}(w)}{\varepsilon} \qquad (w \in W_T).$$

Then, by the assumption,

$$|G(w,\ell)| \leqq C\|\ell\|_{H_T}.$$

for any $w \in W_T$ and $\ell \in W_T^*$.

Let $\{\ell_k\}_{k=1}^\infty \subset W_T^*$ be an orthonormal basis of H_T and set

$$\mathscr{K} = \Big\{\sum_{j=1}^n q_j\ell_j\,;\ q_j \in \mathbb{Q}\ (j = 1,\dots,n),\ n \in \mathbb{N}\Big\}.$$

For $\ell = \sum_{j=1}^n q_j\ell_j \in \mathscr{K}$ and $\phi \in \mathscr{P}$, we have by Lemma 5.1.2

$$\begin{aligned}\int_{W_T} G(\cdot,\ell)\phi\,\mathrm{d}\mu_T &= \lim_{\varepsilon\to 0}\int_{W_T} \frac{\widetilde{F}(\cdot+\varepsilon\ell) - \widetilde{F}(\cdot)}{\varepsilon}\phi\,\mathrm{d}\mu_T = \int_{W_T} \widetilde{F}\partial_\ell\phi\,\mathrm{d}\mu_T \\ &= \sum_{j=1}^n q_j \int_{W_T} \widetilde{F}\partial_{\ell_j}\phi\,\mathrm{d}\mu_T = \int_{W_T}\sum_{j=1}^n q_j G(\cdot,\ell_j)\phi\,\mathrm{d}\mu_T \qquad (5.3.7)\end{aligned}$$

and, for any $\ell \in \mathscr{K}$,

$$G(\cdot,\ell) = \sum_{j=1}^\infty \langle \ell,\ell_j\rangle_{H_T} G(\cdot,\ell_j), \quad \mu_T\text{-a.s.}$$

Hence, setting

$$B = \Big\{w \in \bigcap_{j=1}^\infty A(\ell_j)\,;\ G(w,\ell) = \sum_{j=1}^\infty \langle \ell,\ell_j\rangle_{H_T} G(w,\ell_j),\ \ell \in \mathscr{K}\Big\},$$

we have $\mu_T(B) = 1$. If $w \in B$, then

$$\Big|\sum_{j=1}^{\infty} \langle \ell, \ell_j \rangle_{H_T} G(w, \ell_j)\Big| = |G(w, \ell)| \leqq C \|\ell\|_{H_T} \qquad (\ell \in \mathscr{K}).$$

Hence, letting $N \in \mathbb{N}$ and taking $k_n \in \mathscr{K}$ with

$$\lim_{n\to\infty} \Big\| k_n - \sum_{j=1}^{N} G(w, \ell_j)\ell_j \Big\|_{H_T} = 0 \quad \text{and} \quad \langle k_n, \ell_j \rangle_{H_T} = 0 \; (j \geqq N+1),$$

we obtain

$$\begin{aligned} \sum_{j=1}^{N} G(w, \ell_j)^2 &= \lim_{n\to\infty} \sum_{j=1}^{N} \langle k_n, \ell_j \rangle_{H_T} G(w, \ell_j) \\ &\leqq \limsup_{n\to\infty} C \|k_n\|_{H_T} = C \Big\{ \sum_{j=1}^{N} G(w, \ell_j)^2 \Big\}^{\frac{1}{2}}. \end{aligned}$$

Letting $N \to \infty$, we obtain

$$\sum_{j=1}^{\infty} G(w, \ell_j)^2 \leqq C^2 < \infty. \tag{5.3.8}$$

If we set

$$G(w) = \mathbf{1}_B(w) \sum_{j=1}^{\infty} G(w, \ell_j)\ell_j,$$

then $\|G(w)\|_{H_T} \leqq C$ $(w \in W_T)$ by (5.3.8). Moreover, by (5.3.7),

$$\int_{W_T} F \nabla^*(\phi\ell)\, d\mu_T = \int_{W_T} \langle G, \phi\ell \rangle_{H_T}\, d\mu_T$$

for $\ell \in \mathscr{K}$ and $\phi \in \mathscr{P}$. Since $\mathscr{K}$ is dense in H_T,

$$\int_{W_T} F \nabla^* K\, d\mu_T = \int_{W_T} \langle G, K \rangle_{H_T}\, d\mu_T \quad (K \in \mathscr{P}(H_T)).$$

Therefore, by Theorem 5.3.1, $F \in \mathbb{D}^{1,\infty-}$ and $\nabla F = G$. □

Corollary 5.3.7 *The norm* $\|\theta\| = \max_{0\leqq t\leqq T} |\theta(t)|$ *belongs to* $\mathbb{D}^{1,\infty-}$.

Proof Since $\big|\|w+h\| - \|w\|\big| \leqq \sqrt{T}\, \|h\|_{H_T}$ $(h \in H_T)$, Theorem 5.3.6 implies the assertion. □

Using the following proposition, we can prove that, when $d = 1$, the derivative of the norm $\|\theta\|$ is given by

$$\overbrace{(\nabla\|\theta\|)}^{\cdot} = \mathrm{sgn}(\theta(\tau))\mathbf{1}_{[0,\tau]}, \quad \mu_T\text{-a.s.}, \tag{5.3.9}$$

where $\tau(w) = \inf\{t \in [0,T]; |w(t)| = \|w\|\}$.

Proposition 5.3.8 (1) *Let $F \in \mathbb{D}^{1,p}$. Then, $F^+ = \max\{F,0\} \in \mathbb{D}^{1,p}$ and*

$$\nabla F^+ = \mathbf{1}_{(0,\infty)}(F)\nabla F, \quad \mu_T\text{-a.s.}$$

(2) *Let $F_1,\dots,F_n \in \mathbb{D}^{1,p}$. Then, $\max_{1\leqq i\leqq n} F_i \in \mathbb{D}^{1,p}$ and*

$$\nabla \max_{1\leqq i\leqq n} F_i = \sum_{i=1}^{n} \mathbf{1}_{A_i}\nabla F_i, \quad \mu_T\text{-a.s.},$$

where $A_i = \{w;\ F_j(w) \leqq F_i(w)\,(j<i),\ F_j(w) < F_i(w)\,(j>i)\}$.
(3) *Let $d = 1$. Then, $\max_{0\leqq s\leqq T}\theta(s) \in \mathbb{D}^{1,\infty-}$ and*

$$\overbrace{(\nabla \max_{0\leqq s\leqq T}\theta(s))}^{\cdot} = \mathbf{1}_{[0,\sigma]}, \quad \mu_T\text{-a.s.}, \tag{5.3.10}$$

where $\sigma(w) = \inf\{t \in [0,T];\ w(t) = \max_{0\leqq s\leqq T} w(s)\}$.
(4) (5.3.9) *holds.*

Proof (1) Take $\varphi(x) \in C^\infty(\mathbb{R})$ so that $\varphi(x) = 1$ $(x \geqq 1)$ and $\varphi(x) = 0$ $(x \leqq 0)$. Set $\varphi_n(x) = \varphi(nx)$ and define $\psi_n(x) = \int_0^x \varphi_n(y)\,dy$. By the same arguments as in the proof of Corollary 5.3.2, we can show $\psi_n(F) \in \mathbb{D}^{1,p}$. Letting $n \to \infty$, we obtain the conclusion.
(2) If $n = 2$, then the assertion follows from (1) because $\max\{F_1,F_2\} = (F_1 - F_2)^+ + F_2$. By induction we obtain the assertion for general n.
(3) From (2) we have $\max_{0\leqq k\leqq 2^n}\theta(\frac{k}{2^n}) \in \mathbb{D}^{1,\infty-}$ and

$$\overbrace{\Big(\nabla \max_{0\leqq k\leqq 2^n}\theta\Big(\frac{k}{2^n}\Big)\Big)}^{\cdot} = \sum_{k=0}^{2^n} \mathbf{1}_{A_k^n}\mathbf{1}_{[0,\frac{k}{2^n}]}, \tag{5.3.11}$$

where $A_k^n = \{\theta(\frac{j}{2^n}) \leqq \theta(\frac{k}{2^n})\,(j<k),\ \theta(\frac{j}{2^n}) < \theta(\frac{k}{2^n})\,(j>k)\}$. Since

$$\mu_T(\theta(\sigma) > \theta(t),\ t \neq \sigma) = 1$$

(see [56, p.102]), letting $n \to \infty$ in (5.3.11) yields (5.3.10).
(4) Since $\|\theta\| = \max\big\{\max_{0\leqq s\leqq T}\theta(s), \max_{0\leqq s\leqq T}(-\theta(s))\big\}$, (1) and (3) yield the conclusion. □

5.4 Integration by Parts Formula

In this section we show an integration by parts formula and, by applying it, we introduce the composition of distributions on $\mathbb{R}^N$ and Wiener functionals.

Definition 5.4.1 $F = (F^1, \dots, F^N) \in \mathbb{D}^{\infty,\infty-}(\mathbb{R}^N)$ is called **non-degenerate** if

$$\left(\det\left[\left(\langle \nabla F^i, \nabla F^j \rangle_{H_T}\right)_{i,j=1,\dots,N}\right]\right)^{-1} \in L^{\infty-}(\mu_T) = \bigcap_{p\in(1,\infty)} L^p(\mu_T). \tag{5.4.1}$$

Example 5.4.2 For $\ell_1, \dots, \ell_N \in W_T^*$, suppose that

$$\det\left[(\langle \ell_i, \ell_j \rangle_{H_T})_{i,j=1,\dots,N}\right] \neq 0$$

and set $F = (\ell_1, \dots, \ell_N)$. Then, $F \in \mathscr{P}(\mathbb{R}^N)$ and $\nabla F^i(w) = \ell_i$. Hence, F is non-degenerate.

In particular, for $t > 0$, $N = d$ and $\ell_i(w) = w^i(t)$ $(i = 1, \dots, d,\ w \in W_T)$, we have

$$\det\left[(\langle \ell_i, \ell_j \rangle_{H_T})_{i,j=1,\dots,N}\right] = t^d > 0.$$

Hence, $F = \theta(t)$ is non-degenerate.

Example 5.4.3 Let $\{h_n\}_{n=1}^\infty$ be an orthonormal basis of H_T and $\{a_j\}_{j=1}^\infty \subset \mathbb{R}$ satisfy $\sum_{j=1}^\infty a_j^2 < \infty$. Set

$$F_n = \sum_{j=1}^n a_j\{(\nabla^* h_j)^2 - 1\}.$$

Since $\{\nabla^* h_j\}$ is a sequence of independent identically distributed normal Gaussian random variables, $\|F_n - F_m\|_2^2 = 2\sum_{j=m+1}^n a_j^2$ for $n > m$. Hence, as the limit of F_n in $L^2(\mu_T)$, a random variable

$$F = \sum_{j=1}^\infty a_j\{(\nabla^* h_j)^2 - 1\}$$

is defined.

We have

$$\int_{W_T} e^{\lambda F_n} d\mu_T = \prod_{j=1}^n \int_{\mathbb{R}} e^{\lambda a_j(x^2-1)} \frac{1}{\sqrt{2\pi}} e^{-\frac{x^2}{2}} dx = \left\{\prod_{j=1}^n (1 - 2\lambda a_j) e^{2\lambda a_j}\right\}^{-\frac{1}{2}}$$

for any $\lambda \in \mathbb{R}$ with $a|\lambda| < \frac{1}{2}$, where $a = \sup_{j\in\mathbb{N}} |a_j|$. Since $\log(1 - x) + x = -\frac{x^2}{2} + o(x^2)$ as $x \to 0$, the infinite product $\prod_{j=1}^\infty (1 - 2\lambda a_j) e^{2\lambda a_j}$ converges and is not 0. Since $e^{|y|} \leqq e^y + e^{-y}$, applying Fatou's lemma to the sequences

$\{\int_{W_T} \mathrm{e}^{\pm\lambda F_n} \mathrm{d}\mu_T\}_{n=1}^{\infty}$ (subsequences if necessary), we obtain $\int_{W_T} \mathrm{e}^{|\lambda||F|} \mathrm{d}\mu_T < \infty$. In particular, $F \in L^{\infty-}(\mu_T)$.

Set

$$F' = 2\sum_{j=1}^{\infty} a_j(\nabla^* h_j)h_j.$$

Then, since $\nabla F_n = 2\sum_{j=1}^{n} a_j(\nabla^* h_j)h_j$, we obtain by (5.1.9) and Corollary 5.3.2,

$$\left\|\nabla F_n - F'\right\|_2^2 = \left\|4\sum_{j=n+1}^{\infty} a_j^2(\nabla^* h_j)^2\right\|_1 = 4\sum_{j=n+1}^{\infty} a_j^2 \to 0 \quad (n \to \infty).$$

Hence, for any $G \in \mathscr{P}(H_T)$,

$$\begin{aligned}\int_{W_T} F\nabla^* G \,\mathrm{d}\mu_T &= \lim_{n\to\infty} \int_{W_T} F_n \nabla^* G \,\mathrm{d}\mu_T \\ &= \lim_{n\to\infty} \int_{W_T} \langle \nabla F_n, G\rangle_{H_T} \mathrm{d}\mu_T = \int_{W_T} \langle F', G\rangle_{H_T} \mathrm{d}\mu_T.\end{aligned}$$

Moreover, for $\lambda \in \mathbb{R}$ with $a^2|\lambda| < \frac{1}{2}$, the integrability of $\exp(|\lambda| \left\|F'\right\|_{H_T}^2)$ is shown by a similar argument to that in the preceding paragraph, and $F' \in L^{\infty-}(\mu_T; H_T)$. Thus $F \in \mathbb{D}^{1,\infty-}$ and $\nabla F = F'$.

Furthermore, since $\nabla^2 F_n = 2\sum_{j=1}^{n} a_j h_j \otimes h_j$ and $\nabla^3 F_n = 0$, Theorem 5.2.1 implies

$$F \in \mathbb{D}^{\infty,\infty-}, \qquad \nabla F = 2\sum_{j=1}^{\infty} a_j(\nabla^* h_j)h_j,$$

$$\nabla^2 F = 2\sum_{j=1}^{\infty} a_j h_j \otimes h_j, \qquad \nabla^k F = 0 \quad (k \geqq 3).$$

Finally, we present a sufficient condition for F to be non-degenerate. Suppose that $a_j \neq 0$ for infinitely many js and set $\{j\,;\, a_j \neq 0\} = \{j(1) < j(2) < \cdots\}$. Putting $m_n = \min\{a_{j(k)}^2\,;\, k = 1,\ldots,n\}$, we have $m_n > 0$ and

$$\left\|\nabla F\right\|_{H_T}^2 = \sum_{k=1}^{\infty} a_{j(k)}^2 (\nabla^* h_{j(k)})^2 \geqq m_n \sum_{k=1}^{n} (\nabla^* h_{j(k)})^2 \qquad (n \in \mathbb{N}).$$

Since $\nabla^* h_{j(k)}$ $(k \in \mathbb{N})$ form a sequence of independent identically distributed normal Gaussian random variables, $\left(\sum_{k=1}^{n} (\nabla^* h_{j(k)})^2\right)^{-\frac{1}{2}} \in L^p(\mu_T)$ for $n > p$. Therefore, $\left\|\nabla F\right\|_{H_T}^{-1} \in \bigcap_{p\in(1,\infty)} L^p(\mu_T)$ and F is non-degenerate.

Next we introduce an **integration by parts formula** associated with non-degenerate Wiener functionals. For this purpose we note the following.

Lemma 5.4.4 *For $G \in \mathbb{D}^{\infty,\infty-}$, assume that $G \geqq 0$, μ_T-a.s., and $\frac{1}{G} \in L^{\infty-}(\mu_T)$. Then, $\frac{1}{G} \in \mathbb{D}^{\infty,\infty-}$. In particular, if $F \in \mathbb{D}^{\infty,\infty-}(\mathbb{R}^N)$ is non-degenerate, then*

$$\Big(\langle \nabla F^i, \nabla F^j\rangle_{H_T}\Big)^{-1}_{i,j=1,\dots,N} \in \mathbb{D}^{\infty,\infty-}(\mathbb{R}^N \otimes \mathbb{R}^N).$$

Proof By Corollary 5.3.2, $(G+\varepsilon)^{-1} \in \mathbb{D}^{\infty,\infty-}$ for any $\varepsilon > 0$. Moreover, we have

$$\nabla^n\Big(\frac{1}{G+\varepsilon}\Big) = \sum_{k=0}^{n} \frac{\phi_k(G)}{(G+\varepsilon)^{k+1}},$$

where $\phi_k(G)$ is a polynomial determined by the tensor products of $\nabla G, \dots, \nabla^n G$. Let $\varepsilon \to 0$ to see $\frac{1}{G} \in \mathbb{D}^{\infty,\infty-}$. □

Theorem 5.4.5 *Suppose that $F \in \mathbb{D}^{\infty,\infty-}(\mathbb{R}^N)$ is non-degenerate and set*

$$\gamma = (\gamma_{ij})_{i,j=1,\dots,N} = \Big(\langle \nabla F^i, \nabla F^j\rangle_{H_T}\Big)^{-1}_{i,j=1,\dots,N}.$$

Define the linear mapping $\xi_{i_1\dots i_n} : \mathbb{D}^{\infty,\infty-} \to \mathbb{D}^{\infty,\infty-}$ $(i_1, \dots, i_n \in \{1, \dots, N\})$ by

$$\xi_i[G] = \sum_{j=1}^{N} \nabla^*(\gamma_{ij} G \nabla F^j), \qquad \xi_{i_1\dots i_n}[G] = \xi_{i_n}[\xi_{i_1\dots i_{n-1}}[G]].$$

Then, for any $p > 1$,

$$\sup\Big\{\int_{W_T} |\xi_{i_1\dots i_n}[G]|\, \mathrm{d}\mu_T;\ G \in \mathbb{D}^{\infty,\infty-},\ \|G\|_{n,p} \leqq 1\Big\} < \infty. \tag{5.4.2}$$

Moreover, for $f \in C^n_{\nearrow}(\mathbb{R}^N)$ and $G \in \mathbb{D}^{\infty,\infty-}$,

$$\int_{W_T} \frac{\partial^n f}{\partial x^{i_1} \cdots \partial x^{i_n}}(F)\, G\, \mathrm{d}\mu_T = \int_{W_T} f(F)\, \xi_{i_1\dots i_n}[G]\, \mathrm{d}\mu_T. \tag{5.4.3}$$

Proof (5.4.2) follows from Theorem 5.2.8 and Corollary 5.4.4. We only prove (5.4.3). Let $f \in C^1_{\nearrow}(\mathbb{R}^N)$. Since

$$\nabla(f(F)) = \sum_{i=1}^{N} \frac{\partial f}{\partial x^i}(F) \nabla F^i,$$

we have

$$\frac{\partial f}{\partial x^i}(F) = \Big\langle \nabla(f(F)), \sum_{j=1}^{N} \gamma_{ij} \nabla F^j \Big\rangle_{H_T}.$$

This implies

$$\int_{W_T} \frac{\partial f}{\partial x^i}(F)\, G\, \mathrm{d}\mu_T = \int_{W_T} f(F)\, \xi_i[G]\, \mathrm{d}\mu_T.$$

By induction we obtain (5.4.3). □

As an application of the integration by parts formula, we show that a composition of a distribution on $\mathbb{R}^N$ and a non-degenerate Wiener functional is realized as a generalized Wiener functional. By using this result, we present representations as expectations for probability densities and conditional expectations.

Let $\mathscr{S}(\mathbb{R}^N)$ be the space of rapidly decreasing functions on $\mathbb{R}^N$ and $\mathscr{S}'(\mathbb{R}^N)$ be the space of tempered distributions on $\mathbb{R}^N$. For $k \in \mathbb{Z}$, denote by $\mathscr{S}_{2k}(\mathbb{R}^N)$ the completion of $\mathscr{S}(\mathbb{R}^N)$ by the norm

$$\|f\|_{2k} = \sup_{x\in\mathbb{R}^N}\Big|\{I + |x|^2 - \tfrac{1}{2}\Delta\}^k f(x)\Big|,$$

where $\Delta = \sum_{i=1}^N (\frac{\partial}{\partial x^i})^2$. Then, $\mathscr{S}_{2k}(\mathbb{R}^N) \supset \mathscr{S}_{2k+2}(\mathbb{R}^N)$ and $\mathscr{S}_0(\mathbb{R}^N)$ is the space of continuous functions on $\mathbb{R}^N$ satisfying $\lim_{|x|\to\infty} |f(x)| = 0$. Moreover,

$$\mathscr{S}(\mathbb{R}^N) = \bigcap_{k=1}^{\infty} \mathscr{S}_{2k}(\mathbb{R}^N) \quad \text{and} \quad \mathscr{S}'(\mathbb{R}^N) = \bigcup_{k=1}^{\infty} \mathscr{S}_{-2k}(\mathbb{R}^N).$$

Theorem 5.4.6 *Let $p > 1$ and $k \in \mathbb{Z}_+$, and suppose that $F \in \mathbb{D}^{\infty,\infty-}(\mathbb{R}^N)$ is non-degenerate. Then, there exists a constant C such that*

$$\|f(F)\|_{-2k,p} \leqq C\|f\|_{-2k}$$

for any $f \in \mathscr{S}(\mathbb{R}^N)$.

Proof Define $\eta : \mathbb{D}^{\infty,\infty-} \to \mathbb{D}^{\infty,\infty-}$ by

$$\eta[G] = G + |F|^2 G - \frac{1}{2}\sum_{i=1}^N \xi_{ii}[G].$$

By Theorem 5.4.5,

$$\int_{W_T} (\{I + |x|^2 - \tfrac{1}{2}\Delta\}^k f)(F) G \, d\mu_T = \int_{W_T} f(F)\eta^k[G] \, d\mu_T.$$

This implies

$$\int_{W_T} f(F) G \, d\mu_T = \int_{W_T} (\{I + |x|^2 - \tfrac{1}{2}\Delta\}^{-k} f)(F)\eta^k[G] \, d\mu_T$$

for any $G \in \mathbb{D}^{\infty,\infty-}$. As we have shown in the proof of Theorem 5.1.10, we have

$$\|f(F)\|_{-2k,p} = \sup\Big\{\int_{W_T} f(F) G \, d\mu_T \, ; \, G \in \mathbb{D}^{\infty,\infty-}, \, \|G\|_{2k,q} \leqq 1\Big\},$$

where $q = \frac{p}{p-1}$. Combining this with (5.4.2), we obtain the conclusion. □

Corollary 5.4.7 *If $F \in \mathbb{D}^{\infty,\infty-}(\mathbb{R}^N)$ is non-degenerate, then, for any $p > 1$ and $k \in \mathbb{Z}_+$, the mapping $\mathscr{S}(\mathbb{R}^N) \ni f \mapsto f(F) \in \mathbb{D}^{k,p}$ is extended to a continuous linear mapping $\Phi_F : \mathscr{S}_{-2k} \to \mathbb{D}^{-2k,p}$.*

Definition 5.4.8 Suppose that $F \in \mathbb{D}^{\infty,\infty-}(\mathbb{R}^N)$ is non-degenerate. For $u \in \mathscr{S}_{-2k}(\mathbb{R}^N)$, the generalized Wiener functional $\Phi_F(u) \in \mathbb{D}^{-2k,p}$ in Corollary 5.4.7 is denoted by $u(F)$ and called the **pull-back** of u by F.

By Corollary 5.4.7 we obtain the following.

Corollary 5.4.9 *Assume that $F \in \mathbb{D}^{\infty,\infty-}(\mathbb{R}^N)$ is non-degenerate. Let $p > 1$ and $U \subset \mathbb{R}^n$ be an open set.*
(1) Let $m \in \mathbb{Z}_+$. If the mapping $U \ni z \mapsto u_z \in \mathscr{S}_{-2k}$ is of C^m-class, then so is the mapping $U \ni z \mapsto u_z(F) \in \mathbb{D}^{-2k,p}$.
(2) Assume that $U \ni z \mapsto u_z \in \mathscr{S}_{-2k}$ is continuous and admits the Bochner integral $\int_U u_z \mathrm{d}z$. Then, $z \mapsto u_z(F)$ is Bochner integrable as a $\mathbb{D}^{-2k,p}$-valued function and

$$\Big(\int_U u_z\,\mathrm{d}z\Big)(F) = \int_U u_z(F)\,\mathrm{d}z.$$

Remark 5.4.10 In the above, for a Banach space E, the derivative of an E-valued function $\psi : U \to E$ at $z \in U$ is, by definition, a continuous linear mapping $\psi'(z) : \mathbb{R}^n \to E$ such that $\|\frac{1}{\varepsilon}\{\psi(z + \varepsilon\xi) - \psi(z)\} - [\psi'(z)](\xi)\|_E \to 0$ $(\varepsilon \to 0)$ for any $\xi \in \mathbb{R}^n$. The higher order derivatives are defined inductively. For the Bochner integral, see [133].

Using a composition of a non-degenerate functional and a distribution, we have the following expression of the probability density via a generalized Wiener functional. Let δ_x be the **Dirac measure** on $\mathbb{R}^N$ concentrated at $x \in \mathbb{R}^N$.

Theorem 5.4.11 *Suppose that $F \in \mathbb{D}^{\infty,\infty-}(\mathbb{R}^N)$ is non-degenerate.*
(1) Let $p_F(x)$ be the value of $\delta_x(F) \in \mathbb{D}^{-\infty,1+}$ at $1 \in \mathbb{D}^{\infty,\infty-}$;

$$p_F(x) = \mathbf{E}[\delta_x(F)] = [\delta_x(F)](1).$$

Then, p_F is of C^∞-class and the probability density of F:

$$\mu_T(F \in A) = \int_A p_F(x)\,\mathrm{d}x \qquad (A \in \mathscr{B}(\mathbb{R}^N)).$$

(2) Let $G \in \mathbb{D}^{\infty,\infty-}$ and set $p_{G|F}(x) = \mathbf{E}[\delta_x(F)G]$. Then,

$$\int_{W_T} f(F)G\,\mathrm{d}\mu_T = \int_{\mathbb{R}^N} f(x)p_{G|F}(x)\,\mathrm{d}x$$

for any $f \in \mathscr{S}(\mathbb{R}^N)$. *In particular,* $p_{G|F}(x) = p_F(x)\mathbf{E}[G|F = x]$ *holds for almost all* $x \in \mathbb{R}^N$ *with* $p_F(x) > 0$.

Proof For $k \in \mathbb{Z}_+$, the mapping $\mathbb{R}^N \ni x \mapsto \delta_x \in \mathscr{S}_{-2([\frac{N}{2}]+1+k)}$ is of C^{2k}-class (see [45, Lemma V-9.1]). Hence, by Corollary 5.4.9, both p_F and $p_{G|F}$ are of C^∞-class.

For $f \in \mathscr{S}(\mathbb{R}^N)$, the integral $\int_{\mathbb{R}^N} f(x)\delta_x \mathrm{d}x$ of the $\mathscr{S}_{-2([\frac{N}{2}]+1+k)}$-valued function $x \mapsto f(x)\delta_x$ coincides with f. By Corollary 5.4.9 again,

$$\int_{\mathbb{R}^N} f(x)\delta_x(F)\,\mathrm{d}x = f(F).$$

Hence, by Corollary 5.4.9 and the commutativity between Bochner integrals and linear continuous operators, we obtain, for $G \in \mathbb{D}^{\infty,\infty-}$,

$$\int_{W_T} f(F)G\,\mathrm{d}\mu_T = \int_{\mathbb{R}^N} f(x)\mathbf{E}[\delta_x(F)G]\,\mathrm{d}x. \tag{5.4.4}$$

Setting $G = 1$ in (5.4.4), we see that p_F is the probability density of F. Moreover, since the left hand side of (5.4.4) is equal to $\int_{\mathbb{R}^N} f(x)\mathbf{E}[G|F = x]p_F(x)\,\mathrm{d}x$, the identity

$$p_{G|F}(x) = p_F(x)\mathbf{E}[G|F = x]$$

holds for almost all $x \in \mathbb{R}^N$ with $p_F(x) > 0$. □

Positive distributions on $\mathbb{R}^n$ are realized by measures ([36]). Similar facts holds for generalized Wiener functionals.

Definition 5.4.12 $\Phi \in \mathbb{D}^{-\infty,1+}$ is said to be positive ($\Phi \geqq 0$ in notation) if

$$\int_{W_T} F\Phi\,\mathrm{d}\mu_T \geqq 0$$

holds for any non-negative $F \in \mathbb{D}^{\infty,\infty-}$.

If $\Phi \in L^p(\mu_T)$, then the condition in the above definition is equivalent to $\Phi \geqq 0$, μ_T-a.s.

Set

$$\mathscr{F}C_{\mathrm{b}}^\infty = \{F\,;\, F = f(\ell_1,\ldots,\ell_n),\ f \in C_{\mathrm{b}}^\infty(\mathbb{R}^n),\ \ell_1,\ldots,\ell_n \in W_T{}^*,\ n \in \mathbb{N}\}.$$

Since $\mathscr{F}C_{\mathrm{b}}^\infty$ is dense in $\mathbb{D}^{r,p}$, Φ is positive if and only if $\int_{W_T} F\Phi\,\mathrm{d}\mu_T \geqq 0$ for any non-negative $F \in \mathscr{F}C_{\mathrm{b}}^\infty$.

Proposition 5.4.13 *If* $F \in \mathbb{D}^{\infty,\infty-}(\mathbb{R}^N)$ *is non-degenerate,* $\delta_x(F) \in \mathbb{D}^{-\infty,1+}$ *is positive.*

Proof The assertion follows from the identity

$$\int_{W_T} G\delta_x(F)\,\mathrm{d}\mu_T = \mathbf{E}[G|F=x]p_F(x) \quad (G \in \mathbb{D}^{\infty,\infty-}). \qquad \square$$

Lemma 5.4.14 *Let $\Phi \in \mathbb{D}^{-\infty,1+}$ be positive. Then, $\Phi = 0$ if and only if $\int_{W_T} \Phi\,\mathrm{d}\mu_T = 0$.*

Proof Obviously $\int_{W_T} \Phi\,\mathrm{d}\mu_T = 0$ if $\Phi = 0$. We show the converse. Suppose that $F \in \mathscr{F}C_{\mathrm{b}}^{\infty}$ is non-negative and set $M = \sup_{w \in W_T} F(w)$. Since $M - F \in \mathscr{F}C_{\mathrm{b}}^{\infty}$ is non-negative, we have

$$0 \leqq \int_{W_T} (M-F)\Phi\,\mathrm{d}\mu_T = -\int_{W_T} F\Phi\,\mathrm{d}\mu_T \leqq 0.$$

Hence $\int_{W_T} F\Phi\,\mathrm{d}\mu_T = 0$. Since F is arbitrary, we obtain $\Phi = 0$. $\square$

Theorem 5.4.15 *For any positive $\Phi \in \mathbb{D}^{-\infty,1+}$, there exists a finite measure ν_Φ on W_T such that*

$$\int_{W_T} F\Phi\,\mathrm{d}\mu_T = \int_{W_T} F\,\mathrm{d}\nu_\Phi \tag{5.4.5}$$

for any $F \in \mathscr{F}C_{\mathrm{b}}^{\infty}$.

Remark 5.4.16 If $p > 1$ and $\Phi \in L^p(\mu_T)$, then $\mathrm{d}\nu_\Phi = \Phi\,\mathrm{d}\mu_T$.

Proof Let $\Phi \in \mathbb{D}^{-r,p}$, $\Phi \neq 0$ ($r \in \mathbb{R}$, $p > 1$). By Lemma 5.4.14, we may assume $\int_{W_T} \Phi\,\mathrm{d}\mu_T = 1$. Set

$$\mathbf{D} = \Big\{\frac{k}{2^n}\,;\, n \in \mathbb{Z}_+,\, k \in \mathbb{Z}_+,\, k \leqq 2^n T\Big\}.$$

For $t_j \in \mathbf{D}$ ($j = 1, \dots, n$) with $0 \leqq t_1 < \cdots < t_n$, define $u_{t_1\dots t_n} : \mathscr{S}((\mathbb{R}^d)^n) \to \mathbb{R}$ by

$$u_{t_1\dots t_n}(f) = \int_{W_T} f(\theta(t_1), \dots, \theta(t_n))\Phi\,\mathrm{d}\mu_T,$$

where $\{\theta(t)\}_{t\in[0,T]}$ is the coordinate process. Then $u_{t_1\dots t_n}$ is a positive distribution. Hence, there exists a probability measure $\nu_{t_1\dots t_n}$ on $(\mathbb{R}^d)^n$ such that

$$\int_{W_T} f(\theta(t_1), \dots, \theta(t_n))\Phi\,\mathrm{d}\mu_T = \int_{(\mathbb{R}^d)^n} f(x_1, \dots, x_n)\nu_{t_1\dots t_n}(\mathrm{d}x_1 \cdots \mathrm{d}x_n)$$

for any $f \in \mathscr{S}((\mathbb{R}^d)^n)$. Since $\{\nu_{t_1\dots t_n};\, t_1,\dots,t_n \in \mathbf{D}, n \in \mathbb{N}\}$ is consistent, by Kolmogorov's extension theorem (see, e.g., [56, 114]), there exists a probability measure ν_Φ on $(\mathbb{R}^d)^{\mathbf{D}}$ such that

$$\nu_\Phi((X(t_1),\dots,X(t_n)) \in A) = \nu_{t_1\dots t_n}(A). \tag{5.4.6}$$

for any $A \in \mathscr{B}((\mathbb{R}^d)^n)$ $(t_1 < \cdots < t_n \in \mathbf{D})$, where $X(t) : (\mathbb{R}^d)^{\mathbf{D}} \to \mathbb{R}^d$ is given by $X(t,\phi) = \phi(t)$ $(\phi \in (\mathbb{R}^d)^{\mathbf{D}})$.

By Lemma 5.2.6, there exists a constant C such that

$$\|G\|_{r,q} \leqq C\|G\|_q$$

for any $G \in \bigoplus_{n=0}^4 \mathscr{H}_n$, where $q = \frac{p}{p-1}$. Since $|\theta(t)-\theta(s)|^4 \in \bigoplus_{n=0}^4 \mathscr{H}_n$ for any $t, s \in \mathbf{D}$, we have

$$\begin{aligned}\int_{(\mathbb{R}^d)^{\mathbf{D}}} |X(t)-X(s)|^4 \mathrm{d}\nu_\Phi &= \int_{W_T} |\theta(t)-\theta(s)|^4 \Phi \,\mathrm{d}\mu_T \\ &\leqq C\|\Phi\|_{-r,p}\Big(\int_{\mathbb{R}^d} |x|^{4q} \frac{1}{(2\pi)^{\frac{d}{2}}} \mathrm{e}^{-\frac{|x|^2}{2}} \mathrm{d}x\Big)^{\frac{1}{q}} |t-s|^2.\end{aligned}$$

Hence, by Kolmogorov's continuity theorem (Theorem A.5.1), $\{X(t)\}_{t\in\mathbf{D}}$ is extended to a stochastic process $\{X(t)\}_{t\in[0,T]}$, which is continuous almost surely with respect to ν_Φ. Therefore, ν_Φ is regarded as a probability measure on W_T.

Let $f \in C_{\mathrm{b}}^\infty((\mathbb{R}^d)^n)$. By (5.4.6), we have for $t_1 < \cdots < t_n \in \mathbf{D}$

$$\int_{W_T} f(\theta(t_1),\dots,\theta(t_n))\Phi \,\mathrm{d}\mu_T = \int_{W_T} f(\theta(t_1),\dots,\theta(t_n))\,\mathrm{d}\nu_\Phi.$$

Since $\mathbf{D}$ is dense in $[0,T]$, this identity continues to hold for any $t_1 < \cdots < t_n \in [0,T]$. We have now proved the conclusion because the elements of the form $f(\theta(t_1),\dots,\theta(t_n))$ form a dense subset in $\mathbb{D}^{\infty,\infty-}$. □

Example 5.4.17 Let $\eta_1,\dots,\eta_n \in W_T^*$ form an orthonormal system in H_T. Then, by Example 5.4.2, $\boldsymbol{\eta} = (\eta_1,\dots,\eta_n) \in \mathbb{D}^{\infty,\infty-}(\mathbb{R}^n)$ is non-degenerate and $\delta_x(\boldsymbol{\eta})$ is positive by Proposition 5.4.13.

Take $\varphi \in C_0^\infty(\mathbb{R}^n)$ so that $\varphi(y) = 1$ for $|y| \leqq 1$ and set $\varphi_m(y) = m^n\varphi(\frac{y-x}{m})$. By Theorem 5.4.11 we have

$$\begin{aligned}\frac{1}{\sqrt{2\pi}^n}\mathrm{e}^{-\frac{|x|^2}{2}} &= \int_{W_T} \delta_x(\boldsymbol{\eta})\,\mathrm{d}\mu_T = \int_{W_T} \varphi_m(\boldsymbol{\eta})\delta_x(\boldsymbol{\eta})\,\mathrm{d}\mu_T \\ &= \int_{W_T} \varphi_m(\boldsymbol{\eta})\,\mathrm{d}\nu_{\delta_x(\boldsymbol{\eta})} \to \nu_{\delta_x(\boldsymbol{\eta})}(\{\boldsymbol{\eta} = x\}) \quad (m \to \infty).\end{aligned}$$

Hence

$$\nu_{\delta_x(\eta)}(\{\eta = y\}) = \begin{cases} \dfrac{1}{(2\pi)^{\frac{n}{2}}} e^{-\frac{|x|^2}{2}} & (y = x), \\ 0 & (y \neq x). \end{cases}$$

Thus $\nu_{\delta_x(\eta)}$ is a measure concentrated on the "hyperplane" $\{w\,;\,\eta(w) = x\}$ on the Wiener space. Since $\mu_T(\eta = x) = 0$, $\nu_{\delta_x(\eta)}$ is singular with respect to μ_T.

5.5 Application to Stochastic Differential Equations

We present applications of the Malliavin calculus to stochastic differential equations. Throughout this section, let $V_0, V_1, \dots, V_d : \mathbb{R}^N \to \mathbb{R}^N$ be C^∞ functions on $\mathbb{R}^N$ with bounded derivatives of all orders.

In this section, we think of $\{\theta(t)\}_{t\in[0,T]}$ as an $\{\mathscr{F}_t\}$-Brownian motion as described in Theorem 5.3.3. However, as mentioned in the remark after the theorem, all random variables are $\mathscr{B}(W_T)$-measurable.

Denote by $\{X(t,x)\}_{t\in[0,T]}$ the unique strong solution of the stochastic differential equation

$$dX(t) = \sum_{\alpha=1}^{d} V_\alpha(X(t))\, d\theta^\alpha(t) + V_0(X(t))\, dt, \quad X(0) = x \tag{5.5.1}$$

(Theorem 4.4.5). By Theorem 4.10.8, $X(t,\cdot)$ is of C^∞-class and the Jacobian matrix $Y(t,x) = (\frac{\partial X^i(t,x)}{\partial x^j})_{i,j=1,\dots,N}$ satisfies the stochastic differential equation

$$\begin{aligned} dY(t,x) &= \sum_{\alpha=1}^{d} V'_\alpha(X(t,x))Y(t,x)\, d\theta^\alpha(t) + V'_0(X(t,x))Y(t,x)\, dt, \\ Y(0,x) &= I, \end{aligned} \tag{5.5.2}$$

where $V'_\alpha(x) = (\frac{\partial V^i_\alpha}{\partial x^j}(x))_{i,j=1,\dots,N}$ $(\alpha = 0, 1, \dots, d)$. Moreover, $Y(t,x)$ is non-degenerate and the inverse matrix

$$Z(t,x) = Y(t,x)^{-1}$$

satisfies the stochastic differential equation

$$\begin{aligned} dZ(t,x) = &-\sum_{\alpha=1}^{d} Z(t,x)V'_\alpha(X(t,x))\, d\theta^\alpha(t) - Z(t,x)V'_0(X(t,x))\, dt \\ &+ \sum_{\alpha=1}^{d} Z(t,x)(V'_\alpha(X(t,x)))^2\, dt. \end{aligned} \tag{5.5.3}$$

From these observations, we have, in particular,

$$\sup_{x\in\mathbb{R}^N}\int_{W_T}\sup_{0\leqq t\leqq T}\{|Y(t,x)|^p+|Z(t,x)|^p\}\,\mathrm{d}\mu_T<\infty \tag{5.5.4}$$

for any $p>1$, where $|A|=\left(\sum_{i,j=1}^N a_{ij}^2\right)^{\frac{1}{2}}$ for a matrix $A=(a_{ij})_{i,j=1,\dots,N}$.

Theorem 5.5.1 *Let $t\in[0,T]$. Then, $X(t,x)\in\mathbb{D}^{\infty,\infty-}(\mathbb{R}^N)$ and*

$$(\nabla X^i(t,x))^\alpha(u)=\sum_{j,k=1}^N Y^i_j(t,x)\int_0^{t\wedge u} Z^j_k(v,x)V^k_\alpha(X(v,x))\,\mathrm{d}v \qquad (\alpha=1,\dots,d), \tag{5.5.5}$$

where, for $h\in H_T$, $h^\alpha(u)$ is the value of the α-th component $h^\alpha:[0,T]\to\mathbb{R}$ of h at time $u\in[0,T]$.

Proof For $n\in\mathbb{N}$, set $[s]_n=\frac{[2^n s]}{2^n}$ and define $\{X_n(s)\}_{s\in[0,T]}$ by

$$X_n(0)=x,$$

$$X_n(s)=X_n([s]_n)+\sum_{\alpha=1}^d V_\alpha(X_n([s]_n))\{\theta^\alpha(s)-\theta^\alpha([s]_n)\} + V_0(X_n([s]_n))\{s-[s]_n\}.$$

By definition,

$$X_n(t)\in\mathbb{D}^{\infty,\infty-}(\mathbb{R}^N).$$

Moreover, using the expression

$$\mathrm{d}X_n(s)=\sum_{\alpha=1}^d V_\alpha(X_n([s]_n))\,\mathrm{d}\theta^\alpha(s)+V_0(X_n([s]_n))\,\mathrm{d}s, \tag{5.5.6}$$

we observe

$$\int_{W_T}\sup_{0\leqq s\leqq T}|X_n(s)-X(s,x)|^2\mathrm{d}\mu_T\to 0 \quad (n\to\infty). \tag{5.5.7}$$

To see this, set

$$R_n(s)=\sum_{\alpha=1}^d\int_0^s\{V_\alpha(X_n(u))-V_\alpha(X_n([u]_n))\}\,\mathrm{d}\theta^\alpha(u) + \int_0^s\{V_0(X_n(u))-V_0(X_n([u]_n))\}\,\mathrm{d}u.$$

In the same way as for Theorem 4.3.9, we obtain from (5.5.6)

$$\sup_{n\in\mathbb{N}}\int_{W_T}\sup_{0\leqq s\leqq T}|X_n(s)|^p\mathrm{d}\mu_T<\infty$$

for any $p>1$. Hence, by the definition of $X_n(s)$, there exists a constant C_1 such that

$$\int_{W_T}|X_n(u)-X_n([u]_n)|^2\mathrm{d}\mu_T\leqq C_1 2^{-n}.$$

By this estimate, the Lipschitz continuity of V_α and the Burkholder–Davis–Gundy inequality (Theorem 2.4.1), there exists a constant C_2 such that

$$\int_{W_T}\sup_{0\leqq s\leqq T}|R_n(s)|^2\mathrm{d}\mu_T\leqq C_2 2^{-n}\qquad(n=1,2,\ldots).$$

Moreover, since

$$\begin{aligned}X(s)-X_n(s)=\sum_{\alpha=1}^{n}\int_0^s\{V_\alpha(X(u))-V_\alpha(X_n(u))\}\,\mathrm{d}\theta^\alpha(u)\\+\int_0^s\{V_0(X(u))-V_0(X_n(u))\}\,\mathrm{d}u+R_n(s),\end{aligned}$$

we see, by using the Burkholder–Davis–Gundy inequality again, that there exist constants C_3 and C_4 such that

$$\begin{aligned}&\int_{W_T}\sup_{0\leqq u\leqq s}|X_n(u)-X(u)|^2\mathrm{d}\mu_T\\&\quad\leqq C_3 2^{-n}+C_4\int_0^s\Big(\int_{W_T}\sup_{0\leqq u\leqq v}|X_n(u)-X(u)|^2\mathrm{d}\mu_T\Big)\mathrm{d}v\quad(n=1,2,\ldots).\end{aligned}$$

Hence, by Gronwall's inequality, we obtain (5.5.7).

Let $h\in H_T$ and set

$$J_{n,h}(s)=\langle\nabla X_n(s),h\rangle_{H_T}\qquad(s\in[0,T]).$$

By the definition of $X_n(s)$,

$$\begin{aligned}\mathrm{d}J_{n,h}(s)=\sum_{\alpha=1}^{d}V_\alpha'(X_n([s]_n))J_{n,h}([s]_n)\,\mathrm{d}\theta^\alpha(s)+V_0'(X_n([s]_n))J_{n,h}([s]_n)\,\mathrm{d}s\\+\sum_{\alpha=1}^{d}V_\alpha(X_n([s]_n))\dot{h}^\alpha(s)\,\mathrm{d}s.\end{aligned}$$

Let an $\mathbb{R}^N$-valued stochastic process $\{J_h(s)\}_{s\in[0,T]}$ be the solution of

$$\mathrm{d}J_h(s) = \sum_{\alpha=1}^{d} V'_\alpha(X(s,x))J_h(s)\,\mathrm{d}\theta^\alpha(s) + V'_0(X(s,x))J_h(s)\,\mathrm{d}s$$
$$+ \sum_{\alpha=1}^{d} V_\alpha(X(s,x))\dot{h}^\alpha(s)\,\mathrm{d}s \tag{5.5.8}$$

satisfying $J_h(0) = 0$ and set

$$R'_{n,h}(s) = \sum_{\alpha=1}^{d} \int_0^s \{V'_\alpha(X_n([u]_n))J_{n,h}([u]_n) - V'_\alpha(X(u,x))J_{n,h}(u)\}\,\mathrm{d}\theta^\alpha(u)$$
$$+ \int_0^s \{V'_0(X_n([u]_n))J_{n,h}([u]_n) - V'_0(X(u,x))J_{n,h}(u)\}\,\mathrm{d}u$$
$$+ \sum_{\alpha=1}^{d} \int_0^s \{V_\alpha(X_n([u]_n)) - V_\alpha(X(u,x))\}\dot{h}^\alpha(u)\,\mathrm{d}u.$$

Rewriting as

$$\begin{aligned} &V'_\alpha(X_n([u]_n))J_{n,h}([u]_n) - V'_\alpha(X(u,x))J_{n,h}(u) \\ &\quad = \{V'_\alpha(X_n([u]_n)) - V'_\alpha(X(u,x))\}J_{n,h}([u]_n) \\ &\qquad\qquad + V'_\alpha(X(u,x))\{J_{n,h}([u]_n) - J_{n,h}(u)\} \end{aligned}$$

and using the estimate

$$\sup_{n\in\mathbb{N}} \int_{W_T} \sup_{0\leqq s\leqq T} |J_{n,h}(s)|^p \mathrm{d}\mu_T < \infty$$

for any $p > 1$, we obtain

$$\lim_{n\to\infty} \int_{W_T} \sup_{0\leqq s\leqq T} |R'_{n,h}(s)|^2 \mathrm{d}\mu_T = 0.$$

Hence, by the expression

$$J_{n,h}(s) - J_h(s) = \sum_{\alpha=1}^{d} \int_0^s V'_\alpha(X(u,x))\{J_{n,h}(u) - J_h(u)\}\,\mathrm{d}\theta^\alpha(u)$$
$$+ \int_0^s V'_0(X(u,x))\{J_{n,h}(u) - J_h(u)\}\,\mathrm{d}u + R'_{n,h}(s),$$

a similar argument to that in (5.5.7) yields

$$\int_{W_T} |J_{n,h}(t) - J_h(t)|^2 \mathrm{d}\mu_T \to 0 \quad (n\to\infty). \tag{5.5.9}$$

Define an $H_T \otimes \mathbb{R}^N$-valued random variable $F(t) = (F^1(t), \ldots, F^N(t)) \in \mathbb{D}^{0,\infty-}(H_T \otimes \mathbb{R}^N)$ by

$$\langle F^i(t), g\rangle_{H_T} = \sum_{j,k=1}^{N} \sum_{\alpha=1}^{d} Y^i_j(t,x) \int_0^t Z^j_k(v,x) V^k_\alpha(X(v,x))\, \dot{g}^\alpha(v)\,\mathrm{d}v \quad (g \in H_T)$$

for $i = 1, \ldots, N$. Then, by (5.5.8), we have

$$\langle F(t), h\rangle_{H_T} = J_h(t). \tag{5.5.10}$$

Let $\phi \in \mathscr{P}$, $h \in H_T$, $i = 1, \ldots, N$. Then, by (5.5.7),

$$\int_{W_T} X^i(t,x)\nabla^*(\phi \cdot h)\,\mathrm{d}\mu_T = \lim_{n\to\infty} \int_{W_T} X^i_n(t)\nabla^*(\phi \cdot h)\,\mathrm{d}\mu_T$$
$$= \lim_{n\to\infty} \int_{W_T} \langle \nabla X^i_n(t), \phi \cdot h\rangle_{H_T} \mathrm{d}\mu_T = \lim_{n\to\infty} \int_{W_T} J^i_{n,h}(t)\phi\,\mathrm{d}\mu_T.$$

By (5.5.9) and (5.5.10), the right hand side coincides with

$$\int_{W_T} J^i_h(t)\phi\,\mathrm{d}\mu_T = \int_{W_T} \langle F^i(t), \phi \cdot h\rangle_{H_T} \mathrm{d}\mu_T,$$

and hence we obtain from Theorem 5.3.1,

$$X(t,x) \in \mathbb{D}^{1,\infty-}(\mathbb{R}^N) \quad \text{and} \quad \nabla X(t,x) = F(t).$$

Using the result $\nabla X(t,x) = F(t)$ and repeating a similar argument to the above, we can show $X(t,x) \in \mathbb{D}^{\infty,\infty-}(\mathbb{R}^N)$. We omit the details and refer to [104]. □

On the non-degeneracy of $X(t,x)$, we have the following.

Theorem 5.5.2 *Set* $a^{ij}(y) = \sum_{\alpha=1}^d V^i_\alpha(y)V^j_\alpha(y)$. *If* $a(x) = (a^{ij}(x))_{i,j=1,\ldots,N}$ *is positive definite at the starting point* x *of* $\{X(t,x)\}_{t\in[0,T]}$, *then* $X(t,x)$ *is non-degenerate for any* $t \in (0,T]$.

We give a lemma for the proof.

Lemma 5.5.3 *Let* $\{u_\alpha(t)\}_{t\in[0,T]}$ $(\alpha = 0, 1, \ldots, d)$ *be* $\{\mathscr{F}_t\}$*-predictable and bounded (see Theorem 5.3.3 for* $\mathscr{F}_t$*) and set*

$$M = \sup\{|u_\alpha(t,w)|;\ t \in [0,T],\ w \in W_T,\ \alpha = 0, 1, \ldots, d\}.$$

Define a stochastic process $\{\xi(t)\}_{t\in[0,T]}$ *by*

$$\xi(t) = x + \sum_{\alpha=1}^d \int_0^t u_\alpha(s)\,\mathrm{d}\theta^\alpha(s) + \int_0^t u_0(s)\,\mathrm{d}s$$

and, for $\varepsilon > 0$, set

$$\sigma_\varepsilon = \inf\{t \geqq 0; |\xi(t) - x| > \varepsilon\}.$$

Then,

$$\mu_T(\sigma_\varepsilon \leqq t) \leqq \int_{\frac{\varepsilon}{2\sqrt{dM^2 t}}}^{\infty} \sqrt{\frac{2}{\pi}}\, \mathrm{e}^{-\frac{y^2}{2}}\,\mathrm{d}y$$

holds for $t < \frac{\varepsilon}{2M}$. In particular, $\frac{1}{\sigma_\varepsilon} \in \bigcap_{p\in(1,\infty)} L^p(\mu_T)$.

Proof Let $t < \frac{\varepsilon}{2M}$. Then, since

$$\Big|\sum_{\alpha=1}^{d}\int_0^s u_\alpha(v)\,\mathrm{d}\theta^\alpha(v)\Big| \geqq |\xi(s) - x| - Mt$$

for any $s \in [0, t]$, we have

$$\{\sigma_\varepsilon < t\} \subset \Big\{\sup_{0\leqq s\leqq t}\Big|\sum_{\alpha=1}^{d}\int_0^s u_\alpha(v)\,\mathrm{d}\theta^\alpha(v)\Big| > \frac{\varepsilon}{2}\Big\}.$$

By Theorem 2.5.5, there exists a Brownian motion $\{\beta(t)\}_{t\geqq 0}$ such that[2]

$$\sum_{\alpha=1}^{d}\int_0^s u_\alpha(v)\,\mathrm{d}\theta^\alpha(v) = \beta(\phi(s)),$$

where

$$\phi(s) = \sum_{\alpha=1}^{d}\int_0^s (u_\alpha(v))^2 \mathrm{d}v.$$

Since $\phi(s) \leqq dM^2 s$,

$$\Big\{\sup_{0\leqq s\leqq t}\Big|\sum_{\alpha=1}^{d}\int_0^s u_\alpha(v)\,\mathrm{d}\theta^\alpha(v)\Big| > \frac{\varepsilon}{2}\Big\} \subset \Big\{\max_{0\leqq s\leqq dM^2 t} |\beta(s)| > \frac{\varepsilon}{2}\Big\}.$$

Applying Corollary 3.1.8, we obtain the conclusion. □

Proof of Theorem 5.5.2 Let $t \in (0, T]$ and set

$$A(t, x) = \int_0^t Z(s, x)a(X(s, x))Z(s, x)^*\mathrm{d}s,$$

where Z^* is the transposed matrix of Z. By Theorem 5.5.1,

$$\Big(\langle \nabla X^i(t, x), \nabla X^j(t, x)\rangle_{H_T}\Big)_{i,j=1,\ldots,N} = Y(t, x)A(t, x)Y(t, x)^*.$$

[2] Strictly speaking, we need to extend the probability space. We suppose here that the probability space is already extended and we do not write it explicitly. For details, see Theorem 2.5.5.

Since $\frac{1}{\det Y(t,x)} = \det Z(t,x)$, it suffices to show

$$\frac{1}{\det A(t)} \in \bigcap_{p\in(1,\infty)} L^p(\mu_T) \tag{5.5.11}$$

because of (5.5.4).

Fix a sufficiently small $\varepsilon > 0$. By the positivity of $a(x)$, there exists a $\delta > 0$ such that

$$a(y) \geqq \varepsilon I \quad (y \in B(x,\delta) = \{y; |y-x| < \delta\}).$$

For $\eta > 0$, define stopping times τ_η and σ_η by

$$\tau_\eta = \inf\{s > 0; |X(t,x) - x| > \eta\} \text{ and } \sigma_\eta = \inf\{s > 0; |Z(s,x) - I| > \eta\}.$$

By the definitions, we have

$$A(t) \geqq \varepsilon \int_0^{t\wedge\tau_\delta\wedge\sigma_{\frac{1}{4}}} Z(s,x)Z(s,x)^* \mathrm{d}s \geqq \frac{9\varepsilon}{16}(t\wedge\tau_\delta\wedge\sigma_{\frac{1}{4}})I.$$

In particular,

$$\det A(t) \geqq \Big(\frac{9\varepsilon}{16}(t\wedge\tau_\delta\wedge\sigma_{\frac{1}{4}})\Big)^N.$$

Applying Lemma 5.5.3 to $\{|X(s\wedge\tau_1,x) - x|^2\}_{s\in[0,T]}$ and $\{|Z(s\wedge\sigma_1,x)|^2\}_{s\in[0,T]}$, we obtain $(t\wedge\tau_\delta\wedge\sigma_{\frac{1}{4}})^{-1} \in \bigcap_{p\in(1,\infty)} L^p(\mu_T)$ and (5.5.11). □

As will be mentioned below, the non-degeneracy in Theorem 5.5.2 holds under weaker conditions. Denote the space of $\mathbb{R}^N$-valued C^∞ functions on $\mathbb{R}^N$ by $C^\infty(\mathbb{R}^N;\mathbb{R}^N)$ and identify each element U of $C^\infty(\mathbb{R}^N;\mathbb{R}^N)$ with the differential operator $\sum_{i=1}^N U^i(x)\frac{\partial}{\partial x^i}$. For $U, V \in C^\infty(\mathbb{R}^N;\mathbb{R}^N)$, let $[U,V]$ be the Lie bracket of U and V : $[U,V] = U\circ V - V\circ U$. By the identification mentioned above, $[U,V] \in C^\infty(\mathbb{R}^N;\mathbb{R}^N)$.

Theorem 5.5.4 *Let $x \in \mathbb{R}^N$ and $\mathscr{L}(x)$ be the subspace of $\mathbb{R}^N$ spanned by $V_\alpha(x)$, $[V_{k_1},[V_{k_2},\dots,[V_{k_n},V_\alpha]\dots]](x)$ ($\alpha = 1,\dots,d$, $k_j = 0,1,\dots,d$, $j = 1,\dots,n$, $n \geqq 1$). If $\dim\mathscr{L}(x) = N$, then $X(t,x)$ is non-degenerate for any $t \in [0,T]$.*

As Theorem 5.5.2, this theorem is proven by showing the integrability of $\frac{1}{\det A(t)}$. The condition in the theorem is called **Hörmander's condition**. For details, see [104].

Example 5.5.5 Let $d = 1$ and $N = 2$. Define the vector fields V_1 and V_2 on $\mathbb{R}^2$ by

$$V_1(x) = \begin{pmatrix} 1 \\ 0 \end{pmatrix} \quad \text{and} \quad V_0(x) = \begin{pmatrix} 0 \\ x \end{pmatrix}.$$

A diffusion process on $\mathbb{R}^2$ defined by the solution of the stochastic differential equation

$$\mathrm{d}X^1(t) = \mathrm{d}\theta^1(t), \quad \mathrm{d}X^2(t) = X^1(t)\,\mathrm{d}t, \quad X(0,x) = (x^1, x^2)$$

is called the **Kolmogorov diffusion**. By Theorem 5.5.1, $X(t,x) \in \mathbb{D}^{\infty,\infty-}(\mathbb{R}^2)$. Moreover, since $[V_0, V_1] = \begin{pmatrix} 0 \\ -1 \end{pmatrix}$, $\dim \mathscr{L}(x) = 2$ and $X(t,x)$ is non-degenerate by Theorem 5.5.4. Hence, $\mathbf{E}[\delta_y(X(t,x))]$ gives the transition density $p(t,x,y)$ of the diffusion process $\{X(t,x)\}_{t \geqq 0}$.

The above results can be seen in a more straightforward manner. In fact, the solution of this stochastic differential equation is explicitly given by

$$X(t,x) = \begin{pmatrix} x^1 + \theta^1(t) \\ x^2 + x^1 t + \int_0^t \theta^1(s)\,\mathrm{d}s \end{pmatrix}.$$

This immediately implies $X(t,x) \in \mathbb{D}^{\infty,\infty-}(\mathbb{R}^2)$. Moreover, since

$$(\langle \nabla X^i(t,x), \nabla X^j(t,x) \rangle_{H_T})_{i,j=1,2} = \begin{pmatrix} t & \frac{t^2}{2} \\ \frac{t^2}{2} & \frac{t^3}{3} \end{pmatrix},$$

the non-degeneracy of $X(t,x)$ follows.

Furthermore, $p(t,x,y)$ admits an explicit expression. In fact, the distribution of $X(t,x)$ is Gaussian with mean $\begin{pmatrix} x^1 \\ x^2 + x^1 t \end{pmatrix}$ and covariance matrix $\begin{pmatrix} t & \frac{t^2}{2} \\ \frac{t^2}{2} & \frac{t^3}{2} \end{pmatrix}$. Hence

$$p(t,x,y) = \frac{\sqrt{3}}{\pi t^2} \exp\{-2t^{-1}(y^1 - x^1)^2 + 6t^{-2}(y^1 - x^1)(y^2 - x^2 - tx^1) \\ - 6t^{-3}(y^2 - x^2 - tx^1)^2\},$$

where $x = (x^1, x^2)$ and $y = (y^1, y^2)$.

The generator $\frac{1}{2}\frac{\partial^2}{\partial (x^1)^2} + x^2 \frac{\partial}{\partial x^2}$ is called the **Kolmogorov operator** and it is referred to as a typical degenerate and hypoelliptic operator in the original paper by Hörmander [37]. See [43] and [113] for recent related studies.

Example 5.5.6 Let $d = 2$ and $N = 3$. Define the vector fields V_0, V_1, and V_2 on $\mathbb{R}^3$ by $V_0 = 0$,

$$V_1(x) = \begin{pmatrix} 1 \\ 0 \\ -\frac{x^2}{2} \end{pmatrix}, \quad \text{and} \quad V_2(x) = \begin{pmatrix} 0 \\ 1 \\ \frac{x^1}{2} \end{pmatrix} \qquad (x = (x^1, x^2, x^3) \in \mathbb{R}^3).$$

The solution $X(t, x)$ of the corresponding stochastic differential equation

$$\mathrm{d}X^1(t) = \mathrm{d}\theta^1(t), \qquad \mathrm{d}X^2(t) = \mathrm{d}\theta^2(t),$$
$$\mathrm{d}X^3(t) = \frac{1}{2}X^1(t)\,\mathrm{d}\theta^2(t) - \frac{1}{2}X^2(t)\,\mathrm{d}\theta^1(t)$$

belongs to $\mathbb{D}^{\infty,\infty-}(\mathbb{R}^3)$. Moreover, since

$$[V_1, V_2](x) = \begin{pmatrix} 0 \\ 0 \\ 1 \end{pmatrix},$$

$X(t, x)$ is non-degenerate. $X(t, x)$ is explicitly written as

$$X^\alpha(t, x) = x^\alpha + \theta^\alpha(t) \qquad (\alpha = 1, 2),$$
$$X^3(t, x) = x^3 + \frac{1}{2}\{x^1\theta^2(t) - x^2\theta^1(t)\}$$
$$+ \frac{1}{2}\Big\{\int_0^t \theta^1(s)\,\mathrm{d}\theta^2(s) - \int_0^t \theta^2(s)\,\mathrm{d}\theta^1(s)\Big\}.$$

The stochastic process

$$\mathfrak{s}(t) = \frac{1}{2}\Big\{\int_0^t \theta^1(s)\,\mathrm{d}\theta^2(s) - \int_0^t \theta^2(s)\,\mathrm{d}\theta^1(s)\Big\}$$

which appears in the expression for $X^3(t, x)$ is called **Lévy's stochastic area** and plays an important role in various fields related to stochastic analysis. The explicit form of the characteristic function of $\mathfrak{s}(T)$ is well known (Theorem 5.8.4) and is called **Lévy's formula**.

Next we apply the Malliavin calculus to Schrödinger operators on $\mathbb{R}^d$. First we consider Brownian motions, that is the case where $N = d$ and $X(t, x) = x + \theta(t)$ ($x \in \mathbb{R}^d$). We presented a probabilistic representation for the corresponding heat equations in Chapter 3. We here consider Schrödinger operators with magnetic fields and give representations for the **fundamental solutions** by using the results in the previous section.

Let $V, \Theta_1, \ldots, \Theta_d \in C^\infty_{\exp}(\mathbb{R}^d)$ and assume that

$$\inf_{x \in \mathbb{R}^d} V(x) > -\infty. \tag{5.5.12}$$

The differential operator H given by

$$H = -\frac{1}{2}\sum_{\alpha=1}^{d}\Big(\frac{\partial}{\partial x^\alpha} + \mathrm{i}\,\Theta_\alpha\Big)^2 + V$$

is called a Schrödinger operator with vector potential $\Theta = (\Theta_1, \dots, \Theta_d)$ and scalar potential V. The fundamental solution for the heat equation

$$\frac{\partial u}{\partial t} = -Hu, \quad u(0,\cdot) = f \in C^\infty_{\exp}(\mathbb{R}^d) \tag{5.5.13}$$

associated with H is a function $p(t, x, y)$ such that

$$u(t, x) = \int_{\mathbb{R}^d} f(y)p(t, x, y)\,\mathrm{d}y$$

is a solution of (5.5.13). We construct the fundamental solution by applying the Malliavin calculus. It is easy to see $\int_0^t V(x + \theta(s))\,\mathrm{d}s \in \mathbb{D}^{\infty,\infty-}$ and, from the assumption (5.5.12), we have

$$\exp\Big(-\int_0^t V(x + \theta(s))\,\mathrm{d}s\Big) \in \mathbb{D}^{\infty,\infty-} \qquad (t \in [0, T],\ x \in \mathbb{R}^d).$$

Set

$$L(t, x; \Theta) = \sum_{\alpha=1}^{d}\int_0^t \Theta_\alpha(x + \theta(s)) \circ \mathrm{d}\theta^\alpha(s).$$

By Theorem 5.3.3, $L(t, x; \Theta) \in \mathbb{D}^{\infty,\infty-}$. Hence, by Corollary 5.3.2,

$$e(t, x) = \exp\Big(\mathrm{i}\,L(t, x; \Theta) - \int_0^t V(x + \theta(s))\,\mathrm{d}s\Big) \in \mathbb{D}^{\infty,\infty-}.$$

Theorem 5.5.7 *The function $p(t, x, y)$ $(t > 0,\ x, y \in \mathbb{R}^d)$ defined by*

$$p(t, x, y) = \mathbf{E}[e(t, x)\delta_y(x + \theta(t))] = \int_{W_T} e(t, x)\delta_y(x + \theta(t))\,\mathrm{d}\mu_T$$

is the fundamental solution for the heat equation (5.5.13) *associated with the Schrödinger operator H.*

Proof Let $f \in C^\infty_{\exp}(\mathbb{R}^d)$. Then $Hf \in C^\infty_{\exp}(\mathbb{R}^d)$. Setting

$$v(t, x; f) = \int_{W_T} f(x + \theta(t))e(t, x)\,\mathrm{d}\mu_T,$$

we can prove, by Lebesgue's convergence theorem, that $v(t, \cdot\,; f) \in C^\infty_{\exp}(\mathbb{R}^d)$. By Itô's formula,

$$v(t, x; f) = f(x) + \int_0^t v(s, x; -Hf)\,\mathrm{d}s.$$

Hence, we obtain

$$\frac{\partial v(t,x;f)}{\partial t} = v(t,x;-Hf) \tag{5.5.14}$$

and, by the Markov property of Brownian motions,

$$v(t,x;f) = v(s,x;v(t-s,\cdot\,;f)) \qquad (s \leqq t).$$

Differentiate both sides with respect to s. Then, since the mapping $f \mapsto v(s,x;f)$ is linear, by (5.5.14), we obtain

$$0 = v(s,x;-Hv(t-s,\cdot\,;f)) - v(s,x;v(t-s,\cdot\,;-Hf)).$$

Setting $s=0$, we see

$$-Hv(t,x;f) = v(t,x;-Hf).$$

Hence, by (5.5.14),

$$\frac{\partial v(t,x;f)}{\partial t} = -Hv(t,x;f). \tag{5.5.15}$$

By Theorem 5.4.11,

$$v(t,x;f) = \int_{\mathbb{R}^d} f(y)p(t,x,y)\,\mathrm{d}y \tag{5.5.16}$$

for any $f \in \mathscr{S}(\mathbb{R}^d)$. For $\boldsymbol{a} = (a_1,\dots,a_d) \in \mathbb{R}^d$, set $g_{\boldsymbol{a}}(x) = \cosh\bigl(\sum_{\alpha=1}^d a_\alpha x^\alpha\bigr)$. Moreover, take $\phi_n \in C_0^\infty(\mathbb{R}^d)$ such that $\phi_n(x) = 1$ for $|x| \leqq n$ and $\phi_n(x) = 0$ for $|x| > n+1$, and set $g_{\boldsymbol{a},n} = g_{\boldsymbol{a}}\phi_n$. Since $g_{\boldsymbol{a}} \in C_{\exp}^\infty(\mathbb{R}^d)$, by the monotone convergence theorem and (5.5.16),

$$\begin{aligned}\int_{\mathbb{R}^d} g_{\boldsymbol{a}}(y)p(t,x,y)\,\mathrm{d}y &= \lim_{n\to\infty}\int_{\mathbb{R}^d} g_{\boldsymbol{a},n}(y)p(t,x,y)\,\mathrm{d}y \\ &= \lim_{n\to\infty} v(t,x;g_{\boldsymbol{a},n}) = v(t,x;g_{\boldsymbol{a}}) < \infty.\end{aligned}$$

If $f \in C_{\exp}^\infty(\mathbb{R}^d)$, there exists an $\boldsymbol{a} = (a_1,\dots,a_d) \in \mathbb{R}^d$ such that $|f| \leqq g_{\boldsymbol{a}}$. Hence, by Lebesgue's convergence theorem and (5.5.16), we obtain

$$\begin{aligned}v(t,x;f) &= \lim_{n\to\infty} v(t,x;f\phi_n) \\ &= \lim_{n\to\infty}\int_{\mathbb{R}^d}(f\phi_n)(y)p(t,x,y)\,\mathrm{d}y = \int_{\mathbb{R}^d} f(y)p(t,x,y)\,\mathrm{d}y.\end{aligned}$$

Combining this with (5.5.15), we see that $p(t,x,y)$ is the fundamental solution for the heat equation (5.5.13). □

Remark 5.5.8 By Corollary 5.4.9, $p \in C^\infty((0,T] \times \mathbb{R}^d \times \mathbb{R}^d)$. Moreover, by Theorem 5.4.11(2), we have

$$p(t,x,y) = \mathbf{E}[e(t,x)|x+\theta(t)=y] \times \frac{1}{(2\pi t)^{\frac{d}{2}}} \mathrm{e}^{-\frac{|x-y|^2}{2t}}.$$

The expression in Theorem 5.5.7 above is essentially a conditional expectation.

The above result is naturally extended to solutions of general stochastic differential equations. Let $\{X(t,x)\}_{t\in[0,\infty)}$ be the solution of the stochastic differential equation (5.5.1). Assume that the functions $V, \Theta_1, \dots, \Theta_d \in C^\infty_{\nearrow}(\mathbb{R}^N)$ satisfy (5.5.12). Define the Schrödinger operator $\widetilde{H}$ by

$$\begin{aligned}\widetilde{H}f &= \frac{1}{2}\sum_{i,j=1}^N a^{ij}\frac{\partial^2 f}{\partial x^i \partial x^j} + \sum_{i=1}^N \Big(V_0^i + \mathrm{i}\sum_{j=1}^N a^{ij}\Theta_j\Big)\frac{\partial f}{\partial x^i} \\ &\quad + \Big\{\mathrm{i}\Big(\frac{1}{2}\sum_{i,j=1}^N a^{ij}\frac{\partial \Theta_i}{\partial x^j} + \sum_{i=1}^N V_0^i\Theta_i\Big) - V - \frac{1}{2}\sum_{i,j=1}^N a^{ij}\Theta_i\Theta_j\Big\}f,\end{aligned}$$

where $a^{ij} = \sum_{\alpha=1}^N V_\alpha^i V_\alpha^j$. Set

$$\widetilde{e}(t,x) = \exp\Big(\mathrm{i}\sum_{i=1}^N \int_0^t \Theta_i(X(s,x)) \circ \mathrm{d}X^i(s,x) - \int_0^t V(X(s,x))\,\mathrm{d}s\Big).$$

Then, $\widetilde{e}(t,x) \in \mathbb{D}^{\infty,\infty-}$ and the following holds as in the case of Brownian motions.

Theorem 5.5.9 *Suppose that Hörmander's condition holds at every $x \in \mathbb{R}^N$. Then the function $q(t,x,y) = \int_{W_T} \widetilde{e}(t,x)\delta_y(X(t,x))\,\mathrm{d}\mu_T$ is the fundamental solution of the heat equation*

$$\frac{\partial u}{\partial t} = \widetilde{H}u, \quad u(0,\cdot) = f \in C^\infty_{\nearrow}(\mathbb{R}^N)$$

associated with the Schrödinger operator $\widetilde{H}$. That is, the function $u(t,x) = \int_{\mathbb{R}^N} f(y)q(t,x,y)\,\mathrm{d}y$ is the solution of this heat equation.

Proof For $\beta = (\beta_1, \dots, \beta_N) \in \mathbb{Z}_+^N$, let ∂^β be the differential operator

$$\partial^\beta = \Big(\frac{\partial}{\partial x^1}\Big)^{\beta_1} \cdots \Big(\frac{\partial}{\partial x^N}\Big)^{\beta_N}.$$

Since the mapping

$$x \mapsto \int_{W_T} |\partial^\beta X(t,x)|^p\,\mathrm{d}\mu_T$$

is at most of polynomial growth for any $p > 1$, a repetition of the arguments in the proof of Theorem 5.5.7 yields the conclusion. □

Another application of the Malliavin calculus to a study of Greeks in mathematical finance will be discussed in the next chapter.

5.6 Change of Variables Formula

The integration by parts formula and the change of variables formula are fundamental in calculus. We have discussed the integration by parts formula on Wiener spaces and its applications. In this section we investigate a change of variables formula on a Wiener space.

Let E be a real separable Hilbert space. For $A \in E^{\otimes 2}$, we define the regularized determinant $\det_2(I + A)$ of $I + A$ so that

$$\det_2(I + A) = \det(I + A)\mathrm{e}^{-\mathrm{tr}\, A}$$

if A is of trace class, where I is the identity mapping of E. For details, see [19, XI.9] and [107, Chapter 9].

With the eigenvalues $\{\lambda_j\}_{j=1}^{\infty}$ of A, repeated according to multiplicity, the regularized determinant is written as

$$\det_2(I + A) = \prod_{j=1}^{\infty}(1 + \lambda_j)\mathrm{e}^{-\lambda_j}. \tag{5.6.1}$$

The following change of variables formula holds on W_T.

Theorem 5.6.1 *Let $F \in \mathbb{D}^{\infty,\infty-}(H_T)$. Suppose that there exists a $q > \frac{1}{2}$ such that*

$$\mathrm{e}^{-\nabla^* F + q\|\nabla F\|^2_{H_T^{\otimes 2}}} \in L^{1+}(\mu_T) = \bigcup_{p\in(1,\infty)} L^p(\mu_T). \tag{5.6.2}$$

Then, for any $f \in C_{\mathrm{b}}(W_T)$,

$$\int_{W_T} f(\iota + F)\det_2(I + \nabla F)\mathrm{e}^{-\nabla^* F - \frac{1}{2}\|F\|^2_{H_T}}\mathrm{d}\mu_T = \int_{W_T} f\,\mathrm{d}\mu_T, \tag{5.6.3}$$

where $\iota(w) = w$ ($w \in W_T$).

The left hand side is well defined because

$$\left|\det_2(I + A)\right| \leqq \exp\Bigl(\frac{1}{2}\|A\|^2_{E^{\otimes 2}}\Bigr). \tag{5.6.4}$$

This estimate is obtained by combining (5.6.1) with the inequality

$$\left|(1+x)\mathrm{e}^{-x}\right|^2 \leqq \mathrm{e}^{x^2} \qquad (x \in \mathbb{R}).$$

Remark 5.6.2 (1) By (5.6.3), if $\det_2(I+\nabla F) \geqq 0$, μ_T-a.s., the measure on W_T with density

$$\det{}_2(I+\nabla F)\mathrm{e}^{-\nabla^* F - \frac{1}{2}\|F\|_{H_T}^2}$$

with respect to μ_T is a probability measure and the distribution of the W_T-valued function $\iota + F$ under this probability measure is the Wiener measure. In this case, (5.6.3) also holds for any bounded measurable $f : W_T \to \mathbb{R}$.
(2) Suppose that $G : W_T \to H_T$ is continuous. If there exists an $F \in \mathbb{D}^{\infty,\infty-}(H_T)$ such that the conditions of Theorem 5.6.1 are fulfilled and $(\iota+G)\circ(\iota+F) = \iota$, then

$$\int_{W_T} f(\iota+G)\,\mathrm{d}\mu_T = \int_{W_T} f \det{}_2(I+\nabla F)\mathrm{e}^{-\nabla^* F - \frac{1}{2}\|F\|_{H_T}^2}\,\mathrm{d}\mu_T$$

for any $f \in C_b(W_T)$, that is, the distribution of $\iota + G$ under μ_T coincides with the probability measure $\widehat{\mu_T}$ given by

$$\widehat{\mu_T}(A) = \int_A \det{}_2(I+\nabla F)\mathrm{e}^{-\nabla^* F - \frac{1}{2}\|F\|_{H_T}^2}\,\mathrm{d}\mu_T \qquad (A \in \mathscr{B}(W_T)).$$

(3) The Cameron–Martin theorem (Theorem 1.7.2) is a special case of this theorem. In fact, if F is an H_T-valued constant function, say $F = h$ ($h \in H_T$), then $\nabla F = 0$ and $\|F\|_{H_T} = \|h\|_{H_T}$. Moreover, by Example 5.1.5, $\nabla^* F = \mathscr{I}(h)$. Hence, by Theorem 5.6.1, we have

$$\int_{W_T} f(w+h)\mathrm{e}^{-\mathscr{I}(h) - \frac{1}{2}\|h\|_{H_T}^2}\mu_T(\mathrm{d}w) = \int_{W_T} f\,\mathrm{d}\mu_T$$

for any $f \in C_b(W_T)$.[3]
(4) Girsanov's theorem (Theorem 4.6.2) is also derived from Theorem 5.6.1. To show this, let $\{u(t) = (u_1(t), \ldots, u_d(t))\}_{t\in[0,T]}$ be an $\{\mathscr{F}_t\}$-predictable and bounded $\mathbb{R}^d$-valued stochastic process. As in Theorem 5.3.3, we define $\Phi_u : W_T \to H_T$ by $\dot{\Phi}_u(t) = u(t)$ ($t \in [0,T]$). Assume that $\Phi_u \in \mathbb{D}^{\infty,\infty-}(H_T)$. Since, by Theorem 5.3.3,

$$\nabla^*\Phi_u = \sum_{\alpha=1}^d \int_0^T u_\alpha(t)\,\mathrm{d}\theta^\alpha(t),$$

[3] Research on the change of variables formula on W_T started from a series of studies by Cameron and Martin in the 1940s, including this Cameron–Martin theorem.

we can rewrite Girsanov's theorem as

$$\int_{W_T} f(\iota + \Phi_u)e^{-\nabla^*\Phi_u - \frac{1}{2}\|\Phi_u\|_{H_T}^2}\,\mathrm{d}\mu_T = \int_{W_T} f\,\mathrm{d}\mu_T. \tag{5.6.5}$$

We take $\mathbb{R}^d$-valued stochastic processes $\{u_n(t) = (u_n^1(t), \ldots, u_n^d(t))\}_{t\in[0,T]}$ with components $\{u_n^\alpha(t)\}_{t\in[0,T]} \in \mathscr{L}^0$ $(\alpha = 1, \ldots, d)$ (see Definition 2.2.4) such that

$$\lim_{n\to\infty} \int_{W_T} \Big(\int_0^T |u_n(t) - u(t)|^2\mathrm{d}t\Big)\mathrm{d}\mu_T = 0 \quad \text{and} \quad M := \sup_{n\in\mathbb{N}, w\in W_T} |u_n(t, w)| < \infty.$$

Moreover, we may assume that each $u_n^\alpha(t)$ is written as

$$u_n^\alpha(t) = \xi_{n,k}^\alpha, \quad t_k^n < t \leqq t_{k+1}^n,\ k = 0, 1, \ldots, m_n - 1 \qquad (\alpha = 1, \ldots, d),$$

where the random variables $\xi_{n,k}^\alpha$ are given by

$$\xi_{n,k}^\alpha = \phi_{n,k}^\alpha(\theta(s_1^{k,n}), \ldots, \theta(s_{j_{k,n}}^{k,n}))$$

for a monotone increasing sequence $0 = t_0^n < t_1^n < \cdots < t_k^n < \cdots < t_{m_n}^n = T$ and $0 < s_1^{k,n} < \cdots < s_{j_{k,n}}^{k,n} \leqq t_k^n$ and $\phi_{n,k}^\alpha \in C_{\mathrm{b}}^\infty((\mathbb{R}^d)^{j_{k,n}})$ (see the proof of Theorem 5.3.3). Then, we have

$$\Phi_{u_n} = \sum_{k=0}^{m_n-1}\sum_{\alpha=1}^{d} \xi_{n,k}^\alpha \ell_{(t_k^n, t_{k+1}^n]}^\alpha,$$

where $e_\alpha = (\overbrace{0, \ldots, 0}^{\alpha-1}, 1, 0, \ldots, 0) \in \mathbb{R}^d$ and $\ell_{(s,t]}^\alpha \in H_T$ is defined by $\dot{\ell}_{(s,t]}^\alpha(v) = \mathbf{1}_{(s,t]}(v)e_\alpha$ $(v \in [0, T])$. By Corollary 5.3.2,

$$\nabla\Phi_{u_n} = \sum_{k=0}^{m_n-1}\sum_{i=1}^{j_{k,n}}\sum_{\alpha,\beta=1}^{d} \frac{\partial\phi_{n,k}^\alpha}{\partial x_i^\beta}(\theta(s_1^{k,n}), \ldots, \theta(s_{j_{k,n}}^{k,n}))\ell_{(0,s_i^{k,n}]}^\beta \otimes \ell_{(t_k^n, t_{k+1}^n]}^\alpha,$$

where the coordinate of $(\mathbb{R}^d)^{j_{k,n}}$ is $(x_1^1, \ldots, x_1^d, \ldots, x_{j_{k,n}}^1, \ldots, x_{j_{k,n}}^d)$. Hence, there exist an $N \in \mathbb{N}$, an orthonormal system $g_1, \ldots, g_N$ of H_T, and random variables a_{ij} $(i, j = 1, \ldots, N)$ such that

$$\nabla\Phi_{u_n} = \sum_{1\leqq i<j\leqq N} a_{ij}g_i \otimes g_j.$$

Since all the eigenvalues of upper triangular matrices are zero, $\det_2(I+\nabla\Phi_{u_n}) = 1$. Hence, by Theorem 5.6.1, (5.6.5) holds for $u = u_n$.

Since $\{e^{-2\sum_{\alpha=1}^d \int_0^t u_{n,\alpha}(s)\mathrm{d}\theta^\alpha(s) - 2\int_0^t |u_n(s)|^2\mathrm{d}s}\}_{t\in[0,T]}$ is a martingale,

$$\int_{W_T} e^{2\{-\nabla^*\Phi_n - \frac{1}{2}\|\Phi_{u_n}\|_{H_T}^2\}}\mathrm{d}\mu_T \leqq e^{M^2T}$$

and $\mathrm{e}^{-\nabla^*\Phi_n-\frac{1}{2}\|\Phi_{u_n}\|_{H_T}^2}$ $(n \in \mathbb{N})$ is uniformly integrable. Hence, setting $u = u_n$ in (5.6.5) and letting $n \to \infty$, we obtain (5.6.5) for a bounded stochastic process $\{u(t)\}_{t\in[0,T]}$.
(5) The change of variables formula via the regularized determinant $\det_2$ and the derivative ∇ on Wiener spaces as in the theorem was first studied by Kusuoka [66] and his result was applied to the degree theorem on Wiener spaces in [30]. The proof of the theorem below is based on the arguments in Üstünel and Zakai [120].

For a proof of Theorem 5.6.1, we show a change of variables formula for the integrals with respect to Gaussian measures on $\mathbb{R}^n$. Denote the inner product in $\mathbb{R}^n$ by $\langle\cdot,\cdot\rangle$ and the norm on $\mathbb{R}^n$ by $\|\cdot\|$, where the norm was denoted by $|\cdot|$ in the previous chapters. This change is to make notation analogous to that for H_T. The gradient operator on $\mathbb{R}^n$ is also denoted by ∇, that is, $\nabla f = (\frac{\partial f}{\partial x^1},\dots,\frac{\partial f}{\partial x^n})$. Let ν_n be the probability measure on $\mathbb{R}^n$ defined by

$$\nu_n(\mathrm{d}x) = (2\pi)^{-\frac{n}{2}}\mathrm{e}^{-\frac{\|x\|^2}{2}}\mathrm{d}x \tag{5.6.6}$$

and ∇^* be the formal adjoint operator of ∇ with respect to ν_n,

$$\nabla^* F(x) = \langle x, F(x)\rangle - \mathrm{tr}(\nabla F(x)) \qquad (F \in C^\infty(\mathbb{R}^n;\mathbb{R}^n)),$$

where, for $F = (F^1,\dots,F^n) \in C^\infty(\mathbb{R}^n;\mathbb{R}^n)$,

$$\nabla F = \Bigl(\frac{\partial F^i}{\partial x^j}\Bigr)_{i,j=1,\dots,n}.$$

Moreover, the space of $\mathbb{R}^k$-valued C^∞ functions F on $\mathbb{R}^n$ such that $\nabla^j F \in L^{\infty-}(\nu_n;\mathbb{R}^{kn^j})$ for any $j \in \mathbb{Z}_+$ is denoted by $\mathscr{D}^{\infty,\infty-}(\mathbb{R}^n;\mathbb{R}^k)$. When $k = 1$, we simply write $\mathscr{D}^{\infty,\infty-}(\mathbb{R}^n)$.

For $F \in \mathscr{D}^{\infty,\infty-}(\mathbb{R}^n;\mathbb{R}^n)$, set

$$\Lambda_F = \det{}_2(I+\nabla F)\mathrm{e}^{-\nabla^* F-\frac{1}{2}\|F\|^2} = \det(I+\nabla F)\mathrm{e}^{-\langle F,\cdot\rangle-\frac{1}{2}\|F\|^2}.$$

For a matrix $A \in \mathbb{R}^n\otimes\mathbb{R}^n$, let $A^\sim$ be its cofactor matrix, that is, letting $\widehat{a}_{ij}$ be the (i,j)-cofactor of A, $A^\sim = (\widehat{a}_{ji})_{i,j=1,\dots,n}$. The product of A and $A^\sim$ is

$$A\,(A^\sim) = \det A \times I.$$

By this identity we can define $\Lambda_F(I+\nabla F)^{-1}(x) \in \mathbb{R}^n\otimes\mathbb{R}^n$ by

$$\Lambda_F(I+\nabla F)^{-1}(x) = \mathrm{e}^{-\langle x,F(x)\rangle-\frac{1}{2}\|F(x)\|^2}(I+\nabla F(x))^\sim \tag{5.6.7}$$

regardless of the regularity of the matrix $I + \nabla F(x) \in \mathbb{R}^n\otimes\mathbb{R}^n$.

Lemma 5.6.3 *For $x, v \in \mathbb{R}^n$,*

$$\nabla^*(\Lambda_F(I+\nabla F)^{-1}v)(x) = \Lambda_F(x)\langle v, x+F(x)\rangle.$$

Proof Let $v = (v_1, \ldots, v_n) \in \mathbb{R}^n$. By (5.6.7) and the definition of ∇^*,

$$\nabla^*(\Lambda_F(I+\nabla F)^{-1}v) = \Lambda_F\langle v, \cdot + F\rangle - \mathrm{e}^{-\langle F,\cdot\rangle - \frac{1}{2}\|F\|^2} \sum_{i,j=1}^n \frac{\partial}{\partial x^i}(I+\nabla F)^{\sim}_{ij} v_j.$$

Hence it suffices to show

$$\sum_{i,j=1}^n \frac{\partial}{\partial x^i}(I+\nabla F)^{\sim}_{ij} v_j = 0. \tag{5.6.8}$$

For $x \in \mathbb{R}^n$ and $\zeta \in \mathbb{C}$, set

$$f(x,\zeta) = \sum_{i,j=1}^n \frac{\partial}{\partial x^i}(I+\zeta\nabla F)^{\sim}_{ij} v_j.$$

Suppose that $x \in \mathbb{R}^n$ and $\zeta \in \mathbb{C}$ satisfy $\det(I+\zeta\nabla F(x)) \neq 0$. Since $\det(I + \zeta\nabla F(\cdot)) \neq 0$ in a neighborhood of x,

$$(I+\zeta\nabla F)^{\sim}_{ij} = \det(I+\zeta\nabla F)((I+\zeta\nabla F)^{-1})_{ij}.$$

Since $\frac{\partial}{\partial a_{pq}} \det A = \det A\,(A^{-1})_{qp}$ for $A = (a_{ij})_{i,j=1,\ldots,n}$, a straightforward computation yields

$$\sum_{i=1}^n \frac{\partial}{\partial x^i}(I+\zeta\nabla F)^{\sim}_{ij} = 0.$$

Hence, if $\det(I+\zeta\nabla F(x)) \neq 0$, then $f(x,\zeta) = 0$. For each $x \in \mathbb{R}^n$ there are at most n ζs such that $\det(I+\zeta\nabla F(x)) = 0$. Therefore we obtain $f(x,\zeta) \equiv 0$ and (5.6.8). □

Lemma 5.6.4 *Let $F \in \mathscr{D}^{\infty,\infty-}(\mathbb{R}^n;\mathbb{R}^n)$. Suppose that there exist $\gamma > 0$ and $q > \frac{1}{2}$ such that*

$$\mathrm{e}^{-\nabla^* F + q\|\nabla F\|^2} \in L^{1+\gamma}(\nu_n). \tag{5.6.9}$$

Then, for any $v \in \mathbb{R}^n$,

$$\Lambda_F(I+\nabla F)^{-1}v \in L^{1+\gamma}(\nu_n), \quad \Lambda_F\langle \cdot + F, v\rangle \in \bigcap_{p\in(1,1+\gamma)} L^p(\nu_n).$$

Moreover, for any $G \in \mathscr{D}^{\infty,\infty-}(\mathbb{R}^n)$,

$$\int_{\mathbb{R}^n} \langle \nabla G, \Lambda_F(I+\nabla F)^{-1}v\rangle\,\mathrm{d}\nu_n = \int_{\mathbb{R}^n} G\Lambda_F\langle \cdot + F, v\rangle\,\mathrm{d}\nu_n.$$

Proof By Lemma 5.6.3, it suffices to show the integrability of the first two Wiener functionals.

First we show, for $A \in \mathbb{R}^n \otimes \mathbb{R}^n$,

$$\left\|\det_2(I+A)\{(I+A)^{-1}-I\}\right\| \leqq \exp\Big(\frac{1}{2}(\|A\|+1)^2\Big). \tag{5.6.10}$$

Since $\zeta \mapsto \det_2(I+A+\zeta B)$ ($B \in \mathbb{R}^{n\times n}$) is holomorphic and

$$\frac{\mathrm{d}}{\mathrm{d}\zeta}\Big|_{\zeta=0}\det_2(I+A+\zeta B) = \det_2(I+A)\mathrm{tr}[\{(I+A)^{-1}-I\}B],$$

by Cauchy's integral formula and (5.6.4), we obtain

$$\begin{aligned}
\left|\det_2(I+A)\mathrm{tr}[\{(I+A)^{-1}-I\}B]\right| &= \Big|\frac{1}{2\pi}\int_0^{2\pi}\frac{\det_2(I+A+\mathrm{e}^{\mathrm{i}s}B)}{\mathrm{e}^{\mathrm{i}s}}\,\mathrm{d}s\Big| \\
\leqq \frac{1}{2\pi}\int_0^{2\pi}\exp\Big(\frac{1}{2}\|A+\mathrm{e}^{\mathrm{i}s}B\|^2\Big)\mathrm{d}s &\leqq \exp\Big(\frac{1}{2}(\|A\|+\|B\|)^2\Big).
\end{aligned}$$

Hence, since $\|T\| = \sup\{|\mathrm{tr}(TB)| \mid \|B\| \leqq 1\}$, we have (5.6.10).

Second, by using (5.6.4), (5.6.10), and an elementary inequality $\frac{1}{2}(a+1)^2 \leqq qa^2 + \frac{q}{2q-1}$ $(a > 0)$, we obtain

$$\begin{aligned}
\|\Lambda_F(I+\nabla F)^{-1}v\| &\leqq 2\mathrm{e}^{-\nabla^*F-\frac{1}{2}\|F\|^2+q\|\nabla F\|^2+\frac{q}{2q-1}}\|v\| \\
&\leqq 2\mathrm{e}^{-\nabla^*F+q\|\nabla F\|^2+\frac{q}{2q-1}}\|v\|.
\end{aligned}$$

This implies the first assertion.

Since $\langle\cdot+F, v\rangle \in L^{\infty-}(\nu_n)$, the second assertion follows from (5.6.4) and the assumption. □

Lemma 5.6.5 *If* $F \in \mathscr{D}^{\infty,\infty-}(\mathbb{R}^n;\mathbb{R}^n)$ *satisfies* (5.6.9), *then*

$$\int_{\mathbb{R}^n} f(x+F(x))\Lambda_F(x)\nu_n(\mathrm{d}x) = \int_{\mathbb{R}^n}\Lambda_F\,\mathrm{d}\nu_n\int_{\mathbb{R}^n} f\,\mathrm{d}\nu_n \tag{5.6.11}$$

for any $f \in C_b(\mathbb{R}^n)$.

Proof For $v \in \mathbb{R}^n$ and $\lambda \in \mathbb{R}$, set $f_\lambda(x) = \exp(\mathrm{i}\,\lambda\langle x, v\rangle)$. Since

$$\langle\nabla(f_\lambda(\cdot+F)), k\rangle = \mathrm{i}\,\lambda\langle v, (I+\nabla F)k\rangle f_\lambda(\cdot+F) \qquad (k \in \mathbb{R}^n),$$

setting $k = \Lambda_F(I+\nabla F)^{-1}v$, we have

$$\langle\nabla(f_\lambda(\cdot+F)), \Lambda_F(I+\nabla F)^{-1}v\rangle = \mathrm{i}\,\lambda\|v\|^2 f_\lambda(\cdot+F)\Lambda_F.$$

Combining this identity with Lemma 5.6.3, we obtain

$$\frac{1}{\mathrm{i}}\frac{\mathrm{d}}{\mathrm{d}\lambda}\int_{\mathbb{R}^n} f_\lambda(x+F(x))\Lambda_F(x)\nu_n(\mathrm{d}x)$$
$$=\int_{\mathbb{R}^n} f_\lambda(x+F(x))\Lambda_F(x)\langle x+F(x),v\rangle\nu_n(\mathrm{d}x)$$
$$=\mathrm{i}\,\lambda\|v\|^2\int_{\mathbb{R}^n} f_\lambda(x+F(x))\Lambda_F(x)\nu_n(\mathrm{d}x).$$

Solving this ordinary differential equation, we arrive at

$$\int_{\mathbb{R}^n} f_\lambda(x+F(x))\Lambda_F(x)\nu_n(\mathrm{d}x)=\mathrm{e}^{-\frac{1}{2}\lambda^2\|v\|^2}\int_{\mathbb{R}^n}\Lambda_F(x)\nu_n(\mathrm{d}x).$$

Since $\mathrm{e}^{-\frac{1}{2}\lambda^2\|v\|^2}=\int_{\mathbb{R}^n} f_\lambda\,\mathrm{d}\nu_n$, we obtain (5.6.11).

For general $f\in C_\mathrm{b}(\mathbb{R}^n)$, approximating it by elements in $\mathscr{S}(\mathbb{R}^n)$ and expressing elements of $\mathscr{S}(\mathbb{R}^n)$ in terms of Fourier transforms, we obtain the assertion from the identity above. □

Lemma 5.6.6 *Suppose that $F\in\mathscr{D}^{\infty,\infty-}(\mathbb{R}^n;\mathbb{R}^n)$ satisfies* (5.6.9). *Then,*

$$\int_{\mathbb{R}^n} f(x+F(x))\mathrm{det}_2(I+\nabla F(x))\mathrm{e}^{-\nabla^*F(x)-\frac{1}{2}\|F(x)\|^2}\nu_n(\mathrm{d}x)=\int_{\mathbb{R}^n} f\,\mathrm{d}\nu_n \quad (5.6.12)$$

for any $f\in C_\mathrm{b}(\mathbb{R}^n)$.

Proof Let $t\in[0,1]$. Since $a^t\leqq 1+a$ $(a\geqq 0)$,

$$\mathrm{e}^{-\nabla^*(tF)+q\|\nabla(tF)\|^2}\leqq \mathrm{e}^{t(-\nabla^*F+q\|\nabla F\|^2)}\leqq 1+\mathrm{e}^{-\nabla^*F+q\|\nabla F\|^2}.$$

Hence tF also satisfies (5.6.9) and the mapping $t\mapsto\int_{\mathbb{R}^n}\Lambda_{tF}\mathrm{d}\nu_n$ is continuous. If $\int_{\mathbb{R}^n}\Lambda_{tF}\mathrm{d}\nu_n\in\mathbb{Z}$ $(t\in[0,1])$, then $\int_{\mathbb{R}^n}\Lambda_{tF}\mathrm{d}\nu_n=\int_{\mathbb{R}^n}\Lambda_0\mathrm{d}\nu_n=1$ and we obtain (5.6.12) by Lemma 5.6.5. Hence we show $\int_{\mathbb{R}^n}\Lambda_F\mathrm{d}\nu_n\in\mathbb{Z}$.

Let $\{E_k\}_{k=1}^\infty$ be a sequence of disjoint Borel sets such that

$$\bigcup_{k=1}^\infty E_k=\{x\in\mathbb{R}^n\,;\,\det(I+\nabla F(x))\neq 0\}$$

and, in a neighborhood of each E_k, the mapping $x\mapsto T(x)=x+F(x)$ is a diffeomorphism. By the change of variables formula with respect to the Lebesgue measure, we have

$$\int_{E_k} f(T(x))|\Lambda_F(x)|\nu_n(\mathrm{d}x)$$
$$=\int_{E_k} f(T(x))|\det(\nabla T(x))|(2\pi)^{-\frac{n}{2}}\mathrm{e}^{-\frac{1}{2}\|T(x)\|^2}\mathrm{d}x$$
$$=\int_{T(E_k)} f(x)\nu_n(\mathrm{d}x)\qquad (k=1,2,\dots,\ f\in C_\mathrm{b}(\mathbb{R}^n)). \quad (5.6.13)$$

Since $\Lambda_F(x) = 0$ if $\det(I + \nabla F(x)) = 0$, this implies

$$\int_{\mathbb{R}^n} f(T)|\Lambda_F|\, \mathrm{d}\nu_n = \int_{\{\det(I+\nabla F)\neq 0\}} f(T)|\Lambda_F|\, \mathrm{d}\nu_n = \sum_{k=1}^{\infty} \int_{T(E_k)} f\, \mathrm{d}\nu_n$$

for any $f \in C_{\mathrm{b}}(\mathbb{R}^n)$. Setting $f = 1$, we obtain

$$\int_{\mathbb{R}^n} \Big(\sum_{k=1}^{\infty} \mathbf{1}_{T(E_k)}\Big) \mathrm{d}\nu_n < \infty. \tag{5.6.14}$$

Denote by $s_k \in \{\pm 1\}$ the signature of $\det(I + \nabla F)$ on E_k. By Lemma 5.6.5 and (5.6.13), we have

$$\int_{\mathbb{R}^n} \Lambda_F \mathrm{d}\nu_n \int_{\mathbb{R}^n} f\, \mathrm{d}\nu_n = \int_{\mathbb{R}^n} f(\cdot + F)\Lambda_F \mathrm{d}\nu_n = \sum_{k=1}^{\infty} s_k \int_{T(E_k)} f\, \mathrm{d}\nu_n$$

for any $f \in C_{\mathrm{b}}(\mathbb{R}^n)$. The sum $\sum_{k=1}^{\infty} s_k \mathbf{1}_{T(E_k)}$ is dominated by $\sum_{k=1}^{\infty} \mathbf{1}_{T(E_k)}$ and, by (5.6.14), converges absolutely ν_n-a.e. Hence we have

$$\int_{\mathbb{R}^n} \Lambda_F \mathrm{d}\nu_n \int_{\mathbb{R}^n} f\, \mathrm{d}\nu_n = \int_{\mathbb{R}^n} \Big(\sum_{k=1}^{\infty} s_k \mathbf{1}_{T(E_k)}(x)\Big) f(x)\, \nu_n(\mathrm{d}x)$$

for any $f \in C_{\mathrm{b}}(\mathbb{R}^n)$ and

$$\int_{\mathbb{R}^n} \Lambda_F \mathrm{d}\nu_n = \sum_{k=1}^{\infty} s_k \mathbf{1}_{T(E_k)}, \quad \nu_n\text{-a.e.}$$

In particular, $\int_{\mathbb{R}^n} \Lambda_F \mathrm{d}\nu_n \in \mathbb{Z}$. □

We extend the identity (5.6.12) on $\mathbb{R}^n$ to that on W_T. For this purpose we prepare some notation. Let $\{\ell_i\}_{i=1}^{\infty} \subset W_T^*$ be an orthonormal basis of H_T. For each $n \in \mathbb{N}$, let $\mathscr{G}_n$ be the σ-field generated by the random variables $\ell_1, \ldots, \ell_n$: $\mathscr{G}_n = \sigma(\ell_1, \ldots, \ell_n)$. Define the projection $\pi_n : W_T \to W_T^* \subset H_T \subset W_T$ by

$$\pi_n w = \sum_{j=1}^{n} \ell_j(w)\ell_j \qquad (w \in W_T).$$

Moreover, for $j \in \mathbb{N}$, define $\pi_n^{\otimes j} : H_T^{\otimes j} \to H_T^{\otimes j}$ by

$$\pi_n^{\otimes j}(h_1 \otimes \cdots \otimes h_j) = \pi_n h_1 \otimes \cdots \otimes \pi_n h_j.$$

Denote by $\mathbf{E}_n$ the conditional expectation given $\mathscr{G}_n$, $\mathbf{E}_n(F) = \mathbf{E}[F|\mathscr{G}_n]$, and extend it to the $H_T^{\otimes j}$-valued random variable $G \in L^2(\mu_T; H_T^{\otimes j})$ by

$$\mathbf{E}_n(G) = \sum_{i=1}^{\infty} \mathbf{E}_n(\langle G, \psi_i \rangle_{H_T^{\otimes j}})\psi_i, \tag{5.6.15}$$

where $\{\psi_i\}_{i=1}^{\infty}$ is an orthonormal basis of $H_T^{\otimes j}$. The above infinite sum converges in $H_T^{\otimes j}$ almost surely and in the L^2-sense. Specifically, since $\{\mathbf{E}_n(G)\}_{n=1}^{\infty}$ is a discrete time martingale, by the monotone convergence theorem, Doob's inequality, and Jensen's inequality, we have

$$\begin{aligned}
\int_{W_T} \sup_{n\in\mathbb{N}}(\mathbf{E}_n(\langle G,\psi_i\rangle_{H_T^{\otimes j}}))^2 d\mu_T &= \lim_{m\to\infty}\int_{W_T}\max_{n\leqq m}(\mathbf{E}_n(\langle G,\psi_i\rangle_{H_T^{\otimes j}}))^2 d\mu_T \\
&\leqq 4\limsup_{m\to\infty}\int_{W_T}(\mathbf{E}_m(\langle G,\psi_i\rangle_{H_T^{\otimes j}}))^2 d\mu_T \\
&\leqq 4\limsup_{m\to\infty}\int_{W_T}\mathbf{E}_m(\langle G,\psi_i\rangle_{H_T^{\otimes j}}^2)\, d\mu_T \\
&= 4\int_{W_T}\langle G,\psi_i\rangle_{H_T^{\otimes j}}^2 d\mu_T \qquad (i\in\mathbb{N}).
\end{aligned}$$

Hence we obtain

$$\int_{W_T}\sum_{i=1}^{\infty}\Big(\sup_{n\in\mathbb{N}}\big(\mathbf{E}_n(\langle G,\psi_i\rangle_{H_T^{\otimes j}})\big)^2\Big)d\mu_T \leqq 4\int_{W_T}\|G\|_{H_T^{\otimes j}}^2 d\mu_T. \tag{5.6.16}$$

The almost sure and L^2-convergence of the right hand side of (5.6.15) follows from this estimate.

Lemma 5.6.7 (1) *$\mathbf{E}_n(G)$ is an $H_T^{\otimes j}$-valued random variable, unique up to μ_T-null sets, such that*

$$\int_{W_T}\langle G,G'\rangle_{H_T^{\otimes j}} d\mu_T = \int_{W_T}\langle \mathbf{E}_n(G),G'\rangle_{H_T^{\otimes j}} d\mu_T$$

for any $\mathscr{G}_n$-measurable $G'\in L^2(\mu_T;H_T^{\otimes j})$. In particular, $\mathbf{E}_n(G)$ is independent of the choice of the orthonormal basis $\{\psi_i\}_{i=1}^{\infty}$.
(2) *For any $G,K\in L^2(\mu_T;H_T^{\otimes j})$,*

$$\begin{aligned}
\int_{W_T}\langle \mathbf{E}_n(G),K\rangle_{H_T^{\otimes j}} d\mu_T &= \int_{W_T}\langle \mathbf{E}_n(G),\mathbf{E}_n(K)\rangle_{H_T^{\otimes j}} d\mu_T \\
&= \int_{W_T}\langle G,\mathbf{E}_n(K)\rangle_{H_T^{\otimes j}} d\mu_T.
\end{aligned} \tag{5.6.17}$$

Moreover, for $G\in L^p(\mu_T;H_T^{\otimes j})$, $p\geqq 1$,

$$\|\mathbf{E}_n(G)\|_{H_T^{\otimes j}}^p \leqq \mathbf{E}_n(\|G\|_{H_T^{\otimes j}}^p). \tag{5.6.18}$$

(3) *The following convergence holds:*

$$\lim_{n\to\infty}\int_{W_T}\|\mathbf{E}_n(\pi_n G)-G\|_{H_T^{\otimes j}}^2 d\mu_T = 0.$$

(4) *Let* $G \in \mathscr{P}$. *Then, for* μ_T*-a.s.* $w \in W_T$,

$$\mathbf{E}_n(G)(w) = \int_{W_T} G(\pi_n w + (1-\pi_n)w')\mu_T(\mathrm{d}w'). \tag{5.6.19}$$

Proof (1) Let $G' \in L^2(\mu_T; H_T^{\otimes j})$ be $\mathscr{G}_n$-measurable. By the expansion with respect to $\{\psi_i\}_{i=1}^\infty$,

$$\langle \mathbf{E}_n(G), G'\rangle_{H_T^{\otimes j}} = \sum_{i=1}^\infty \mathbf{E}_n(\langle G, \psi_i\rangle_{H_T^{\otimes j}})\langle G', \psi_i\rangle_{H_T^{\otimes j}}.$$

By the identities

$$\sum_{i=1}^\infty \langle G', \psi_i\rangle^2_{H_T^{\otimes j}} = \|G'\|^2_{H_T^{\otimes j}}, \quad \sum_{i=1}^\infty \langle G, \psi_i\rangle^2_{H_T^{\otimes j}} = \|G\|^2_{H_T^{\otimes j}}$$

and (5.6.16), we can apply Lebesgue's convergence theorem to obtain the desired equality

$$\begin{aligned}\int_{W_T} \langle \mathbf{E}_n(G), G'\rangle_{H_T^{\otimes j}} \mathrm{d}\mu_T &= \sum_{i=1}^\infty \int_{W_T} \mathbf{E}_n(\langle G, \psi_i\rangle_{H_T^{\otimes j}})\langle G', \psi_i\rangle_{H_T^{\otimes j}} \mathrm{d}\mu_T \\ &= \sum_{i=1}^\infty \int_{W_T} \langle G, \psi_i\rangle_{H_T^{\otimes j}}\langle G', \psi_i\rangle_{H_T^{\otimes j}} \mathrm{d}\mu_T \\ &= \int_{W_T} \langle G, G'\rangle_{H_T^{\otimes j}} \mathrm{d}\mu_T.\end{aligned}$$

because $\langle G', \psi_i\rangle_{H_T^{\otimes j}}$ is $\mathscr{G}_n$-measurable.

The uniqueness is shown in the same way as the usual conditional expectation.

(2) Set $G = K$ and $G' = \mathbf{E}_n(G)$ in (1) to obtain the first identity of (5.6.17). The second identity is obtained by changing G and K in the first one.

Next we show (5.6.18). It suffices to prove it in the case when $p = 1$ because the general case is obtained by Jensen's inequality $(\mathbf{E}_n(X))^p \leqq \mathbf{E}_n(X^p)$ for conditional expectations.

Let $g \in H_T^{\otimes j}$. For any $\mathscr{G}_n$-measurable $\phi \in L^2(\mu_T)$, set $G' = \phi \cdot g$ in the identity described in (1). Then we have

$$\int_{W_T} \langle \mathbf{E}_n(G), g\rangle_{H_T^{\otimes j}} \phi \, \mathrm{d}\mu_T = \int_{W_T} \langle G, g\rangle_{H_T^{\otimes j}} \phi \, \mathrm{d}\mu_T.$$

Hence

$$\langle \mathbf{E}_n(G), g\rangle_{H_T^{\otimes j}} = \mathbf{E}_n(\langle G, g\rangle_{H_T^{\otimes j}}).$$

In particular, since $|\langle G, g\rangle_{H_T^{\otimes j}}| \leqq \|G\|_{H_T^{\otimes j}}\|g\|_{H_T^{\otimes j}}$, we obtain

$$|\langle \mathbf{E}_n(G), g\rangle_{H_T^{\otimes j}}| \leqq \mathbf{E}_n(\|G\|_{H_T^{\otimes j}})\|g\|_{H_T^{\otimes j}}. \tag{5.6.20}$$

Since the space $H_T^{\otimes j}$ is separable, there exists a countable sequence $\{g_i\}_{i=1}^{\infty}$ with $\|g_i\|_{H_T^{\otimes j}} \leqq 1$ such that

$$\|\xi\|_{H_T^{\otimes j}} = \sup_{i\in\mathbb{N}} |\langle \xi, g_i\rangle_{H_T^{\otimes j}}| \qquad (\xi \in H_T^{\otimes j}).$$

Combining this with (5.6.20), we obtain (5.6.18) when $p = 1$.
(3) By the linearity of $\mathbf{E}_n$,

$$\int_{W_T} \left\|\mathbf{E}_n(\pi_n G) - G\right\|_{H_T^{\otimes j}}^2 d\mu_T$$
$$\leqq 2\int_{W_T} \left\|\mathbf{E}_n(\pi_n G - G)\right\|_{H_T^{\otimes j}}^2 d\mu_T + 2\int_{W_T} \left\|\mathbf{E}_n(G) - G\right\|_{H_T^{\otimes j}}^2 d\mu_T.$$

Using (5.6.18) for $p = 2$, we obtain

$$\int_{W_T} \left\|\mathbf{E}_n(\pi_n G - G)\right\|_{H_T^{\otimes j}}^2 d\mu_T \leqq \int_{W_T} \left\|\pi_n G - G\right\|_{H_T^{\otimes j}}^2 d\mu_T \to 0 \quad (n \to \infty).$$

By (5.6.16), the martingale convergence theorem (Theorem 1.4.21) and the dominated convergence theorem, we have

$$\int_{W_T} \left\|\mathbf{E}_n(G) - G\right\|_{H_T^{\otimes j}}^2 d\mu_T = \int_{W_T} \sum_{i=1}^{\infty} \{\mathbf{E}_n(\langle G, \psi_i\rangle_{H_T^{\otimes j}}) - \langle G, \psi_i\rangle_{H_T^{\otimes j}}\}^2 d\mu_T$$
$$\longrightarrow 0 \quad (n \to \infty).$$

(4) Let $G \in \mathscr{P}$ be of the form

$$G(w) = f(\eta_1(w), \dots, \eta_m(w)) \qquad (w \in W_T)$$

with a polynomial $f : \mathbb{R}^m \to \mathbb{R}$ and $\eta_1, \dots, \eta_m \in W_T^*$. By the embedding $W_T^* \subset H_T^* = H_T \subset W_T$, we have

$$\eta_i(\ell_j) = \langle \eta_i, \ell_j\rangle_{H_T} = \ell_j(\eta_i).$$

Hence, for any $w \in W_T$,

$$\eta_i(\pi_n w) = (\pi_n \eta_i)(w), \quad \eta_i((I - \pi_n)w) = ((I - \pi_n)\eta_i)(w). \tag{5.6.21}$$

Since $\langle \pi_n\eta_i, (I-\pi_n)\eta_j\rangle_{H_T} = 0$, $\{\eta_i \circ \pi_n\}_{i=1}^m$ and $\{\eta_i \circ (I-\pi_n)\}_{i=1}^m$ are independent. Define a polynomial $\widetilde{f}$ by

$$\widetilde{f}(x_1, \dots, x_m) = \int_{W_T} f(x_1 + \eta_1((I-\pi_n)w'), \dots, x_m + \eta_m((I-\pi_n)w'))\mu_T(dw').$$

Then, since $\{\eta_i \circ \pi_n\}_{i=1}^m$ is $\mathscr{G}_n$-measurable, by Proposition 1.4.2, we obtain

$$\begin{aligned}\mathbf{E}_n(G) &= \mathbf{E}[f(\eta_1 \circ \pi_n + \eta_1 \circ (I - \pi_n), \dots, \eta_m \circ \pi_n + \eta_m \circ (I - \pi_n))|\mathscr{G}_n] \\ &= \widetilde{f}(\eta_1 \circ \pi_n, \dots, \eta_m \circ \pi_n) = \int_{W_T} G(\pi_n \cdot +(I - \pi_n)w')\mu_T(\mathrm{d}w'). \qquad \square\end{aligned}$$

Lemma 5.6.8 *Let $F \in \mathbb{D}^{\infty,\infty-}(H_T^{\otimes j})$ and $F_1 \in \mathbb{D}^{\infty,\infty-}(H_T)$. Then $\mathbf{E}_n(\pi_n^{\otimes j} F) \in \mathbb{D}^{\infty,\infty-}(H_T^{\otimes j})$ and*

$$\nabla(\mathbf{E}_n(\pi_n^{\otimes j} F)) = \pi_n^{\otimes j+1}(\mathbf{E}_n(\nabla F)) = \mathbf{E}_n(\pi_n^{\otimes j+1}(\nabla F)), \tag{5.6.22}$$

$$\nabla^*(\mathbf{E}_n(\pi_n F_1)) = \mathbf{E}_n(\nabla^* F_1), \tag{5.6.23}$$

$$\|\nabla(\mathbf{E}_n(\pi_n F_1))\|_{H_T^{\otimes 2}}^2 \leqq \mathbf{E}_n(\|\nabla F_1\|_{H_T^{\otimes 2}}^2). \tag{5.6.24}$$

Proof First let $F \in \mathscr{P}(H_T^{\otimes j})$. We prove $\mathbf{E}_n(\pi_n^{\otimes j} F)$ belongs to $\mathscr{P}(H_T^{\otimes j})$ and (5.6.22) holds. To do this, it suffices to show it in the case where $F = Ge$ for the same $G \in \mathscr{P}$ as in the proof of Lemma 5.6.7 (4) and $e \in H_T^{\otimes j}$. We use the same notation as in the proof of Lemma 5.6.7 (4).

Then, since $\mathbf{E}_n(G) = \widetilde{f}(\eta_1 \circ \pi_n, \dots, \eta_m \circ \pi_n) \in \mathscr{P}$, we have $\mathbf{E}_n(\pi_n^{\otimes j} F) = \mathbf{E}_n(G)\pi_n^{\otimes j} e \in \mathscr{P}(H_T^{\otimes j})$. Hence, by Lemma 5.6.7 (4),

$$\begin{aligned}\nabla(\mathbf{E}_n(\pi_n^{\otimes j} F)) &= \nabla(\mathbf{E}_n(G))\pi_n^{\otimes j} e \\ &= \sum_{i=1}^m \frac{\partial \widetilde{f}}{\partial x^i}(\eta_1 \circ \pi_n, \dots, \eta_m \circ \pi_n)(\eta_i \circ \pi_n) \otimes \pi_n^{\otimes j} e.\end{aligned}$$

On the other hand, since $\pi_n^{\otimes j+1}(\nabla F) = \sum_{i=1}^m \frac{\partial f}{\partial x_i}(\eta_1, \dots, \eta_m)(\pi_n \eta_i) \otimes \pi_n^{\otimes j} e$, using again Lemma 5.6.7 (4), we obtain

$$\mathbf{E}_n(\pi_n^{\otimes j+1}(\nabla F)) = \sum_{i=1}^m \frac{\partial \widetilde{f}}{\partial x^i}(\eta_1 \circ \pi_n, \dots, \eta_m \circ \pi_n)(\pi_n \eta_i) \otimes \pi_n^{\otimes j} e.$$

By (5.6.21), (5.6.22) holds for $F \in \mathscr{P}(H_T^{\otimes j})$.

Second, let $F \in \mathbb{D}^{\infty,\infty-}(H_T^{\otimes j})$. For $k \in \mathbb{N}$, $p > 1$, take $F_m \in \mathscr{P}(H_T^{\otimes j})$ so that $\lim_{m\to\infty} \|F_m - F\|_{(k,p)} = 0$ (see Definition 5.1.4). Then, applying (5.6.22) to F_m and using (5.6.18), we obtain for $\ell \leqq k$

$$\begin{aligned}&\|\nabla^\ell(\mathbf{E}_n(\pi_n^{\otimes j} F_m)) - \mathbf{E}_n(\pi_n^{\otimes j+\ell}(\nabla^\ell F))\|_p \\ &\leqq \|\mathbf{E}_n(\pi_n^{\otimes j+\ell}(\nabla^\ell F_m)) - \mathbf{E}_n(\pi_n^{\otimes j+\ell}(\nabla^\ell F))\|_p \\ &\leqq \|\nabla^\ell F_m - \nabla^\ell F\|_p \longrightarrow 0 \quad (m \to \infty).\end{aligned}$$

Hence, we have $\mathbf{E}_n(\pi_n^{\otimes j} F) \in \mathbb{D}^{k,p}(H_T^{\otimes j})$ and (5.6.22). Since k and p are arbitrary, $F \in \mathbb{D}^{\infty,\infty-}(H_T^{\otimes j})$.

Third, we show (5.6.23). Let $K \in \mathscr{P}$. By the symmetry of $\mathbf{E}_n$ mentioned in (5.6.17), the symmetry of π_n in H_T, the commutativity of $\mathbf{E}_n$ and π_n, and (5.6.22) for $j = 0$, we have

$$\int_{W_T} K\nabla^*(\mathbf{E}_n(\pi_n F_1))\,\mathrm{d}\mu_T = \int_{W_T} \langle \nabla K, \mathbf{E}_n(\pi_n F_1)\rangle_{H_T}\mathrm{d}\mu_T$$
$$= \int_{W_T} \langle \pi_n(\mathbf{E}_n(\nabla K)), F_1\rangle_{H_T}\mathrm{d}\mu_T = \int_{W_T} \langle \mathbf{E}_n(\pi_n(\nabla K)), F_1\rangle_{H_T}\mathrm{d}\mu_T$$
$$= \int_{W_T} \langle \nabla(\mathbf{E}_n K), F_1\rangle_{H_T}\mathrm{d}\mu_T = \int_{W_T} K\mathbf{E}_n(\nabla^* F_1)\,\mathrm{d}\mu_T.$$

Thus we obtain (5.6.23).

Finally, we show (5.6.24). By (5.6.22) and (5.6.18), we obtain

$$\big\|\nabla(\mathbf{E}_n(\pi_n F_1))\big\|^2_{H_T^{\otimes 2}} = \big\|\mathbf{E}_n(\pi_n^{\otimes 2}(\nabla F_1))\big\|^2_{H_T^{\otimes 2}}$$
$$\leqq \mathbf{E}_n\big(\big\|\pi_n^{\otimes 2}(\nabla F_1)\big\|^2_{H_T^{\otimes 2}}\big) \leqq \mathbf{E}_n\big(\big\|\nabla F_1\big\|^2_{H_T^{\otimes 2}}\big). \qquad \square$$

Proof of Theorem 5.6.1 By (5.6.2), there exist $\gamma > 0$ and $q > \frac{1}{2}$ such that

$$\mathrm{e}^{-\nabla^* F + q\|\nabla F\|^2_{H_T^{\otimes 2}}} \in L^{1+\gamma}(\mu_T).$$

Let $\mathbf{E}_n$ and π_n be as above and set $F_n = \mathbf{E}_n(\pi_n F)$. By (5.6.23), (5.6.24), and Jensen's inequality for conditional expectations, we have

$$\int_{W_T} \mathrm{e}^{(1+\gamma)(-\nabla^* F_n + q\|\nabla F_n\|^2_{H_T^{\otimes 2}})}\mathrm{d}\mu_T \leqq \int_{W_T} \mathrm{e}^{\mathbf{E}_n((1+\gamma)\{-\nabla^* F + q\|\nabla F\|^2_{H_T^{\otimes 2}}\})}\mathrm{d}\mu_T$$
$$\leqq \int_{W_T} \mathbf{E}_n\Big(\mathrm{e}^{(1+\gamma)\{-\nabla^* F + q\|\nabla F\|^2_{H_T^{\otimes 2}}\}}\Big)\mathrm{d}\mu_T$$
$$= \int_{W_T} \mathrm{e}^{(1+\gamma)\{-\nabla^* F + q\|\nabla F\|^2_{H_T^{\otimes 2}}\}}\mathrm{d}\mu_T. \tag{5.6.25}$$

Identify $\pi_n(W_T)$ with $\mathbb{R}^n$ in a natural way. By Lemma 5.6.8, applying the Sobolev embedding theorem ([1]), we may regard $F_n \in \mathscr{D}^{\infty,\infty-}(\mathbb{R}^n;\mathbb{R}^n)$. By (5.6.25), F_n satisfies (5.6.9). Thus, by Lemma 5.6.6, (5.6.3) holds for $F = F_n$.

By the martingale convergence theorem (Theorem 1.4.21), Lemma 5.6.7 (3), (5.6.22), and (5.6.23), we may suppose that F_n, ∇F_n, and $\nabla^* F_n$ converges almost surely to F, ∇F, and $\nabla^* F$, respectively, taking a subsequence if necessary. By (5.6.4), we have

$$|\det_2(I + \nabla F_n)|\mathrm{e}^{-\nabla^* F_n - \frac{1}{2}\|F_n\|^2_{H_T}} \leqq \mathrm{e}^{-\nabla^* F_n + \frac{1}{2}\|\nabla F_n\|^2_{H_T^{\otimes 2}}}.$$

By (5.6.25),

$$\Big\{f(\iota + F_n)\det_2(I + \nabla F_n)\exp\Big(-\nabla^* F_n - \frac{1}{2}\big\|F_n\big\|^2_{H_T}\Big)\Big\}_{n\in\mathbb{N}}$$

is uniformly integrable (Theorem A.3.4). Hence, letting $n \to \infty$ in (5.6.3) for $F = F_n$, we obtain the desired identity (5.6.3) for F. □

5.7 Quadratic Forms

As in other fields of analysis, quadratic forms on the Wiener space play fundamental roles in stochastic analysis. In this section we show a general theory on quadratic Wiener functionals and, in the next section, we present concrete examples.

Definition 5.7.1 Regarding a symmetric $A \in H_T^{\otimes 2}$ as an $H_T^{\otimes 2}$-valued constant function on W_T, set

$$Q_A = (\nabla^*)^2 A, \quad L_A = \nabla^* A.$$

Q_A is called a **quadratic form** associated with A.

By Theorem 5.2.1, $Q_A \in \mathbb{D}^{\infty,\infty-}$ and $L_A \in \mathbb{D}^{\infty,\infty-}(H_T)$.

Since the Hilbert–Schmidt operators are compact operators, by the spectral decomposition for compact operators ([58]), A is diagonalized as

$$A = \sum_{n=1}^{\infty} a_n h_n \otimes h_n, \tag{5.7.1}$$

where $\{h_n\}_{n=1}^{\infty}$ is an orthonormal basis of H_T and $\{a_n\}_{n=1}^{\infty}$ is a sequence of real numbers with $\sum_{n=1}^{\infty} a_n^2 < \infty$. By using this decomposition, Q_A and L_A are represented as infinite sums.

Lemma 5.7.2 (1) *The following convergence in the L^2-sense holds:*

$$Q_A = \sum_{n=1}^{\infty} a_n\{(\nabla^* h_n)^2 - 1\} \quad \text{and} \quad L_A = \sum_{n=1}^{\infty} a_n(\nabla^* h_n)h_n.$$

(2) *Set $\|A\|_{\mathrm{op}} = \sup_n |a_n|$. For $\lambda \in \mathbb{R}$ with $|\lambda|\,\|A\|_{\mathrm{op}} < \frac{1}{2}$, $\mathrm{e}^{\lambda Q_A} \in L^{1+}(\mu_T)$.*
(3) *For $\lambda \in \mathbb{R}$ with $|\lambda|\,\|A\|_{\mathrm{op}} < \frac{1}{2}$,*

$$\int_{W_T} \mathrm{e}^{\lambda Q_A} \mathrm{d}\mu_T = \{\det{}_2(I - 2\lambda A)\}^{-\frac{1}{2}}.$$

The aim of this section is to extend the assertion (3) to general integrals of the form $\int_{W_T} \mathrm{e}^{\lambda Q_A} f\, \mathrm{d}\mu_T$ ($f \in C_{\mathrm{b}}(W_T)$) (Theorem 5.7.6), applying the change of variables formula on W_T as shown in the previous section.

Remark 5.7.3 Since $\{\nabla^* h_n\}_{n=1}^{\infty}$ is a sequence of independent standard Gaussian random variables, by the Itô–Nisio theorem (Theorem 1.2.5), we have

$$\theta = \sum_{n=1}^{\infty} (\nabla^* h_n) h_n.$$

Combining this with (5.7.1), we have a formal expression for Lemma 5.7.2(1):

$$Q_A = \langle \theta, A\theta \rangle_{H_T} - \mathrm{tr}(A).$$

While this expression is "$\infty - \infty$" in general because $H_T \subsetneq W_T$ and A is not necessarily of trace class, it suggests the origin of the name of quadratic forms associated with A.

Proof of Lemma 5.7.2 (1) First we show

$$(\nabla^*)^2 (h \otimes g) = (\nabla^* h)(\nabla^* g) - \langle h, g \rangle_{H_T} \qquad (h, g \in H_T). \tag{5.7.2}$$

For this purpose, let E be a separable Hilbert space, $G \in \mathbb{D}^{\infty,\infty-}(H_T \otimes E)$ and $e \in E$. We have $\langle G, e \rangle_E \in \mathbb{D}^{\infty,\infty-}(H_T)$. Since

$$\begin{aligned}\int_{W_T} \langle \nabla^* G, e \rangle_E \phi \, \mathrm{d}\mu_T &= \int_{W_T} \langle G, \nabla(\phi \cdot e) \rangle_{H_T \otimes E} \mathrm{d}\mu_T \\ &= \int_{W_T} \langle G, (\nabla \phi) \otimes e \rangle_{H_T \otimes E} \mathrm{d}\mu_T = \int_{W_T} \langle \langle G, e \rangle_E, \nabla \phi \rangle_{H_T} \mathrm{d}\mu_T\end{aligned}$$

for any $\phi \in \mathscr{P}$,

$$\langle \nabla^* G, e \rangle_E = \nabla^* (\langle G, e \rangle_E).$$

Using this identity with $E = H_T$, we obtain

$$\langle \nabla^* (h \otimes g), h_n \rangle_{H_T} = (\nabla^* h) \langle g, h_n \rangle_{H_T} \qquad (n = 1, 2, \ldots).$$

Hence we have

$$\nabla^* (h \otimes g) = (\nabla^* h) g. \tag{5.7.3}$$

Since $\nabla(\nabla^* h) = h$ (Example 5.1.5), by Theorem 5.2.8, we obtain (5.7.2).

Setting $K_m = \sum_{n=1}^{m} a_n h_n \otimes h_n$, by (5.7.2) and (5.7.3), we have

$$\nabla^* K_m = \sum_{n=1}^{m} a_n (\nabla^* h_n) h_n \quad \text{and} \quad (\nabla^*)^2 K_m = \sum_{n=1}^{m} a_n \{(\nabla^* h_n)^2 - 1\}.$$

By the continuity of ∇^* and a similar argument to that in Example 5.4.3, we obtain the conclusion.

(2) It suffices to show $e^{\lambda Q_A} \in L^1(\mu_T)$ for $\lambda \in \mathbb{R}$ with $|\lambda| \, \|A\|_{\text{op}} < \frac{1}{2}$. By (1), there exists a subsequence $\{m_n\}_{n=1}^\infty$ such that

$$F_n = \sum_{k=1}^{m_n} a_k\{(\nabla^* h_k)^2 - 1\} \to Q_A, \quad \mu_T\text{-a.s.}$$

Since $\{\nabla^* h_n\}$ is a sequence of independent standard Gaussian random variables by (5.1.8), we have

$$\begin{aligned}\int_{W_T} e^{\lambda Q_A} d\mu_T &\leqq \liminf_{n\to\infty} \int_{W_T} e^{\lambda F_n} d\mu_T \\ &= \liminf_{n\to\infty} \prod_{k=1}^{m_n} (1 - 2\lambda a_k)^{-\frac{1}{2}} e^{-\lambda a_k} = \det_2(1 - 2\lambda A)^{-\frac{1}{2}} < \infty.\end{aligned}$$

(3) By the proof of (2), $\{e^{\lambda F_n}\}_{n=1}^\infty$ is uniformly integrable. Hence, in the same way as in (2), we have

$$\begin{aligned}\int_{W_T} e^{\lambda Q_A} d\mu_T &= \lim_{n\to\infty} \int_{W_T} e^{\lambda F_n} d\mu_T \\ &= \lim_{n\to\infty} \prod_{k=1}^{m_n} (1 - 2\lambda a_k)^{-\frac{1}{2}} e^{-\lambda a_k} = \det_2(1 - 2\lambda A)^{-\frac{1}{2}}.\end{aligned}$$ □

By Lemma 5.7.2 (1), $\nabla^3 Q_A = 0$. The converse is also true.

Proposition 5.7.4 *Let $F \in \mathbb{D}^{\infty,\infty-}$. If $\nabla^3 F = 0$, then there exist a symmetric operator $A \in H_T^{\otimes 2}$, $h \in H_T$, and $c \in \mathbb{R}$ such that*

$$F = c + \nabla^* h + \frac{1}{2} Q_A. \tag{5.7.4}$$

Moreover,

$$c = \int_{W_T} F \, d\mu_T \quad \text{and} \quad h = \int_{W_T} \nabla F \, d\mu_T. \tag{5.7.5}$$

Proof By Proposition 5.2.9, there exists an $A \in H_T^{\otimes 2}$ such that $\nabla^2 F = A$. In particular, A is symmetric.

Set $F_1 = F - \frac{1}{2} Q_A$. Then, by Lemma 5.7.2, $\nabla^2 F_1 = 0$. By Proposition 5.2.9 again, there exists an $h \in H_T$ such that $\nabla F_1 = h$.

Next set $F_2 = F_1 - \nabla^* h$. By Example 5.1.5, $\nabla F_2 = 0$. Hence, there exists a $c \in \mathbb{R}$ such that $F_2 = c$. From these observations, we obtain (5.7.4).

By Lemma 5.7.2 and Example 5.1.5, we have

$$\int_{W_T} Q_A \, d\mu_T = 0, \quad \int_{W_T} \nabla Q_A \, d\mu_T = 0, \quad \int_{W_T} \nabla^* h \, d\mu_T = 0.$$

Hence, (5.7.5) follows from (5.7.4). □

Remark 5.7.5 From the expression (5.7.4), we can prove that the distribution of F is infinitely divisible ([101]) and can compute the corresponding Lévy measure. For details, see [82].

Develop a symmetric operator $A \in H_T^{\otimes 2}$ as (5.7.1). For $\lambda \in \mathbb{R}$ with $2|\lambda|\,\|A\|_{\mathrm{op}} < 1$, set

$$s_n^{A,\lambda} = (1 - 2\lambda a_n)^{-\frac{1}{2}} - 1 \quad \text{and} \quad S^{A,\lambda} = \sum_{n=1}^{\infty} s_n^{A,\lambda} h_n \otimes h_n.$$

Since

$$|s_n^{A,\lambda}| \leqq \frac{|2\lambda a_n|}{\sqrt{1 - 2|\lambda|\,\|A\|_{\mathrm{op}}}}, \tag{5.7.6}$$

$S^{A,\lambda}$ is a symmetric Hilbert–Schmidt operator. If $|\lambda|$ is sufficiently small, for example, if $|\lambda|\,\|A\|_{\mathrm{op}} < \frac{3}{16}$, then $\|S^{A,\lambda}\|_{\mathrm{op}} < \frac{1}{2}$.

In connection with quadratic forms, the following change of variables formula holds.

Theorem 5.7.6 *For $\lambda \in \mathbb{R}$ with $|\lambda|\,\|A\|_{\mathrm{op}} < \frac{3}{16}$ and $f \in C_{\mathrm{b}}(W_T)$,*

$$\int_{W_T} \mathrm{e}^{\lambda Q_A} f \, \mathrm{d}\mu_T = \{\det{}_2(I - 2\lambda A)\}^{-\frac{1}{2}} \int_{W_T} f(\iota + L_{S^{A,\lambda}}) \, \mathrm{d}\mu_T. \tag{5.7.7}$$

For a proof, we give a lemma.

Lemma 5.7.7 (1) *For each $h \in H_T$, there exists an H_T-invariant $X_h \in \mathscr{B}(W_T)$ with $\mu_T(X_h) = 1$ such that*

$$(\nabla^* h)(w + g) = (\nabla^* h)(w) + \langle h, g \rangle_{H_T} \tag{5.7.8}$$

for any $w \in X_h$ and $g \in H_T$, where the H_T-invariance of X_h means that $X_h + g = X_h$ for any $g \in H_T$.
(2) *For any symmetric operator $A \in H_T^{\otimes 2}$, there exists an H_T-invariant $X_A \in \mathscr{B}(W_T)$ with $\mu(X_A) = 1$ such that*

$$Q_A(w + g) = Q_A(w) + 2\langle L_A(w), g \rangle_{H_T} + \langle Ag, g \rangle_{H_T} \tag{5.7.9}$$

for any $w \in X_A$ and $g \in H_T$.

$\nabla^* h$, L_A and Q_A are defined up to null sets and the assertions of the lemma include the problem of the choice of modifications. We give an answer in the proof below. In order to expand $Q_A(w + F(w))$, which appears in the change of variables formula on W_T, for each w, we need to consider the H_T-invariant sets as in the lemma.

Proof (1) Let $\{\ell_n\}_{n=1}^\infty \subset W_T^*$ be an orthonormal basis of H_T. Set

$$\widetilde{h}_n = \sum_{k=1}^n \langle h, \ell_k\rangle_{H_T}\ell_k.$$

By (5.1.4), if $\ell_n \in W_T^*$, then $\nabla^*\ell_n = \ell_n$, μ_T-a.s. Hence, we can take the modification of $\nabla^*\widetilde{h}_n$ so that

$$\nabla^*\widetilde{h}_n = \widetilde{h}_n.$$

Moreover, we may assume that $\widetilde{h}_n \to \nabla^* h$, μ_T-a.s., choosing a subsequence if necessary. Set

$$X_h = \Big\{w \in W_T;\ \lim_{n,m\to\infty} |\widetilde{h}_n(w) - \widetilde{h}_m(w)| = 0\Big\}.$$

X_h is H_T-invariant because $\lim_{n\to\infty}\|\widetilde{h}_n - h\|_{H_T} = 0$. Moreover, since $\widetilde{h}_n \to \nabla^* h$, μ_T-a.s., $\mu_T(X_h) = 1$. Define a modification of $\nabla^* h$ by

$$(\nabla^* h)(w) = \begin{cases} \lim_{n\to\infty}\widetilde{h}_n(w) & (w \in X_h) \\ 0 & (w \notin X_h). \end{cases}$$

Then, since

$$\widetilde{h}_n(w+g) = \widetilde{h}_n(w) + \langle \widetilde{h}_n, g\rangle_{H_T} \qquad (w \in W_T,\ g \in H_T),$$

we obtain (5.7.8) by letting $n \to \infty$.
(2) Develop A as in (5.7.1) with an orthonormal basis $\{h_n\}_{n=1}^\infty$ of H_T. For each h_n, define X_{h_n} and $\nabla^* h_n$ by (1). Let X_A be the set of $w \in \bigcap_{n=1}^\infty X_{h_n}$ such that $\sum_{n=1}^\infty a_n^2(\nabla^* h_n)^2(w) < \infty$ and $\sum_{n=1}^\infty a_n\{(\nabla^* h_n)^2(w) - 1\}$ converges. By (5.7.8), X_A is H_T-invariant. Since $\sum_{n=1}^\infty a_n\{(\nabla^* h_n)^2 - 1\}$ converges in L^2 and $\sum_{n=1}^\infty a_n^2(\nabla^* h_n)^2$ is integrable, by the assertion (1), $\mu_T(X_A) = 1$.

Let $w \in X_A$ and $g \in H_T$. Recalling Lemma 5.7.2, set

$$Q_A(w) = \begin{cases} \sum_{n=1}^\infty a_n\{(\nabla^* h_n)^2(w) - 1\} & (w \in X_A), \\ 0 & (w \notin X_A), \end{cases}$$

$$L_A(w) = \begin{cases} \sum_{n=1}^\infty a_n(\nabla^* h_n)(w)h_n & (w \in X_A), \\ 0 & (w \notin X_A). \end{cases}$$

Then, we have (5.7.9). □

Proof of Theorem 5.7.6 For simplicity of notation, write $S = S^{A,\lambda}$ and $s_n = s_n^{A,\lambda}$, and set $F = L_S$. Then, $\nabla F = S$ and $\nabla^* F = Q_S$.

If $|\lambda|\,\|A\|_{\mathrm{op}} < \frac{3}{16}$, then $\|S\|_{\mathrm{op}} < \frac{1}{2}$ by (5.7.6). Lemma 5.7.2(2) implies that $\mathrm{e}^{-\nabla^* F + \|\nabla F\|^2_{H_T^{\otimes 2}}} \in L^{1+}(\mu_T)$. Hence, the conditions of Theorem 5.6.1 hold with $q = 1$. Moreover, by Remark 5.6.2(1), we have

$$\det{}_2(I + S)\int_{W_T} f(\iota + F)\mathrm{e}^{\lambda Q_A \circ(\iota+F)}\mathrm{e}^{-\nabla^* F - \frac{1}{2}\|F\|^2_{H_T}}\mathrm{d}\mu_T = \int_{W_T} f\mathrm{e}^{\lambda Q_A}\mathrm{d}\mu_T. \tag{5.7.10}$$

Since $(1 + s_n)^2(1 - 2\lambda a_n) = 1$,

$$\begin{aligned} &\lambda a_n + 2\lambda a_n s_n + a_n s_n^2 - s_n - \frac{1}{2}s_n^2 = 0, \\ &\lambda A + 2\lambda AS + AS^2 - S - \frac{1}{2}S^2 = 0. \end{aligned} \tag{5.7.11}$$

From these identities, it follows that

$$\det{}_2(I + S) = \left\{\prod_{n=1}^{\infty}(1 + s_n)^2\mathrm{e}^{-2s_n}\right\}^{\frac{1}{2}} = \{\det{}_2(I - 2\lambda A)\}^{-\frac{1}{2}}\mathrm{e}^{\mathrm{tr}(\lambda A - S)}. \tag{5.7.12}$$

By Lemma 5.7.7,

$$Q_A \circ (\iota + F) = Q_A + 2\langle L_A, F\rangle_{H_T} + \langle AF, F\rangle_{H_T}.$$

By Lemma 5.7.2,

$$\begin{aligned} &\langle L_A, F\rangle_{H_T} = Q_{AS} + \mathrm{tr}\,AS, \qquad \langle AF, F\rangle_{H_T} = Q_{AS^2} + \mathrm{tr}\,AS^2, \\ &\|F\|^2_{H_T} = Q_{S^2} + \mathrm{tr}\,S^2. \end{aligned}$$

Thus, by the linearity $pQ_B + qQ_C = Q_{pB+qC}$ and (5.7.11),

$$\lambda Q_A \circ (\iota + F) - \nabla^* F - \frac{1}{2}\|F\|^2_{H_T} = \mathrm{tr}(S - \lambda A).$$

Plugging this and (5.7.12) into (5.7.10), we obtain (5.7.7). □

Corollary 5.7.8 *For $\lambda \in \mathbb{R}$ with $|\lambda|\,\|A\|_{\mathrm{op}} < \frac{1}{2}$ and $g \in H_T$,*

$$\int_{W_T} \mathrm{e}^{\lambda Q_A + \nabla^* g}\mathrm{d}\mu_T = \{\det{}_2(I - 2\lambda A)\}^{-\frac{1}{2}}\mathrm{e}^{\frac{1}{2}\langle (I-2\lambda A)^{-1}g, g\rangle_{H_T}}. \tag{5.7.13}$$

Proof It suffices to show (5.7.13) when $|\lambda|\,\|A\|_{\mathrm{op}} < \frac{3}{16}$, because, by analytic continuation and Lemma 5.7.2(2), (5.7.13) holds also when $|\lambda|\,\|A\|_{\mathrm{op}} < \frac{1}{2}$.

Develop A as in (5.7.1) and set $g_n = \sum_{k=1}^n \langle g, h_k\rangle_{H_T} h_k$. By Lemmas 5.7.7 and 5.7.2, we have

$$\begin{aligned} \nabla^* g_n \circ (\iota + L_{S^{A,\lambda}}) &= \nabla^* g_n + \langle g_n, L_{S^{A,\lambda}}\rangle_{H_T} \\ &= \sum_{k=1}^{n}\langle g, h_k\rangle_{H_T}(1 - 2\lambda a_k)^{-\frac{1}{2}}\nabla^* h_k. \end{aligned}$$

Since $\{\nabla^* h_k\}$ is a sequence of independent standard Gaussian random variables,

$$\int_{W_T} e^{\nabla^* g_n \circ (\iota + L_{SA,\lambda})} d\mu_T = e^{\frac{1}{2}\langle (I-2\lambda A)^{-1} g_n, g_n\rangle_{H_T}}.$$

Hence, by Theorem 5.7.6, we obtain

$$\int_{W_T} e^{\lambda Q_A + \nabla^* g_n} d\mu_T = \{\det_2(I - 2\lambda A)\}^{-\frac{1}{2}} e^{\frac{1}{2}\langle (I-2\lambda A)^{-1} g_n, g_n\rangle_{H_T}}. \tag{5.7.14}$$

Since the distribution of $\nabla^* g_n$ is Gaussian with mean 0 and variance $\|g_n\|_{H_T}^2$, for any $p > 0$

$$\int_{W_T} e^{p\nabla^* g_n} d\mu_T = e^{\frac{1}{2}p^2\|g_n\|_{H_T}^2} \leqq e^{\frac{1}{2}p^2\|g\|_{H_T}^2}.$$

Hence, by Lemma 5.7.2(2), $\{e^{\lambda Q_A + \nabla^* g_n}\}_{n=1}^{\infty}$ is uniformly integrable. Since $\nabla^* g_n$ converges to $\nabla^* g$ in L^2, we obtain (5.7.13) by letting $n \to \infty$ in (5.7.14). □

Corollary 5.7.9 *Let $\eta_1, \dots, \eta_n \in W_T^*$ be an orthonormal system in H_T. Define $\pi : W_T \to W_T$ by $\pi(w) = \sum_{i=1}^n \eta_i(w)\eta_i$, and A_0 and $A_1 : H_T \to H_T$ by $A_0 = (I - \pi)A(I - \pi)$ and $A_1 = \pi A \pi$, respectively. Write $\delta_0(\boldsymbol{\eta})d\mu_T$ for the measure $\nu_{\delta_0(\boldsymbol{\eta})}$ in Theorem 5.4.15. Then, for $\lambda \in \mathbb{R}$ with $|\lambda|\,\|A\|_{\mathrm{op}} < \frac{1}{2}$ and $g \in H_T$,*

$$\int_{W_T} e^{\lambda Q_A + \nabla^* g}\delta_0(\boldsymbol{\eta}) d\mu_T$$
$$= \frac{1}{(2\pi)^{\frac{n}{2}}}\{\det_2(I - 2\lambda A_0)\}^{-\frac{1}{2}} e^{-\lambda tr(A_1)} e^{\frac{1}{2}\langle (I-2\lambda A_0)^{-1}(I-\pi)g, (I-\pi)g\rangle_{H_T}}. \tag{5.7.15}$$

Proof Write $w = (I - \pi)(w) + \sum_{i=1}^n \eta_i(w)\eta_i$. Then, by Example 5.4.17,

$$\eta_i(w) = 0, \quad \delta_0(\boldsymbol{\eta})d\mu_T\text{-a.e. } w \in W_T, \quad \eta_i \circ (I - \pi) = 0 \quad (i = 1, \dots, n). \tag{5.7.16}$$

Hence, for $F = f(\ell_1, \dots, \ell_m)$ with $\ell_1, \dots, \ell_m \in W_T^*$ and $f \in C_{\nearrow}^{\infty}(\mathbb{R}^m)$, we have

$$F = F \circ (I - \pi), \quad \delta_0(\boldsymbol{\eta})d\mu_T\text{-a.e.} \tag{5.7.17}$$

On the other hand, since $I - \pi$ and $\boldsymbol{\eta}$ are independent under μ_T,

$$\int_{W_T} (F \circ (I - \pi))\varphi(\boldsymbol{\eta})\, d\mu_T = \int_{W_T} (F \circ (I - \pi))\, d\mu_T \times \int_{W_T} \varphi(\boldsymbol{\eta})\, d\mu_T$$
$$= \int_{W_T} (F \circ (I - \pi))\, d\mu_T \times \int_{\mathbb{R}^n} \varphi(x) \frac{1}{(2\pi)^{\frac{n}{2}}} e^{-\frac{1}{2}|x|^2} dx$$

for any $\varphi \in \mathscr{S}(\mathbb{R}^n)$. Hence, by taking $\{\varphi_k\}_{k=1}^{\infty} \in \mathscr{S}(\mathbb{R}^n)$ converging to δ_0 and letting $k \to \infty$, by Corollary 5.4.7 and (5.7.16), we obtain

$$\int_{W_T} F\delta_0(\boldsymbol{\eta}) d\mu_T = \frac{1}{(2\pi)^{\frac{n}{2}}} \int_{W_T} (F \circ (I - \pi))\, d\mu_T. \tag{5.7.18}$$

Extend $\eta_1,\dots,\eta_n$ to an orthonormal basis $\{\eta_i\}_{i=1}^{\infty}$ of H_T. Set $c_i = \langle g,\eta_i\rangle_{H_T}$, $a_{ij} = \langle \eta_i, A\eta_j\rangle_{H_T}$,

$$g_N = \sum_{i=1}^{N} c_i\eta_i \quad \text{and} \quad A_N = \sum_{i,j=1}^{N} a_{ij}\eta_i\otimes\eta_j \qquad (N \geqq n).$$

Then, we have

$$Q_{A_N} = Q_{(I-\pi)A_N(I-\pi)} - \mathrm{tr}(\pi A_N\pi), \quad \delta_0(\boldsymbol{\eta})\mathrm{d}\mu_T\text{-a.e.}, \tag{5.7.19}$$

$$Q_{(I-\pi)A_N(I-\pi)}\circ(I-\pi) = Q_{(I-\pi)A_N(I-\pi)}, \quad \mu_T\text{-a.s.} \tag{5.7.20}$$

In fact, by (5.7.2),

$$Q_{A_N} = \sum_{i,j=1}^{N} a_{ij}\{\eta_i\eta_j - \delta_{ij}\} \quad \text{and} \quad Q_{(I-\pi)A_N(I-\pi)} = \sum_{n<i,j\leqq N} a_{ij}\{\eta_i\eta_j-\delta_{ij}\}.$$

Then we obtain (5.7.19) from $\mathrm{tr}(\pi A_n\pi) = \sum_{i=1}^{n} a_{ii}$ and (5.7.16). Moreover, the identity

$$Q_{(I-\pi)A_N(I-\pi)} = \sum_{n<i,j\leqq N} a_{ij}\{((I-\pi)\eta_i)((I-\pi)\eta_j) - \langle (I-\pi)\eta_i,\eta_j\rangle_{H_T}\}$$

yields (5.7.20).

Since $\nabla^* g_N = \sum_{i=1}^{N} c_i\eta_i$ and $\nabla^*((I-\pi)g_N) = \sum_{n<i\leqq N} c_i\eta_i$, by (5.7.16) again, we have

$$\nabla^* g_N = \nabla^*((I-\pi)g_N), \quad \delta_0(\boldsymbol{\eta})\mathrm{d}\mu_T\text{-a.e.},$$

$$\nabla^*((I-\pi)g_N) = \nabla^*((I-\pi)g_N)\circ(I-\pi), \quad \mu_T\text{-a.s.}$$

Recalling the inequality $\|(I-\pi)A_N(I-\pi)\|_{\mathrm{op}} \leqq \|A\|_{\mathrm{op}}$ and then applying (5.7.18), we obtain

$$\begin{aligned}
&\int_{W_T} \mathrm{e}^{\lambda Q_{A_N}+\nabla^* g_N}\delta_0(\boldsymbol{\eta})\mathrm{d}\mu_T \\
&\quad = \int_{W_T} \mathrm{e}^{\lambda\{Q_{(I-\pi)A_N(I-\pi)}-\mathrm{tr}(\pi A_N\pi)\}+\nabla^*((I-\pi)g_N)}\delta_0(\boldsymbol{\eta})\mathrm{d}\mu_T \\
&\quad = \frac{\mathrm{e}^{-\lambda\mathrm{tr}(\pi A_N\pi)}}{(2\pi)^{\frac{n}{2}}}\int_{W_T} \mathrm{e}^{\lambda Q_{(I-\pi)A_N(I-\pi)}+\nabla^*((I-\pi)g_N)}\mathrm{d}\mu_T.
\end{aligned}$$

The uniform integrability of the integrands can be seen in the same way as in the proof of Lemma 5.7.2(2). Therefore, letting $N\to\infty$, we obtain

$$\int_{W_T} \mathrm{e}^{\lambda Q_A+\nabla^* g}\delta_0(\boldsymbol{\eta})\mathrm{d}\mu_T = \frac{\mathrm{e}^{-\lambda\mathrm{tr}(\pi A\pi)}}{(2\pi)^{\frac{n}{2}}}\int_{W_T} \mathrm{e}^{\lambda Q_{A_0}+\nabla^*((I-\pi)g)}\mathrm{d}\mu_T.$$

In conjunction with Corollary 5.7.8 the conclusion follows. □

Remark 5.7.10 By analytic continuation, Corollaries 5.7.8 and 5.7.9 hold for $\lambda \in \mathbb{C}$ with $|\mathrm{Re}(\lambda)| \, \|A\|_{\mathrm{op}} < \frac{1}{2}$.

5.8 Examples of Quadratic Forms

In this section, the results in the previous section are applied to concrete examples: harmonic oscillators, Lévy's stochastic area, and sample variance.

5.8.1 Harmonic Oscillators

Let $d = 1$ and W_T be the one-dimensional Wiener space. Set

$$\mathfrak{h}_T(w) = \int_0^T w(t)^2 \mathrm{d}t \qquad (w \in W_T).$$

The functional $\mathfrak{h}_T$ is closely related to the **harmonic oscillator** $-\frac{1}{2}\frac{\mathrm{d}^2}{\mathrm{d}x^2} + \lambda x^2$, which is one of the fundamental Schrödinger operators.

First we present the Laplace transforms of the probability law of $\mathfrak{h}_T$.

Theorem 5.8.1 *For* $\lambda > -\frac{\pi^2}{4T^2}$,

$$\int_{W_T} \mathrm{e}^{-\frac{1}{2}\lambda\mathfrak{h}_T} \mathrm{d}\mu_T = \sqrt{\frac{1}{\cosh(\sqrt{\lambda}\,T)}}, \tag{5.8.1}$$

$$\int_{W_T} \mathrm{e}^{-\frac{1}{2}\lambda\mathfrak{h}_T} \delta_0(\theta(T)) \mathrm{d}\mu_T = \frac{1}{\sqrt{2\pi T}} \sqrt{\frac{\sqrt{\lambda}\,T}{\sinh(\sqrt{\lambda}\,T)}}. \tag{5.8.2}$$

Proof First we show that $\mathfrak{h}_T \in \mathbb{D}^{\infty,\infty-}$ and

$$\mathfrak{h}_T = Q_A + \frac{T^2}{2}, \tag{5.8.3}$$

where $A : H_T \to H_T$ is given by

$$(\dot{A h})(t) = \int_t^T h(s)\,\mathrm{d}s \qquad (t \in [0,T],\ h \in H_T). \tag{5.8.4}$$

For $n \in \mathbb{N}$, set

$$\mathfrak{h}_T^{(n)} = \frac{T}{n} \sum_{i=0}^{n-1} \theta\Big(\frac{i}{n}T\Big)^2.$$

$\mathfrak{h}_T^{(n)} \in \mathscr{P}$ and, by (5.1.1),

$$\langle \nabla \mathfrak{h}_T^{(n)}, h \rangle_{H_T} = \frac{2T}{n} \sum_{i=0}^{n-1} \theta\Big(\frac{i}{n}T\Big) h\Big(\frac{i}{n}T\Big) \qquad (h \in H_T).$$

Hence, defining $\ell_{[0,t]} \in W_T^* \subset H_T$ by $\ell_{[0,t]}(w) = w(t)$ $(w \in W_T)$, we have

$$\nabla \mathfrak{h}_T^{(n)} = \frac{2T}{n} \sum_{i=0}^{n-1} \theta\Big(\frac{i}{n}T\Big)\ell_{[0,\frac{i}{n}T]}.$$

From this expression, using (5.1.1) again, we obtain

$$\nabla^2 \mathfrak{h}_T^{(n)} = \frac{2T}{n} \sum_{i=0}^{n-1} \ell_{[0,\frac{i}{n}T]} \otimes \ell_{[0,\frac{i}{n}T]}, \quad \nabla^3 \mathfrak{h}_T^{(n)} = 0.$$

Let $n \to \infty$. Then, for any $p > 1$, we have the following L^p-convergence:

$$\mathfrak{h}_T^{(n)} \to \mathfrak{h}_T, \quad \nabla \mathfrak{h}_T^{(n)} \to 2\int_0^T \theta(t)\ell_{[0,t]}\,\mathrm{d}t, \quad \nabla^2 \mathfrak{h}_T^{(n)} \to 2\int_0^T \ell_{[0,t]} \otimes \ell_{[0,t]}\,\mathrm{d}t.$$

Hence, $\mathfrak{h}_T \in \mathbb{D}^{\infty,\infty-}$ and

$$\nabla \mathfrak{h}_T = 2\int_0^T \theta(t)\ell_{[0,t]}\,\mathrm{d}t, \quad \nabla^2 \mathfrak{h}_T = 2\int_0^T \ell_{[0,t]} \otimes \ell_{[0,t]}\,\mathrm{d}t, \quad \nabla^3 \mathfrak{h}_T = 0.$$

Since the integration by parts yields

$$\Big\langle\Big(\int_0^T \ell_{[0,t]} \otimes \ell_{[0,t]}\,\mathrm{d}t\Big)[h], g\Big\rangle_{H_T} = \int_0^T h(t)g(t)\,\mathrm{d}t = \int_0^T \Big(\int_t^T h(s)\mathrm{d}s\Big)\dot{g}(t)\,\mathrm{d}t,$$

we have

$$\nabla^2 \mathfrak{h}_T = 2A.$$

Moreover, by the above expression, we have

$$\int_{W_T} \nabla \mathfrak{h}_T\,\mathrm{d}\mu_T = 0 \quad \text{and} \quad \int_{W_T} \mathfrak{h}_T\,\mathrm{d}\mu_T = \frac{T^2}{2}.$$

Thus, by Proposition 5.7.4, (5.8.3) holds.

Second, we compute the eigenvalues and eigenfunctions of the Hilbert–Schmidt operator A. By (5.8.4), the equation $\lambda h = Ah$ is equivalent to

$$\lambda \dot{h}(t) = \int_t^T h(s)\,\mathrm{d}s \quad (t \in [0,T]) \quad \text{and} \quad h(0) = 0.$$

From this, A does not have a zero eigenvalue since $\lambda = 0$ implies $h = 0$. For $\lambda \neq 0$, we rewrite the above equation as

$$\lambda \ddot{h} + h = 0, \quad \dot{h}(T) = 0, \quad h(0) = 0.$$

The solution of this second order ordinary differential equation is given by a linear combination of $\cos(\lambda^{-\frac{1}{2}}t)$ and $\sin(\lambda^{-\frac{1}{2}}t)$. By the initial condition $\dot{h}(T) = 0,\ h(0) = 0$, we obtain

$$\lambda^{-\frac{1}{2}} = \frac{(n+\frac{1}{2})\pi}{T} \quad \text{and} \quad h(t) = c\sin\Big(\frac{(n+\frac{1}{2})\pi t}{T}\Big),$$

where c is a non-zero constant. Set

$$h_n(t) = \frac{\sqrt{2T}}{(n+\frac{1}{2})\pi} \sin\Big(\frac{(n+\frac{1}{2})\pi t}{T}\Big).$$

Then, $\{h_n\}_{n=0}^{\infty}$ is an orthonormal basis of H_T and each h_n is an eigenfunction of A corresponding to the eigenvalue $\frac{T^2}{(n+\frac{1}{2})^2\pi^2}$. Hence, A is diagonalized as

$$A = \sum_{n=0}^{\infty} \frac{T^2}{(n+\frac{1}{2})^2\pi^2} h_n \otimes h_n.$$

We have $\|A\|_{\text{op}} = \frac{4T^2}{\pi^2}$. Moreover, since $\sum_{n=0}^{\infty} \frac{1}{(2n+1)^2} = \frac{\pi^2}{8}$, $\text{tr}(A) = \frac{T^2}{2}$.

By Corollary 5.7.8, for $\lambda \in \mathbb{R}$ with $|\lambda| < \frac{\pi^2}{4T^2}$,

$$\int_{W_T} \mathrm{e}^{-\frac{1}{2}\lambda \mathfrak{b}_T} \mathrm{d}\mu_T = \int_{W_T} \mathrm{e}^{-\frac{1}{2}\lambda Q_A} \mathrm{d}\mu_T \cdot \mathrm{e}^{-\frac{1}{4}\lambda T^2} = \{\det_2(I+\lambda A)\}^{-\frac{1}{2}} \mathrm{e}^{-\frac{1}{4}\lambda T^2}$$
$$= \Big\{\prod_{n=0}^{\infty}\Big(1 + \frac{4\lambda T^2}{(2n+1)^2\pi^2}\Big)\Big\}^{-\frac{1}{2}}.$$

Combining this with the identity

$$\cosh x = \prod_{n=0}^{\infty}\Big(1 + \frac{4x^2}{(2n+1)^2\pi^2}\Big),$$

we see that (5.8.1) holds for $\lambda \in \mathbb{R}$ with $|\lambda| < \frac{\pi^2}{4T^2}$. By analytic continuation, (5.8.1) holds for $\lambda \in \mathbb{R}$ with $\lambda > -\frac{\pi^2}{4T^2}$.

Next we show (5.8.2). As above, it suffices to show (5.8.2) for $\lambda \in \mathbb{R}$ with $|\lambda| < \frac{\pi^2}{4T^2}$. Define $\eta \in H_T$ by $\eta(t) = \frac{t}{\sqrt{T}}$ ($t \in [0,T]$) and π, A_0, A_1 as in Corollary 5.7.9. A_0 is given by

$$(A_0\dot{h})(t) = \int_t^T h(s)\,\mathrm{d}s - \frac{1}{T}\int_0^T \Big(\int_s^T h(u)\,\mathrm{d}u\Big)\mathrm{d}s$$

for $h \in H_T$ with $\pi h = 0$ or $h(T) = 0$. By a similar argument to the above, A_0 is developed as

$$A_0 = \sum_{n=1}^{\infty} \frac{T^2}{n^2\pi^2} k_n \otimes k_n, \qquad k_n(t) = \frac{\sqrt{2T}}{n\pi} \sin\Big(\frac{n\pi t}{T}\Big).$$

Since $\delta_0(\theta(T)) = \frac{1}{\sqrt{T}}\delta_0(\boldsymbol{\eta})$ and $\text{tr}(A) = \text{tr}(A_0) + \text{tr}(A_1)$, by Corollary 5.7.9, we obtain

$$\int_{W_T} \mathrm{e}^{-\frac{1}{2}\lambda \mathfrak{b}_T} \delta_0(\theta(T)) \mathrm{d}\mu_T = \int_{W_T} \mathrm{e}^{-\frac{1}{2}\lambda Q_A} \delta_0(\theta(T)) \mathrm{d}\mu_T \mathrm{e}^{-\frac{1}{2}\lambda \text{tr}(A)}$$

$$= \frac{1}{\sqrt{2\pi T}}\{\det_2(I + \lambda A_0)\}^{-\frac{1}{2}} e^{-\frac{1}{2}\lambda\{\mathrm{tr}(A) - \mathrm{tr}(A_1)\}}$$

$$= \frac{1}{\sqrt{2\pi T}}\Big\{\prod_{n=1}^{\infty}\Big(1 + \frac{\lambda T^2}{n^2\pi^2}\Big)^2\Big\}^{-\frac{1}{2}}.$$

By the identity

$$\sinh x = x\prod_{n=1}^{\infty}\Big(1 + \frac{x^2}{n^2\pi^2}\Big), \tag{5.8.5}$$

we arrive at (5.8.2). □

By using Theorem 5.8.1, we show the explicit formula for the heat kernel of the Schrödinger operator investigated in Theorem 5.5.7 when $d = 1$, $\Theta = 0$, and $V(x) = x^2$.

Theorem 5.8.2 *Fix $\lambda > 0$. Then, for $x, y \in \mathbb{R}$ and $T > 0$,*

$$\int_{W_T} \exp\Big(-\frac{\lambda^2}{2}\int_0^T (x + \theta(t))^2 \mathrm{d}t\Big)\delta_y(x + \theta(T))\,\mathrm{d}\mu_T \tag{5.8.6}$$
$$= \frac{1}{\sqrt{2\pi T}}\sqrt{\frac{\lambda T}{\sinh(\lambda T)}}\exp\Big(-\frac{\lambda}{2}\coth(\lambda T)\{x^2 - 2xy\,\mathrm{sech}(\lambda T) + y^2\}\Big).$$

Proof Let $\phi : [0, T] \to \mathbb{R}$ be the unique solution for the ordinary differential equation[4]

$$\phi'' - \lambda^2\phi = 0, \quad \phi(0) = x, \ \ \phi(T) = y. \tag{5.8.7}$$

Define $h \in H_T$ by $h(t) = \phi(t) - x$. Since

$$\int_0^T \phi'(t)\,\mathrm{d}\theta(t) = \theta(T)\phi'(T) - \int_0^T \phi''(t)\theta(t)\,\mathrm{d}t$$

and ϕ satisfies (5.8.7), applying the Cameron–Martin theorem (Theorem 1.7.2), we obtain

$$\int_{W_T} \exp\Big(-\frac{\lambda^2}{2}\int_0^T (x + \theta(t))^2 \mathrm{d}t\Big)\delta_y(x + \theta(T))\,\mathrm{d}\mu_T$$
$$= \int_{W_T} \exp\Big(-\frac{\lambda^2}{2}\int_0^T (x + \theta(t) + h(t))^2 \mathrm{d}t\Big)\, e^{-\nabla^* h - \frac{1}{2}\|h\|_{H_T}^2}$$
$$\times \delta_y(x + \theta(T) + h(T))\,\mathrm{d}\mu_T$$
$$= \exp\Big(-\frac{1}{2}\int_0^T \{\lambda^2\phi(t)^2 + (\phi'(t))^2\}\,\mathrm{d}t\Big)\int_{W_T} e^{-\frac{1}{2}\lambda^2\mathfrak{h}_T}\delta_0(\theta(T))\,\mathrm{d}\mu_T.$$

[4] This equation is the Lagrange equation corresponding to the action integral $S_T(\phi) = \int_0^T L(\phi(t), \dot{\phi}(t))\,\mathrm{d}t$ associated with the Lagrangian $L(p, q) = \frac{1}{2}\{p^2 + q^2\}$. See Section 7.1.

By (5.8.7), we have

$$\int_0^T \{\lambda^2\phi(t)^2 + (\phi'(t))^2\}\,\mathrm{d}t = y\phi'(T) - x\phi'(0).$$

Then, plugging in the explicit form of ϕ,

$$\phi(t) = \frac{y - \mathrm{e}^{-\lambda T}x}{\mathrm{e}^{\lambda T} - \mathrm{e}^{-\lambda T}}\mathrm{e}^{\lambda t} - \frac{y - \mathrm{e}^{\lambda T}x}{\mathrm{e}^{\lambda T} - \mathrm{e}^{-\lambda T}}\mathrm{e}^{-\lambda t} \qquad (t \in [0,T])$$

and using Theorem 5.8.1, we obtain (5.8.6). □

Remark 5.8.3 We have derived (5.8.6) by applying (5.8.2) in Theorem 5.8.1. The identity (5.8.6) can be shown in a direct and functional analytical way associated with the Schrödinger operator $H_\lambda = -\frac{1}{2}(\frac{\mathrm{d}}{\mathrm{d}x})^2 + \frac{\lambda^2}{2}x^2$ ($\lambda > 0$) on $\mathbb{R}$. The method is as follows.

Realize H_λ as a self-adjoint operator on $L^2(\mathbb{R})$, the Hilbert space of square-integrable functions with respect to the Lebesgue measure. The spectrum of H_λ consists only of the eigenvalues $\{\lambda(n+\frac{1}{2})\}_{n=0}^\infty$ with multiplicity one and the corresponding normalized eigenfunction ϕ_n is given by

$$\phi_n(x) = \sqrt{n!}\Big(\frac{\lambda}{\pi}\Big)^{\frac{1}{4}}\mathrm{e}^{-\frac{1}{2}\lambda x^2}H_n(\sqrt{2\lambda}\,x),$$

where $H_n(x)$ is a Hermite polynomial. Since $p(t,x,y)$ admits the eigenfunction expansion

$$p(t,x,y) = \sum_{n=0}^\infty \mathrm{e}^{-\lambda(n+\frac{1}{2})t}\phi_n(x)\phi_n(y),$$

a well-known formula for the Hermite polynomials

$$\sum_{n=0}^\infty n!H_n(x)H_n(y)t^n = \frac{1}{\sqrt{1-t^2}}\exp\Big(-\frac{1}{2}\frac{1}{1-t^2}(t^2x^2 - 2txy + t^2y^2)\Big)$$

yields (5.8.6). For this identity, see [67].

5.8.2 Lévy's Stochastic Area

Let W_T be the two-dimensional Wiener space and consider Lévy's stochastic area $\mathfrak{s}(T)$ (Example 5.5.6).

Theorem 5.8.4 *For $\lambda \in \mathbb{R}$ with $|\lambda| < \frac{\pi}{T}$,*

$$\int_{W_T} \mathrm{e}^{\lambda\mathfrak{s}(T)}\mathrm{d}\mu_T = \frac{1}{\cos(\frac{1}{2}\lambda T)}, \tag{5.8.8}$$

$$\int_{W_T} e^{\lambda \mathfrak{s}(T)} \delta_0(\theta(T)) d\mu_T = \frac{1}{2\pi T} \frac{\frac{1}{2}\lambda T}{\sin(\frac{1}{2}\lambda T)}. \tag{5.8.9}$$

Proof First we show $\mathfrak{s}(T) \in \mathbb{D}^{\infty,\infty-}$ and the expression

$$\mathfrak{s}(T) = \frac{1}{2} Q_A, \tag{5.8.10}$$

where $A : H_T \to H_T$ is given by

$$(\dot{Ah})(t) = J\Big(h(t) - \frac{1}{2}h(T)\Big) \qquad (t \in [0,T],\ h \in H_T)$$

and $J = \begin{pmatrix} 0 & -1 \\ 1 & 0 \end{pmatrix}$. For $n \in \mathbb{N}$, define $\mathfrak{s}^{(n)}(T) \in \mathscr{P}$ by

$$\mathfrak{s}^{(n)}(T) = \frac{1}{2} \sum_{i=0}^{n-1} \Big\langle J\theta\Big(\frac{i}{n}T\Big), \theta\Big(\frac{i+1}{n}T\Big) - \theta\Big(\frac{i}{n}T\Big)\Big\rangle_{\mathbb{R}^2}.$$

By (5.1.1), we have for $h \in H_T$

$$\begin{aligned}\langle \nabla \mathfrak{s}^{(n)}(T), h\rangle_{H_T} &= \frac{1}{2} \sum_{i=0}^{n-1} \Big\langle Jh\Big(\frac{i}{n}T\Big), \theta\Big(\frac{i+1}{n}T\Big) - \theta\Big(\frac{i}{n}T\Big)\Big\rangle_{\mathbb{R}^2} \\ &\quad + \frac{1}{2} \sum_{i=0}^{n-1} \Big\langle J\theta\Big(\frac{i}{n}T\Big), h\Big(\frac{i+1}{n}T\Big) - h\Big(\frac{i}{n}T\Big)\Big\rangle_{\mathbb{R}^2}.\end{aligned}$$

Since $\theta(0) = h(0) = 0$, an algebraic manipulation yields

$$\begin{aligned}\langle \nabla \mathfrak{s}^{(n)}(T), h\rangle_{H_T} &= \frac{1}{2} \sum_{i=1}^{n-1} \Big\langle J\theta\Big(\frac{i}{n}T\Big), h\Big(\frac{i+1}{n}T\Big) - h\Big(\frac{i-1}{n}T\Big)\Big\rangle_{\mathbb{R}^2} \\ &\quad - \frac{1}{2}\Big\langle J\theta(T), h\Big(\frac{n-1}{n}T\Big)\Big\rangle_{\mathbb{R}^2}.\end{aligned}$$

Using (5.1.1) again, we obtain for $h, g \in H_T$

$$\begin{aligned}\langle [\nabla^2 \mathfrak{s}^{(n)}(T)](g), h\rangle_{H_T} &= \frac{1}{2} \sum_{i=1}^{n-1} \Big\langle Jg\Big(\frac{i}{n}T\Big), h\Big(\frac{i+1}{n}T\Big) - h\Big(\frac{i-1}{n}T\Big)\Big\rangle_{\mathbb{R}^2} \\ &\quad - \frac{1}{2}\Big\langle Jg(T), h\Big(\frac{n-1}{n}T\Big)\Big\rangle_{\mathbb{R}^2}.\end{aligned}$$

Letting $n \to \infty$, we see that $\mathfrak{s}(T) \in \mathbb{D}^{\infty,\infty-}$,

$$\langle \nabla \mathfrak{s}(T), h\rangle_{H_T} = \int_0^T \langle J\theta(t), \dot{h}(t)\rangle_{\mathbb{R}^2}\, dt - \frac{1}{2}\langle J\theta(T), h(T)\rangle_{\mathbb{R}^2}$$

and

$$\begin{aligned}\langle[\nabla^2\mathfrak{s}(T)](g),h\rangle_{H_T} &= \int_0^T \langle Jg(t),\dot{h}(t)\rangle_{\mathbb{R}^2}\,\mathrm{d}t - \frac{1}{2}\langle Jg(T),h(T)\rangle_{\mathbb{R}^2} \\ &= \int_0^T \Big\langle J\Big(g(t)-\frac{1}{2}g(T)\Big),\dot{h}(t)\Big\rangle_{\mathbb{R}^2}\mathrm{d}t \qquad (h,g\in H_T).\end{aligned}$$

From these observations we obtain

$$\nabla^2\mathfrak{s}(T)=A, \qquad \int_{W_T}\nabla\mathfrak{s}(T)\,\mathrm{d}\mu_T=0, \qquad \int_{W_T}\mathfrak{s}(T)\,\mathrm{d}\mu_T=0.$$

By Proposition 5.7.4, (5.8.10) holds.

Second, we compute the eigenvalues and eigenfunctions of A. The equation $Ah=\lambda h$ is equivalent to

$$\lambda\ddot{h}=J\dot{h}, \quad h(0)=0, \quad \lambda\dot{h}(0)=-\frac{1}{2}Jh(T).$$

Solving this ordinary differential equation, we see that the following functions h_n and $\widehat{h_n}$ are the eigenfunctions corresponding to the eigenvalue $\lambda_n=\frac{T}{(2n+1)\pi}$:

$$h_n(t)=\frac{\sqrt{T}}{(2n+1)\pi}\begin{pmatrix}\cos(\frac{(2n+1)\pi t}{T})-1\\ \sin(\frac{(2n+1)\pi t}{T})\end{pmatrix} \quad\text{and}\quad \widehat{h_n}=Jh_n \qquad (n\in\mathbb{Z}).$$

Moreover, $\{h_n,\widehat{h_n}\}_{n\in\mathbb{Z}}$ is an orthonormal basis of H_T. Hence, we have the expansion

$$A=\sum_{n\in\mathbb{Z}}\frac{T}{(2n+1)\pi}\{h_n\otimes h_n+\widehat{h_n}\otimes\widehat{h_n}\}.$$

The multiplicity of each eigenvalue $\frac{T}{(2n+1)\pi}$ is two and $\|A\|_{\mathrm{op}}=\frac{T}{\pi}$.

If $|\lambda|<\frac{\pi}{T}$, then, by Corollary 5.7.8,

$$\begin{aligned}\int_{W_T} \mathrm{e}^{\lambda\mathfrak{s}(T)}\mathrm{d}\mu_T &= \{\det_2(I-\lambda A)\}^{-\frac{1}{2}} \\ &= \Big\{\prod_{n\in\mathbb{Z}}\Big(1-\frac{\lambda T}{(2n+1)\pi}\Big)\mathrm{e}^{\frac{\lambda T}{(2n+1)\pi}}\Big\}^{-1} = \Big\{\prod_{n=0}^{\infty}\Big(1-\frac{\lambda^2T^2}{(2n+1)^2\pi^2}\Big)\Big\}^{-1}.\end{aligned}$$

By the identity

$$\cos x=\prod_{n=0}^{\infty}\Big(1-\frac{4x^2}{(2n+1)^2\pi^2}\Big),$$

we obtain (5.8.8).

Next we show (5.8.9). Let $e_1 = (1,0)$ and $e_2 = (0,1) \in \mathbb{R}^2$. Define $\eta_i \in H_T$ by $\eta_i(t) = \frac{t}{\sqrt{T}} e_i$ $(i = 1, 2)$. Moreover, define π, A_0, and A_1 as in Corollary 5.7.9. For $h \in H_T$ with $\pi h = 0$ or $h(T) = 0$, we have

$$(\dot{A_0 h})(t) = J(h(t) - \overline{h}),$$

where $\overline{h} = T^{-1} \int_0^T h(s)\,\mathrm{d}s$. Hence, in the same way as above,

$$A_0 = \sum_{n\in\mathbb{Z}\setminus\{0\}}^{\infty} \frac{T}{2n\pi}\{k_n \otimes k_n + \widehat{k}_n \otimes \widehat{k}_n\},$$

where

$$k_n(t) = \frac{\sqrt{T}}{2n\pi}\begin{pmatrix} \cos(\frac{2n\pi t}{T}) - 1 \\ \sin(\frac{2n\pi t}{T}) \end{pmatrix} \quad \text{and} \quad \widehat{k}_n = Jk_n.$$

Furthermore, since

$$\operatorname{tr}(A_1) = \sum_{i=1}^{2} \langle \eta_i, A\eta_i \rangle_{H_T} = \sum_{i=1}^{2} \int_0^T \frac{t-T}{T} \langle e_i, Je_i \rangle_{\mathbb{R}^2} \mathrm{d}t = 0$$

and $\delta_0(\theta(T)) = \frac{1}{T}\delta_0(\boldsymbol{\eta})$, by (5.7.9), we obtain

$$\begin{aligned}\int_{W_T} \mathrm{e}^{\lambda \mathfrak{s}(T)} \delta_0(\theta(T))\mathrm{d}\mu_T &= \int_{W_T} \mathrm{e}^{\frac{1}{2}\lambda Q_A} \delta_0(\theta(T))\mathrm{d}\mu_T = \frac{1}{2\pi T}\{\det{}_2(I - \lambda A_0)\}^{-\frac{1}{2}} \\ &= \frac{1}{2\pi T}\Big\{\prod_{n\in\mathbb{Z}\setminus\{0\}} \Big(1 - \frac{\lambda T}{2n\pi}\Big)\Big\}^{-1} = \frac{1}{2\pi T}\Big\{\prod_{n=1}^{\infty}\Big(1 - \frac{\lambda^2 T^2}{4n^2\pi^2}\Big)\Big\}^{-1}.\end{aligned}$$

By the identity

$$\sin x = x \prod_{n=1}^{\infty}\Big(1 - \frac{x^2}{n^2\pi^2}\Big)$$

we obtain (5.8.9). □

As Theorem 5.8.2, Theorem 5.8.4 is applicable to compute the heat kernel.

Theorem 5.8.5 *Let $\Theta(x) = (-\frac{x^2}{2}, \frac{x^1}{2})$ $(x = (x^1, x^2) \in \mathbb{R}^2)$. Define $L(t, x; \Theta)$ as in Theorem 5.5.7. Then, for $\lambda \in \mathbb{R}$,*

$$\begin{aligned}&\int_{W_T} \mathrm{e}^{\mathrm{i}\,\lambda L(T,x;\Theta)} \delta_y(x + \theta(T))\,\mathrm{d}\mu_T \\ &\qquad = \frac{\lambda}{4\pi \sinh(\frac{1}{2}\lambda T)} \exp\Big(\frac{\mathrm{i}\lambda}{2}\langle Jx, y\rangle_{\mathbb{R}^2} - \frac{\lambda}{4}\coth\Big(\frac{1}{2}\lambda T\Big)|y - x|^2\Big).\end{aligned}$$

Proof Let $x, y \in \mathbb{R}^2$. If we show

$$\int_{W_T} e^{\alpha L(T,x;\Theta)} \delta_y(x+\theta(T))\, d\mu_T = \frac{\alpha}{4\pi \sin(\frac{1}{2}\alpha T)} \exp\Big(\frac{\alpha}{2}\langle Jx, y\rangle_{\mathbb{R}^2} - \frac{\alpha}{4}\cot\Big(\frac{1}{2}\alpha T\Big)|y-x|^2\Big) \quad (5.8.11)$$

for sufficiently small $\alpha \in \mathbb{R}$, we obtain the conclusion by analytic continuation.

Let $\phi : [0, T] \to \mathbb{R}^2$ be the solution of the ordinary differential equation

$$\ddot{\phi} - \alpha J \dot{\phi} = 0, \quad \phi(0) = x, \ \ \phi(T) = y \quad (5.8.12)$$

and define $h \in H_T$ by $h(t) = \phi(t) - x$. Since

$$L(t, x; \Theta) = \frac{1}{2}\int_0^T \langle J(x+\theta(t)), d\theta(t)\rangle_{\mathbb{R}^2}$$

and

$$L(t, x; \Theta)(\cdot + h) = \mathfrak{s}(T) + \int_0^T \langle J\theta(t), \phi'(t)\rangle_{\mathbb{R}^2} dt - \frac{1}{2}\langle J\theta(T), \phi(T)\rangle_{\mathbb{R}^2} + \frac{1}{2}\int_0^T \langle J\phi(t), \phi'(t)\rangle_{\mathbb{R}^2} dt,$$

by the Cameron–Martin theorem, we obtain

$$\int_{W_T} e^{\alpha L(T,x;\Theta)} \delta_y(x+\theta(T))\, d\mu_T \quad (5.8.13)$$
$$= \exp\Big(\frac{1}{2}\int_0^T \{\langle \alpha J\phi(t), \phi'(t)\rangle_{\mathbb{R}^2} - |\phi'(t)|^2\}\, dt\Big) \int_{W_T} e^{\alpha \mathfrak{s}(T)} \delta_0(\theta(T))\, d\mu_T.$$

By integration by parts on $[0, T]$ and (5.8.12),

$$\int_0^T \{\langle \alpha J\phi(t), \phi'(t)\rangle_{\mathbb{R}^2} - |\phi'(t)|^2\}\, dt = \langle \phi'(0), x\rangle_{\mathbb{R}^2} - \langle \phi'(T), y\rangle_{\mathbb{R}^2}.$$

The solution of (5.8.12) is explicitly given by

$$\phi(t) = x + \frac{1}{\alpha} J(I - e^{\alpha t J})\gamma \qquad (t \in [0, T]),$$

where

$$\gamma = \frac{\alpha}{2\sin(\frac{1}{2}\alpha T)} \begin{pmatrix} \cos(\frac{1}{2}\alpha T) & \sin(\frac{1}{2}\alpha T) \\ -\sin(\frac{1}{2}\alpha T) & \cos(\frac{1}{2}\alpha T) \end{pmatrix} (y - x).$$

Hence

$$\phi'(0) = \gamma \quad \text{and} \quad \phi'(T) = e^{\alpha T J}\gamma = \alpha J(y - x) + \gamma.$$

Moreover, we have

$$\langle \phi'(0), x\rangle_{\mathbb{R}^2} - \langle \phi'(T), y\rangle_{\mathbb{R}^2} = \alpha\langle Jx, y\rangle_{\mathbb{R}^2} + \langle \gamma, x-y\rangle_{\mathbb{R}^2}$$
$$= \alpha\langle Jx, y\rangle_{\mathbb{R}^2} - \frac{\alpha\cos(\frac{1}{2}\alpha T)}{2\sin(\frac{1}{2}\alpha T)}|x-y|^2.$$

Plugging this and (5.8.9) into (5.8.13), we obtain (5.8.11). □

Remark 5.8.6 Lévy [70] showed the results in this section by using the Fourier expansion of Brownian motion. Moreover, several proofs are known (see [4]).

5.8.3 Sample Variance

Let W_T be the one-dimensional Wiener space and set

$$\mathfrak{v}_T(w) = \int_0^T (w(t) - \overline{w})^2 \mathrm{d}t \qquad (w \in W_T),$$

where $\overline{w} = \frac{1}{T}\int_0^T w(t)\,\mathrm{d}t$.

Theorem 5.8.7 *For $\lambda \in \mathbb{R}$ with $\lambda > -\frac{\pi^2}{T^2}$,*

$$\int_{W_T} \mathrm{e}^{-\frac{1}{2}\lambda\mathfrak{v}_T}\mathrm{d}\mu_T = \Big(\frac{\sqrt{\lambda}\,T}{\sinh(\sqrt{\lambda}\,T)}\Big)^{\frac{1}{2}}, \tag{5.8.14}$$

$$\int_{W_T} \mathrm{e}^{-\frac{1}{2}\lambda\mathfrak{v}_T}\delta_0(\theta(T))\,\mathrm{d}\mu_T = \frac{\frac{1}{2}\sqrt{\lambda}\,T}{\sinh(\frac{1}{2}\sqrt{\lambda}\,T)}. \tag{5.8.15}$$

Proof First we show (5.8.14). Define $A : H_T \to H_T$ by

$$(\dot{A h})(t) = \int_t^T (h(s) - \overline{h})\,\mathrm{d}s \qquad (t \in [0,T],\ h \in H_T).$$

In the same way as in Theorem 5.8.1, we can show for $h, g \in H_T$

$$\langle \nabla\mathfrak{v}_T, h\rangle_{H_T} = 2\int_0^T (\theta(t) - \overline{\theta})(h(t) - \overline{h})\,\mathrm{d}t$$

and

$$\langle [\nabla\mathfrak{v}_T](g), h\rangle_{H_T} = 2\int_0^T (g(t) - \overline{g})(h(t) - \overline{h})\,\mathrm{d}t = 2\int_0^T (g(t) - \overline{g})h(t)\,\mathrm{d}t$$
$$= 2\int_0^T \Big(\int_t^T (g(s) - \overline{g})\,ds\Big)\dot{h}(t)\,\mathrm{d}t.$$

From these identities, we obtain

$$\nabla^2 \mathfrak{v}_T = 2A, \quad \int_{W_T} \nabla \mathfrak{v}_T \, d\mu_T = 0, \quad \int_{W_T} \mathfrak{v}_T \, d\mu_T = \frac{T^2}{6}.$$

Hence, by Proposition 5.7.4, we obtain

$$\mathfrak{v}_T = Q_A + \frac{T^2}{6}.$$

The equation $Ah = \lambda h$ is equivalent to

$$\lambda \dddot{h} = -\dot{h}, \quad h(0) = 0, \quad \dot{h}(0) = \dot{h}(T) = 0.$$

Solving this equation, we obtain the expansion

$$A = \sum_{n=1}^{\infty} \Big(\frac{T}{n\pi}\Big)^2 h_n \otimes h_n$$

where $h_n(t) = \frac{\sqrt{2T}}{n\pi}\Big\{\cos\Big(\frac{n\pi t}{T}\Big) - 1\Big\}$. In particular, we have $\|A\|_{\text{op}} = \frac{T^2}{\pi^2}$. If $|\lambda| < \frac{\pi^2}{T^2}$, then

$$\int_{W_T} e^{-\frac{1}{2}\lambda \mathfrak{v}_T} d\mu_T = \{\det_2(I + \lambda A)\}^{-\frac{1}{2}} = \Big\{\prod_{n=1}^{\infty}\Big(1 + \frac{\lambda T^2}{n^2\pi^2}\Big)\Big\}^{-\frac{1}{2}}$$

by Corollary 5.7.8. Combining this with (5.8.5), we obtain (5.8.14) for $\lambda \in \mathbb{R}$ with $|\lambda| < \frac{\pi^2}{T^2}$. By analytic continuation, (5.8.14) holds also for $\lambda \in \mathbb{R}$ with $\lambda > -\frac{\pi^2}{T^2}$.

Second, we show (5.8.15). It suffices to show it when $|\lambda| < \frac{\pi^2}{T^2}$.

Define $\eta \in H_T$ by $\eta(t) = \frac{t}{\sqrt{T}}$ $(t \in [0, T])$. Define π, A_0, and A_1 as in Corollary 5.7.9. For $h \in H_T$ with $\pi h = 0$ or $h(T) = 0$, we have

$$(A_0 \dot{h})(t) = \int_t^T (h(s) - \overline{h})\, ds - \frac{1}{T}\int_0^T \Big(\int_s^T (h(u) - \overline{h})\, du\Big) ds.$$

From this, by a similar argument to the above, we obtain

$$A_0 = \sum_{n=1}^{\infty}\Big(\frac{T}{2n\pi}\Big)^2 \{k_n \otimes k_n + \widehat{k}_n \otimes \widehat{k}_n\},$$

where the eigenfunctions are given by

$$k_n(t) = \frac{\sqrt{2T}}{2n\pi}\sin\Big(\frac{2n\pi t}{T}\Big), \qquad \widehat{k}_n(t) = \frac{\sqrt{2T}}{2n\pi}\Big\{\cos\Big(\frac{2n\pi t}{T}\Big) - 1\Big\}.$$

Hence, since $\delta_0(\theta(T)) = \frac{1}{\sqrt{T}}\delta(\boldsymbol{\eta})$ and $\operatorname{tr}(A) = \operatorname{tr}(A_0) + \operatorname{tr}(A_1)$, by Corollary 5.7.9, we have

$$\int_{W_T} e^{-\frac{1}{2}\lambda \mathfrak{v}_T} \delta_0(\theta(T))\, d\mu_T = \Big\{\prod_{n=1}^{\infty}\Big(1 + \frac{\lambda T^2}{(2n\pi)^2}\Big)^2\Big\}^{-\frac{1}{2}}.$$

By (5.8.5), we obtain (5.8.15) for $\lambda \in \mathbb{R}$ with $|\lambda| < \frac{\pi^2}{T^2}$. □

5.9 Abstract Wiener Spaces and Rough Paths

An **abstract Wiener space** is a triplet $(\mathscr{W}, \mathscr{H}, \nu)$ such that

(i) $\mathscr{W}$ is a real separable Banach space,
(ii) $\mathscr{H}$ is a real separable Hilbert space embedded in $\mathscr{W}$ densely and continuously,
(iii) ν is a probability measure on $(\mathscr{W}, \mathscr{B}(\mathscr{W}))$ under which every $\ell \in \mathscr{W}^*$ is a Gaussian random variable with mean 0 and variance $\|\ell\|_{\mathscr{H}}^2$, where we have used the inclusion $\mathscr{W}^* \subset \mathscr{H}^* = \mathscr{H} \subset \mathscr{W}$.

Example 5.9.1 Let (W_T, H_T, μ_T) be the d-dimensional Wiener space.
(1) By Lemma 1.7.1, (W_T, H_T, μ_T) is an abstract Wiener space.
(2) Let $W_{T,0} = \{w \in W_T; w(T) = 0\}$, $H_{T,0} = \{h \in H_T; h(T) = 0\}$ and $\mu_{T,0} = (2\pi T)^{\frac{d}{2}}\delta_0(\theta(T))$, where we have thought of the positive generalized Wiener functional $\delta_0(\theta(T))$ as a Borel measure on W_T by Theorem 5.4.15. Then $(W_{T,0}, H_{T,0}, \mu_{T,0})$ is an abstract Wiener space. To see this, define $\ell_\alpha \in W_T^*$ $(\alpha = 1, \ldots, d)$ by $\ell_\alpha(w) = \frac{1}{\sqrt{T}}w^\alpha(T)$ $(w = (w^1, \ldots, w^d) \in W_T)$. Observe that $\|\ell_\alpha\|_{H_T} = 1$. Take $\ell_j \in W_T^*$ $(j \geqq d+1)$ so that $\{\ell_n\}_{n=1}^\infty$ is an orthonormal basis of H_T. Then, thinking of $\{\ell_n\}_{n=1}^\infty$ as a sequence of independent Gaussian random variables with mean 0 and variance 1, μ_T can be decomposed as a product measure of a d-dimensional standard normal distribution and $\mu_{T,0}$. This implies the desired result. We left the details to the reader.
(3) Let γ be the distribution on W_T of a continuous d-dimensional Gaussian process with mean 0. For $Z \in \mathbf{H}$, the $L^2(\gamma)$-closure of the span of $\theta^\alpha(t)$ $(\alpha = 1, \ldots, d, 0 \leqq t \leqq T)$, define $h_Z \in W_T$ by $h_Z(t) = \int_{W_T} Z\theta(t)\mathrm{d}\gamma$. Let $\mathscr{H} = \{h_Z; Z \in \mathbf{H}\}$ and $\mathbf{W}$ be the closure of $\mathscr{H}$ in W_T with respect to the uniform norm. Then, $(\mathbf{W}, \mathbf{H}, \gamma)$ becomes an abstract Wiener space. For details, see [7, 68].

Repeating the arguments in the preceding sections with an abstract Wiener space $(\mathscr{W}, \mathscr{H}, \nu)$ instead of (W_T, H_T, μ_T), we can define the Sobolev spaces $\mathbb{D}^{p,k}(E)$, the operator ∇, and other things on $\mathscr{W}$ similarly. All assertions and proofs there, except the proof of Theorem 5.4.15, continue to be true without any changes. For the proof of Theorem 5.4.15, an additional observation is necessary (see [116]).

As an extension of Theorem 5.4.11, we have the following assertion on absolute continuity.

Proposition 5.9.2 *Let $(\mathscr{W}, \mathscr{H}, \nu)$ be an abstract Wiener space and $F : \mathscr{W} \to \mathbb{R}$ be of class $\mathbb{D}^{\infty,\infty-}$. Then the distribution of F under $\|\nabla F\|_{\mathscr{H}}^2\mathrm{d}\nu$ has a density* *function* $p(x) = \mathbf{E}[\mathbf{1}_{[x,\infty)}(F)\nabla^*\nabla F]$ *with respect to the Lebesgue measure. In particular, if* $\nabla F \neq 0$ ν*-a.s., then the distribution of* F *is absolutely continuous with respect to the Lebesgue measure.*

Proof Let $\hat{\mathscr{W}} = \mathscr{W} \times W_1^1$, $\hat{\mathscr{H}} = \mathscr{H} \times H_1^1$, and $\hat{\nu} = \nu \times \mu_1^1$. Then $(\hat{\mathscr{W}}, \hat{\mathscr{H}}, \hat{\nu})$ is an abstract Wiener space. Denote by $\hat{\nabla}$ and $\hat{\mathbb{D}}^{\infty,\infty-}$ the gradient operator and the $\mathbb{D}^{\infty,\infty-}$-space on $\hat{\mathscr{W}}$, respectively. By a natural inclusion, $\mathbb{D}^{\infty,\infty-} \subset \hat{\mathbb{D}}^{\infty,\infty-}$, $\hat{\nabla}|_{\mathbb{D}^{\infty,\infty-}} = \nabla$, and $\hat{\nabla}^*|_{\mathbb{D}^{\infty,\infty-}(\mathscr{H})} = \nabla^*$.

For $\varepsilon > 0$, define $\hat{F}_\varepsilon : \hat{\mathscr{W}} \to \mathbb{R}$ by $\hat{F}_\varepsilon(w, w') = F(w) + \varepsilon e^{\xi(w')}$ $((w, w') \in \hat{\mathscr{W}})$, where $\xi(w') = w'(1)$. Then, $\hat{\nabla}\hat{F}_\varepsilon = \nabla F + \varepsilon e^{\xi}\nabla'\xi$, where ∇' stands for the gradient operator on W_1^1. In particular,

$$\|\hat{\nabla}\hat{F}_\varepsilon\|^2_{\hat{\mathscr{H}}} = \|\nabla F\|^2_{\mathscr{H}} + \varepsilon^2 e^{2\xi} \quad \text{and} \quad \hat{\nabla}^*\hat{\nabla}\hat{F}_\varepsilon = \nabla^*\nabla F + \varepsilon(\xi - 1)e^{\xi},$$

where we have used Theorem 5.2.8 to see the second identity. Thus, $\hat{F}_\varepsilon$ is of class $\mathbb{D}^{\infty,\infty-}$ and non-degenerate. By Theorem 5.4.11, the integration by parts formula and Theorem 5.2.1,

$$\begin{aligned}\mathbf{E}[f(\hat{F}_\varepsilon)\|\hat{\nabla}\hat{F}_\varepsilon\|^2_{\hat{\mathscr{H}}}] &= \int_{\mathbb{R}} f(x)\mathbf{E}[\delta_x(\hat{F}_\varepsilon)\|\hat{\nabla}\hat{F}_\varepsilon\|^2_{\hat{\mathscr{H}}}]\mathrm{d}x \\ &= \int_{\mathbb{R}} f(x)\mathbf{E}[\mathbf{1}_{[x,\infty)}(\hat{F}_\varepsilon)\hat{\nabla}^*\hat{\nabla}\hat{F}_\varepsilon]\mathrm{d}x \quad (f \in C_b(\mathscr{W})).\end{aligned}$$

Letting $\varepsilon \to 0$, we arrive at

$$\mathbf{E}[f(F)\|\nabla F\|^2_{\mathscr{H}}] = \int_{\mathbb{R}} f(x)p(x)\mathrm{d}x.$$

This implies the first assertion. The second assertion is an immediate consequence of the first one. □

It was shown by Bouleau and Hirsch [8] that the second assertion continues to hold for $\mathbb{R}^n$-valued Wiener functionals. See also [92].

Proposition 5.9.3 *Let* $F = (F^1, \ldots, F^n) : \mathscr{W} \to \mathbb{R}^n$ *be of class* $\mathbb{D}^{1,p}$ *for some* $p > 1$. *Suppose* $\det[(\langle\nabla F^i, \nabla F^j\rangle)_{i,j=1,\ldots,n}] \neq 0$ ν*-a.s. Then, the distribution of* F *on* $\mathbb{R}^n$ *is absolutely continuous with respect to the Lebesgue measure.*

An application of the Malliavin calculus on abstract Wiener spaces is the one to stochastic differential equations extended by the theory of rough paths. The theory of rough paths was initiated by Lyons in the 1990s, and developed widely to produce several monographs [26, 27, 71, 72]. In the remainder of this section, we shall give a glance at the theory of rough paths, following [26].

For a while, we work in the deterministic setting. Let V be a Banach space. A rough path $\boldsymbol{X} = (X, \mathbb{X})$ is a pair of continuous functions $X : [0, T] \to V$ and $\mathbb{X} : [0, T]^2 \to V \otimes V$, satisfying

$$\mathbb{X}(s,t) - \mathbb{X}(s,u) - \mathbb{X}(u,t) = X(s,u) \otimes X(u,t), \quad \text{where } X(s,t) = X(t) - X(s).$$

For $\frac{1}{3} < \alpha \leqq \frac{1}{2}$, $\mathcal{C}^\alpha = \mathcal{C}^\alpha([0,T],V)$ denotes the space of rough paths $\boldsymbol{X} = (X, \mathbb{X})$ such that

$$\|X\|_\alpha = \sup_{s \neq t \in [0,T]} \frac{|X(s,t)|}{|t-s|^\alpha} < \infty, \quad \|\mathbb{X}\|_{2\alpha} = \sup_{s \neq t \in [0,T]} \frac{|\mathbb{X}(s,t)|}{|t-s|^{2\alpha}} < \infty.$$

Moreover, $\mathcal{C}^\alpha_{\mathrm{g}}$ stands for the space of rough paths $\boldsymbol{X} = (X, \mathbb{X}) \in \mathcal{C}^\alpha$ such that $\mathrm{Sym}(\mathbb{X}(s,t)) = \frac{1}{2}X(s,t) \otimes \frac{1}{2}X(s,t)$, where Sym is the symmetrizing operator.

For $X \in C^\alpha([0,T],V)$, $Y \in C^\alpha([0,T],\bar{W})$, $\bar{W}$ being a Banach space, is said to be controlled by X if there exists $Y' \in C^\alpha([0,T],\mathcal{L}(V,\bar{W}))$, where $\mathcal{L}(V,\bar{W})$ is the space of continuous linear mappings of V to $\bar{W}$, such that $R^Y(s,t) = Y(s,t) - Y'(s)X(s,t)$ satisfies $\|R^Y\|_{2\alpha} < \infty$. The space of such pairs (Y, Y') is denoted by $\mathcal{D}^{2\alpha}_X([0,T],\bar{W})$.

Let $\alpha > \frac{1}{3}$. If $\boldsymbol{X} = (X, \mathbb{X}) \in \mathcal{C}^\alpha([0,T],V)$ and $(Y,Y') \in \mathcal{D}^{2\alpha}_X([0,T],\mathcal{L}(V,W))$, then, for every $s < t \leqq T$, $\lim_{|\mathcal{P}|\to 0} \sum_{[u,v]\in\mathcal{P}} (Y(u)X(u,v) + Y'(u)\mathbb{X}(u,v))$ exists, where $\mathcal{P}$ is a partition of $[s,t]$. The limit is called the integration of Y against the rough path $\boldsymbol{X}$, and denoted by $\int_s^t Y(r)\mathrm{d}\boldsymbol{X}(r)$.

Using integrations against rough paths, a differential equation driven by a rough path, say a rough differential equation, can be defined; the rough differential equation

$$\mathrm{d}Y = f(Y)\mathrm{d}\boldsymbol{X}, \quad Y_0 = \xi$$

means the integral equation

$$Y(t) = \xi + \int_0^t f(Y(s))\mathrm{d}\boldsymbol{X}(s).$$

We now proceed to the stochastic setting. First we shall see that a rough differential equation extends a stochastic differential equation. To do this, let $B = \{B(t)\}_{t \geqq 0}$ be a d-dimensional standard Brownian motion. Set

$$\mathbb{B}^{\text{Itô}} = \int_s^t B(s,r) \otimes \mathrm{d}B(r) \in \mathbb{R}^d \otimes \mathbb{R}^d.$$

Then $\boldsymbol{B}^{\text{Itô}} = (B, \mathbb{B}^{\text{Itô}}) \in \mathcal{C}^\alpha([0,T],\mathbb{R}^d)$ a.s. for any $\alpha \in (\frac{1}{3}, \frac{1}{2})$ and $T > 0$. Similarly, if we set

$$\mathbb{B}^{\text{Strat}} = \int_s^t B(s,r) \otimes \circ\mathrm{d}B(r) \in \mathbb{R}^d \otimes \mathbb{R}^d,$$

then $\boldsymbol{B}^{\text{Strat}} = (B, \mathbb{B}^{\text{Strat}}) \in C_{\text{g}}^{\alpha}([0,T], \mathbb{R}^d)$ a.s. for any $\alpha \in (\frac{1}{3}, \frac{1}{2})$ and $T > 0$. It may be worthwhile to notice that, if $s = 0$, then the anti-symmetric parts of $\mathbb{B}^{\text{Itô}}$ and $\mathbb{B}^{\text{Strat}}$ coincide with Lévy's stochastic area.

If $(Y(\omega), Y'(\omega)) \in \mathcal{D}_{B(\omega)}^{2\alpha}$ for a.a. ω, and Y, Y' are both predictable, then

$$\int_0^T Y \mathrm{d}\boldsymbol{B}^{\text{Itô}} = \int_0^T Y(t)\mathrm{d}B(t) \quad \text{and} \quad \int_0^T Y \mathrm{d}\boldsymbol{B}^{\text{Strat}} = \int_0^T Y(t) \circ \mathrm{d}B(t).$$

Moreover, for $f \in C_{\text{b}}^3(\mathbb{R}^e, \mathcal{L}(\mathbb{R}^d, \mathbb{R}^e))$, Lipschitz continuous $f_0 : \mathbb{R}^e \to \mathbb{R}^e$, and $\xi \in \mathbb{R}^e$, (i) for a.a. ω, there is a unique solution $(Y(\omega), f(Y(\omega))) \in \mathcal{D}_{B(\omega)}^{2\alpha}$ to the rough differential equation

$$\mathrm{d}Y(t,\omega) = f_0(Y(t,\omega))\mathrm{d}t + f(Y(t,\omega))\mathrm{d}\boldsymbol{B}^{\text{Itô}}(t,\omega), \quad Y(0,\omega) = \xi,$$

and (ii) $Y = \{Y(t,\omega)\}_{t \geq 0}$ is a strong solution to the Itô stochastic differential equation

$$\mathrm{d}Y(t) = f_0(Y(t))\mathrm{d}t + f(Y(t))\mathrm{d}B(t), \quad Y(0) = \xi.$$

A similar assertion holds with "Strat" instead of "Itô".

We now investigate rough paths arising from Gaussian processes, for which the Malliavin calculus on abstract Wiener spaces works. Let X be a continuous, centered Gaussian process with values in $\mathbb{R}^d$. In what follows, we work on the abstract Wiener space given in Example 5.9.1 (3). The rectangle increment of the covariance is defined by

$$R\begin{pmatrix} s,t \\ s',t' \end{pmatrix} = \mathbb{E}[X(s,t) \otimes X(s',t')].$$

Its ρ-variation on a rectangle $I \times I'$, where I and I' are both rectangles in $\mathbb{R}^d$, is given by

$$\|R\|_{\rho, I \times I'} = \Bigg(\sup_{\substack{\mathcal{P} \subset I, \\ \mathcal{P}' \subset I'}} \sum_{\substack{[s,t] \in \mathcal{P} \\ [s',t'] \in \mathcal{P}'}} \left| R\begin{pmatrix} s,t \\ s',t' \end{pmatrix} \right|^{\rho} \Bigg)^{\frac{1}{\rho}},$$

where $\mathcal{P}$ (resp. $\mathcal{P}'$) is a partition of I (resp. I').

Let $\rho \in [1, \frac{3}{2})$, $\alpha \in (\frac{1}{3}, \frac{1}{2\rho})$, and $\{X(t)\}_{0 \leqq t \leqq T}$ be a d-dimensional, continuous, centered Gaussian process with independent components such that

$$\sup_{0 \leqq s < t \leqq T} \frac{\|R_{X^i}\|_{\rho, [s,t]^2}}{|t-s|^{\frac{1}{\rho}}} < \infty \quad (i = 1, \ldots, d).$$

Define

$$\mathbb{X}^{i,j}(s,t) = \begin{cases} L^2 - \lim\limits_{\substack{\mathcal{P}\in\Pi_{[s,t]} \\ |\mathcal{P}|\to 0}} \sum\limits_{[u,v]\in\mathcal{P}} X^i(s,u)X^j(u,v), & \text{if } i < j, \\ \frac{1}{2}(X^i_{s,t})^2, & \text{if } i = j, \\ -\mathbb{X}^{j,i} + X^i(s,t)X^j(s,t), & \text{if } i > j, \end{cases}$$

where $\Pi_{[s,t]}$ is the set of partitions of $[s,t]$. Then $\boldsymbol{X} = (X, \mathbb{X}) \in \mathscr{C}^{\alpha}_{\mathrm{g}}$. For this $\boldsymbol{X}$ and $V_1, \ldots, V_d \in C^{\infty}_{\mathrm{b}}(\mathbb{R}^e, \mathbb{R}^e)$, let $\{Y(t)\}_{0\leqq t\leqq T}$ be the solution to the rough differential equation

$$\mathrm{d}Y = V(Y)\mathrm{d}\boldsymbol{X}, \quad Y(0) = y_0 \in \mathbb{R}^e,$$

where $V = (V_1, \ldots, V_d)$. As an application of Proposition 5.9.3, we have the following.

Theorem 5.9.4 *Assume that*
(1) *For* $f \in C^{\alpha}([0,t], \mathbb{R}^d)$, $f = 0$ *if and only if* $\sum_{j=1}^d \int_0^t f_j \mathrm{d}h^j = 0$ *for all* $h \in \mathscr{H}$.
(2) *For a.a.* ω, $X(\omega)$ *is truly rough, at least in a right neighborhood of* 0, *that is, there is a dense subset* A *of a right neighborhood of* 0 *such that for any* $s \in A$,

$$\limsup_{t\downarrow s} \frac{|\langle v, X(s,t)\rangle|}{|t-s|^{2\alpha}} = \infty, \quad \textit{for any } v \in \mathbb{R}^d \setminus \{0\}.$$

Moreover, suppose that $\mathrm{Lie}(V_1, \ldots, V_d)\big|_{y_0} = \mathbb{R}^e$. *Then, for any* $t > 0$, *the distribution of* $Y(t)$ *is absolutely continuous with respect to the Lebesgue measure on* $\mathbb{R}^e$.

6

The Black–Scholes Model

Throughout this chapter, we fix $T > 0$ and a 1-dimensional Brownian motion $\{B(t)\}_{0 \leqq t \leqq T}$ on a complete probability space $(\Omega, \mathscr{F}, \mathbf{P})$. As in Section 2.6, let $\mathscr{F}_t^B = \sigma\{B(s); s \leqq t\} \vee \mathscr{N}$, $\mathscr{N}$ being the totality of $\mathbf{P}$-null sets. We assume $\mathscr{F} = \mathscr{F}_T^B$. In what follows, we omit the prefix "$\{\mathscr{F}_t^B\}$-", and say simply "predictable", "martingale" and so on. Moreover, $\mathbf{E}$ stands for the expectation with respect to $\mathbf{P}$, and the expectation with respect to another probability measure $\mathbf{Q}$ will be denoted by $\mathbf{E}_{\mathbf{Q}}$. Finally, for $p = 1, 2$, $\mathscr{L}^p_{\text{loc}}$ denotes the space of predictable processes $\{f(t)\}_{0 \leqq t \leqq T}$ with $\int_0^T |f(t)|^p \mathrm{d}t < \infty$, $\mathbf{P}$-a.s.

6.1 The Black–Scholes Model

Let $r, \mu, \sigma > 0$, and define the two-dimensional stochastic process $X = \{X(t) = (\rho(t), S(t))\}_{0 \leqq t \leqq T}$ by

$$\begin{cases} \mathrm{d}\rho(t) = r\rho(t)\mathrm{d}t, & \rho(0) = 1, \\ \mathrm{d}S(t) = \mu S(t)\mathrm{d}t + \sigma S(t)\mathrm{d}B(t), & S(0) = s_0 > 0. \end{cases} \tag{6.1.1}$$

Then

$$\rho(t) = \mathrm{e}^{rt} \quad \text{and} \quad S(t) = s_0 \exp\Big(\sigma B(t) + \Big(\mu - \frac{\sigma^2}{2}\Big)t\Big).$$

In particular, $\rho(t), S(t) > 0$.

$\rho(t)$ represents the price of a safe security, like a bond, at time t. r is an interest rate and $\rho(t)$ represents the amount of continuous compounding at time t. $S(t)$ represents the price of a risky security, like a stock. As (6.1.1) indicates, it is an amount of continuous compounding perturbed by the noise driven by a Brownian motion. The parameter σ corresponds to how much $S(t)$ fluctuates, and is called **volatility**. The stochastic process $X = \{X(t) = (\rho(t), S(t))\}_{0 \leqq t \leqq T}$ is called the **Black–Scholes model**. We restrict ourselves to the Black–Scholes model, while general market models can be investigated via stochastic differential equations with more general coefficients than (6.1.1). In what follows, we call $\{S(t)\}_{0\leqq t\leqq T}$ the **stock price process**.

Definition 6.1.1 (1) A **portfolio** is a two-dimensional predictable process $\varphi = \{\varphi(t) = (\varphi^0(t), \varphi^1(t))\}_{0\leqq t\leqq T}$. The totality of portfolios is denoted by $\mathcal{P}$.
(2) Given $\varphi = \{\varphi(t) = (\varphi^0(t), \varphi^1(t))\}_{0\leqq t\leqq T} \in \mathcal{P}$, set

$$V(t;\varphi) = \langle\varphi(t), X(t)\rangle = \varphi^0(t)\rho(t) + \varphi^1(t)S(t).$$

The process $\{V(t;\varphi)\}_{0\leqq t\leqq T}$ is called the **value process** of φ.
(3) $\varphi \in \mathcal{P}$ is said to be **self-financing** in the market X if $\{\varphi^0(t)\}_{0\leqq t\leqq T} \in \mathscr{L}^1_{\text{loc}}$, $\{\varphi^1(t)\}_{0\leqq t\leqq T} \in \mathscr{L}^2_{\text{loc}}$, and

$$\begin{aligned} V(t;\varphi) &= V_0(\varphi) + \int_0^t \varphi^0(s)\mathrm{d}\rho(s) + \int_0^t \varphi^1(s)\mathrm{d}S(s) \\ &= V_0(\varphi) + \int_0^t (r\varphi^0(s)\rho(s) + \mu\varphi^1(s)S(s))\mathrm{d}s + \int_0^t \sigma\varphi^1(s)S(s)\mathrm{d}B(s). \end{aligned} \tag{6.1.2}$$

The totality of self-financing portfolios is denoted by $\mathcal{P}_{\text{sf}}$.

Remark 6.1.2 (1) By the continuity of $\{\rho(t)\}_{0\leqq t\leqq T}$ and $\{S(t)\}_{0\leqq t\leqq T}$,

$$\{r\varphi^0(t)\rho(t) + \mu\varphi^1(t)S(t)\}_{0\leqq t\leqq T} \in \mathscr{L}^1_{\text{loc}} \quad \text{and} \quad \{\sigma\varphi^1(t)S(t)\}_{0\leqq t\leqq T} \in \mathscr{L}^2_{\text{loc}}$$

if $\{\varphi^0(t)\}_{0\leqq t\leqq T} \in \mathscr{L}^1_{\text{loc}}$ and $\{\varphi^1(t)\}_{0\leqq t\leqq T} \in \mathscr{L}^2_{\text{loc}}$. Hence the integrals appearing in (6.1.2) are well defined.
(2) By discretizing time, it can be seen that being self-financing means no inflow and outflow of money. In fact, suppose a next trade after time t occurs at $t + \Delta$. In this case, the portfolio $\varphi(t + \Delta)$ is the trading strategy taken at time t. If there is no inflow or outflow of money, then the wealth $V(t;\varphi)$ of the trader at time t must coincide with the amount of investment: $\langle\varphi(t), X(t)\rangle = \langle\varphi(t + \Delta), X(t)\rangle$. Hence

$$V(t + \Delta;\varphi) - V(t;\varphi) = \langle\varphi(t + \Delta), X(t + \Delta) - X(t)\rangle.$$

Letting $\Delta \to 0$, we obtain

$$\mathrm{d}V(t;\varphi) = \langle\varphi(t), \mathrm{d}X(t)\rangle = \varphi^0(t)\mathrm{d}\rho(t) + \varphi^1(t)\mathrm{d}S(t).$$

Thus (6.1.2) holds.
(3) Since $B(0) = 0$, $\mathscr{F}_0^B = \{\emptyset, \Omega\} \cup \mathscr{N}$. Hence every $\mathscr{F}_0^B$-measurable function is a constant function. In particular, $\varphi^0(0)$, $\varphi^1(0)$, and $V(0;\varphi)$ are all constants.

Example 6.1.3 A constant portfolio, that is, a portfolio such that $\varphi(t) = \varphi(0)$ $(0 \leqq t \leqq T)$ is self-financing.

Definition 6.1.4 Set

$$\xi(t) = \frac{1}{\rho(t)} = \mathrm{e}^{-rt}, \quad \overline{S}(t) = \xi(t)S(t), \quad \overline{X}(t) = \xi(t)X(t) = (1, \overline{S}(t)).$$

$\{\overline{S}(t)\}_{0 \leqq t \leqq T}$ is called a stock price process discounted by numeraire $\{\rho(t)\}_{0 \leqq t \leqq T}$.

Since $\mathrm{d}\xi(t) = -r\xi(t)\mathrm{d}t$, by Itô's formula,

$$\mathrm{d}\overline{S}(t) = S(t)\mathrm{d}\xi(t) + \xi(t)\mathrm{d}S(t) = (\mu - r)\overline{S}(t)\mathrm{d}t + \sigma\overline{S}(t)\mathrm{d}B(t). \tag{6.1.3}$$

Being "self-financing" is common in both markets X and $\overline{X}$ as follows.

Lemma 6.1.5 *Let $\overline{\mathcal{P}}_{\mathrm{sf}}$ be the totality of portfolios which are self-financing in the market $\overline{X}$. Then $\overline{\mathcal{P}}_{\mathrm{sf}} = \mathcal{P}_{\mathrm{sf}}$. In particular, for $\varphi \in \mathcal{P}_{\mathrm{sf}}$, $\overline{V}(t;\varphi) = \xi(t)V(t;\varphi)$ satisfies*

$$\overline{V}(t;\varphi) = \overline{V}_0(\varphi) + \int_0^t \varphi^1(s)\mathrm{d}\overline{S}(s). \tag{6.1.4}$$

Proof Let $\varphi \in \mathcal{P}_{\mathrm{sf}}$. By Itô's formula and (6.1.3),

$$\begin{aligned}\mathrm{d}\overline{V}(t;\varphi) &= \xi(t)\mathrm{d}V(t;\varphi) + V(t;\varphi)\mathrm{d}\xi(t) \\ &= \xi(t)\{(r\varphi^0(t)\rho(t) + \mu\varphi^1(t)S(t))\mathrm{d}t + \sigma\varphi^1(t)S(t)\mathrm{d}B(t)\} \\ &\quad - r\xi(t)\{\varphi^0(t)\rho(t) + \varphi^1(t)S(t)\}\mathrm{d}t \\ &= \xi(t)S(t)\varphi^1(t)\{(\mu - r)\mathrm{d}t + \sigma\mathrm{d}B(t)\} = \varphi^1(t)\mathrm{d}\overline{S}(t).\end{aligned}$$

Thus $\varphi \in \overline{\mathcal{P}}_{\mathrm{sf}}$.

Conversely, suppose $\varphi \in \overline{\mathcal{P}}_{\mathrm{sf}}$. Since $V(t;\varphi) = \rho(t)\overline{V}(t;\varphi)$, by Itô's formula, (6.1.4), and (6.1.3),

$$\begin{aligned}\mathrm{d}V(t;\varphi) &= \rho(t)\mathrm{d}\overline{V}(t;\varphi) + \overline{V}(t;\varphi)\mathrm{d}\rho(t) \\ &= \rho(t)\varphi^1(t)\{(\mu - r)\overline{S}(t)\mathrm{d}t + \sigma\overline{S}(t)\mathrm{d}B(t)\} + r\rho(t)\{\varphi^0(t) + \varphi^1(t)\overline{S}(t)\}\mathrm{d}t \\ &= r\varphi^0(t)\rho(t)\mathrm{d}t + \varphi^1(t)S(t)\{\mu\mathrm{d}t + \sigma\mathrm{d}B(t)\} = \varphi^0(t)\mathrm{d}\rho(t) + \varphi^1(t)\mathrm{d}S(t).\end{aligned}$$

Hence $\varphi \in \mathcal{P}_{\mathrm{sf}}$. □

We close this section with a remark on $\mathcal{P}_{\mathrm{sf}}$.

Lemma 6.1.6 *Given $a \in \mathbb{R}$ and $\{\varphi^1(t)\}_{0 \leqq t \leqq T} \in \mathscr{L}^2_{\mathrm{loc}}$, define*

$$\varphi^0(t) = a + \int_0^t \varphi^1(s)\mathrm{d}\overline{S}(s) - \varphi^1(t)\overline{S}(t).$$

Then $\varphi = (\varphi^0, \varphi^1) \in \mathcal{P}_{\rm sf}$ *and* $V(0;\varphi) = a$.

Proof By Lemma 6.1.5, $\varphi = (\varphi^0, \varphi^1) \in \mathcal{P}$ is self-financing if and only if

$$\varphi^0(t) + \varphi^1(t)\overline{S}(t) = V(0;\varphi) + \int_0^t \varphi^1(s)\mathrm{d}\overline{S}(s) \quad (0 \leqq t \leqq T).$$

Solving this in $\varphi^0(t)$, we obtain the desired assertion. □

Example 6.1.7 Suppose $\varphi^1(t) = b$ $(0 \leqq t \leqq T)$ for some $b \in \mathbb{R}$. If we apply the lemma to this φ^1, then $\varphi^0(t) = a - bs_0$ $(0 \leqq t \leqq T)$. Thus, we arrive at a constant portfolio.

6.2 Arbitrage Opportunity, Equivalent Martingale Measures

Definition 6.2.1 (1) $\varphi \in \mathcal{P}_{\rm sf}$ is said to be admissible if

$$\mathbf{P}(V(t;\varphi) \geqq C \text{ for any } t \in [0,T]) = 1$$

for some $C \in \mathbb{R}$. The totality of admissible portfolios is denoted by $\mathcal{P}_{\rm adm}$.
(2) A portfolio $\varphi \in \mathcal{P}_{\rm adm}$ is called an **arbitrage opportunity** if $V(0;\varphi) = 0$, $V(t;\varphi) \geqq 0$, $\mathbf{P}$-a.s. for every $t \in [0,T]$, and $\mathbf{P}(V(t;\varphi) > 0) > 0$. Denote by $\mathcal{P}_{\rm arb}$ the totality of arbitrage opportunities.
(3) An **equivalent martingale measure** is a probability measure $\mathbf{Q}$ on $(\Omega, \mathscr{F})$ satisfying

(i) $\mathbf{Q}$ is equivalent to $\mathbf{P}$, that is, $A \in \mathscr{F}$ is $\mathbf{Q}$-null if and only if it is $\mathbf{P}$-null.
(ii) Under $\mathbf{Q}$, $\{\overline{S}(t)\}_{0\leqq t\leqq T}$ is a local martingale.

The totality of equivalent martingale measures is denoted by $\mathcal{EMM}$.

Remark 6.2.2 (1) $\varphi \in \mathcal{P}_{\rm adm}$ is also called a tame portfolio. The condition $V(t;\varphi) \geqq C$ means that an investor has to keep the debt within manageable limits.
(2) The trading strategy $\varphi \in \mathcal{P}_{\rm arb}$ enables an investor to start with no asset, invest without running into debt, and end up with a profit with positive probability.
(3) There exists $\varphi \in \mathcal{P}_{\rm sf} \setminus \mathcal{P}_{\rm adm}$ such that $V(0;\varphi) = 0$ and $V(T;\varphi) > 0$ $\mathbf{P}$-a.s. To see this, we assume $r = \mu = 0$. Let

$$Y(t) = \int_0^t \frac{1}{\sqrt{T-s}}\mathrm{d}B(s),$$

$a > 0$, and $\tau_a = \inf\{t \geqq 0\,;\, Y(t) \geqq a\}$. By Theorem 2.5.3, there exists a Brownian motion $\hat{B}$ such that

$$Y(t) = \hat{B}\Big(\int_0^t \frac{1}{T-s}\mathrm{d}s\Big) \tag{6.2.1}$$

Then Theorem 3.2.1 implies $\mathbf{P}(\tau_a < T) = 1$.

Set $\varphi^1(t) = \frac{1}{\sigma S(t)\sqrt{T-t}}\mathbf{1}_{[0,\tau_a)}(t)$. On account of Lemma 6.1.6, there exists a $\{\varphi^0(t)\}_{0\leqq t\leqq T} \in \mathscr{L}^1_{\mathrm{loc}}$ such that $\varphi = (\varphi^0, \varphi^1) \in \mathcal{P}_{\mathrm{sf}}$ and $V(0;\varphi) = 0$. Since $r = 0$,

$$V(t;\varphi) = \int_0^t \sigma S(s)\varphi^1(s)\mathrm{d}B(s) = Y_{t\wedge\tau_a}. \tag{6.2.2}$$

In conjunction with (6.2.1) and Theorem 1.5.20, this yields

$$\mathbf{P}(\inf_{0\leqq t\leqq T} V(t;\varphi) \geqq C) < 1$$

for any $C \in \mathbb{R}$. Thus $\varphi \notin \mathcal{P}_{\mathrm{adm}}$. Moreover, by (6.2.2) and the definition of τ_a, $V(T;\varphi) = a > 0$.

Theorem 6.2.3 *Set* $\alpha = \frac{r-\mu}{\sigma}$ *and define*

$$\widehat{\mathbf{P}}(A) = \mathbf{E}[\mathrm{e}^{\alpha B(t) - \frac{\alpha^2}{2}T}; A] \quad (A \in \mathscr{F}) \quad \textit{and} \quad \widehat{B}(t) = B(t) - \alpha t \quad (0 \leqq t \leqq T).$$

(1) $\widehat{\mathbf{P}}$ *is a probability measure on* $(\Omega, \mathscr{F})$*. Moreover,* $\{\widehat{B}(t)\}_{0\leqq t\leqq T}$ *is an* $(\mathscr{F}^B_t)$*-Brownian motion under* $\widehat{\mathbf{P}}$.
(2) $\mathcal{EMM} = \{\widehat{\mathbf{P}}\}$.
(3) *Let* $\varphi \in \mathcal{P}_{\mathrm{sf}}$*. Under* $\widehat{\mathbf{P}}$*,* $\{\overline{V}(t;\varphi)\}_{0\leqq t\leqq T}$ *is a local martingale.*
(4) $\mathcal{P}_{\mathrm{arb}} = \emptyset$.

Proof (1) While the assertion is a direct application of the Girsanov theorem (Theorem 4.6.2), we give an elementary proof.

Let $0 \leqq s_1 < \cdots < s_n \leqq s < t$, $f \in C_{\mathrm{b}}(\mathbb{R}^n)$ and $\lambda \in \mathbb{R}$. Then, by the independence of $B(t) - B(s)$ and $\mathscr{F}^B_s$,

$$\begin{aligned}
&\mathbf{E}_{\widehat{\mathbf{P}}}[f(\widehat{B}(s_1), \dots, \widehat{B}(s_n))\exp(\mathrm{i}\lambda(\widehat{B}(t) - \widehat{B}(s)))] \\
&= \mathbf{E}\Big[f(\widehat{B}(s_1), \dots, \widehat{B}(s_n))\exp\Big(\alpha B(s) - \frac{1}{2}\alpha^2 s\Big) \\
&\quad \times \exp\Big((\mathrm{i}\lambda + \alpha)(B(t) - B(s)) - \mathrm{i}\lambda\alpha(t-s) - \frac{\alpha^2}{2}(t-s)\Big)\Big] \\
&= \mathbf{E}_{\widehat{\mathbf{P}}}[f(\widehat{B}(s_1), \dots, \widehat{B}(s_n))]\mathrm{e}^{-\frac{1}{2}\lambda^2(t-s)}.
\end{aligned}$$

This implies that $\{\widehat{B}(t)\}_{0\leqq t\leqq T}$ is an $(\mathscr{F}^{\widehat{B}}_t)$-Brownian motion under $\widehat{\mathbf{P}}$, where $\mathscr{F}^{\widehat{B}}_t = \sigma\{\widehat{B}(s); s \leqq T\} \vee \{\widehat{\mathbf{P}}\text{-null sets}\}$. Since $\widehat{B}(t) = B(t) - \alpha t$ and $\widehat{\mathbf{P}}$ and $\mathbf{P}$ are equivalent, $\mathscr{F}^{\widehat{B}}_t = \mathscr{F}^B_t$. Thus $\{\widehat{B}(t)\}_{0\leqq t\leqq T}$ is an $(\mathscr{F}^B_t)$-Brownian motion under $\widehat{\mathbf{P}}$.

(2) By (6.1.3), we have

$$\mathrm{d}\overline{S}(t) = \sigma\overline{S}(t)\mathrm{d}\widehat{B}(t). \tag{6.2.3}$$

Then, by (1), $\{\overline{S}(t)\}_{0\leqq t\leqq T}$ is a local martingale under $\widehat{\mathbf{P}}$. Thus $\widehat{\mathbf{P}} \in \mathcal{EMM}$.

Next let $\mathbf{Q} \in \mathcal{EMM}$. The proof of the second assertion is completed once we have shown $\mathbf{Q} = \widehat{\mathbf{P}}$. Since $\mathbf{Q}$ is equivalent to $\mathbf{P}$, there exists a $Z \in L^1(\mathbf{P})$ such that $Z > 0$, $\mathbf{P}$-a.s. and

$$\mathbf{Q}(A) = \mathbf{E}[Z; A] \quad (A \in \mathscr{F}). \tag{6.2.4}$$

Define $Z(t) = \mathbf{E}[Z|\mathscr{F}_t^B]$ $(0 \leqq t \leqq T)$. By Corollary 2.6.6 and a standard stopping time argument, $\{Z(t)\}_{0\leqq t\leqq T}$ is a continuous martingale.

Set $\tau = \inf\{t \leqq T \,|\, Z(t) = 0\}$, where $\tau = \infty$ if $\{\cdots\} = \emptyset$. Since $\{\tau \leqq T\} \in \mathscr{F}^B_{\tau\wedge T}$, by the optional sampling theorem (Theorem 1.5.11), we have

$$0 = \mathbf{E}[Z(\tau); \tau \leqq T] = \mathbf{E}[Z(\tau \wedge T); \tau \leqq T] = \mathbf{E}[Z(T); \tau \leqq T].$$

This implies that $\mathbf{P}(\tau \leqq T) = 0$, for $Z(T) = Z > 0$, $\mathbf{P}$-a.s. Hence

$$\inf_{0\leqq t\leqq T} Z(t) > 0, \quad \mathbf{P}\text{-a.s.} \tag{6.2.5}$$

As remarked after Theorem 2.6.2, there is an $\{f(t)\}_{0\leqq t\leqq T} \in \mathscr{L}^2_{\mathrm{loc}}$ satisfying

$$Z(t) = 1 + \int_0^t f(s)\mathrm{d}B(s) \quad (0 \leqq t \leqq T).$$

If we set $g(t) = \frac{f(t)}{Z(t)}$, then by (6.2.5), $\{g(t)\}_{0\leqq t\leqq T} \in \mathscr{L}^2_{\mathrm{loc}}$. Furthermore,

$$Z(t) = 1 + \int_0^t g(s)Z(s)\mathrm{d}B(s) \quad (0 \leqq t \leqq T). \tag{6.2.6}$$

Set $\tau_n = \inf\{t \geqq 0 \,|\, \overline{S}(t) \geqq n\}$ $(n = 1, 2, \ldots)$, where $\tau_n = \infty$ if $\{\cdots\} = \emptyset$. Since $\{\overline{S}(t)\}_{0\leqq t\leqq T}$ is a local martingale under $\mathbf{Q}$, $\{\overline{S}^{\tau_n}(t)\}_{0\leqq t\leqq T}$ is a bounded martingale under $\mathbf{Q}$. Hence for $s < t$ and $A \in \mathscr{F}_s^B$,

$$\mathbf{E}_{\mathbf{Q}}[\overline{S}^{\tau_n}(t); A] = \mathbf{E}_{\mathbf{Q}}[\overline{S}^{\tau_n}(s); A].$$

From this and the identity

$$\mathbf{E}_{\mathbf{Q}}[\overline{S}^{\tau_n}(u); C] = \mathbf{E}[\overline{S}^{\tau_n}(u)Z; C] = \mathbf{E}[\overline{S}^{\tau_n}(u)Z(u); C] \quad (0 \leqq u \leqq T, C \in \mathscr{F}_u^B),$$

it follows that $\{\overline{S}^{\tau_n}(t)Z(t)\}_{0\leqq t\leqq T}$ is a martingale under $\mathbf{P}$.

By Itô's formula, (6.1.3), and (6.2.6), we obtain

$$\begin{aligned}\overline{S}(t)Z(t) = & s + \int_0^t \overline{S}(s)g(s)Z(s)\mathrm{d}B(s) + \int_0^t \sigma\overline{S}(s)Z(s)\mathrm{d}B(s) \\ & + \int_0^t \overline{S}(s)Z(s)\{(\mu - r) + \sigma g(s)\}\mathrm{d}s.\end{aligned}$$

Since $\{\overline{S}^{\tau_n}(t)Z(t)\}_{0\leqq t\leqq T}$ is a martingale under $\mathbf{P}$, this yields

$$\int_0^t \overline{S}(s)Z(s)\{(\mu - r) + \sigma g(s)\}\mathrm{d}s = 0 \quad (0 \leqq t \leqq T).$$

Hence $g(s) = \alpha$ $(\leqq s \leqq T)$. Substitute this into (6.2.6) to see

$$Z(t) = 1 + \int_0^t \alpha Z(t)\mathrm{d}B(t) \quad (0 \leqq t \leqq T).$$

Thus $Z(t) = \mathrm{e}^{\alpha B(t) - \frac{\alpha^2}{2}t}$ $(t \leqq T)$, and $Z = Z(T) = \mathrm{e}^{\alpha B(T) - \frac{\alpha^2}{2}T}$. In conjunction with the definition (6.2.4) of $\mathbf{Q}$, we have $\mathbf{Q} = \widehat{\mathbf{P}}$.
(3) Since $\{\overline{S}(t)\}_{0\leqq t\leqq T}$ is a local martingale under $\widehat{\mathbf{P}}$, so is $\{\overline{V}(t;\varphi)\}_{0\leqq t\leqq T}$ by (6.1.4).
(4) It suffices to show that if $\varphi \in \mathcal{P}_{\mathrm{adm}}$ satisfies $V_0(\varphi) = 0$ and $V(t;\varphi) \geqq 0$ $(0 \leqq t \leqq T)$ $\mathbf{P}$-a.s., then $\varphi \notin \mathcal{P}_{\mathrm{arb}}$.

Let φ be as above, and put $\sigma_n = \inf\{t \geqq 0 \,|\, \overline{V}(t;\varphi) \geqq n\}$ $(n = 1, 2, \dots)$, where $\sigma_n = \infty$ if $\{\cdots\} = \emptyset$. By (3), $\{\overline{V}(t \wedge \sigma_n;\varphi)\}_{0\leqq t\leqq T}$ is a martingale under $\widehat{\mathbf{P}}$.

By the equivalence of $\widehat{\mathbf{P}}$ and $\mathbf{P}$, $V(0;\varphi) = 0$ and $V(T;\varphi) \geqq 0$, $\widehat{\mathbf{P}}$-a.s. and there exists a $C \in \mathbb{R}$ such that $\widehat{\mathbf{P}}(\overline{V}(t;\varphi) \geqq C\,(0 \leqq t \leqq T)) = 1$. Applying Fatou's lemma, we obtain

$$0 \leqq \mathbf{E}_{\widehat{\mathbf{P}}}[\overline{V}(T;\varphi)] \leqq \liminf_{n\to\infty} \mathbf{E}_{\widehat{\mathbf{P}}}[\overline{V}(T \wedge \tau_n;\varphi)] = \mathbf{E}_{\widehat{\mathbf{P}}}[\overline{V}(0;\varphi)] = 0.$$

This implies $\overline{V}(T;\varphi) = 0$, $\widehat{\mathbf{P}}$-a.s. and hence also $\mathbf{P}$-a.s. Hence $\varphi \notin \mathcal{P}_{\mathrm{arb}}$. □

Remark 6.2.4 By (6.2.3),

$$\overline{S}(t) = \exp\Big(\sigma\widehat{B}(t) - \frac{\sigma^2}{2}t\Big) \quad (0 \leqq t \leqq T).$$

Thus, $\{\overline{S}(t)\}_{0\leqq t\leqq T}$ is a martingale under $\widehat{\mathbf{P}}$.

6.3 Pricing Formula

In this section, we shall give a formula to give the price of a claim by using the equivalent martingale measure $\widehat{\mathbf{P}}$ defined in Theorem 6.2.3.

Consider a derivative of the stock price process $S = \{S(t)\}_{0\leqq t\leqq T}$, which is paid off at maturity T. It is an $\mathscr{F}_T^S$-measurable function. Since

$$B(t) = \frac{1}{\sigma}\Big\{\log S(t) - \Big(\mu - \frac{\sigma^2}{2}\Big)t\Big\},$$

the $\mathscr{F}_T^S$-measurability coincides with $\mathscr{F}$-measurability. Thus, such a claim is an $\mathscr{F}$-measurable function.

Definition 6.3.1 (1) C_{E} denotes the totality of $\mathscr{F}$-measurable functions F with $\mathbf{P}(F \geqq C) = 1$ for some $C \in \mathbb{R}$. An element of C_{E} is called a **European contingent claim**.
(2) $F \in C_{\mathrm{E}}$ is said to be **replicable** if there exists a $\varphi \in \mathcal{P}_{\mathrm{adm}}$ such that $V(T;\varphi) = F$. In this case, we say φ replicates (hedges) F.

A natural question is "which $F \in C_{\mathrm{E}}$ can be replicated?", and the following is an answer.

Theorem 6.3.2 *Given $F \in L^1(\widehat{\mathbf{P}}) \cap C_{\mathrm{E}}$, there exists a $\varphi \in \mathcal{P}_{\mathrm{adm}}$ which replicates F and satisfies $V(0;\varphi) = \mathbf{E}_{\widehat{\mathbf{P}}}[\xi(T)F]$.*

Remark 6.3.3 The theorem asserts that $\mathbf{E}_{\widehat{\mathbf{P}}}[\xi(T)F]$ is the price of F at time 0 since there is no arbitrage opportunity. In fact, suppose we buy F at the price $x < \mathbf{E}_{\widehat{\mathbf{P}}}[\xi(T)F]$. Then we take the strategy $-\varphi$, where φ is the portfolio described in the theorem. By this investment, we still have $\mathbf{E}_{\widehat{\mathbf{P}}}[\xi(T)F]-x$ at the beginning, and invest it into ρ by a constant portfolio. Receiving the payoff F, we end up with $(\mathbf{E}_{\widehat{\mathbf{P}}}[\xi(T)F] - x)\mathrm{e}^{rT} > 0$ remaining. Thus an arbitrage opportunity occurs. Next if we sell F at the price $y > \mathbf{E}_{\widehat{\mathbf{P}}}[\xi(T)F]$, then taking the strategy φ and investing $y-\mathbf{E}_{\widehat{\mathbf{P}}}[\xi(T)F]$ into ρ, we earn $(y-\mathbf{E}_{\widehat{\mathbf{P}}}[\xi(T)F])\mathrm{e}^{rT} > 0$ at maturity. This is also an arbitrage opportunity. Thus, trades at a price different from $\mathbf{E}_{\widehat{\mathbf{P}}}[\xi(T)F]$ cause an arbitrage opportunity.

Proof As was seen in the proof of Theorem 6.2.3, $\widehat{\mathscr{F}_t^B}$, the σ-field constructed as $\mathscr{F}_t^B$ with $\widehat{B}$ and $\widehat{\mathbf{P}}$ instead of B and $\mathbf{P}$, coincides with $\mathscr{F}_t^B$. Moreover, by the equivalence of $\mathbf{P}$ and $\widehat{\mathbf{P}}$, the space defined as $\mathscr{L}^2_{\mathrm{loc}}$ with $\widehat{\mathbf{P}}$ instead of $\mathbf{P}$ coincides with $\mathscr{L}^2_{\mathrm{loc}}$ itself. Due to these observations, as an application of Itô's representation theorem (Theorem 2.6.2) to $\{\widehat{B}(t)\}_{0\leqq t\leqq T}$, there is an $\{f(t)\}_{0\leqq t\leqq T} \in \mathscr{L}^2_{\mathrm{loc}}$ such that

$$\mathbf{E}_{\widehat{\mathbf{P}}}[\xi(T)F|\mathscr{F}_t^B] = \mathbf{E}_{\widehat{\mathbf{P}}}[\xi(T)F] + \int_0^t f(s)\mathrm{d}\widehat{B}(s) \quad (0 \leqq t \leqq T).$$

Define $\{\varphi^1(t)\}_{0\leqq t\leqq T} \in \mathscr{L}^2_{\mathrm{loc}}$ by $\varphi^1(t) = \frac{f(t)}{\sigma\overline{S}(t)}$. By virtue of Lemma 6.1.6, using this $\{\varphi^1(t)\}_{0\leqq t\leqq T}$, we construct a $\varphi = (\varphi^0(t), \varphi^1(t)) \in \mathcal{P}_{\mathrm{sf}}$ such that $V(0;\varphi) = \mathbf{E}_{\widehat{\mathbf{P}}}[\xi(T)F]$. Then, by (6.1.4) and (6.2.3),

$$\overline{V}(t;\varphi) = \mathbf{E}_{\widehat{\mathbf{P}}}[\xi(T)F] + \int_0^t \sigma\varphi^1(s)\overline{S}(s)\mathrm{d}\widehat{B}(s) = \mathbf{E}_{\widehat{\mathbf{P}}}[\xi(T)F|\mathscr{F}_t^B] \quad (0 \leqq t \leqq T).$$

Since $\xi(T)F$ is bounded from below, $\varphi \in \mathcal{P}_{\mathrm{adm}}$. Moreover, substitute $t = T$ to see $\overline{V}(T;\varphi) = \xi(T)F$, which means φ replicates F. □

As was seen in the above proof, finding a replicating portfolio reduces to finding a process $\{f(t)\}_{0\leqq t\leqq T}$ satisfying

$$\xi(T)F = \mathbf{E}_{\widehat{\mathbf{P}}}[\xi(T)F] + \int_0^T f(t)\mathrm{d}\widehat{B}(t). \tag{6.3.1}$$

By the Clark–Ocone formula (Theorem 5.3.5), if F is in $\mathbb{D}^{1,2}$, then the desired process is given by

$$f(t) = \xi(T)\mathbf{E}_{\widehat{\mathbf{P}}}[\overbrace{(\nabla F)}^{\cdot}(t)|\mathscr{F}_t^B].$$

If F is of the form $g(\overline{S}(T))$, then a more precise expression is available.

Proposition 6.3.4 *Let $g \in C^1(\mathbb{R})$ be of polynomial growth order, and $F = g(\overline{S}(T))$. Set*

$$f(t) = \sigma\xi(T)\overline{S}(t)\int_{\mathbb{R}} \frac{1}{\sqrt{2\pi\sigma(T-t)}} xg'(\overline{S}(t)\mathrm{e}^x)\exp\Big(-\frac{(x+\frac{\sigma^2}{2}(T-t))^2}{2\sigma(T-t)}\Big)\mathrm{d}x.$$

Then $\{f(t)\}_{0\leqq t\leqq T}$ satisfies (6.3.1).

In particular, let

$$\varphi^1(t) = \xi(T)\int_{\mathbb{R}} \frac{1}{\sqrt{2\pi\sigma(T-t)}} xg'(\overline{S}(t)\mathrm{e}^x)\exp\Big(-\frac{(x+\frac{\sigma^2}{2}(T-t))^2}{2\sigma(T-t)}\Big)\mathrm{d}x,$$

$$\varphi^0(t) = \mathbf{E}_{\widehat{\mathbf{P}}}[\xi(T)F] + \int_0^t \varphi^1(s)\mathrm{d}\overline{S}(s) - \varphi^1(t)\overline{S}(t).$$

Then $\varphi = (\varphi^0, \varphi^1)$ replicates F.

Proof Since $\overline{S}(T) = \mathrm{e}^{\sigma\widehat{B}(T)-\frac{\sigma^2}{2}T}$, by Corollary 5.3.2,

$$\overbrace{(\nabla F)}^{\cdot}(t) = \sigma g'(\overline{S}(T))\overline{S}(T).$$

Rewriting as $\overline{S}(T) = \overline{S}(t)\widehat{S}(t,T)$, where $\widehat{S}(t,T) = \mathrm{e}^{\sigma(\widehat{B}(T)-\widehat{B}(t))-\frac{\sigma^2}{2}(T-t)}$, and using the independence of $\widehat{B}(T) - \widehat{B}(t)$ and $\mathscr{F}_t^B$, we obtain

$$\mathbf{E}_{\widehat{\mathbf{P}}}[g'(\overline{S}(T))\overline{S}(T)|\mathscr{F}_t^B] = \overline{S}(t)\mathbf{E}_{\widehat{\mathbf{P}}}[g'(y\widehat{S}(t,T))\widehat{S}(t,T)]\big|_{y=\overline{S}(t)}.$$

Since $\sigma(\widehat{B}(T) - \widehat{B}(t)) - \frac{\sigma^2}{2}(T-t)$ obeys the normal distribution with mean $-\frac{\sigma^2}{2}(T-t)$ and variance $\sigma^2(T-t)$, this implies the desired expression of $f(t)$.

The second assertion has been seen in the proof of Theorem 6.3.2. □

We now proceed to determining the price of a European contingent claim $F \in C_{\mathrm{E}}$. For this purpose, we introduce two candidates of price as follows. Let

$$\mathcal{P}_B(F) = \{\varphi \in \mathcal{P}_{\text{adm}} \mid V(T;\varphi) + F \geqq 0\},$$
$$\mathcal{P}_S(F) = \{\psi \in \mathcal{P}_{\text{adm}} \mid V(T;\psi) - F \geqq 0\},$$
$$\pi_B(F) = \sup\{y \mid V(0;\varphi) = -y \text{ for some } \varphi \in \mathcal{P}_B\},$$
$$\pi_S(F) = \inf\{z \mid V(0;\psi) = z \text{ for some } \psi \in \mathcal{P}_S\}.$$

Remark 6.3.5 Suppose a buyer buys F at the price y at time 0, hence starts with initial value $-y$, and then invests by a portfolio φ. What he/she expects is that the sum $V(T;\varphi) + F$, the total assets at maturity, is non-negative, otherwise he/she suffers a loss. Hence $\pi_B(F)$ is the supremum of prices which a buyer accepts.

Contrarily, suppose a seller sells F at the price z at time 0. By a portfolio ψ with initial value z, he/she earns $V(T;\psi)$. After paying F, what remains is $V(T;\psi) - F$, which he/she hopes to be non-negative. Thus $\pi_S(F)$ is the infimum of prices which a seller sets.

Theorem 6.3.6 *Let* $F \in C_{\mathrm{E}}$.
(1) $\pi_B(F) \leqq \pi_S(F)$.
(2) *If* $F \in L^1(\widehat{\mathbf{P}})$ *in addition, then* $\pi_B(F) = \pi_S(F) = \mathbf{E}_{\widehat{\mathbf{P}}}[\xi(T)F]$.

Proof (1) Let $\phi \in \mathcal{P}_{\text{adm}}$. By Theorem 6.2.3(3), $\{\overline{V}(t;\phi)\}_{0 \leqq t \leqq T}$ is a local martingale under $\widehat{\mathbf{P}}$. Since $\widehat{\mathbf{P}}$ and $\mathbf{P}$ are equivalent, $\widehat{\mathbf{P}}(\overline{V}(t;\phi) \geqq C\ (0 \leqq t \leqq T)) = 1$ for some $C \in \mathbb{R}$. By a similar argument as in the proof of Theorem 6.2.3(4), we have

$$\mathbf{E}_{\widehat{\mathbf{P}}}[\overline{V}(t;\phi)] \leqq \mathbf{E}_{\widehat{\mathbf{P}}}[\overline{V}(0;\phi)]. \tag{6.3.2}$$

Since $\overline{V}(0;\phi)$ is a constant, in conjunction with the boundedness from below, this implies $\overline{V}(t;\phi) \in L^1(\widehat{\mathbf{P}})$.

Take $\varphi \in \mathcal{P}_B$ and $\psi \in \mathcal{P}_S$. Set $V(0;\varphi) = -y$ and $V(0;\psi) = z$. Then $\varphi + \psi \in \mathcal{P}_{\text{adm}}$ and $V(t;\varphi) + V(t;\psi) = \{V(t;\varphi) + F\} + \{V(t;\psi) - F\} \geqq 0$. By (6.3.2),

$$0 \leqq \mathbf{E}_{\widehat{\mathbf{P}}}[\overline{V}(t;\varphi) + \overline{V}(t)(\psi)] \leqq -y + z.$$

Hence $y \leqq z$. Taking the supremum over y and the infimum over z, we obtain $\pi_B(F) \leqq \pi_S(F)$.

(2) Since $F \in L^1(\widehat{\mathbf{P}})$, by Theorem 6.3.2, there exists a $\varphi \in \mathcal{P}_{\text{adm}}$, which replicates F and satisfies $V(0;\varphi) = \mathbf{E}_{\widehat{\mathbf{P}}}[\xi(T)F]$. This φ belongs to $\mathcal{P}_S(F)$, because $V(T;\varphi) - F = 0$. Hence

$$\pi_S(F) \leqq \mathbf{E}_{\widehat{\mathbf{P}}}[\xi(T)F]. \tag{6.3.3}$$

Next set $F_n = F \wedge n$ $(n = 1, 2, \ldots)$. By Theorem 6.3.2, for each n, take a $\varphi_n \in \mathcal{P}_{\text{adm}}$ replicating $-F_n$ and satisfying $V(0;\varphi_n) = -\mathbf{E}_{\widehat{\mathbf{P}}}[\xi(T)F_n]$.

Since $F_n \leqq F$, $V(T;\varphi_n) + F \geqq V(T;\varphi_n) + F_n = 0$. Thus $\varphi_n \in \mathcal{P}_B$. Hence

$$\pi_B(F) \geqq \mathbf{E}_{\widehat{\mathbf{P}}}[\xi(T)F_n].$$

By the boundedness from below of F and the monotone convergence theorem, letting $n \to \infty$, we obtain

$$\pi_B(F) \geqq \mathbf{E}_{\widehat{\mathbf{P}}}[\xi(T)F].$$

In conjunction with the assertion (1) and (6.3.3), this implies the desired identity. □

In Theorem 6.3.6 (2), the identity means the expectation is an amount accepted by both buyer and seller. Hence the **price** of $F \in C_{\mathrm{E}} \cap L^1(\widehat{\mathbf{P}})$ is given by

$$\pi(F) = \mathbf{E}_{\widehat{\mathbf{P}}}[\xi(T)F].$$

Theorem 6.3.7 *Let $f \in C(\mathbb{R})$ be bounded from below and at most of polynomial growth. Then $F = f(S(T)) \in C_{\mathrm{E}} \cap L^1(\widehat{\mathbf{P}})$, and*

$$\pi(F) = \mathrm{e}^{-rT} \int_{\mathbb{R}} f(s_0 \mathrm{e}^{(r-\frac{\sigma^2}{2})T} \mathrm{e}^x) \frac{1}{\sqrt{2\sigma^2 T}} \mathrm{e}^{-\frac{x^2}{2\sigma^2 T}} \,\mathrm{d}x. \tag{6.3.4}$$

It should be noted that in the pricing formula (6.3.4), only r and σ are involved, and μ is not. The stock price process $\{S(t)\}_{0 \leqq t \leqq T}$ reflects on prices via only σ, which indicates how risky the market is.

Proof By (6.2.3), $\overline{S}(t) = s_0 \exp(\sigma \widehat{B}(t) - \frac{\sigma^2}{2} T)$. Hence

$$S(T) = s_0 \mathrm{e}^{rT} \exp\Big(\sigma \widehat{B}(T) - \frac{\sigma^2}{2} T\Big).$$

Since $\{\widehat{B}(t)\}_{0 \leqq t \leqq T}$ is a Brownian motion starting at 0 under $\widehat{\mathbf{P}}$, $\sigma \widehat{B}(T)$ obeys the normal distribution with mean 0 and variance $\sigma^2 T$. Thus $F \in L^1(\widehat{\mathbf{P}})$ and

$$\begin{aligned}
\pi(F) &= \mathbf{E}_{\widehat{\mathbf{P}}}[\xi(T) f(S(T))] = \mathrm{e}^{-rT} \mathbf{E}_{\widehat{\mathbf{P}}}\Big[f\Big(s_0 \mathrm{e}^{rT} \exp\Big(\sigma \widehat{B}(T) - \frac{\sigma^2}{2} T\Big)\Big)\Big] \\
&= \mathrm{e}^{-rT} \int_{\mathbb{R}} f(s_0 \mathrm{e}^{rT} \mathrm{e}^{x - \frac{\sigma^2}{2} T}) \frac{1}{\sqrt{2\sigma^2 T}} \mathrm{e}^{-\frac{x^2}{2\sigma^2 T}} \,\mathrm{d}x.
\end{aligned}$$

This implies (6.3.4). □

A **European call option** is a contingent claim whose payoff at maturity is $C = (S(T) - K)^+$. The holder of this claim is allowed to buy a unit of stock at the **exercise price** K at maturity. The payoff is computed as follows. If the price of the stock is more than K, then the holder exercises the option; buys a stock at the price K, immediately sells it at the market price $S(T)$, and earns a profit $S(T) - K$. If $K \geqq S(T)$, the holder never exercises the option, and makes no profit. Thus the payoff is $(S(T) - K)^+$. There is an exact expression of the price of the European call option, called the **Black–Scholes formula**.

Proposition 6.3.8 *The price of the European call option is given by*

$$\pi(C) = s_0\Phi(d_+) - K\mathrm{e}^{-rT}\Phi(d_-), \tag{6.3.5}$$

where

$$\Phi(x) = \int_{-\infty}^{x} \frac{1}{\sqrt{2\pi}}\mathrm{e}^{-\frac{y^2}{2}}\,\mathrm{d}y \quad (x \in \mathbb{R})$$

and

$$d_\pm = \frac{1}{\sigma\sqrt{T}}\Big(\log\Big(\frac{s_0}{K}\Big) + \Big(r \pm \frac{\sigma^2}{2}\Big)T\Big) \quad \textit{(the double signs correspond).}$$

Proof By Theorem 6.3.7,

$$\begin{aligned}
\mathrm{e}^{rT}\pi(C) &= \int_{\mathbb{R}} (s_0\mathrm{e}^{(r-\frac{\sigma^2}{2})T}\mathrm{e}^x - K)^+ \frac{1}{\sqrt{2\sigma^2 T}}\mathrm{e}^{-\frac{x^2}{2\sigma^2 T}}\,\mathrm{d}x \\
&= s_0\mathrm{e}^{(r-\frac{\sigma^2}{2})T}\int_{\log(\frac{K}{s_0})-(r-\frac{\sigma^2}{2})T}^{\infty} \mathrm{e}^x\frac{1}{\sqrt{2\sigma^2 T}}\mathrm{e}^{-\frac{x^2}{2\sigma^2 T}}\,\mathrm{d}x \\
&\quad - K\int_{\log(\frac{K}{s_0})-(r-\frac{\sigma^2}{2})T} \frac{1}{\sqrt{2\sigma^2 T}}\mathrm{e}^{-\frac{x^2}{2\sigma^2 T}}\,\mathrm{d}x.
\end{aligned}$$

By the change of variables $x = \sigma\sqrt{T}\,y$ in both terms in the last equation, and then $z = y - \sigma\sqrt{T}$ in the first term, the first term turns into $s_0\mathrm{e}^{rT}\Phi(d_+)$ and the second one becomes $K\Phi(d_-)$, because $\int_a^\infty \frac{1}{\sqrt{2\pi}}\mathrm{e}^{-\frac{x^2}{2}}\,\mathrm{d}x = \Phi(-a)$. □

A **European put option** allows the holder to sell at maturity a unit of stock at the exercise price K. Its payoff is $P = (K - S(T))^+$. By the same method as above, its price is computed as

$$\pi(P) = K\mathrm{e}^{-rT}\Phi(-d_-) - s_0\Phi(-d_+). \tag{6.3.6}$$

From this and (6.3.5), the **put-call parity** follows:

$$\pi(P) - \pi(C) = K\mathrm{e}^{-rT} - s_0.$$

This identity is also derived by taking advantage of the equivalent martingale measure $\widehat{\mathbf{P}}$. Indeed, as was seen in Remark 6.2.4, $\{\overline{S}(T)\}_{0\leqq t\leqq T}$ is a martingale under $\widehat{\mathbf{P}}$. Since $(K - S(T))^+ - (S(T) - K)^+ = K - S(T)$,

$$\pi(P) - \pi(C) = \mathbf{E}[\mathrm{e}^{-rT}K - \overline{S}(T)] = K\mathrm{e}^{-rT} - s_0.$$

To apply the Black–Scholes model to a real market, it is indispensable to estimate the volatility σ. Thinking of the price of the European call option as a function of σ, we write $\pi(C;\sigma)$ for $\pi(C)$. Since $d_+ = d_- + \sigma\sqrt{T}$ and

$$\Phi'(d_- + \sigma\sqrt{T}) = \frac{K}{s_0}e^{-rT}\Phi'(d_-),$$

it follows from Proposition 6.3.8 that

$$\begin{aligned}\frac{d}{d\sigma}\pi(C;\sigma) &= s_0\Phi'(d_- + \sigma\sqrt{T})\{d_-' + \sqrt{T}\} - Ke^{-rT}\Phi'(d_-)d_-' \\ &= s_0\Phi'(d_- + \sigma\sqrt{T})\sqrt{T} > 0.\end{aligned}$$

Thus, the mapping $\sigma \mapsto \pi(C;\sigma)$ is strictly increasing, and hence for each $\gamma \geqq 0$, there exists a unique solution to the identity $\pi(C;\sigma) = \gamma$. The solution σ_γ is called an **implied volatility**. Hence, substituting the real price of the European call option into γ, we estimate the volatility σ.

6.4 Greeks

Partial derivatives of $\pi(F)$ are called **Greeks**. They indicate the sensitivity of $\pi(F)$ to changes of parameters. The Greeks **Delta**, Gamma, Vega, and Rho are defined to be $\frac{\partial}{\partial s_0}\pi(F)$, $\frac{\partial^2}{\partial^2 s_0}\pi(F)$, $\frac{\partial}{\partial\sigma}\pi(F)$, and $\frac{\partial}{\partial r}\pi(F)$, respectively.

By (6.3.4), the Greeks are directly computed. Moreover, by the integration by parts on $\mathbb{R}$, they can be represented as weighted integrals of $f(s_0e^{r-\frac{\sigma^2}{2})T}e^x)$. This can be done with the help of the Malliavin calculus. Such an application of the Malliavin calculus to Greeks was first described in [24]. See also [76].

We shall see this for the Delta.

Proposition 6.4.1 *Let $f \in C(\mathbb{R})$ be bounded from below. Suppose f and its derivative are at most of polynomial growth. Set $F = f(S(T))$. Then,*

$$\frac{\partial}{\partial s_0}\pi(F) = e^{-rT}\mathbf{E}_{\widehat{\mathbf{P}}}\Big[f(S(T))\frac{\widehat{B}(T)}{s_0\sigma T}\Big]. \tag{6.4.1}$$

Proof By Theorem 6.3.7, $\pi(F)$ is differentiable in s_0. Moreover, due to the same theorem, we may assume that f is of C^1-class.

Let $(W_T, \mathscr{B}(W_T), \mu_T)$ be the one-dimensional Wiener space on $[0,T]$, and $\{\varphi(t)\}_{0\leqq t\leqq T}$ be the coordinate process on it. By Theorem 6.2.3,

$$\pi(F) = e^{-rT}\int_{W_T} f(s_0e^{(r-\frac{\sigma^2}{2})T}e^{\sigma\varphi(T)})d\mu_T.$$

Then

$$\frac{\partial}{\partial s_0}\pi(F) = \mathrm{e}^{-rT}\int_{W_T} f'(s_0\mathrm{e}^{(r-\frac{\sigma^2}{2})T}\mathrm{e}^{\sigma\varphi(T)})\mathrm{e}^{(r-\frac{\sigma^2}{2})T}\mathrm{e}^{\sigma\varphi(T)}\mathrm{d}\mu_T.$$

By Corollary 5.3.2,

$$\nabla(f(s_0\mathrm{e}^{(r-\frac{\sigma^2}{2})T}\mathrm{e}^{\sigma\varphi(T)})) = f'(s_0\mathrm{e}^{(r-\frac{\sigma^2}{2})T}\mathrm{e}^{\sigma\varphi(T)})s_0\mathrm{e}^{(r-\frac{\sigma^2}{2})T}\mathrm{e}^{\sigma\varphi(T)}\sigma\ell_{[0,T]},$$

where $\ell_{[0,T]} \in W_T^*$ is defined by $\ell_{[0,T]}(w) = w(T)$ $(w \in W_T)$. Since $\|\ell_{[0,T]}\|^2 = T$, by Remark 5.1.3(2),

$$\begin{aligned}\frac{\partial}{\partial s_0}\pi(F) &= \frac{\mathrm{e}^{-rT}}{s_0\sigma T}\int_{W_T}\langle\nabla(f(s_0\mathrm{e}^{(r-\frac{\sigma^2}{2})T}\mathrm{e}^{\sigma\varphi(T)})), \ell_{[0,T]}\rangle_{H_T}\mathrm{d}\mu_T \\ &= \frac{\mathrm{e}^{-rT}}{s_0\sigma T}\int_{W_T} f(s_0\mathrm{e}^{(r-\frac{\sigma^2}{2})T}\mathrm{e}^{\sigma\varphi(T)})\varphi(T)\mathrm{d}\mu_T.\end{aligned}$$

This means (6.4.1) holds. □

The other Greeks possess similar expectation expressions. The proofs are carried out in exactly the same manner as above via the Malliavin calculus.

In general mathematical models in finance, stock prices are realized as solutions of stochastic differential equations with general coefficients. For such general models, Greeks are defined similarly. For example, the Delta is defined as partial derivatives of the price of European contingent claims with respect to the initial value of risky assets. Thus it relates to partial derivatives of Wiener integrals with respect to the initial conditions of the stochastic differential equations. To be precise, let $\{X(t,x)\}_{t\in[0,T]}$ be the solution of the stochastic differential equation (5.5.1) and f be a function on $\mathbb{R}^N$. The Delta corresponds to the derivative of the expectation of $f(X(t,x))$ with respect to the initial value $x = X(0,x)$,

$$\frac{\partial}{\partial x^i}\int_{W_T} f(X(t,x))\,\mathrm{d}\mu_T,$$

where $(W_T, \mathscr{B}(W_T), \mu_T)$ is the d-dimensional Wiener space. We present an expression for the derivative by using the Malliavin calculus. In the following we assume that $X(t,x)$ is non-degenerate for all $t \in (0,T]$ and $x \in \mathbb{R}^N$. For example, the reader may assume Hörmander's condition at every $x \in \mathbb{R}^N$.

Let $Y(t,x) = (Y^i_j(t,x))_{1\leqq i,j\leqq N}$ be the Jacobian matrix of the mapping $x \mapsto X(t,x)$ as before and $Z(t,x) = (Z^i_j(t,x))_{1\leqq i,j\leqq N}$ be its inverse matrix. Define $G^i(t,x) \in \mathbb{D}^{\infty,\infty-}(H_T)$ by

$$\overbrace{G^i(t,x)}^{\cdot\,\alpha}(s) = \sum_{j=1}^N Z^i_j(s,x)V^j_\alpha(X(s,x))\mathbf{1}_{[0,t]}(s) \qquad (0 \leqq s \leqq T)$$

and $A(t,x) = (A^{ij}(t,x))_{1 \leqq i,j \leqq N}$ by

$$A^{ij}(t,x) = \langle G^i(t,x), G^j(t,x) \rangle_{H_T}.$$

As mentioned after Theorem 5.5.4, $(\det A(t,x))^{-1} \in L^{\infty-}(\mu_T)$ and $\det A(t,x) \neq 0$, μ_T-a.s. Denote the inverse matrix of $A(t,x)$ by $B(t,x)$:

$$B(t,x) = (B_{ij}(t,x))_{1 \leqq i,j \leqq N} = A(t,x)^{-1}.$$

Theorem 6.4.2 *For any* $f \in C^1_{\nearrow}(\mathbb{R}^N)$,

$$\frac{\partial}{\partial x^i} \int_{W_T} f(X(t,x))\, \mathrm{d}\mu_T = \int_{W_T} f(X(t,x)) \Phi_i(t,x)\, \mathrm{d}\mu_T, \tag{6.4.2}$$

where Φ_i *is given by*

$$\Phi_i(t,x) = \sum_{k=1}^N B_{ik}(t,x) \sum_{j=1}^N \sum_{\alpha=1}^d \int_0^t Z_j^k(s,x) V_\alpha^j(X(s,x))\, \mathrm{d}\theta^\alpha(s) \\ + 2 \sum_{j,k,m=1}^N B_{ik}(t,x) \langle \nabla G^k, G^m \otimes G^j \rangle_{H_T^{\otimes 2}} B_{mj}(t,x).$$

Proof By Theorem 4.10.8,

$$\frac{\partial}{\partial x^i} \int_{W_T} f(X(t,x))\, \mathrm{d}\mu_T = \sum_{j=1}^N \int_{W_T} \frac{\partial f}{\partial x^j}(X(t,x)) Y_i^j(t,x)\, \mathrm{d}\mu_T. \tag{6.4.3}$$

By Theorem 5.5.1,

$$\nabla X^k(t,x) = \sum_{i=1}^N Y_i^k(t,x) G^i(t,x),$$
$$(\langle \nabla X^i(t,x), \nabla X^j(t,x) \rangle_{H_T})_{1 \leqq i,j \leqq N} = Y(t,x) A(t,x) Y(t,x)^*,$$
$$(\langle \nabla X^i(t,x), \nabla X^j(t,x) \rangle_{H_T})^{-1}_{1 \leqq i,j \leqq N} = Z^*(t,x) B(t,x) Z(t,x).$$

Then Theorem 5.4.5 yields

$$\sum_{j=1}^N \int_{W_T} \frac{\partial f}{\partial x^j}(X(t,x)) Y_i^j(t,x)\, \mathrm{d}\mu_T \\ = \int_{W_T} f(X(t,x)) \sum_{k=1}^N \nabla^*(B_{ik}(t,x) G^k(t,x))\, \mathrm{d}\mu_T. \tag{6.4.4}$$

By Theorem 5.3.3,

$$\nabla^* G^k(t,x) = \sum_{j=1}^N \sum_{\alpha=1}^d \int_0^t Z_j^k(s,x) V_\alpha^j(X(s,x))\, \mathrm{d}\theta^\alpha(s). \tag{6.4.5}$$

Moreover, applying ∇ to both sides of the identity

$$\sum_{k=1}^{N} B_{ik}(t,x)\langle G^k(t,x), G^j(t,x)\rangle_{H_T} = \delta_{ij},$$

we obtain

$$\sum_{k=1}^{N} \nabla B_{ik}(t,x)\langle G^k(t,x), G^j(t,x)\rangle_{H_T} + \sum_{k=1}^{N} B_{ik}(t,x)\{\langle \nabla G^k(t,x), G^j(t,x)\rangle_{H_T} + \langle G^k(t,x), \nabla G^j(t,x)\rangle_{H_T}\} = 0.$$

Since $B(t,x)$ is a symmetric matrix, we have

$$\nabla B_{ij}(t,x) = -2 \sum_{k,m=1}^{N} B_{ik}(t,x)\langle \nabla G^k(t,x), G^m(t,x)\rangle_{H_T} B_{mj}(t,x).$$

Combining this identity with (6.4.3)–(6.4.5) and Theorem 5.2.8, we obtain (6.4.2). □

Remark 6.4.3 By Remark 5.5.8, the mapping

$$x \mapsto \int_{W_T} f(X(t,x))\,\mathrm{d}\mu_T$$

is smooth for any Borel measurable function f with at most polynomial growth. We can show that the assertion of the theorem holds for such a function f by approximating f by smooth functions. In fact, let $p(t,x,y)$ be the transition density of $X(t,x)$. Then $p(t,x,y)$ is smooth in $(t,x,y) \in (0,\infty) \times \mathbb{R}^N \times \mathbb{R}^N$ and we have

$$\frac{\partial}{\partial x^i} \int_{W_T} f(X(t,x))\,\mathrm{d}\mu_T = \int_{\mathbb{R}^N} \frac{\partial}{\partial x^i} p(t,x,y) f(y)\,\mathrm{d}y.$$

The right hand side of (6.4.2) is an analytical expression for the right hand side of this identity.

Those who are interested in the mathematical finance may proceed to books by Duffie [18], Elliott and Kopp [21], Karatzas and Shreve [57], Musiela and Rutkowski [88], and Shreve [105].

7

The Semiclassical Limit

Kac (see [52, 54, 55]) found a similarity between the Feynman path integral and the Wiener integral. The Feynman–Kac formula is a typical example. He also applied the Wiener integral to analysis of differential operators. Among these his contribution to the "problem of drums" is well known.

In the first half of this chapter, we introduce Van Vleck's formula [121], which asserts that, if the Hamiltonian of the classical mechanics is given by a quadratic polynomial, the corresponding propagator (the fundamental solution for the Schrödinger operator) is represented in terms of action integrals, and we show its analogue to heat equations. As an application, a probabilistic representation of soliton solutions for the KdV equation and other related topics will be discussed. In the second half, by investigating the semiclassical approximations for the eigenvalues of Schrödinger operators and Selberg's trace formula on the upper half plane, we present applications of stochastic analysis to differential operators and see the correspondence to classical mechanics.

7.1 Van Vleck's Result and Quadratic Functionals

In chapter 5, the explicit forms of the heat kernels of the harmonic oscillators and the Schrödinger operators with constant magnetic fields were derived. For these Schrödinger operators on $\mathbb{R}^d$, whose Hamiltonians of the corresponding classical mechanics are given by quadratic polynomials, Van Vleck ([121]) showed that the propagators are given by means of action integrals of classical mechanics.

For simplicity we study on $\mathbb{R}^2$. Let k and ℓ be non-negative constants and K and J be the 2×2 constant matrices given by

$$K = \begin{pmatrix} k^2 & 0 \\ 0 & \ell^2 \end{pmatrix} \quad \text{and} \quad J = \begin{pmatrix} 0 & -1 \\ 1 & 0 \end{pmatrix},$$

respectively. For $\gamma \in \mathbb{R}$, define $V : \mathbb{R}^2 \to [0, \infty)$ and $\Theta : \mathbb{R}^2 \to \mathbb{R}^2$ by

$$V(x) = \langle Kx, x \rangle \quad \text{and} \quad \Theta(x) = \begin{pmatrix} \Theta_1(x) \\ \Theta_2(x) \end{pmatrix} = \frac{1}{2}\gamma Jx = \frac{1}{2}\gamma \begin{pmatrix} -x^2 \\ x^1 \end{pmatrix},$$

respectively, where $x = (x^1, x^2) \in \mathbb{R}^2$ and $\langle x, y \rangle$ is the inner product of $x, y \in \mathbb{R}^2$.

Let $H = H_{(V,\Theta)}$ be the Schrödinger operator defined by

$$H = \frac{1}{2}\sum_{\alpha=1}^{2}\Big(\frac{1}{\mathrm{i}}\frac{\partial}{\partial x^\alpha} + \Theta_\alpha(x)\Big)^2 + V(x).$$

The **Hamiltonian** of the corresponding classical mechanics is

$$H(p, q) = \frac{1}{2}\sum_{\alpha=1}^{2}(p^\alpha + \Theta_\alpha(q))^2 + V(q) \qquad (p, q \in \mathbb{R}^2).$$

The **classical path** $(p(s), q(s))$ $(s \geqq 0)$ satisfies the Hamilton equation

$$\dot{q}^\alpha(s) = \frac{\partial H}{\partial p^\alpha}(p(s), q(s)), \quad \dot{p}^\alpha(s) = -\frac{\partial H}{\partial q^\alpha}(p(s), q(s)) \qquad (\alpha = 1, 2).$$

The corresponding Lagrangian is given by

$$L(q, \dot{q}) = \frac{1}{2}((\dot{q}^1)^2 + (\dot{q}^2)^2) + \frac{1}{2}\gamma(q^1\dot{q}^2 - q^2\dot{q}^1) - \frac{1}{2}\langle Kq, q \rangle$$

and the classical path also satisfies the Lagrange equation

$$\frac{\mathrm{d}}{\mathrm{d}s}\Big(\frac{\partial L}{\partial \dot{q}^\alpha}(q(s), \dot{q}(s))\Big) - \frac{\partial L}{\partial q^\alpha}(q(s), \dot{q}(s)) = 0 \qquad (\alpha = 1, 2).$$

Moreover, for each path $\phi = \{\phi(s)\}_{0 \leqq s \leqq t}$, the integral

$$S(\phi) = \int_0^t L(\phi(s), \dot{\phi}(s))\, \mathrm{d}s$$

is called the **action integral** of ϕ. For classical mechanics, see for example Arnold [3].

For fixed $x, y \in \mathbb{R}^2$ and $t > 0$, the classical path $\phi_{\mathrm{cl}} = \{\phi_{\mathrm{cl}}(s)\}_{0 \leqq s \leqq t}$ with $\phi(0) = x$ and $\phi(t) = y$ is uniquely determined. Denote the action integral of ϕ_{cl} by $S_{\mathrm{cl}}(t, x, y)$.

Let $q(t, x, y)$ be the propagator of the Schrödinger equation

$$\frac{1}{\mathrm{i}}\frac{\partial u}{\partial t} = Hu.$$

Van Vleck showed that it is given by

$$q(t, x, y) = \frac{1}{2\pi}\Big(\det\Big(\frac{\partial^2 S_{\mathrm{cl}}(t, x, y)}{\partial x^\alpha \partial y^\beta}\Big)_{\alpha,\beta=1,2}\Big)^{\frac{1}{2}} \mathrm{e}^{\mathrm{i}S_{\mathrm{cl}}(t,x,y)}.$$

In order to achieve an analogous expression for the heat equation, we define a formal Lagrangian $\widetilde{L}$ by

$$\widetilde{L}(x, \dot{x}) = \frac{1}{2}|\dot{x}|^2 - \frac{\mathrm{i}}{2}\gamma\langle Jx, \dot{x}\rangle + \frac{1}{2}\langle Kx, x\rangle.$$

Denote by $\widetilde{\phi}_{\mathrm{cl}} = \{\widetilde{\phi}_{\mathrm{cl}}(s)\}_{0\leqq s\leqq t}$ the solution of the Lagrange equation

$$\frac{\mathrm{d}}{\mathrm{d}s}\Big(\frac{\partial \widetilde{L}}{\partial \dot{x}^\alpha}(\phi(s), \dot{\phi}(s))\Big) - \frac{\partial \widetilde{L}}{\partial x^\alpha}(\phi(s), \dot{\phi}(s)) = 0 \qquad (\alpha = 1, 2)$$

satisfying $\phi(0) = x$ and $\phi(t) = y$. Define its action integral by

$$\widetilde{S}_{\mathrm{cl}}(t, x, y) = \int_0^t \widetilde{L}(\widetilde{\phi}_{\mathrm{cl}}(s), \dot{\widetilde{\phi}}_{\mathrm{cl}}(s))\,\mathrm{d}s.$$

Note that $\widetilde{\phi}_{\mathrm{cl}}$ is a path in $\mathbb{C}^2$ and $\widetilde{S}_{\mathrm{cl}}(t, x, y) \in \mathbb{C}$.

Let $p(t, x, y)$ be the fundamental solution of the heat equation

$$\frac{\partial u}{\partial t} = -Hu.$$

Let $(W, \mathscr{B}(W), \mu)$ be the two-dimensional Wiener space and set

$$\mathfrak{h}(t)(w) = \frac{1}{2}\int_0^t \{(kw^1(s))^2 + (\ell w^2(s))^2\}\,\mathrm{d}s,$$
$$\mathfrak{s}(t)(w) = \frac{1}{2}\int_0^t (w^1(s)\,\mathrm{d}w^2(s) - w^2(s)\,\mathrm{d}w^1(s)) \qquad (w \in W).$$

Then we have (see Theorem 5.5.9)

$$p(t, x, y) = \int_W \mathrm{e}^{\mathrm{i}\gamma\mathfrak{s}(t)(w_x) - \mathfrak{h}(t)(w_x)}\delta_y(w_x(t))\,\mathrm{d}\mu,$$

where $w_x = \{w_x(s)\}_{s\geqq 0}$ is given by $w_x(s) = x + w(s)$.

The following is known ([40, 41]).

Theorem 7.1.1 *For all $x, y \in \mathbb{R}^2$ and $t > 0$,*

$$p(t, x, y) = \frac{1}{2\pi}\Big(\det\Big(\frac{\partial^2 \widetilde{S}_{\mathrm{cl}}(t, x, y)}{\partial x^\alpha \partial y^\beta}\Big)_{\alpha,\beta=1,2}\Big)^{\frac{1}{2}} \mathrm{e}^{-\widetilde{S}_{\mathrm{cl}}(t,x,y)}. \qquad (7.1.1)$$

(7.1.1) holds for $\gamma \in \mathbb{C}$ and for a more general symmetric matrix K, which is not non-negative. However, in general, the corresponding (formal) classical path has conjugate points and (7.1.1) holds before conjugacy. See the above cited references together with the proofs.

Example 7.1.2 (stochastic area) For $\gamma \in \mathbb{R} \setminus \{0\}$, let H be the Schrödinger operator with a constant magnetic field

$$H = -\frac{1}{2}\Big(\frac{\partial}{\partial x^1} - \mathrm{i}\,\frac{\gamma}{2}x^2\Big)^2 - \frac{1}{2}\Big(\frac{\partial}{\partial x^2} + \mathrm{i}\,\frac{\gamma}{2}x^1\Big)^2.$$

As was shown in Theorem 5.8.5, the fundamental solution of the heat equation is given by

$$\begin{aligned} p(t,x,y) &= \int_W \mathrm{e}^{\mathrm{i}\gamma\mathfrak{s}(t)(w_x)}\delta_y(w_x(t))\mu(\mathrm{d}w) \\ &= \mathrm{e}^{\frac{\mathrm{i}\gamma\langle Jx,y\rangle}{2}}\frac{\gamma}{4\pi\sinh(\frac{\gamma t}{2})}\exp\Big(-\frac{\gamma}{4}\coth\Big(\frac{\gamma t}{2}\Big)|y-x|^2\Big). \end{aligned} \tag{7.1.2}$$

To derive (7.1.2) from (7.1.1), we consider the formal Lagrangian

$$\widetilde{L}(q,\dot{q}) = \frac{1}{2}|\dot{q}|^2 - \frac{\mathrm{i}\gamma}{2}(q^1\dot{q}^2 - q^2\dot{q}^1).$$

The Lagrange equation is written as

$$\ddot{q}^1(s) = -\mathrm{i}\,\gamma\dot{q}^2(s), \quad \ddot{q}^2(s) = \mathrm{i}\,\gamma\dot{q}^1(s)$$

and the classical path $\{\widetilde{\phi}_{\mathrm{cl}}(s)\}_{0\leqq s\leqq t}$ with $\widetilde{\phi}_{\mathrm{cl}}(0) = x$, $\widetilde{\phi}_{\mathrm{cl}}(t) = y$ is given by

$$\begin{aligned} \widetilde{\phi}^1_{\mathrm{cl}}(s) &= x^1 + \Big(\frac{1}{2}\coth\Big(\frac{\gamma t}{2}\Big)(y^1 - x^1) + \frac{\mathrm{i}\,(y^2 - x^2)}{2}\Big)\sinh(\gamma s) \\ &\quad - \Big(\frac{y^1 - x^1}{2} + \frac{\mathrm{i}}{2}\coth\Big(\frac{\gamma t}{2}\Big)(y^2 - x^2)\Big)(\cosh(\gamma s) - 1), \\ \widetilde{\phi}^2_{\mathrm{cl}}(s) &= x^2 + \Big(-\frac{\mathrm{i}\,(y^1 - x^1)}{2} + \frac{1}{2}\coth\Big(\frac{\gamma t}{2}\Big)(y^2 - x^2)\Big)\sinh(\gamma s) \\ &\quad + \Big(\frac{\mathrm{i}}{2}\coth\Big(\frac{\gamma t}{2}\Big)(y^1 - x^1) - \frac{y^2 - x^2}{2}\Big)(\cosh(\gamma s) - 1). \end{aligned}$$

Plugging these representations into $\int_0^t \widetilde{L}(\widetilde{\phi}_{\mathrm{cl}}(s), \dot{\widetilde{\phi}}_{\mathrm{cl}}(s))\,\mathrm{d}s$, we obtain (7.1.2).

Computing the action integrals of the classical paths in the same way as above, we can show the following explicit representations in general. The Lagrange equation in the general setting is

$$\ddot{q}_1 + \mathrm{i}\,\gamma\dot{q}_2 - k^2 q_1 = 0 \quad \text{and} \quad \ddot{q}_2 - \mathrm{i}\,\gamma\dot{q}_1 - \ell^2 q_2 = 0.$$

Set

$$\begin{aligned} m_1 &= \{(k+\ell)^2 + \gamma^2\}, \quad m_2 = \{(k-\ell)^2 + \gamma^2\}, \\ s_1 &= \frac{m_1 + m_2}{2}, \quad s_2 = \frac{-m_1 + m_2}{2}. \end{aligned}$$

Then the classical paths $\widetilde{\phi}_{\mathrm{cl}}(s)$ are written as a linear combination of $\mathrm{e}^{s_1 s}$ and $\mathrm{e}^{s_2 s}$. Solving the Lagrange equation with the conditions $\widetilde{\phi}_{\mathrm{cl}}(0) = x$ and $\widetilde{\phi}_{\mathrm{cl}}(t) = y$ and substituting the solution in the action integral

$$\int_0^t \widetilde{L}(\widetilde{\phi}_{\mathrm{cl}}(s), \dot{\widetilde{\phi}}_{\mathrm{cl}}(s))\,\mathrm{d}s,$$

we obtain the following explicit expression for the right hand side of (7.1.1).

Corollary 7.1.3 ([80]) *For $x = (x^1, x^2)$, $y = (y^1, y^2) \in \mathbb{R}^2$ and $t > 0$,*

$$p(t, x, y) = \frac{1}{2\pi}\Big(\frac{k\ell m_1^2 m_2^2}{K(t)}\Big)^{\frac{1}{2}} \exp(-\widetilde{S}_{\mathrm{cl}}(t, x, y)),$$

where $K(t)$ and $\widetilde{S}_{\mathrm{cl}}(t, x, y)$ are given by

$$\begin{aligned}
K(t) &= 2k\ell\gamma^2(\cosh(s_1 t)\cosh(s_2 t) - 1) \\
&\qquad - \{\gamma^2(k^2 + \ell^2) + (k - \ell)^2\}\sinh(s_1 t)\sinh(s_2 t), \\
\widetilde{S}_{\mathrm{cl}}(t, x, y) &= \frac{m_1 m_2}{K(t)}\alpha_1(t)\{(x^1)^2 + (y^1)^2\} + \frac{m_1 m_2}{K(t)}\beta_1(t) x^1 y^1 \\
&\quad + \frac{m_1 m_2}{K(t)}\alpha_2(t)\{(x^2)^2 + (y^2)^2\} + \frac{m_1 m_2}{K(t)}\beta_2(t) x^2 y^2 \\
&\quad + \frac{\mathrm{i}\gamma(k^2 - \ell^2)}{2K(t)}\eta(t)(x^1 x^2 - y^1 y^2) \\
&\quad - \frac{\mathrm{i}k\ell m_1 m_2 \gamma}{K(t)}\{\cosh(s_1 t) - \cosh(s_2 t)\}(x^1 y^2 - x^2 y^1)
\end{aligned}$$

with

$$\begin{aligned}
\alpha_1(t) &= s_1(s_2^2 - k^2)\cosh(s_1 t)\sinh(s_2 t) - s_2(s_1^2 - k^2)\sinh(s_1 t)\cosh(s_2 t), \\
\alpha_2(t) &= s_1(s_2^2 - \ell^2)\cosh(s_1 t)\sinh(s_2 t) - s_2(s_1^2 - \ell^2)\sinh(s_1 t)\cosh(s_2 t), \\
\beta_1(t) &= s_2(s_1^2 - k^2)\sinh(s_1 t) - s_1(s_2^2 - k^2)\sinh(s_2 t), \\
\beta_2(t) &= s_2(s_1^2 - \ell^2)\sinh(s_1 t) - s_1(s_2^2 - \ell^2)\sinh(s_2 t), \\
\eta(t) &= 2k\ell\{\cosh(s_1 t)\cosh(s_2 t) - 1\} + (m_1^2 - 2k\ell)\sinh(s_1 t)\sinh(s_2 t).
\end{aligned}$$

Corollary 7.1.4 *For $\gamma, k \in \mathbb{R}$, set $m_1 = (\gamma^2 + 4k^2)^{\frac{1}{2}}$. Then,*

$$\begin{aligned}
&\int_W \exp\Big(\mathrm{i}\gamma\mathfrak{s}(t) - \frac{k^2}{2}\int_0^t |\theta(s)|^2 \mathrm{d}s\Big)\delta_y(\theta(t))\,\mathrm{d}\mu \\
&\qquad = \frac{1}{2\pi t}\frac{\frac{m_1 t}{2}}{\sinh(\frac{m_1 t}{2})}\exp\Big(-\frac{1}{2t}\frac{\frac{m_1 t}{2}}{\tanh(\frac{m_1 t}{2})}|y|^2\Big) \qquad (y \in \mathbb{R}^2,\ t > 0),
\end{aligned}$$

where $\{\theta(t)\}_{t \geqq 0}$ is the coordinate process of W.

Proof Set $k = \ell$ and $x = 0$ in Corollary 7.1.3. Then, $m_2 = |\gamma|$. By the identities

$$\cosh a \cosh b - \sinh a \sinh b = \cosh(a-b) \quad \text{and} \quad \cosh a - 1 = 2\sinh^2\Big(\frac{a}{2}\Big),$$

we have

$$K(t) = 4\gamma^2 k^2 \sinh^2\Big(\frac{m_1 t}{2}\Big).$$

Hence,

$$\frac{k^2 m_1^2 m_2^2}{K(t)} = \frac{(\frac{m_1}{2})^2}{\sinh^2(\frac{m_1 t}{2})}.$$

Moreover, since

$$s_1(s_2^2 - k^2) = s_2(s_1^2 - k^2) = -k^2|\gamma|,$$

by the identity $\sinh a \cosh b - \cosh a \sinh b = \sinh(a-b)$, we obtain

$$\alpha_1(t) = \alpha_2(t) = |\gamma| k^2 \sinh(m_1 t).$$

Thus

$$\widetilde{S}_{\mathrm{cl}}(t,0,y) = \frac{m_1 m_2}{2K(t)}\alpha_1(t)|y|^2 = \frac{1}{2t}\frac{\frac{m_1 t}{2}}{\tanh(\frac{m_1 t}{2})}|y|^2.$$

By Corollary 7.1.3, we obtain the conclusion. □

7.1.1 Soliton Solutions for the KdV Equation

In this subsection, applying Corollary 7.1.4, we give an expectation representation of soliton solutions for the Korteweg–de Vries (KdV) equation, which is one of the fundamental non-linear partial differential equations.

For $p \in \mathbb{R}$, let $\{\xi^p(t)\}_{t \geqq 0}$ be a two-dimensional Ornstein–Uhlenbeck process obtained as the unique strong solution of the stochastic differential equation

$$\mathrm{d}\xi^p(t) = \mathrm{d}\theta(t) + p\xi^p(t)\,\mathrm{d}t, \quad \xi^p(0) = 0.$$

Each component of $\{\xi^p(t) = (\xi^{p,1}(t), \xi^{p,2}(t))\}$ satisfies

$$\mathrm{d}\xi^{p,i}(t) = \mathrm{d}\theta^i(t) + p\xi^{p,i}(t)\,\mathrm{d}t, \quad \xi^{p,i}(0) = 0 \qquad (i = 1,2).$$

We assume $t > 0$ in the following.

Theorem 7.1.5 *Let $\gamma, C \in \mathbb{R}$ and $y \in \mathbb{R}^2$ and set $m_1 = (\gamma^2 + 4p^2)^{\frac{1}{2}}$. Then,*

$$\int_W \exp\Big(\frac{\mathrm{i}\gamma}{2}\int_0^t \langle J\xi^p(s), \mathrm{d}\xi^p(s)\rangle - C|\xi^p(t)|^2\Big)\delta_y(\xi^p(t))\,\mathrm{d}\mu$$
$$= \frac{1}{2\pi t}\frac{\frac{m_1 t}{2}}{\sinh(\frac{m_1 t}{2})}\exp\Big(-\Big\{\frac{1}{2t}\frac{\frac{m_1 t}{2}}{\tanh(\frac{m_1 t}{2})} + \Big(C - \frac{p}{2}\Big)\Big\}|y|^2 - pt\Big). \quad (7.1.3)$$

Moreover, if $C \geqq \frac{p}{2}$*, then*

$$\int_W \exp\left(\frac{\mathrm{i}\gamma}{2}\int_0^t \langle J\xi^p(s), \mathrm{d}\xi^p(s)\rangle - C|\xi^p(t)|^2\right)\mathrm{d}\mu$$
$$= \frac{m_1 \mathrm{e}^{-pt}}{m_1\cosh(\frac{m_1 t}{2}) + (4C-2p)\sinh(\frac{m_1 t}{2})}. \qquad (7.1.4)$$

We give a remark on the left hand side of (7.1.3). Since

$$\xi^{p,i}(t) = \mathrm{e}^{pt}\int_0^t \mathrm{e}^{-ps}\mathrm{d}\theta^i(s) \qquad (i=1,2),$$

the distribution of $\xi^p(t)$ is two-dimensional Gaussian with mean $0 \in \mathbb{R}^2$ and covariance matrix $\frac{\mathrm{e}^{2pt}-1}{2p}I$, where I is the 2×2 unit matrix. Therefore, if

$$C < -\frac{1}{2}\frac{2p}{\mathrm{e}^{2pt}-1},$$

then the integrand $\exp(\cdots)$ of the left hand side of (7.1.3) is not integrable. Even worse, it does not belong to a nice Sobolev space in the sense of the Malliavin calculus. However, by the effect of the localization by $\delta_y(\xi^p(x))$, the left hand side of (7.1.3) is meaningful.

To see this, fix $y \in \mathbb{R}^2$ and take $\varphi \in C_0^\infty(\mathbb{R}^2)$ so that $\varphi(x) = 1$ if $|x-y| \leqq \frac{1}{2}|y|$. If $\{f_n\}_{n=1}^\infty \subset C_0^\infty(\mathbb{R}^2)$ is a sequence in $\mathscr{S}'(\mathbb{R}^2)$ which converges to δ_y, then φf_n also converges to δ_y in $\mathscr{S}'(\mathbb{R}^2)$ and

$$\int_W F\delta_y(\xi^p(t))\,\mathrm{d}\mu = \int_W F\varphi(\xi^p(t))\delta_y(\xi^p(t))\,\mathrm{d}\mu \qquad (F \in \mathbb{D}^{\infty,\infty-}).$$

Combining this with the identity

$$\exp\left(\frac{\mathrm{i}\gamma}{2}\int_0^t \langle J\xi^p(s), \mathrm{d}\xi^p(s)\rangle - C|\xi^p(t)|^2\right)\varphi(\xi^p(t))$$
$$= \exp\left(\frac{\mathrm{i}\gamma}{2}\int_0^t \langle J\xi^p(s), \mathrm{d}\xi^p(s)\rangle - C\psi(|\xi^p(t)|^2)\right)\varphi(\xi^p(t))$$

for $\psi \in C_0^\infty(\mathbb{R})$ such that $\psi \equiv 1$ on the interval $[0, 2|y|^2)$, we see that the left hand side of (7.1.3) is meaningful.

Proof By Girsanov's theorem (Theorem 4.6.2), the stochastic process $\left\{\theta(s) - \int_0^s p\theta(u)\mathrm{d}u\right\}_{s\in[0,t]}$ is a two-dimensional Brownian motion under the probability measure Q given by

$$\mathrm{d}Q = \exp\left(p\int_0^t \langle\theta(s), \mathrm{d}\theta(s)\rangle - \frac{p^2}{2}\int_0^t |\theta(s)|^2\mathrm{d}s\right)\mathrm{d}\mu.$$

Since $\int_0^t \langle \theta(s), \mathrm{d}\theta(s)\rangle = \frac{1}{2}|\theta(t)|^2 - t$,

$$\int_W \Phi(\{\xi^p(s)\}_{s\in[0,t]})\,\mathrm{d}\mu$$
$$= \int_W \Phi(\{\theta(s)\}_{s\in[0,t]}) \exp\left(\frac{p}{2}|\theta(t)|^2 - \frac{p^2}{2}\int_0^t |\theta(s)|^2 \mathrm{d}s\right) \mathrm{e}^{-pt}\,\mathrm{d}\mu$$

for any $\Phi \in C_b(W_t)$. Let $\{f_n\}_{n=1}^\infty \subset C_0^\infty(\mathbb{R}^2)$ be as above and take $\Psi \in C_{\mathrm{b}}(W_t)$ satisfying $\Psi(\{\xi^p(s)\}_{s\in[0,t]})$ and $\Psi(\{\theta(s)\}_{s\in[0,t]}) \in \mathbb{D}^{\infty,\infty-}$. Applying the above identity to $\Phi \in C_{\mathrm{b}}(W_t)$ given by

$$\Psi(\{\theta(s)\}_{s\in[0,t]}) f_n(\theta(t))$$

and letting $n \to \infty$, we obtain

$$\int_W \Psi(\{\xi^p(s)\}_{s\in[0,t]})\delta_y(\xi^p(t))\,\mathrm{d}\mu$$
$$= \int_W \Psi(\{\theta(s)\}_{s\in[0,t]}) \exp\left(-\frac{p^2}{2}\int_0^t |\theta(s)|^2 \mathrm{d}s\right)\delta_y(\theta(t))\,\mathrm{d}\mu\, \mathrm{e}^{p\{\frac{|y|^2}{2}-t\}}.$$

Substitute

$$\Psi = \exp\left(\frac{\mathrm{i}\gamma}{2}\int_0^t \langle J\theta(s), \mathrm{d}\theta(s)\rangle - C|\theta(t)|^2\right)$$

into the above identity, where $\gamma, C \in \mathbb{R}$. Then, for $y \in \mathbb{R}^2$, we obtain

$$\int_W \exp\left(\frac{\mathrm{i}\gamma}{2}\int_0^t \langle J\xi^p(s), \mathrm{d}\xi^p(s)\rangle - C|\xi^p(t)|^2\right)\delta_y(\xi^p(t))\,\mathrm{d}\mu$$
$$= \int_W \exp\left(\frac{\mathrm{i}\gamma}{2}\int_0^t \langle J\theta(s), \mathrm{d}\theta(s)\rangle - C|\theta(t)|^2\right)$$
$$\times \exp\left(-\frac{p^2}{2}\int_0^t |\theta(s)|^2\mathrm{d}s\right)\delta_y(\theta(t))\,\mathrm{d}\mu \times \mathrm{e}^{p(\frac{|y|^2}{2}-t)}$$
$$= \int_W \exp\left(\mathrm{i}\gamma\mathfrak{s}(t) - \frac{p^2}{2}\int_0^t |\theta(s)|^2\mathrm{d}s\right)\delta_y(\theta(t))\,\mathrm{d}\mu \times \mathrm{e}^{(\frac{p}{2}-C)|y|^2 - pt}.$$

This observation and Corollary 7.1.4 yield (7.1.3).

Let $C \geqq \frac{p}{2}$. Since

$$\frac{1}{2t}\frac{\frac{m_1 t}{2}}{\tanh(\frac{m_1 t}{2})} + \left(C - \frac{p}{2}\right) > 0,$$

integrating (7.1.3) in y over $\mathbb{R}^2$, we obtain

$$\int_W \exp\left(\frac{i\gamma}{2}\int_0^t \langle J\xi^p(s), d\xi^p(s)\rangle - C|\xi^p(t)|^2\right)d\mu$$
$$= \frac{\frac{m_1}{2}}{\sinh(\frac{m_1 t}{2})}\left\{2\left(\frac{1}{2t}\frac{\frac{m_1 t}{2}}{\tanh(\frac{m_1 t}{2})} + \left(C - \frac{p}{2}\right)\right)\right\}^{-1} e^{-pt}.$$

Thus (7.1.4) holds. □

The **KdV equation** is the non-linear partial differential equation on $\mathbb{R}^2$ given by

$$\frac{\partial u}{\partial t}(x,t) = \frac{3}{2}u(x,t)\frac{\partial u}{\partial x}(x,t) + \frac{1}{4}\frac{\partial^3 u}{\partial x^3}(x,t) \qquad ((x,t)\in\mathbb{R}^2),$$

which shows the movement of waves on shallow water surfaces. The solution $u(x,t)$ expresses the height of the wave at time t and position $x \in \mathbb{R}$. There has been much research on the solutions of the KdV equation. Among them, the existence of interesting solutions, called the **soliton solutions**, is well known. We briefly discuss this soliton solution. For details, see [87].

Let

$$\mathscr{S} = \{\{\eta_j, m_j\}_{j=1,\dots,n};\ 0 < m_1,\dots,m_n,\ \eta_1 < \cdots < \eta_n,\ n = 1,2,\dots\}.$$

$\mathbf{s} = \{\eta_j, m_j\}_{j=1,\dots,n} \in \mathscr{S}$ is called the **scattering data** of length n. For $\mathbf{s} = \{\eta_j, m_j\}_{j=1,\dots,n} \in \mathscr{S}$, the function $u_{\mathbf{s}}$ defined by

$$u_{\mathbf{s}}(x) = -2\left(\frac{d}{dx}\right)^2 \log\det(I + G_{\mathbf{s}}(x)) \qquad (x \in \mathbf{R})$$

is called the **reflectionless potential** with scattering data $\mathbf{s}$, where $G_{\mathbf{s}}(x)$ is an $n \times n$ matrix given by

$$G_{\mathbf{s}}(x) = \left(\frac{\sqrt{m_i m_j}e^{-(\eta_i+\eta_j)x}}{\eta_i + \eta_j}\right)_{i,j=1,\dots,n}.$$

For $\mathbf{s} = \{\eta_j, m_j\}_{j=1,\dots,n} \in \mathscr{S}$, the function $v(x,t)$ given by

$$v(x,t) = -u_{\mathbf{s}(t)}(x)$$

is a solution of the KdV equation, where $\mathbf{s}(t) = \{\eta_j, m_j\exp(-2\eta_j^3 t)\}_{j=1,\dots,n}$. Such a v is called an n-soliton solution.

Next, applying Theorem 7.1.5, we show probabilistic representations for the reflectionless potential and 1-soliton solution by means of the Ornstein–Uhlenbeck process.

Proposition 7.1.6 *Under the same framework as Theorem 7.1.5, let* $C \in [\frac{p}{2}, \frac{p}{2} + \frac{m_1}{4})$ *and set*

$$Q(x) = \log\Big(\int_W \exp\Big(\frac{\mathrm{i}\gamma}{2}\int_0^x \langle J\xi^p(s), \mathrm{d}\xi^p(s)\rangle - C|\xi^p(x)|^2\Big)\mathrm{d}\mu\Big) \quad (x \geqq 0).$$

Then, on $[0,\infty)$*, the function* $q = 2(\frac{\mathrm{d}}{\mathrm{d}x})^2 Q$ *coincides with the reflectionless potential with scattering data*

$$\Big\{\frac{m_1}{2}, \frac{m_1(m_1 - 4C + 2p)}{m_1 + 4C - 2p}\Big\}.$$

Remark 7.1.7 (1) Since reflectionless potentials are analytic on $\mathbb{R}$, they are uniquely determined by the value on $[0,\infty)$. Hence, the probabilistic representation above determines the reflectionless potential completely.
(2) In the proposition above, if $p \leqq 0$, then we can take $C = 0$. In particular, then the function

$$2\Big(\frac{\mathrm{d}}{\mathrm{d}x}\Big)^2 \log\Big(\int_W \exp\Big(\frac{\mathrm{i}\gamma}{2}\int_0^x \langle J\xi^p(s), \mathrm{d}\xi^p(s)\rangle\Big)\mathrm{d}\mu\Big)$$

is a reflectionless potential.

Proof Define $\delta \geqq 0$ by $m_1 \tanh\delta = 4C - 2p$. By the identity

$$\cosh(t+s) = \cosh t \cosh s + \sinh t \sinh s,$$

we have

$$\begin{aligned} &m_1 \cosh\Big(\frac{m_1 x}{2}\Big) + (4C - 2p)\sinh\Big(\frac{m_1 x}{2}\Big) \\ &= m_1\Big\{\cosh\Big(\frac{m_1 x}{2}\Big) + \tanh\delta \sinh\Big(\frac{m_1 x}{2}\Big)\Big\} \\ &= \frac{m_1}{\cosh\delta}\frac{\mathrm{e}^{(\frac{m_1 x}{2})+\delta}}{2}(1 + \mathrm{e}^{-2\delta}\mathrm{e}^{-m_1 x}). \end{aligned}$$

Combining this with Theorem 7.1.5, we obtain

$$\begin{aligned} q(x) &= -2\Big(\frac{\mathrm{d}}{\mathrm{d}x}\Big)^2\Big\{\log m_1 - px - \log\frac{m_1}{2\cosh\delta} - \frac{m_1 x}{2} + \delta \\ &\qquad - \log(1 + \mathrm{e}^{-2\delta}\,\mathrm{e}^{-m_1 x})\Big\} \\ &= -2\Big(\frac{\mathrm{d}}{\mathrm{d}x}\Big)^2 \log(1 + \mathrm{e}^{-2\delta}\,\mathrm{e}^{-m_1 x}). \end{aligned}$$

Hence, q is the reflectionless potential with scattering data $\{\frac{m_1}{2}, m_1\mathrm{e}^{-2\delta}\}$. □

Corollary 7.1.8 *For* $p \in \mathbb{R}$ *and* $t \geqq 0$*, set*

$$\begin{aligned} V^p(x,t) = \log\Big(\int_W \exp\Big(\frac{\mathrm{i}\gamma}{2}\int_0^x \langle J\xi^p(y), \mathrm{d}\xi^p(y)\rangle \\ - \Big\{\frac{p}{2} + \frac{m_1}{4}\tanh\Big(\frac{m_1^3 t}{8}\Big)\Big\}|\xi^p(x)|^2\Big)\mathrm{d}\mu\Big) \end{aligned}$$

and

$$v^p(x,t) = 2\Big(\frac{\partial}{\partial x}\Big)^2 V^p(x,t).$$

Then, v^p is a 1-soliton solution of the KdV equation with the initial condition

$$u(x,0) = -\frac{(\gamma^2+4p^2)^{\frac{1}{2}}}{2\cosh^2((\gamma^2+4p^2)^{\frac{1}{2}}x)}.$$

Proof By Proposition 7.1.6, $v^p(\cdot,t)$ is a reflectionless potential with scattering data $\{\frac{m_1}{2}, m_1 \mathrm{e}^{-2(\frac{m_1}{2})^3 t}\}$. Hence, v^p is a 1-soliton solution of the KdV equation. □

We can present probabilistic representations for the reflectionless potentials and n-soliton solutions with scattering data of length n. In such a study, it is important to have an explicit representation for Wiener integrals corresponding to the function $V^p(t,x)$ by using a solution for the Riccati equation. For details, see [44, 118].

7.1.2 Euler Polynomials

A sequence of polynomials $\{p_n(x)\}_{n=0}^\infty$ satisfying

$$p_n' = np_{n-1} \qquad (n=1,2,\ldots)$$

is called an **Appell sequence**. $p_n(x) = x^n$ $(n=0,1,\ldots)$ is a typical example. The sequence $\{n!H_n(x)\}_{n=0}^\infty$, H_n being the Hermite polynomial, also satisfies this relationship.

In this subsection, we are concerned with the **Euler polynomial** E_n defined by

$$\frac{2\mathrm{e}^{\zeta x}}{\mathrm{e}^{\zeta}+1} = \sum_{n=0}^\infty E_n(x)\frac{\zeta^n}{n!}. \tag{7.1.5}$$

The sequence $\{E_n\}_{n=0}^\infty$ is also an Appell sequence. The **Euler number** e_n is defined by

$$e_n = 2^n E_n(\tfrac{1}{2}) \qquad (n=0,1,\ldots).$$

For the Euler polynomials, the Euler numbers, and Lévy's stochastic area, the following relationship holds.

Theorem 7.1.9 *For $x\in\mathbb{R}$,*

$$E_n(x) = \mathrm{i}^n \sum_{k=0}^n \binom{n}{k}(1-2x)^k \int_W \Big\{\mathfrak{s}(1)+\frac{\mathrm{i}}{4}|\theta(1)|^2\Big\}^k \mathrm{d}\mu \times \int_W \mathfrak{s}(1)^{n-k}\mathrm{d}\mu, \tag{7.1.6}$$

$$\int_W \mathfrak{s}(1)^n \, \mathrm{d}\mu = \begin{cases} 0 & \textit{(if } n \textit{ is odd),} \\ (-1)^{\frac{n}{2}} 2^{-n} e_n & \textit{(if } n \textit{ is even).} \end{cases} \tag{7.1.7}$$

For a proof, we give a lemma.

Lemma 7.1.10 *For $a, \beta \in \mathbb{R}$ with $|a| \leqq \frac{1}{2}, -a\beta < 1$,*

$$\int_W \mathrm{e}^{\mathrm{i}\beta\{\mathfrak{s}(1)+\mathrm{i}\frac{a}{2}|\theta(1)|^2\}} \mathrm{d}\mu = \left\{\left(\frac{1}{2}+a\right)\mathrm{e}^{\frac{\beta}{2}} + \left(\frac{1}{2}-a\right)\mathrm{e}^{-\frac{\beta}{2}}\right\}^{-1}.$$

Proof Since $-a\beta < \frac{1}{2}$, $\mathrm{e}^{-\frac{a\beta}{2}|\theta(1)|^2} \in \mathbb{D}^{\infty,\infty-}$. Moreover, both the real and imaginary parts of $\mathrm{e}^{\mathrm{i}\beta\mathfrak{s}(1)}$ belong to $\mathbb{D}^{\infty,\infty-}$. Hence, by Theorem 5.4.11,

$$\begin{aligned} \int_W \mathrm{e}^{\mathrm{i}\beta\{\mathfrak{s}(1)+\mathrm{i}\frac{a}{2}|\theta(1)|^2\}} \mathrm{d}\mu &= \int_{\mathbb{R}^2} \left(\int_W \mathrm{e}^{\mathrm{i}\beta\{\mathfrak{s}(1)+\mathrm{i}\frac{a}{2}|\theta(1)|^2\}} \delta_y(\theta(1)) \, \mathrm{d}\mu \right) \mathrm{d}y \\ &= \int_{\mathbb{R}^2} \mathrm{e}^{-\frac{a\beta}{2}|y|^2} \left(\int_W \mathrm{e}^{\mathrm{i}\beta\mathfrak{s}(1)} \delta_y(\theta(1)) \, \mathrm{d}\mu \right) \mathrm{d}y. \end{aligned}$$

By Corollary 7.1.4 (or Theorem 5.8.5), the right hand side coincides with

$$\frac{1}{2\pi} \frac{\frac{\beta}{2}}{\sinh(\frac{\beta}{2})} \int_{\mathbb{R}^2} \exp\left(-\frac{\frac{\beta}{2}}{\sinh(\frac{\beta}{2})} \left\{ \cosh\left(\frac{\beta}{2}\right) + 2a \sinh\left(\frac{\beta}{2}\right) \right\} \frac{|y|^2}{2} \right) \mathrm{d}y.$$

Since

$$\cosh\left(\frac{\beta}{2}\right) + 2a \sinh\left(\frac{\beta}{2}\right) = \left(\frac{1}{2}+a\right)\mathrm{e}^{\frac{\beta}{2}} + \left(\frac{1}{2}-a\right)\mathrm{e}^{-\frac{\beta}{2}} > 0,$$

computing the Gaussian integral yields the conclusion. □

Proof of Theorem 7.1.9 First we show (7.1.6). Rewrite (7.1.5) as

$$\sum_{n=0}^{\infty} E_n(x) \frac{\zeta^n}{n!} = \frac{\mathrm{e}^{-\frac{1}{2}\zeta(1-2x)}}{\cosh(\frac{\zeta}{2})}. \tag{7.1.8}$$

Set $a = \frac{1}{2}$, $\beta = \zeta(1-2x)$ and let $\zeta(x - \frac{1}{2}) < 1$. Then, applying Lemma 7.1.10, we have

$$\mathrm{e}^{-\frac{1}{2}\zeta(1-2x)} = \int_W \exp\left(\mathrm{i}\,\zeta(1-2x) \left\{ \mathfrak{s}(1) + \frac{\mathrm{i}}{4}|\theta(1)|^2 \right\} \right) \mathrm{d}\mu.$$

Plugging this and the identity

$$\int_W \mathrm{e}^{\mathrm{i}\,\zeta\mathfrak{s}(1)} \mathrm{d}\mu = \frac{1}{\cosh(\frac{\zeta}{2})},$$

which is obtained in Theorem 5.8.4, into (7.1.8), we obtain

$$\sum_{n=0}^{\infty} \frac{E_n(x)}{n!}\zeta^n = \int_W \exp\Big(\mathrm{i}\,\zeta(1-2x)\Big\{\mathfrak{s}(1) + \frac{\mathrm{i}}{4}|\theta(1)|^2\Big\}\Big)\mathrm{d}\mu \times \int_W \mathrm{e}^{\mathrm{i}\,\zeta\mathfrak{s}(1)}\mathrm{d}\mu.$$

Considering the series expansion in ζ of the right hand side and comparing the coefficients of ζ^n ($n \in \mathbb{Z}_+$), we obtain (7.1.6).

Second, we show (7.1.7). Since $\widehat{\theta} = (\theta^2, \theta^1)$ is also a Brownian motion under μ, $\mathfrak{s}(1)$ is identical in law with $-\mathfrak{s}(1)$. In particular,

$$\int_W \mathfrak{s}(1)^n \mathrm{d}\mu = \int_W (-\mathfrak{s}(1))^n \mathrm{d}\mu.$$

Hence, if n is odd, then

$$\int_W \mathfrak{s}(1)^n \mathrm{d}\mu = 0.$$

When n is even, let $x = \frac{1}{2}$ in (7.1.6). □

7.2 Asymptotic Distribution of Eigenvalues

Let V be a real-valued function on $\mathbb{R}^d$ satisfying

(V.1) V is continuous and bounded from below. (For simplicity, we assume that $V \geqq 0$.)

Then, the Schrödinger operator

$$H = -\frac{1}{2}\Delta + V$$

is an essentially self-adjoint operator on $C_0^\infty(\mathbb{R}^d)$. Denote also by H the corresponding self-adjoint operator on $L^2(\mathbb{R}^d)$ (see [58, 133]).

By virtue of detailed study on semigroups generated by Schrödinger operators, a part of which is presented in Chapter 3, it is known that H generates a semigroup $\{\mathrm{e}^{-tH}\}_{t\geqq 0}$ on $L^2(\mathbb{R}^d)$ and the heat kernel $p(t,x,y)$ is continuous in (t,x,y) under additional assumptions, which will be described below. We utilize these fundamental facts in functional analysis. See, for example, [108]. We work with these semigroups and the heat kernel in this section.

Under the assumption that $V(x) \to \infty$ as $|x| \to \infty$ and the spectrum of H consists only of eigenvalues with finite multiplicity (discrete spectrum) $\{\lambda_n\}_{n=1}^\infty$, we shall investigate the asymptotic behavior of $N(\lambda)$, the number of λ_ns with $\lambda_n < \lambda$, as $\lambda \to \infty$. If the Tauberian theorem (Theorem A.6.4) is applicable, it suffices to show the asymptotic behavior of the trace $\mathrm{tr}(\mathrm{e}^{-tH})$ of the semigroup $\{\mathrm{e}^{-tH}\}_{t\geqq 0}$ as $t \downarrow 0$. We show this asymptotic behavior as $t \downarrow 0$ via analysis based on the Wiener integral (the Feynman–Kac formula).

This problem is an analogue of Weyl's theorem on the Laplacian in bounded domains ([124]). As was mentioned at the beginning of this chapter, Kac [54] pointed out that a Brownian motion is useful in solving such questions. See also [50, Section 7.6]. The result in this section was first shown by Ray [96].

In the rest of this section we assume the following conditions on V.

(V.2) There exist an $\alpha > 0$ and a slowly increasing function F (see Section A.6) such that

$$\lim_{\lambda\to\infty} \frac{\mathrm{vol}(\{x \in \mathbb{R}^d; V(x) < \lambda\})}{\lambda^\alpha F(\lambda)} = 1,$$

where vol is the Lebesgue measure on the corresponding Euclidean space.

(V.3) There exists a $\delta > 0$ such that

$$\lim_{\lambda\to\infty} \frac{\mathrm{vol}\Big(\Big\{x \in \mathbb{R}^d;\ \max_{|y|<\delta} V(x+y) < \lambda\Big\}\Big)}{\mathrm{vol}(\{x \in \mathbb{R}^d; V(x) < \lambda\})} = 1.$$

The conditions (V.2) and (V.3) cause restrictions on the continuity and the growth of V. If V is a polynomial, then both of the conditions are satisfied.

Theorem 7.2.1 *Under the conditions* (V.1) – (V.3),

$$N(\lambda) = (2\pi)^{-d}\mathrm{vol}\Big(\Big\{(x,p) \in \mathbb{R}^d \times \mathbb{R}^d\ ;\ \frac{1}{2}|p|^2 + V(x) < \lambda\Big\}\Big)(1 + o(1))$$

as $\lambda \to \infty$.

Proof Let $p(t,x,y)$ be the heat kernel of H, that is, the integral kernel of the semigroup $\{\mathrm{e}^{-tH}\}_{t\geqq 0}$. Take an orthonormal basis $\{\varphi_n\}_{n=1}^\infty$ of $L^2(\mathbb{R}^d)$ consisting of the eigenfunctions of H: $H\varphi_n = \lambda_n\varphi_n$. Then we have the eigenfunction expansion

$$p(t,x,y) = \sum_{n=1}^\infty \mathrm{e}^{-\lambda_n t}\varphi_n(x)\varphi_n(y).$$

In particular,

$$\int_{\mathbb{R}^d} p(t,x,x)\,\mathrm{d}x = \mathrm{tr}(\mathrm{e}^{-tH}) = \sum_{n=1}^\infty \mathrm{e}^{-\lambda_n t} = \int_0^\infty \mathrm{e}^{-\lambda t}\mathrm{d}N(\lambda).$$

Hence, to complete the proof, it suffices to show that

$$\int_{\mathbb{R}^d} p(t,x,x)\,\mathrm{d}x = (2\pi t)^{-\frac{d}{2}} \int_{\mathbb{R}^d} \mathrm{e}^{-tV(x)}\mathrm{d}x(1 + o(1))$$

holds as $t \downarrow 0$. In fact, if this holds, then, noting

$$(2\pi t)^{-\frac{d}{2}} \int_{\mathbb{R}^d} \mathrm{e}^{-tV(x)}\mathrm{d}x = (2\pi)^{-d} \int_{\mathbb{R}^d\times\mathbb{R}^d} \mathrm{e}^{-t(\frac{1}{2}|p|^2+V(x))}\mathrm{d}p\mathrm{d}x$$

and applying the Tauberian theorem (Theorem A.6.4) in conjunction with the condition (V.2), we obtain the conclusion of the theorem.

Let $\{B(t)\}_{t\geqq 0}$ be a Brownian motion with $B(0) = 0$ defined on a probability space $(\Omega, \mathscr{F}, \mathbf{P})$. Then we have the following representation for $p(t,x,x)$ (see Remark 5.5.8):

$$p(t,x,x) = (2\pi t)^{-\frac{d}{2}} \mathbf{E}\Big[e^{-\int_0^t V(x+B(s))ds}\Big| B(t) = 0\Big].$$

Applying Jensen's inequality for the integral $\frac{1}{t}\int_0^t$, we obtain

$$p(t,x,x) \leqq (2\pi t)^{-\frac{d}{2}} \mathbf{E}\Big[\frac{1}{t}\int_0^t e^{-tV(x+B(s))}ds \,\Big| B(t) = 0\Big].$$

Moreover, by Fubini's theorem,

$$\begin{aligned}\int_{\mathbb{R}^d} p(t,x,x)\,dx &\leqq (2\pi t)^{-\frac{d}{2}} \mathbf{E}\Big[\frac{1}{t}\int_0^t ds \int_{\mathbb{R}^d} e^{-tV(x+B(s))}dx \,\Big| B(t) = 0\Big] \\ &= (2\pi t)^{-\frac{d}{2}} \int_{\mathbb{R}^d} e^{-tV(x)}dx.\end{aligned}$$

Hence, for any $t > 0$, we have

$$\mathrm{tr}(e^{-tH}) \leqq (2\pi t)^{-\frac{d}{2}} \int_{\mathbb{R}^d} e^{-tV(x)}dx. \tag{7.2.1}$$

Next, set

$$e_V(t) = \int_0^t V(x + B(s))\,ds,$$

and rewrite as

$$p(t,x,x) = (2\pi t)^{-\frac{d}{2}} \mathbf{E}\Big[\int_0^\infty e^{-u}\mathbf{1}_{\{e_V(t)\leqq u\}}du \,\Big| B(t) = 0\Big].$$

Let $M(t) = \max_{0\leqq s\leqq t}|B(s)|$. For $\delta > 0$,

$$p(t,x,x) \geqq (2\pi t)^{-\frac{d}{2}} \int_0^\infty e^{-u}\mathbf{E}[\mathbf{1}_{\{e_V(t)\leqq u\}}\mathbf{1}_{\{M(t)<\delta\}}|B(t) = 0]\,du.$$

Since $e_V(t) \leqq t\max_{|y|<\delta} V(x+y)$ if $M(t) < \delta$,

$$p(t,x,x) \geqq (2\pi t)^{-\frac{d}{2}} \int_0^\infty e^{-u}\mathbf{E}[\mathbf{1}_{\{\max_{|y|<\delta} V(x+y)<\frac{u}{t}\}}\mathbf{1}_{\{M(t)<\delta\}}\big|B(t) = 0]\,du.$$

Integrating both sides with respect to x, we obtain

$$\begin{aligned}\mathrm{tr}(e^{-tH}) \geqq (2\pi t)^{-\frac{d}{2}} \int_0^\infty e^{-u}\mathrm{vol}\Big(\Big\{x;\ \max_{|y|<\delta} V(x+y) < \frac{u}{t}\Big\}\Big)du \\ \times \mathbf{P}(M(t) < \delta \,|B(t) = 0).\end{aligned}$$

Set $M^i(t) = \max_{0\leqq s\leqq t} |B^i(s)|$. Then, by Corollary 3.1.8,

$$\begin{aligned}\mathbf{P}(M(t) < \delta | B(t) = 0) &\geqq \mathbf{P}\Big(|M^1(t)| < \frac{\delta}{\sqrt{d}}, \dots, |M^d(t)| < \frac{\delta}{\sqrt{d}} \Big| B(t) = 0\Big) \\ &= \Big(1 - \mathrm{e}^{-\frac{2\delta^2}{dt}}\Big)^d.\end{aligned}$$

Moreover, by the conditions and Fatou's lemma,

$$\begin{aligned}&\liminf_{t\downarrow 0} \frac{1}{\Gamma(\alpha+1)F(\frac{1}{t})t^{-\alpha}} \int_0^\infty \mathrm{e}^{-u} \mathrm{vol}\Big(\Big\{x;\ \max_{|y|<\delta} V(x+y) < \frac{u}{t}\Big\}\Big)\mathrm{d}u \\ &\geqq \frac{1}{\Gamma(\alpha+1)} \int_0^\infty \mathrm{e}^{-u} \lim_{t\downarrow 0}\Bigg(\frac{\mathrm{vol}\big(\big\{x;\ \max_{|y|<\delta} V(x+y) < \frac{u}{t}\big\}\big)}{F(\frac{u}{t})(\frac{u}{t})^\alpha} \frac{F(\frac{u}{t})}{F(\frac{1}{t})}\Bigg) u^\alpha \mathrm{d}u \\ &= \frac{1}{\Gamma(\alpha+1)} \int_0^\infty \mathrm{e}^{-u} u^\alpha \mathrm{d}u = 1.\end{aligned}$$

Thus we have shown

$$\liminf_{t\to\infty} \frac{\mathrm{tr}(\mathrm{e}^{-tH})}{(2\pi t)^{-\frac{d}{2}}\Gamma(\alpha+1)F(\frac{1}{t})t^{-\alpha}} \geqq 1.$$

On the other hand, applying the Abelian theorem (Theorem A.6.4) together with (V.2), we obtain

$$\int_0^\infty \mathrm{e}^{-tV(x)}\mathrm{d}x = \Gamma(\alpha+1)F\Big(\frac{1}{t}\Big)t^{-\alpha}(1+o(1))$$

as $t \downarrow 0$. Combining these with (7.2.1), we arrive at the conclusion

$$\mathrm{tr}(\mathrm{e}^{-tH}) = (2\pi t)^{-\frac{d}{2}} \int_{\mathbb{R}^d} \mathrm{e}^{-tV(x)}\mathrm{d}x(1+o(1)). \qquad \square$$

For other applications of stochastic analysis to studies of Schrödinger operators and the spectral theory, see Simon [106] and the references therein.

7.3 Semiclassical Approximation of Eigenvalues

Let V be an $\mathbb{R}$-valued function on $\mathbb{R}^d$ and consider the Schrödinger operator

$$H(\hbar) = -\frac{\hbar^2}{2}\Delta + V.$$

$\hbar$ is a positive parameter called the Plank constant and the problem of investigating asymptotic behaviors of characteristic quantities of $H(\hbar)$ as $\hbar \to 0$ is called a problem of semiclassical approximation (limit).

Analyzing the trace of the semigroup $\mathrm{tr}(\mathrm{e}^{-tH(\hbar)})$ as in the previous section, we study a problem of semiclassical approximation on eigenvalues of $H(\hbar)$ when V has a finite number of minimum points. For analytical results, see [34, 109] and so on.

In this section assume that the potential V is a periodic function, and regard $H(\hbar)$ as an operator on the torus. We follow the arguments in Watanabe [123], where a similar problem on a Riemannian manifold was treated by using the Brownian motion on it.

Let $\boldsymbol{e}^1, \ldots, \boldsymbol{e}^d$ be the standard basis of $\mathbb{R}^d$. In the rest of this section, we suppose that V is a non-negative, C^∞, and periodic function such that

$$V(x+\boldsymbol{e}^i) = V(x) \qquad (x \in \mathbb{R}^d,\ i = 1, \ldots, d).$$

In place of $H(\hbar)$, for $\lambda > 0$, we are concerned with the Schrödinger operator H_λ on $\mathbb{R}^d$ defined by

$$H_\lambda = -\frac{1}{2}\Delta + \lambda^2 V.$$

Regard H_λ as an operator acting on periodic functions on $\mathbb{R}^d$. Then, there corresponds a self-adjoint operator on $L^2([0,1)^d)$ or $L^2(\mathbb{T}^d)$, where $\mathbb{T}^d$ is the d-dimensional torus. Identify $[0,1)^d$ with $\mathbb{T}^d$ and denote this operator by H_λ^{per}. H_λ^{per} is a self-adjoint operator on a compact manifold $\mathbb{T}^d$ and the spectrum of H_λ^{per} consists only of eigenvalues $\{\nu_n(\lambda)\}_{n=1}^\infty$ $(\nu_1(\lambda) < \nu_2(\lambda) \leqq \nu_3(\lambda) \leqq \cdots)$ with finite multiplicity (see [58]).

We further suppose the following.

(V) $V(x)$ has a finite number of zeros $a_1, a_2, \ldots, a_m$ in $[0,1)^d$ and the Hessian $(V_{ij}(a_\ell))_{i,j=1,\ldots,d}$ $(\ell = 1, \ldots, m)$ at each zero is positive definite, where $V_{ij}(a_\ell)$ is given by

$$V_{ij}(a_\ell) = \frac{\partial^2 V}{\partial x^i \partial x^j}(a_\ell).$$

Denote the eigenvalues of the Hessian $(V_{ij}(a_\ell))_{i,j=1,\ldots,d}$ at a_ℓ by $\mu_1(a_\ell), \ldots, \mu_d(a_\ell)$ $(\ell = 1, \ldots, m)$ and consider the harmonic oscillator

$$L_\ell = -\frac{1}{2}\Delta + \frac{1}{2}\sum_{i,j=1}^d V_{ij}(a_\ell)x^i x^j \qquad (x = (x^1, \ldots, x^d) \in \mathbb{R}^d).$$

By a change of variables $x \mapsto y = Px$ with an orthogonal matrix P diagonalizing $(V_{ij}(a_\ell))_{i,j=1,\ldots,d}$, L_ℓ is transformed into

$$\widetilde{L_\ell} = -\frac{1}{2}\Delta_y + \frac{1}{2}\sum_{i=1}^d \mu_i(a_\ell)(y^i)^2.$$

Thus the set of eigenvalues of L_ℓ is given by $\left\{\sum_{k=1}^d (n_k+\frac{1}{2})\sqrt{\mu_k(a_\ell)};\ n_k = 0,1,2,\dots\right\}$ (Remark 5.8.1). We rearrange the set

$$\bigcup_{\ell=1}^m \left\{\sum_{k=1}^d \left(n_k+\frac{1}{2}\right)\sqrt{\mu_k(a_\ell)}\,;\ n_k = 0,1,\dots,\ k=1,\dots,d\right\}$$

as $\{\mu_n\}_{n=1}^\infty$ $(\mu_1 \leqq \mu_2 \leqq \cdots)$, repeated according to the multiplicity.

Theorem 7.3.1 *Under the assumption* (V), $\lambda^{-1}\nu_n(\lambda) \to \mu_n$ *as* $\lambda\to\infty$ *for each* $n = 1,2,\dots$.

Remark 7.3.2 If $\mu_1 = \mu_2$, an interesting problem related to the large deviation principle arises from the asymptotic behavior of $\nu_2(\lambda)-\nu_1(\lambda)$ as $\lambda\to\infty$ (exponential decay in λ). Such an equality holds, for example, if there is a pair of the zeros of V at which the eigenvalues of the Hessians coincide. See, for example, [35, 110].

By the continuity of Laplace transforms (Theorem A.6.2), it suffices to prove the following theorem in order to show Theorem 7.3.1.

Theorem 7.3.3 *For any* $t>0$,

$$\lim_{\lambda\to\infty} \mathrm{tr}(\mathrm{e}^{-\frac{t}{\lambda}H_\lambda^{\mathrm{per}}}) = \sum_{\ell=1}^m \mathrm{tr}(\mathrm{e}^{-tL_\ell}).$$

Proof Let $p_\lambda(t,x,y)$ $(t>0,\ x,y\in\mathbb{R}^d)$ be the heat kernel of the semigroup $\{\mathrm{e}^{-\frac{t}{\lambda}H_\lambda}\}_{t\geqq 0}$ on $L^2(\mathbb{R}^d)$ and $q_\lambda(t,x,y)$ $(t>0,\ x,y\in[0,1)^d)$ be that of $\{\mathrm{e}^{-\frac{t}{\lambda}H_\lambda^{\mathrm{per}}}\}_{t\geqq 0}$ on $L^2([0,1)^d)$. Then, we have

$$q_\lambda(t,x,y) = \sum_{\boldsymbol{n}\in\mathbb{Z}^d} p_\lambda(t,x,y+\boldsymbol{n}).$$

Let $(W,\mathscr{B}(W),\mu)$ be the d-dimensional Wiener space and set $\lambda=\varepsilon^{-2}$. By the scaling property of the Wiener measure (Theorem 1.2.8), for $p_\lambda(t,x,x+\boldsymbol{n})$, we have

$$\begin{aligned} p_\lambda(t,x,x+\boldsymbol{n}) &= \int_W \mathrm{e}^{-\lambda^2\int_0^{t/\lambda} V(x+\theta(s))\mathrm{d}s}\delta_{\boldsymbol{n}}\Big(\theta\Big(\frac{t}{\lambda}\Big)\Big)\mathrm{d}\mu \qquad (7.3.1)\\ &= \int_W \mathrm{e}^{-\frac{1}{\varepsilon^2}\int_0^t V(x+\theta(\varepsilon^2 s))\mathrm{d}s}\delta_{\boldsymbol{n}}(\theta(\varepsilon^2 t))\,\mathrm{d}\mu \\ &= \int_W \mathrm{e}^{-\frac{1}{\varepsilon^2}\int_0^t V(x+\varepsilon\theta(s))\mathrm{d}s}\varepsilon^{-d}\delta_{\frac{1}{\varepsilon}\boldsymbol{n}}(\theta(t))\,\mathrm{d}\mu. \qquad (7.3.2)\end{aligned}$$

Since V is non-negative, it is easy to see

$$p_\lambda(t,x,x+\boldsymbol{n}) \leqq \int_W \varepsilon^{-d}\delta_{\frac{1}{\varepsilon}\boldsymbol{n}}(\theta(t))\,\mathrm{d}\mu = (2\pi\varepsilon^2 t)^{-\frac{d}{2}}\mathrm{e}^{-\frac{|\boldsymbol{n}|^2}{2\varepsilon^2 t}} \quad (t>0). \tag{7.3.3}$$

Take $\delta > 0$ so that $O_{\ell,\delta} \subset (0,1)^d$ ($\ell = 1,\dots,m$), $O_{\ell,\delta}$ being the ball with center a_ℓ and radius δ. Divide the trace of the semigroup $\{\mathrm{e}^{-\frac{t}{\lambda}H_\lambda^{\mathrm{per}}}\}_{t\geqq 0}$ into three terms as follows:

$$\mathrm{tr}(\mathrm{e}^{-\frac{t}{\lambda}H_\lambda^{\mathrm{per}}}) = \int_{[0,1)^d} q_\lambda(t,x,x)\,\mathrm{d}x = I_\lambda^1(t) + I_\lambda^2(t) + I_\lambda^3(t),$$

$$I_\lambda^1(t) = \sum_{\ell=1}^m \int_{O_{\ell,\delta}} p_\lambda(t,x,x)\,\mathrm{d}x, \quad I_\lambda^2(t) = \int_{[0,1)^d\setminus(\bigcup_\ell O_{\ell,\delta})} p_\lambda(t,x,x)\,\mathrm{d}x,$$

$$I_\lambda^3(t) = \sum_{\boldsymbol{n}\neq 0}\int_{[0,1)^d} p_\lambda(t,x,x+\boldsymbol{n})\,\mathrm{d}x.$$

By (7.3.3), $I_\lambda^3(t) \to 0$ ($\lambda\to\infty$) for any $t>0$.

We show $I_\lambda^2(t) \to 0$ by using (7.3.1). Let $x \in [0,1)^d \setminus (\bigcup_{\ell=1}^m O_{\ell,\delta})$, that is, suppose that $|x-a_\ell| > \delta$ for every $\ell = 1,\dots,m$. Set $h_\varepsilon = \mathbf{1}_{\{\max_{0\leqq u\leqq\varepsilon^2 t}|\theta(u)|\leqq\frac{\delta}{2}\}}$ and, using the conditional expectation, rewrite $p_\lambda(t,x,x)$ as

$$p_\lambda(t,x,x) = p_\lambda^{(1)}(t,x) + p_\lambda^{(2)}(t,x),$$

$$p_\lambda^{(1)}(t,x) = (2\pi\varepsilon^2 t)^{-\frac{d}{2}}\mathbf{E}\Big[h_\varepsilon \mathrm{e}^{-\frac{1}{\varepsilon^4}\int_0^{\varepsilon^2 t}V(x+\theta(u))\mathrm{d}u}\Big|\,\theta(\varepsilon^2 t)=0\Big],$$

$$p_\lambda^{(2)}(t,x) = (2\pi\varepsilon^2 t)^{-\frac{d}{2}}\mathbf{E}\Big[(1-h_\varepsilon)\mathrm{e}^{-\frac{1}{\varepsilon^4}\int_0^{\varepsilon^2 t}V(x+\theta(u))\mathrm{d}u}\Big|\,\theta(\varepsilon^2 t)=0\Big].$$

If $\max_{0\leqq u\leqq\varepsilon^2 t}|\theta(u)| \leqq \frac{\delta}{2}$, then $|x+\theta(u)-a_\ell| > \frac{\delta}{2}$ for any $u \in [0,\varepsilon^2 t]$ and there exists a positive constant $C = C(\delta)$ such that $V(x+\theta(u)) \geqq C$. Hence, we have

$$p_\lambda^{(1)}(t,x) \leqq (2\pi\varepsilon^2 t)^{-\frac{d}{2}}\mathrm{e}^{-\frac{Ct}{\varepsilon^2}} \qquad (t>0).$$

Next, denote by $\mu(A\,|\,\theta(\varepsilon^2 t)=0)$ the conditional probability of $A \in \mathscr{B}(W)$ given $\theta(\varepsilon^2 t)=0$. By Corollary 3.1.8,

$$\mu\Big(\max_{0\leqq u\leqq\varepsilon^2 t}|\theta(u)| > \frac{\delta}{2}\,\Big|\,\theta(\varepsilon^2 t)=0\Big)$$
$$\leqq d\cdot\mu\Big(\max_{0\leqq u\leqq\varepsilon^2 t}|\theta^1(u)| > \frac{\delta}{2\sqrt{d}}\,\Big|\,\theta(\varepsilon^2 t)=0\Big) = d\mathrm{e}^{-\frac{\delta^2}{2\varepsilon^2 dt}}.$$

Hence, by the non-negativity of V, we obtain

$$p_\lambda^{(2)}(t,x) \leqq d(2\pi\varepsilon^2 t)^{-\frac{d}{2}}\mathrm{e}^{-\frac{\delta^2}{2\varepsilon^2 dt}} \qquad (t>0).$$

Thus, $p_\lambda(t,x,x) \to 0$ uniformly in $x \in [0,1)^d \setminus (\bigcup_\ell O_{\ell,\delta})$ and $I_\lambda^2(t) \to 0$ as $\lambda\to\infty$ (i.e., $\varepsilon\to 0$).

We now turn to the main term $I_\lambda^1(t)$. Fix $\ell = 1, \dots, m$. By (7.3.2), we have

$$\int_{O_{\ell,\delta}} p_\lambda(t,x,x)\,\mathrm{d}x = \int_{|x|<\delta} \mathrm{d}x \int_W \mathrm{e}^{-\frac{1}{\varepsilon^2}\int_0^t V(a_\ell+x+\varepsilon\theta(s))\mathrm{d}s}\varepsilon^{-n}\delta_0(\theta(t))\,\mathrm{d}\mu$$
$$= \int_{|y|<\frac{\delta}{\varepsilon}} \mathrm{d}y \int_W \mathrm{e}^{-\frac{1}{\varepsilon^2}\int_0^t V(a_\ell+\varepsilon(y+\theta(s)))\mathrm{d}s}\delta_0(\theta(t))\,\mathrm{d}\mu.$$

For $N > 0$, we divide this into two terms,

$$J_1(N) = \int_{|y|<\frac{\delta}{\varepsilon}, |y|\leqq N} \mathrm{d}y \int_W \mathrm{e}^{-\frac{1}{\varepsilon^2}\int_0^t V(a_\ell+\varepsilon(y+\theta(s)))\mathrm{d}s}\delta_0(\theta(t))\,\mathrm{d}\mu$$

and

$$J_2(N) = \int_{|y|<\frac{\delta}{\varepsilon}, |y|>N} \mathrm{d}y \int_W \mathrm{e}^{-\frac{1}{\varepsilon^2}\int_0^t V(a_\ell+\varepsilon(y+\theta(s)))\mathrm{d}s}\delta_0(\theta(t))\,\mathrm{d}\mu.$$

For $\eta > 0$, set $\widetilde{h}_\eta = \mathbf{1}_{\{\max_{0\leqq u\leqq t}|\varepsilon\theta(u)|<\eta\}}$, and divide $J_2(N)$ into two terms

$$J_2^{(1)}(N) = \int_{|y|<\frac{\delta}{\varepsilon}, |y|>N} (2\pi t)^{-\frac{d}{2}}\mathbf{E}\Big[\widetilde{h}_\eta \mathrm{e}^{-\frac{1}{\varepsilon^2}\int_0^t V(a_\ell+\varepsilon(y+\theta(s)))\mathrm{d}s}\Big|\,\theta(t)=0\Big]\mathrm{d}y$$

and

$$J_2^{(2)}(N) = \int_{|y|<\frac{\delta}{\varepsilon}, |y|>N} (2\pi t)^{-\frac{d}{2}}\mathbf{E}\Big[(1-\widetilde{h}_\eta)\mathrm{e}^{-\frac{1}{\varepsilon^2}\int_0^t V(a_\ell+\varepsilon(y+\theta(s)))\mathrm{d}s}\Big|\,\theta(t)=0\Big]\mathrm{d}y.$$

By Corollary 3.1.8,

$$\mathbf{E}[1-\widetilde{h}_\eta|\theta(t)=0] \leqq d(2\pi t)^{-\frac{d}{2}}\mathrm{e}^{-\frac{\eta^2}{2\varepsilon^2 dt}}.$$

Then, by the non-negativity of V,

$$J_2^{(2)}(N) \leqq d(2\pi t)^{-\frac{d}{2}}\mathrm{vol}\Big(|y| < \frac{\delta}{\varepsilon}\Big)\mathrm{e}^{-\frac{\eta^2}{2\varepsilon^2 dt}}. \tag{7.3.4}$$

Hence, $J_2^{(2)}(N) \to 0$ as $\varepsilon \to 0$.

By the condition (V), there exists a constant C_1 such that $V(a_\ell+\xi) \geqq C_1|\xi|^2$ if $|\xi|$ is sufficiently small. Hence, there exists a constant C, depending only on η, δ, and the Hessian $(V_{ij}(a_\ell))_{i,j=1,2,\dots,d}$, such that

$$J_2^{(1)}(N) \leqq \int_{|y|<\frac{\delta}{\varepsilon}, |y|>N} (2\pi)^{-\frac{d}{2}}\mathbf{E}\Big[\widetilde{h}_\eta \mathrm{e}^{-C\int_0^t |y+\theta(s)|^2\mathrm{d}s}\Big|\,\theta(t)=0\Big]\mathrm{d}y$$
$$\leqq \int_{|y|>N} (2\pi t)^{-\frac{d}{2}}\mathbf{E}\Big[\mathrm{e}^{-C\int_0^t |y+\theta(s)|^2\mathrm{d}s}\Big|\,\theta(t)=0\Big]\mathrm{d}y$$

for sufficiently small $\eta > 0$. Hence, by Theorem 5.8.2,

$$\sup_{\varepsilon>0} J_2^{(1)}(N) \leqq \int_{|y|>N} \Big(\frac{2C}{\pi \sinh(\sqrt{2C}t)}\Big)^{\frac{d}{2}} \mathrm{e}^{-\sqrt{2C}\tanh(\frac{\sqrt{2C}t}{2})y^2} \mathrm{d}y$$
$$= O\Big(\frac{1}{N}\mathrm{e}^{-N^2}\Big) \qquad (N \to \infty).$$

Combining this with (7.3.4), we obtain

$$\limsup_{\varepsilon\downarrow 0} J_2(N) = O\Big(\frac{1}{N}\mathrm{e}^{-N^2}\Big).$$

Set

$$\mathfrak{h}_\ell(t) = \frac{1}{2}\sum_{i,j=1}^d V_{ij}(a_\ell)\int_0^t (y^i + \theta^i(s))(y^j + \theta^j(s))\,\mathrm{d}s$$

and decompose $J_1(N)$ as

$$J_1(N) = J_1^{(1)}(N) + J_1^{(2)}(N) + J_1^{(3)}(N),$$

where

$$J_1^{(1)}(N) = \int_{|y|<\frac{\delta}{\varepsilon}\wedge N} (2\pi t)^{-\frac{d}{2}} \mathbf{E}[\widetilde{h}_\eta \mathrm{e}^{-\mathfrak{h}_\ell(t)} | \theta(t) = 0]\,\mathrm{d}y,$$
$$J_1^{(2)}(N) = \int_{|y|<\frac{\delta}{\varepsilon}\wedge N} (2\pi t)^{-\frac{d}{2}}$$
$$\times \mathbf{E}\Big[\widetilde{h}_\eta\Big(\mathrm{e}^{-\frac{1}{\varepsilon^2}\int_0^t V(a_\ell+\varepsilon(y+\theta(s)))\mathrm{d}s} - \mathrm{e}^{-\mathfrak{h}_\ell(t)}\Big)\Big| \theta(t) = 0\Big]\mathrm{d}y,$$
$$J_1^{(3)}(N) = \int_{|y|<\frac{\delta}{\varepsilon}\wedge N} (2\pi t)^{-\frac{d}{2}}$$
$$\times \mathbf{E}\Big[(1-\widetilde{h}_\eta)\mathrm{e}^{-\frac{1}{\varepsilon^2}\int_0^t V(a_\ell+\varepsilon(y+\theta(s)))\mathrm{d}s}\Big| \theta(t) = 0\Big]\mathrm{d}y.$$

For $J_1^{(3)}(N)$, we can show in the same way as above

$$J_1^{(3)}(N) \leqq (2\pi t)^{-\frac{d}{2}}\Big(\frac{\delta}{\varepsilon}\Big)^d \mathrm{e}^{-\frac{\eta^2}{\varepsilon^2}}.$$

If $\max_{0\leqq s\leqq t}|\varepsilon\theta(s)| < \eta$ and $|\varepsilon y| < \delta$, then there exists a constant C_2, depending only on η and δ, such that

$$\mathrm{e}^{-\varepsilon^{-2}\int_0^t V(a_\ell+\varepsilon(y+\theta(s)))\mathrm{d}s} - \mathrm{e}^{-\mathfrak{h}_\ell} \leqq C_2\varepsilon.$$

Hence, we obtain

$$J_1^{(2)}(N) \leqq C_2\varepsilon(2\pi t)^{-\frac{d}{2}}\mathrm{vol}(|y| < N).$$

Thus $\limsup_{\varepsilon\downarrow 0} J_1^{(i)}(N) = 0$ $(i = 2, 3)$.

Let $r_\ell(t,x,y)$ $(t>0,\ x,y\in\mathbb{R}^d)$ be the heat kernel for L_ℓ. Then, by Lebesgue's convergence theorem, we have

$$\lim_{\varepsilon\to 0} J_1^{(1)}(N) = \int_{|y|<N} (2\pi t)^{-\frac{d}{2}} \mathbf{E}[\mathrm{e}^{-\mathfrak{b}_\ell} | \theta(t)=0]\,\mathrm{d}y$$
$$= \int_{|y|<N} r_\ell(t,y,y)\,\mathrm{d}y.$$

Summing up the above observations, we obtain constants C_3 and C_4 such that

$$\limsup_{\lambda\to\infty} \left| \mathrm{tr}(\mathrm{e}^{-\frac{t}{\lambda}H_\lambda^{\mathrm{per}}}) - \sum_{\ell=1}^m \int_{|y|<N} r_\ell(t,y,y)\,\mathrm{d}y \right| \leqq \frac{C_3}{N}\mathrm{e}^{-C_4N^2}$$

for any N. Letting $N\to\infty$, we arrive at the desired identity

$$\lim_{\lambda\to\infty} \mathrm{tr}(\mathrm{e}^{-\frac{t}{\lambda}H_\lambda^{\mathrm{per}}}) = \sum_{\ell=1}^m \int_{\mathbb{R}^d} r_\ell(t,y,y)\,\mathrm{d}y = \sum_{\ell=1}^m \mathrm{tr}(\mathrm{e}^{-tL_\ell}).$$ □

7.4 Selberg's Trace Formula on the Upper Half Plane

Let $(W,\mathscr{B}(W),\mu)$ be the two-dimensional Wiener space and $\{(\theta^1(t),\theta^2(t))\}_{t\geqq 0}$ be the coordinate process. Consider the stochastic differential equation

$$\begin{cases} \mathrm{d}X(t) = Y(t)\,\mathrm{d}\theta^1(t), & X(0)=x\in\mathbb{R}, \\ \mathrm{d}Y(t) = Y(t)\,\mathrm{d}\theta^2(t), & Y(0)=y>0. \end{cases} \tag{7.4.1}$$

The unique solution $Z=\{Z(t)=(X(t),Y(t))\}_{t\geqq 0}$ is given by

$$X(t) = x + \int_0^t y\mathrm{e}^{\theta^2(s)-\frac{s}{2}}\mathrm{d}\theta^1(s), \qquad Y(t) = y\mathrm{e}^{\theta^2(t)-\frac{t}{2}}.$$

Z is a diffusion process on the **upper half plane** $\{(x,y)\in\mathbb{R}^2; y>0\}$ generated by

$$L = \frac{1}{2}\Delta, \qquad \Delta = y^2\Big(\frac{\partial^2}{\partial x^2} + \frac{\partial^2}{\partial y^2}\Big).$$

In particular, $\{Y(t)\}_{t\geqq 0}$ is a geometric Brownian motion (see Example 4.2.4).

The above Δ is the Laplace–Beltrami operator on the upper half plane, endowed with the Riemannian metric, called the **Poincaré metric**,

$$\mathrm{d}s^2 = \frac{\mathrm{d}x^2+\mathrm{d}y^2}{y^2}.$$

Because of this, we continue to use the same symbol Δ, which was used to denote the Laplacian on $\mathbb{R}^d$. In the rest of this section we denote by $\mathbf{H}^2$ the upper half plane with the Poincaré metric, which is a Riemannian manifold. $\mathbf{H}^2$ has a constant negative curvature -1 and the volume element is given by $dv(z) = y^{-2}dxdy$ (see for example [33]).

On a Riemannian manifold, a second order elliptic differential operator, called the Laplace–Beltrami operator, is defined from the Riemannian metric and a diffusion process generated by it is called a Brownian motion on the Riemannian manifold. While $\mathbf{H}^2$ admits a global coordinate, under which we shall work, a general Riemannian manifold does not. For this reason, a construction of the Brownian motion on a Riemannian manifold requires an additional geometric observation, and is one of the interesting problems in stochastic analysis. Several constructions are known. For details, see [22, 38, 45, 100] and others.

We continue to explain the geometry of $\mathbf{H}^2$. Identify $\mathbf{H}^2 \ni (x, y)$ with $z = x + iy \in \mathbb{C}$. Then, the **special linear group**

$$\mathrm{SL}_2(\mathbb{R}) = \left\{ g = \begin{pmatrix} a & b \\ c & d \end{pmatrix};\ a, b, c, d \in \mathbb{R},\ \det g = 1 \right\}$$

acts on $\mathbf{H}^2$ by **linear fractional transformations**

$$gz = \frac{az + b}{cz + d},$$

which is an isometry ([11, 33]). In this connection, it may be useful to recall that the distance function $d(z, z')$ associated with the Poincaré metric on $\mathbf{H}^2$ is given by

$$\cosh d(z, z') = \frac{|x - x'|^2 + y^2 + (y')^2}{2yy'}. \tag{7.4.2}$$

Let M be a two-dimensional compact orientable Riemannian manifold with constant sectional curvature -1. Then, the universal covering $\widetilde{M}$ is a manifold which is isometric with $\mathbf{H}^2$ and the fundamental group $\pi_1(M)$ of M is identified with a discrete subgroup of the projective special linear group $\mathrm{PSL}_2(\mathbb{R}) = \mathrm{SL}_2(\mathbb{R})/ \pm I$. Since M is assumed to be compact, the representative of $g \in \pi_1(M)$ in $\mathrm{SL}_2(\mathbb{R})$ is conjugate with an element of the form $\gamma_a = \begin{pmatrix} a & 0 \\ 0 & a^{-1} \end{pmatrix}$ $(a > 0)$.

Using the Brownian motion Z on $\mathbf{H}^2$ and following the argument in [85], we show **Selberg's trace formula** ([103]) for the Laplace–Beltrami operator on M. To present the formula, we need the explicit form of the value $p(t, z, z)$ $(t > 0,\ z \in \mathbf{H}^2)$ on the diagonal for the transition density of Z or the heat kernel of L. We prove the following explicit form of $p(t, z, z')$ in the next section.

Theorem 7.4.1 *The transition density* $p(t,z,z')$ $(t > 0,\ z,z' \in \mathbf{H}^2)$ *of* Z *with respect to the volume element* $\mathrm{d}v(z) = y^{-2}\mathrm{d}x\mathrm{d}y$ *on* $\mathbf{H}^2$ *is given by*

$$p(t,z,z') = \frac{\sqrt{2}\mathrm{e}^{-\frac{t}{8}}}{(2\pi t)^{\frac{3}{2}}} \int_r^\infty \frac{b\mathrm{e}^{-\frac{b^2}{2t}}}{(\cosh b - \cosh r)^{\frac{1}{2}}}\,\mathrm{d}b, \tag{7.4.3}$$

where $r = d(z,z')$.

Identify the fundamental group $\pi_1(M)$ of M with a discrete subgroup of $\mathrm{PSL}_2(\mathbb{R})$ and denote it by Γ. Let Δ_M be the restriction of Δ to the space of smooth functions f which is periodic with respect to the action of Γ, that is, $f(\gamma z) = f(z)$ for any $\gamma \in \Gamma$, $z \in \mathbf{H}^2$. Moreover, let $\mathscr{D}$ be the fundamental domain of Γ, a connected subset of $\mathbf{H}^2$ satisfying

$$\bigcup_{\gamma\in\Gamma} \gamma\mathscr{D} = \mathbf{H}^2 \quad \text{and} \quad \gamma\mathscr{D} \cap \gamma'\mathscr{D} = \emptyset \ \ (\gamma \neq \gamma').$$

It is known that Δ_M is essentially self-adjoint and extended to a unique self-adjoint operator on $L^2(\mathbf{H}^2, \mathrm{d}v)$, say Δ_M again. Δ_M may be regarded as a self-adjoint operator on $L^2(\mathscr{D}, \mathrm{d}v)$ and its spectrum consists only of eigenvalues $\{\lambda_k\}_{k=1}^\infty$ with finite multiplicities.

For $\gamma \in \Gamma$, set

$$\ell(\gamma) = \inf_{z\in\mathbf{H}^2} d(z, \gamma z).$$

$\ell(\gamma)$ is the length of the closed geodesic on M associated with γ and, if γ is conjugate with γ_a $(a > 0)$, satisfies

$$\ell(\gamma) = d(z, a^2 z) = \log a^2 \qquad (z \in \mathbf{H}^2).$$

Moreover, for $n \in \mathbb{N}$, $\ell(\gamma^n) = n\ell(\gamma)$.

We call $\gamma \in \Gamma$ primitive if it is not represented as a positive power of another element of Γ. Let Γ_0 be a set of the primitive elements in Γ whose elements are not conjugate with each other.

Theorem 7.4.2 *For all* $t > 0$,

$$\begin{aligned}\mathrm{tr}(\mathrm{e}^{\frac{t}{2}\Delta_M}) &= \sum_{k=1}^\infty \mathrm{e}^{-\frac{1}{2}\lambda_k t} \\ &= \mathrm{vol}(M)\frac{\mathrm{e}^{-\frac{t}{8}}}{(2\pi t)^{\frac{3}{2}}} \int_0^\infty \frac{b\mathrm{e}^{-\frac{b^2}{2t}}}{\sinh(\frac{b}{2})}\,\mathrm{d}b + \sum_{\gamma\in\Gamma_0}\sum_{n=1}^\infty \frac{\mathrm{e}^{-\frac{t}{8}}}{2\sqrt{2\pi t}} \frac{\ell(\gamma)}{\sinh(\frac{n\ell(\gamma)}{2})}\mathrm{e}^{-\frac{n^2\ell(\gamma)^2}{2t}}.\end{aligned}$$

Proof We show a sketch of a proof. Let $\delta^{(2)}_{z'}$ be the Dirac measure concentrated at $z' \in \mathbf{H}^2$ with respect to the Lebesgue measure. Then, the heat kernel $p(t,z,z')$ for L is written as

$$p(t,z,z') = (y')^2 \int_W \delta^{(2)}_{z'}(Z(t))\,\mathrm{d}\mu.$$

Set $(\widetilde{\theta}^1(s), \widetilde{\theta}^2(s)) = (\theta^1(s), \theta^2(s) - \frac{s}{2})$. By the Cameron–Martin theorem (Theorem 1.7.2, Theorem 4.6.2), $\widetilde{\theta} = \{(\widetilde{\theta}^1(s), \widetilde{\theta}^2(s))\}_{0 \leqq s \leqq t}$ is a two-dimensional Brownian motion under the probability measure $\widetilde{\mu}$ on $(W, \mathscr{B}(W))$ given by

$$\mathrm{d}\widetilde{\mu}\Big|_{\mathscr{B}_t} = \mathrm{e}^{\frac{1}{2}\theta^1(t) - \frac{t}{8}}\,\mathrm{d}\mu\Big|_{\mathscr{B}_t}.$$

Since $\delta_{z'}(\alpha Z(t)) = \alpha^{-2}\delta_{\frac{z'}{\alpha}}(Z(t))$ $(\alpha > 0)$, we have

$$\begin{aligned} p(t,z,z') &= (y')^2 \int_W \mathrm{e}^{-\frac{1}{2}\widetilde{\theta}^2(t) - \frac{t}{8}} \delta^{(2)}_{z'}(x + y\beta(t), y\mathrm{e}^{\widetilde{\theta}^2(t)})\,\mathrm{d}\widetilde{\mu} \\ &= \mathrm{e}^{-\frac{t}{8}} \left(\frac{y'}{y}\right)^{\frac{3}{2}} \int_W \delta^{(2)}_{(\frac{x'-x}{y}, \frac{y'}{y})}(\beta(t), \mathrm{e}^{\widetilde{\theta}^2(t)})\,\mathrm{d}\widetilde{\mu}, \end{aligned}$$

where $\beta(t)$ is given by

$$\beta(t) = \int_0^t \mathrm{e}^{\widetilde{\theta}^2(s)}\mathrm{d}\widetilde{\theta}^1(s).$$

The conditional probability distribution of $\beta(t)$ given $\widetilde{\theta}^2 = \{\widetilde{\theta}^2(s)\}_{0 \leqq s \leqq t}$ is the Gaussian distribution with mean 0 and variance

$$\widetilde{A}(t) = \int_0^t \mathrm{e}^{2\widetilde{\theta}^2(s)}\mathrm{d}s.$$

Taking advantage of the conditional expectation,[1] we obtain

$$\begin{aligned} p(t,z,z') &= \mathrm{e}^{-\frac{t}{8}} \left(\frac{y'}{y}\right)^{\frac{3}{2}} \int_W \mathrm{d}\widetilde{\mu} \int_{\mathbb{R}} \frac{1}{\sqrt{2\pi\widetilde{A}(t)}} \mathrm{e}^{-\frac{\xi^2}{2\widetilde{A}(t)}} \delta^{(2)}_{(\frac{x'-x}{y}, \frac{y'}{y})}(\xi, \mathrm{e}^{\widetilde{\theta}^2(t)})\,\mathrm{d}\xi \\ &= \mathrm{e}^{-\frac{t}{8}} \left(\frac{y'}{y}\right)^{\frac{3}{2}} \int_W \frac{1}{\sqrt{2\pi\widetilde{A}(t)}} \mathrm{e}^{-\frac{(x'-x)^2}{2y^2\widetilde{A}(t)}} \delta_{\frac{y'}{y}}(\mathrm{e}^{\widetilde{\theta}^2(t)})\,\mathrm{d}\widetilde{\mu} \\ &= \mathrm{e}^{-\frac{t}{8}} \left(\frac{y'}{y}\right)^{\frac{1}{2}} \int_W \frac{1}{\sqrt{2\pi\widetilde{A}(t)}} \mathrm{e}^{-\frac{(x'-x)^2}{2y^2\widetilde{A}(t)}} \delta_{\log(\frac{y'}{y})}(\widetilde{\theta}^2(t))\,\mathrm{d}\widetilde{\mu}, \end{aligned} \tag{7.4.4}$$

[1] While we carry out a formal computation on the Dirac measure, we can justify the argument by considering the expectation $\mathbf{E}[f_1(X(t))f_2(Y(t))]$ for bounded continuous functions f_1 and f_2.

where δ_a is the Dirac measure concentrated at $a \in \mathbb{R}$ and, in the last equality, we have used $\delta_a(e^x) = a^{-1}\delta_{\log a}(x)$ $(a > 0)$.

The explicit form of the probability density of the distribution of $(\widetilde{\theta}(t), \widetilde{A}(t))$ for $t > 0$ is known ([132]), which, in conjunction with (7.4.4), leads us to Theorem 7.4.1. In the next section, we firstly study the distribution, and then we shall prove Theorem 7.4.1.

We continue the sketch of the proof of Theorem 7.4.2. For the details of the following argument, see [85].

For $\sigma \in \Gamma$, there exist $n \in \mathbb{N}$, $\kappa \in \Gamma$, and $\gamma \in \Gamma_0$ such that

$$\sigma = \kappa^{-1}\gamma^n\kappa.$$

γ, n and the conjugacy class $[\kappa] \in \Gamma/\Gamma_\gamma$ containing κ are uniquely determined by σ, where Γ_γ is the centralizer of γ. Hence, we obtain

$$\begin{aligned}\operatorname{tr}(e^{-\frac{1}{2}t\Delta_M}) &= \int_{\mathscr{D}} p(t,z,z)\,dv(z) + \sum_{\sigma \neq I}\int_{\mathscr{D}} p(t,z,\sigma z)\,dv(z)\\ &= \operatorname{vol}(M)p(t,z,z) + \sum_{\gamma\in\Gamma_0}\sum_{n=1}^{\infty}\sum_{[\kappa]\in\Gamma/\Gamma_\gamma} I(\gamma,n,[\kappa]),\end{aligned}$$

where $I(\gamma, n, [\kappa])$ is given by

$$I(\gamma,n,[\kappa]) = \int_{\mathscr{D}} p(t,z,\kappa^{-1}\gamma^n\kappa z)\,dv(z).$$

Recall now that $p(t,z,z)$ does not depend on z and, by Theorem 7.4.1, is given by

$$p(t,z,z) = \frac{e^{-\frac{t}{8}}}{(2\pi t)^{\frac{3}{2}}}\int_0^\infty \frac{be^{-\frac{b^2}{2t}}}{\sinh(\frac{b}{2})}\,db.$$

For $\gamma \in \Gamma_0$, take $a = a(\gamma) > 1$ and $\tau \in PSL_2(\mathbb{R})$ so that $\gamma = \tau^{-1}\gamma_a\tau$ and set $\mathscr{D}_\gamma = \bigcup_{[\kappa]\in\Gamma/\Gamma_\gamma}\kappa\mathscr{D}$. $\mathscr{D}_\gamma$ is a fundamental domain of Γ_γ and

$$\bigcup_{m=-\infty}^{\infty}\gamma_a^m\tau\mathscr{D}_\gamma = \tau\Big(\bigcup_{m=-\infty}^{\infty}\gamma^m\mathscr{D}_\gamma\Big) = \mathbf{H}^2$$

Thus, $\tau\mathscr{D}_\gamma$ is a fundamental domain of the cyclic group $\{(\gamma_a)^m\}_{m=-\infty}^{\infty}$ and we may assume that

$$\tau\mathscr{D}_\gamma = \{(x,y) \in \mathbf{H}^2; 1 < y \leqq a^2\}.$$

Moreover, as was mentioned above, $\ell(\gamma) = \ell(\gamma_a) = \log a^2$.

Combining the arguments above, we obtain, for fixed $\gamma \in \Gamma_0$ and $n \in \mathbb{N}$,

$$\sum_{[\kappa]\in\Gamma/\Gamma_\gamma} I(\gamma,n,[\kappa]) = \int_{\mathscr{D}_\gamma} p(t,z,\gamma^n z)\,dv(z).$$

Moreover, since $\tau \in \mathrm{PSL}_2(\mathbb{R})$ is an isometry of $\mathbf{H}^2$ and

$$p(t,z,\gamma^n z) = p(t,z,\tau^{-1}\gamma_a^n \tau z) = p(t,\tau z,\gamma_a^n \tau z)$$

holds, we have

$$\sum_{[\kappa]\in\Gamma/\Gamma_\gamma} I(\gamma,n,[\kappa]) = \int_{\tau\mathscr{D}_\gamma} p(t,z,\gamma_a^n z)\,\mathrm{d}v(z) = \int_1^{a^2} \frac{\mathrm{d}y}{y^2}\int_{\mathbb{R}} p(t,z,a^{2n}z)\,\mathrm{d}x.$$

Using the expression (7.4.4) of the heat kernel, we obtain[2]

$$\begin{aligned}
&\int_1^{a^2} \frac{\mathrm{d}y}{y^2}\int_{\mathbb{R}} p(t,z,a^{2n}z)\,\mathrm{d}x \\
&= \mathrm{e}^{-\frac{t}{8}} a^n \int_1^{a^2} \frac{\mathrm{d}y}{y^2}\int_{\mathbb{R}} \mathrm{d}x \int_W \frac{1}{\sqrt{2\pi\widetilde{A}(t)}} \mathrm{e}^{-\frac{(a^{2n}-1)^2x^2}{2y^2\widetilde{A}(t)}} \delta_{\log(a^{2n})}(\widetilde{\theta^2}(t))\,\mathrm{d}\widetilde{\mu} \\
&= \mathrm{e}^{-\frac{t}{8}} a^n \int_1^{a^2} \frac{\mathrm{d}y}{y^2}\int_W \frac{y}{a^{2n}-1}\delta_{\log(a^{2n})}(\widetilde{\theta^2}(t))\,\mathrm{d}\widetilde{\mu} \\
&= \frac{\mathrm{e}^{-\frac{t}{8}}}{2\sqrt{2\pi t}} \frac{\ell(\gamma)}{\sinh(\frac{n\ell(\gamma)}{2})} \mathrm{e}^{-\frac{n^2\ell(\gamma)^2}{2t}},
\end{aligned}$$

where, to see the last identity, we have used

$$\int_W \delta_x(\widetilde{\theta^2}(t))\,\mathrm{d}\widetilde{\mu} = \frac{1}{\sqrt{2\pi t}}\mathrm{e}^{-\frac{x^2}{2t}}. \qquad \square$$

A probabilistic and simple proof as above is extensible to showing Selberg's trace formula for the Maass Laplacian (see [33]), a generalization of Δ, given by

$$y^2\Big(\frac{\partial}{\partial x} + \mathrm{i}\frac{k}{y}\Big)^2 + y^2\frac{\partial^2}{\partial y^2} \qquad (k \in \mathbb{R}).$$

For details, see [42].

7.5 Integral of Geometric Brownian Motion and Heat Kernel on $\mathbf{H}^2$

To prove Selberg's trace formula, evaluating the diagonal of the heat kernel on $\mathbf{H}^2$ was indispensable. In this section we prove Theorem 7.4.1 which gives

[2] It may seem that we formally apply Fubini's theorem since the δ-function is used for the expression of the heat kernel. In fact, we can apply it because $\{\widetilde{\theta^2}(t)\}_{t\geqq 0}$ is a one-dimensional Brownian motion under $\widetilde{\mu}$ and the generalized Wiener functionals determined by δ-functions are measures on the Wiener space (Theorem 5.4.15).

the explicit form of the heat kernel. We need some formulae for the modified Bessel functions, Legendre functions and so on. For these, see, for example, [31, 67].

Let $\{B(t)\}_{t\geqq 0}$ be a one-dimensional Brownian motion with $B(0) = 0$ defined on a probability space $(\Omega, \mathscr{F}, \mathbf{P})$ and $A(t)$ be a functional given by

$$A(t) = \int_0^t e^{2B(s)} ds.$$

In order to derive the explicit form of the heat kernel from the expression (7.4.4), it is necessary to show the explicit form of the probability density of the distribution of $(A(t), B(t))$ for fixed $t > 0$. This was done by Yor [130, 132].

For $\mu \in \mathbb{R}$, the function

$$I_\mu(z) = \sum_{n=0}^{\infty} \frac{1}{n!\,\Gamma(\mu+n+1)} \left(\frac{z}{2}\right)^{2n+\mu} \qquad (z \in \mathbb{C} \setminus (-\infty, 0))$$

is called the **modified Bessel function** of the first kind of order μ. When $z = v > 0$, the function $[0, \infty) \ni \lambda \mapsto I_{\sqrt{2\lambda}}(v)$ is completely monotone. Set

$$\Theta(v, t) = \frac{v}{(2\pi^3 t)^{\frac{1}{2}}} \int_0^\infty e^{\frac{\pi^2 - \xi^2}{2t}} e^{-v\cosh(\xi)} \sinh(\xi) \sin\left(\frac{\pi\xi}{t}\right) d\xi. \tag{7.5.1}$$

Then it is known that ([130, 132])

$$I_{\sqrt{2\lambda}}(v) = \int_0^\infty e^{-\lambda t} \Theta(v, t)\, dt. \tag{7.5.2}$$

The function K_μ $(\mu \in \mathbb{R})$ defined by

$$K_\mu(z) = \begin{cases} \dfrac{\pi}{2} \dfrac{I_{-\mu}(z) - I_\mu(z)}{\sin \mu\pi} & (\mu \notin \mathbb{Z}), \\[2ex] \dfrac{(-1)^n}{2} \left[\dfrac{\partial I_{-\mu}(z)}{\partial \mu} - \dfrac{\partial I_\mu(z)}{\partial \mu} \right]_{\mu=n} & (\mu = n \in \mathbb{Z}) \end{cases}$$

is called the modified Bessel function of the second kind, or the Macdonald function, of order μ. For $\mu \geqq 0$, I_μ and K_μ are monotone increasing and decreasing functions on $(0, \infty)$, respectively. They satisfy the modified Bessel equation

$$\frac{d^2 u}{dz^2} + \frac{1}{z}\frac{du}{dz} - \left(1 + \frac{\mu^2}{z^2}\right) u = 0.$$

Theorem 7.5.1 *For $t > 0$, the distribution of $(A(t), B(t))$ has the smooth density $a_t(u, b)$ $(u > 0,\ b \in \mathbb{R})$ given by*

$$a_t(u, b) = \frac{1}{u} \exp\left(-\frac{1 + e^{2b}}{2u}\right) \Theta\left(\frac{e^b}{u}, t\right). \tag{7.5.3}$$

Proof $\{(A(t), B(t))\}_{t\geqq 0}$ is a diffusion process on $\mathbb{R}^2$ generated by

$$\frac{1}{2}\frac{\partial^2}{\partial y^2} + e^{2y}\frac{\partial}{\partial x}.$$

Since

$$\left[\frac{\partial}{\partial y}, e^{2y}\frac{\partial}{\partial x}\right] = 2e^{2y}\frac{\partial}{\partial y},$$

it satisfies Hörmander's condition. Therefore, by Theorems 5.4.11 and 5.5.4, $(A(t), B(t))$ has a density of C^∞-class. We denote it by $a_t(u, b)$ and show that it has the expression (7.5.3).

For this purpose, for $\lambda > 0$, set

$$\begin{aligned} q_\lambda(t, x, y) &= \mathbf{E}\Big[e^{-\frac{1}{2}\lambda^2 e^{2x} A(t)}\Big| x + B(t) = y\Big]\frac{1}{\sqrt{2\pi t}}e^{-\frac{(y-x)^2}{2t}} \\ &= \int_0^\infty e^{-\frac{1}{2}\lambda^2 e^{2x} u} a_t(u, y - x)\,du. \end{aligned}$$

$q_\lambda(t, x, y)$ is the heat kernel (with respect to the Lebesgue measure) for the Schrödinger operator on $\mathbb{R}$:

$$H_\lambda = -\frac{1}{2}\frac{d^2}{dx^2} + \frac{1}{2}\lambda^2 e^{2x}.$$

That is, the function u given by

$$u(t, x) = \int_{\mathbb{R}} q_\lambda(t, x, y) f(y)\,dy$$

for $f \in L^2(\mathbb{R})$ satisfies the heat equation

$$\frac{\partial u}{\partial t} = -H_\lambda u, \qquad \lim_{t\downarrow 0} u(t, x) = f(x).$$

The Laplace transform of the heat kernel in t,

$$G_\lambda(x, y; \alpha) := \int_0^\infty e^{-\alpha t} q_\lambda(t, x, y)\,dt \qquad (\alpha > 0)$$

is called the Green function with respect to the Lebesgue measure. Now consider the equation

$$-H_\lambda u = \alpha u.$$

Then, there exist solutions $u_1(x; \alpha)$ and $u_2(x; \alpha)$, which are monotone increasing and decreasing in x, respectively, such that

$$\lim_{x\to-\infty} u_1(x; \alpha) = 0 \quad \text{and} \quad \lim_{x\to\infty} u_2(x; \alpha) = 0.$$

Such solutions are unique up to multiplicative constants. It is known in the general theory of one-dimensional diffusion processes (see, for example, [50, 56]) that the function $\widetilde{G}_\lambda(x, y; \alpha) = \frac{1}{W(u_1,u_2)} u_1(x; \alpha) u_2(y; \alpha)$ $(x \leqq y)$ is the Green function with respect to the speed measure $m(\mathrm{d}y) = 2\mathrm{d}y$ (Section 4.8), where $W(u_1, u_2)$ is the Wronskian given by $W(u_1, u_2) = u_1'(x)u_2(x) - u_1(x)u_2'(x)$. Hence,

$$G_\lambda(x, y; \alpha) = \frac{2}{W(u_1, u_2)} u_1(x; \alpha) u_2(y; \alpha).$$

We can take

$$u_1(x; \lambda) = I_{\sqrt{2\alpha}}(\lambda \mathrm{e}^x) \quad \text{and} \quad u_2(x; \lambda) = K_{\sqrt{2\alpha}}(\lambda \mathrm{e}^x).$$

By the formula $I_\mu'(z)K_\mu(z) - I_\mu(z)K_\mu'(z) = \frac{1}{z}$, we have

$$G_\lambda(x, y; \alpha) = 2I_{\sqrt{2\alpha}}(\lambda \mathrm{e}^x) K_{\sqrt{2\alpha}}(\lambda \mathrm{e}^y) \qquad (x \leqq y,\ \alpha > 0).$$

Moreover, by using the formula

$$I_\mu(x)K_\mu(y) = \frac{1}{2}\int_0^\infty \mathrm{e}^{-\frac{t}{2} - \frac{x^2+y^2}{2t}} I_\mu\Big(\frac{xy}{t}\Big)\frac{\mathrm{d}t}{t},$$

we obtain

$$G_\lambda(x, y; \alpha) = \int_0^\infty \mathrm{e}^{-\frac{t}{2} - \frac{\lambda^2(e^{2x}+e^{2y})}{2t}} I_{\sqrt{2\alpha}}\Big(\frac{\lambda^2 \mathrm{e}^{x+y}}{t}\Big)\frac{\mathrm{d}t}{t},$$

that is,

$$\int_0^\infty \mathrm{e}^{-\alpha t}\mathrm{d}t \int_0^\infty \mathrm{e}^{-\frac{1}{2}\lambda^2 \mathrm{e}^{2x} u} a_t(u, y - x)\,\mathrm{d}u$$
$$= \int_0^\infty \mathrm{e}^{-\frac{t}{2} - \frac{\lambda^2(\mathrm{e}^{2x}+\mathrm{e}^{2y})}{2t}} I_{\sqrt{2\alpha}}\Big(\frac{\lambda^2 \mathrm{e}^{x+y}}{t}\Big)\frac{\mathrm{d}t}{t}.$$

On the other hand, by the characterization (7.5.2) of the function $\Theta(v, t)$, Fubini's theorem yields

$$\int_0^\infty \mathrm{e}^{-\alpha t}\mathrm{d}t \int_0^\infty \mathrm{e}^{-\frac{1}{2}\lambda^2 \mathrm{e}^{2x} u} \frac{1}{u} \mathrm{e}^{-\frac{1+\mathrm{e}^{2(y-x)}}{2u}} \Theta\Big(\frac{\mathrm{e}^{y-x}}{u}, t\Big)\mathrm{d}u$$
$$= \int_0^\infty \mathrm{e}^{-\frac{1}{2}\lambda^2 \mathrm{e}^{2x} u} \mathrm{e}^{-\frac{1+\mathrm{e}^{2(y-x)}}{2u}} I_{\sqrt{2\alpha}}\Big(\frac{\mathrm{e}^{y-x}}{u}\Big)\frac{\mathrm{d}u}{u}$$
$$= \int_0^\infty \mathrm{e}^{-\frac{t}{2}} \mathrm{e}^{-\frac{\lambda^2(\mathrm{e}^{2x}+\mathrm{e}^{2y})}{2t}} I_{\sqrt{2\alpha}}\Big(\frac{\lambda^2 \mathrm{e}^{x+y}}{t}\Big)\frac{\mathrm{d}t}{t},$$

where, for the second identity, we have used the change of variables given by $\lambda^2 \mathrm{e}^{2x} u = t$.

Hence, by a repeated use of the uniqueness of the Laplace transforms, we obtain (7.5.3). □

Plugging (7.5.3) into (7.4.4), we obtain the following ([32]).

Proposition 7.5.2 *Set $r = d(z,z')$. Then*

$$p(t,z,z') = \frac{\mathrm{e}^{-\frac{t}{8}}}{4\pi^{\frac{3}{2}}t^{\frac{1}{2}}} \int_0^\infty \frac{\mathrm{e}^{\frac{\pi^2-\xi^2}{2t}} \sinh\xi \sin(\frac{\pi\xi}{t})}{(\cosh r + \cosh\xi)^{\frac{3}{2}}}\,\mathrm{d}\xi. \tag{7.5.4}$$

Proof By (7.4.2), (7.4.4), and (7.5.3),

$$\begin{aligned}
p(t,z,z') &= \mathrm{e}^{-\frac{t}{8}}\Big(\frac{y'}{y}\Big)^{\frac{1}{2}} \int_0^\infty \frac{1}{\sqrt{2\pi}u^{\frac{3}{2}}}\mathrm{e}^{-\frac{(x-x')^2+y^2+(y')^2}{2y^2u}}\Theta\Big(\frac{y'}{yu},t\Big)\mathrm{d}u \\
&= \mathrm{e}^{-\frac{t}{8}} \int_0^\infty \frac{1}{\sqrt{2\pi}u^{\frac{3}{2}}}\mathrm{e}^{-\frac{y'\cosh r}{yu}}\Theta\Big(\frac{y'}{yu},t\Big)\mathrm{d}u \\
&= \frac{\mathrm{e}^{-\frac{t}{8}}}{\sqrt{2\pi}} \int_0^\infty \mathrm{e}^{-v\cosh r}v^{-\frac{1}{2}}\Theta(v,t)\,\mathrm{d}v.
\end{aligned} \tag{7.5.5}$$

On account of the expression (7.5.1) of $\Theta(v,t)$ and Fubini's theorem, we obtain

$$\begin{aligned}
p(t,z,z') &= \frac{\mathrm{e}^{-\frac{t}{8}}}{\sqrt{2\pi}}\frac{1}{(2\pi^3t)^{\frac{1}{2}}} \int_0^\infty \mathrm{e}^{\frac{\pi^2-\xi^2}{2t}} \sinh\xi \sin\Big(\frac{\pi\xi}{t}\Big)\mathrm{d}\xi \\
&\qquad\qquad \times \int_0^\infty \mathrm{e}^{-v(\cosh r+\cosh\xi)}v^{\frac{1}{2}}\,\mathrm{d}v \\
&= \frac{\mathrm{e}^{-\frac{t}{8}}}{2\pi^2t^{\frac{1}{2}}} \int_0^\infty \frac{\Gamma(\frac{3}{2})\mathrm{e}^{\frac{\pi^2-\xi^2}{2t}} \sinh\xi \sin(\frac{\pi\xi}{t})}{(\cosh r + \cosh\xi)^{\frac{3}{2}}}\,\mathrm{d}\xi.
\end{aligned}$$

Since $\Gamma(\frac{3}{2}) = \frac{\sqrt{\pi}}{2}$, we arrive at (7.5.4). □

Proof of Theorem 7.4.1 We give a proof following [81]. See [11] for another approach. For $\alpha,\beta,\gamma > 0$, let $F(\alpha,\beta,\gamma;z)$ be the Gauss **hypergeometric function** given by

$$F(\alpha,\beta,\gamma;z) = \frac{\Gamma(\gamma)}{\Gamma(\alpha)\Gamma(\beta)}\sum_{n=0}^\infty \frac{\Gamma(\alpha+n)\Gamma(\beta+n)}{\Gamma(\gamma+n)}\frac{z^n}{n!}.$$

Moreover, for $\mu \in \mathbb{R}\setminus\{-1,-2,\ldots\}$, $z \in \mathbb{C}\setminus(-\infty,0)$, let Q_μ be the **Legendre function** of the second kind defined by

$$Q_\mu(z) = \frac{\sqrt{\pi}\Gamma(\mu+1)}{\Gamma(\mu+\frac{3}{2})}\frac{1}{(2z)^{\mu+1}}F\Big(\frac{\mu+1}{2},\frac{\mu}{2}+1,\mu+\frac{3}{2};\frac{1}{z^2}\Big).$$

Q_μ is a solution for the Legendre differential equation

$$(1-z^2)\frac{\mathrm{d}^2u}{\mathrm{d}z^2} - 2z\frac{\mathrm{d}u}{\mathrm{d}z} + \mu(\mu+1)u = 0.$$

In what follows, we use the integral representations of Q_μ ($\mu > -1$) (see for example [67, pp.174, 201])

$$Q_\mu(\cosh a) = \sqrt{\frac{\pi}{2}} \int_0^\infty e^{-(\cosh a)u} I_{\mu+\frac{1}{2}}(u) \frac{du}{\sqrt{u}} \tag{7.5.6}$$

$$= \frac{1}{\sqrt{2}} \int_a^\infty \frac{e^{-(\mu+\frac{1}{2})\rho}}{(\cosh\rho - \cosh a)^{\frac{1}{2}}} d\rho, \tag{7.5.7}$$

where $a > 0$.

Set

$$G_1(t) = \int_0^\infty e^{-v\cosh r} v^{-\frac{1}{2}} \Theta(v, t)\, dv$$

and

$$G_2(t) = \frac{\sqrt{2}}{2\pi t^{\frac{3}{2}}} \int_r^\infty \frac{\rho e^{-\frac{\rho^2}{2t}}}{(\cosh\rho - \cosh r)^{\frac{1}{2}}} d\rho.$$

By (7.5.5), it suffices to show $G_1(t) = G_2(t)$ ($t > 0$) in order to obtain (7.4.3). For this purpose, we show the coincidence of the Laplace transforms.

For $G_1(t)$, by (7.5.2) and (7.5.6),

$$\int_0^\infty e^{-\lambda t} G_1(t)\, dt = \int_0^\infty e^{-v\cosh r} v^{-\frac{1}{2}} I_{\sqrt{2\lambda}}(v)\, dv$$

$$= \sqrt{\frac{2}{\pi}} Q_{\sqrt{2\lambda}-\frac{1}{2}}(\cosh r)$$

for $\lambda > 0$. On the other hand, since

$$\int_0^\infty e^{-\lambda t - \frac{\rho^2}{2t}} t^{-\frac{3}{2}} dt = \frac{\sqrt{2\pi}}{\rho} e^{-\sqrt{2\lambda}\rho} \qquad (\lambda, \rho > 0),$$

(7.5.7) implies

$$\int_0^\infty e^{-\lambda t} G_2(t)\, dt = \frac{1}{\sqrt{\pi}} \int_r^\infty \frac{e^{-\sqrt{2\lambda}\rho}}{(\cosh\rho - \cosh r)^{\frac{1}{2}}} d\rho$$

$$= \sqrt{\frac{2}{\pi}} Q_{\sqrt{2\lambda}-\frac{1}{2}}(\cosh r).$$

Hence, we see the coincidence of the Laplace transforms of G_1 and G_2. □

Appendix

Some Fundamentals

A.1 Gronwall's Inequality

Theorem A.1.1 *Let $f(t)$ and $g(t)$ be non-negative continuous functions on $[0, T]$ and α be a positive number. If*

$$f(t) \leqq g(t) + \alpha \int_0^t f(s)\mathrm{d}s \qquad (0 \leqq t \leqq T),$$

then

$$f(t) \leqq g(t) + \alpha \int_0^t g(s)\mathrm{e}^{\alpha(t-s)}\mathrm{d}s \qquad (0 \leqq t \leqq T).$$

In particular, if $f(t) \leqq \alpha \int_0^t f(s)\mathrm{d}s$ $(0 \leqq t \leqq T)$, then $f(t) \equiv 0$.

A.2 Dynkin Class Theorem, Monotone Class Theorem

Definition A.2.1 (1) A family $\mathscr{A}$ of subsets of a set Ω is called a **multiplicative class** if the following conditions are fulfilled:

(i) $\Omega \in \mathscr{A}$;
(ii) if $A, B \in \mathscr{A}$, then $A \cap B \in \mathscr{A}$.

(2) A family $\mathscr{G}$ of subsets of a set Ω is called a **Dynkin class** if the following conditions are fulfilled:

(i) $\Omega \in \mathscr{G}$;
(ii) if $A, B \in \mathscr{A}, A \subset B$, then $B \setminus A \equiv B \cap A^c \in \mathscr{G}$;
(iii) if $A_n \in \mathscr{G}, A_n \subset A_{n+1}$ $(n = 1, 2, \ldots)$, then $\bigcup_{n=1}^{\infty} A_n \in \mathscr{G}$.

Theorem A.2.2 (Dynkin class theorem) (1) *Let $\mathscr{A}$ be a multiplicative class and $\mathscr{G}$ be a Dynkin class. If $\mathscr{A} \subset \mathscr{G}$, then $\sigma(\mathscr{A}) \subset \mathscr{G}$.*
(2) *The smallest Dynkin class containing a multiplicative class $\mathscr{A}$ is $\sigma(\mathscr{A})$.*

Theorem A.2.3 *Let $\mathscr{A}$ be a multiplicative class and μ and $\widetilde{\mu}$ be measures on $\sigma(\mathscr{A})$. If μ and $\widetilde{\mu}$ coincide on $\mathscr{A}$, then $\mu = \widetilde{\mu}$.*

Definition A.2.4 A family $\mathscr{G}$ of subsets of a set Ω is called a **monotone class** if the following conditions are fulfilled:

(i) if $A_n \in \mathscr{G}, A_n \subset A_{n+1}$, then $\bigcup_n A_n \in \mathscr{G}$;
(ii) if $A_n \in \mathscr{G}, A_n \supset A_{n+1}$, then $\bigcap_n A_n \in \mathscr{G}$.

Theorem A.2.5 (Monotone class theorem) (1) *If a monotone class $\mathscr{G}$ is finitely additive, then $\mathscr{G}$ is a σ-field.*
(2) *Let $\mathscr{A}$ be a finitely additive class and $\mathscr{G}$ be a monotone class. If $\mathscr{A} \subset \mathscr{G}$, then $\sigma(\mathscr{A}) \subset \mathscr{G}$.*

The next theorem for functions (random variables) is useful in applications.

Theorem A.2.6 *Let $\mathscr{A}$ be a multiplicative class on a set Ω and $\mathscr{H}$ be a family of real functions on Ω such that*

(i) *for all $A \in \mathscr{A}$, $\mathbf{1}_A \in \mathscr{H}$;*
(ii) *$\mathbf{1} \in \mathscr{H}$;*
(iii) *$\mathscr{H}$ is a vector space;*
(iv) *let $\{f_n\}_{n=1}^{\infty} \subset \mathscr{H}$ be a convergent sequence of bounded functions such that $f_n \geqq 0$ and $f_n \leqq f_{n+1}$. If $f = \lim_{n\to\infty} f_n$ is also a bounded function, then $f \in \mathscr{H}$.*

Then, $\mathscr{H}$ contains all $\sigma(\mathscr{A})$-measurable real bounded functions.

For proofs, see [5, 6, 47, 114].

A.3 Uniform Integrability

The next elementary proposition is used to prove the uniform integrability of martingales.

Proposition A.3.1 *An integrable random variable X defined on a probability space $(\Omega, \mathscr{F}, \mathbf{P})$ satisfies the following.*
(1) $\lim_{\lambda\to\infty} \mathbf{E}[|X|\mathbf{1}_{\{|X|\geqq\lambda\}}] = 0$.
(2) *For any $\varepsilon > 0$ there exists a $\delta > 0$ such that, if $\mathbf{P}(A) < \delta$ $(A \in \mathscr{F})$, then* $\mathbf{E}[|X|\mathbf{1}_A] < \varepsilon$.

The uniform integrability of a family of random variables is defined in the following way.

Definition A.3.2 Let $\mathscr{I}$ be an index set and $\{X_\iota\}_{\iota\in\mathscr{I}}$ be a family of random variables defined on $(\Omega, \mathscr{F}, \mathbf{P})$. $\{X_\iota\}_{\iota\in\mathscr{I}}$ is said to be **uniformly integrable** if

$$\lim_{\lambda\to\infty} \sup_{\iota\in\mathscr{I}} \mathbf{E}[|X_\iota|\mathbf{1}_{\{|X_\iota|>\lambda\}}] = 0.$$

Proposition A.3.3 *If $\{X_\iota\}_{\iota\in\mathscr{I}}$ is uniformly integrable, then $\sup_{\iota\in\mathscr{I}} \mathbf{E}[|X_\iota|] < \infty$.*

Theorem A.3.4 *A family $\{X_\iota\}_{\iota\in\mathscr{I}}$ of integrable random variables is uniformly integrable if and only if there exists a measurable function ψ such that*

$$\lim_{x\to\infty} \frac{\psi(x)}{x} = \infty \quad \textit{and} \quad \sup_{\iota\in\mathscr{I}} \mathbf{E}[\psi(|X_\iota|)] < \infty.$$

In particular, if $\sup_{\iota\in\mathscr{I}} \mathbf{E}[|X_\iota|^p] < \infty$ for some $p > 1$, then $\{X_\iota\}_{\iota\in\mathscr{I}}$ is uniformly integrable.

Theorem A.3.5 *For a uniformly integrable sequence $\{X_n\}_{n=1}^\infty$ of random variables,*

$$\mathbf{E}[\liminf_{n\to\infty} X_n] \leqq \liminf_{n\to\infty} \mathbf{E}[X_n] \leqq \limsup_{n\to\infty} \mathbf{E}[X_n] \leqq \mathbf{E}[\limsup_{n\to\infty} X_n].$$

Moreover, if there exists the limit $\lim_{n\to\infty} X_n = X$ almost surely, then $X \in L^1$ and

$$\lim_{n\to\infty} \mathbf{E}[|X_n - X|] = 0.$$

Theorem A.3.6 *For a uniformly integrable sequence $\{X_n\}_{n=1}^\infty$ of random variables, X_n converges to a random variable X in probability if and only if it does in L^1.*

Theorem A.3.7 *Suppose that a sequence $\{X_n\}_{n=1}^\infty$ of integrable random variables converges to X almost surely. Then, the following conditions are equivalent:*

(i) *$\{X_n\}_{n=1}^\infty$ is uniformly integrable;*
(ii) *X_n converges to X in L^1;*
(iii) *$X \in L^1$ and $\mathbf{E}[|X_n|]$ converges to $\mathbf{E}[|X|]$.*

For details on the topics in this section, see [126].

A.4 Conditional Probability and Regular Conditional Probability

Let $(\Omega, \mathscr{F}, \mathbf{P})$ be a probability space and $\mathscr{G}$ be a sub-σ-field of $\mathscr{F}$. For an integrable random variable X, there exists a unique, up to null sets, $\mathscr{G}$-measurable integrable random variable Y such that

$$\mathbf{E}[X\mathbf{1}_A] = \mathbf{E}[Y\mathbf{1}_A]$$

for all $A \in \mathscr{G}$. This random variable $Y(\omega)$ is called the **conditional expectation** of X given $\mathscr{G}$ and is denoted by $\mathbf{E}[X|\mathscr{G}](\omega)$. In particular, when $X = \mathbf{1}_B$ for $B \in \mathscr{F}$, $\mathbf{E}[\mathbf{1}_B|\mathscr{G}](\omega)$ is called the **conditional probability** of B given $\mathscr{G}$ and is denoted by $\mathbf{P}(B|\mathscr{G})(\omega)$.

Remark A.4.1 (1) If X is $\mathscr{G}$-measurable, then $\mathbf{E}[X|\mathscr{G}] = X$, $\mathbf{P}$-a.s.
(2) If $\mathscr{G}$ is trivial, that is, if $\mathscr{G} = \{\emptyset, \Omega\}$, then $\mathbf{E}[X|\mathscr{G}] = \mathbf{E}[X]$, $\mathbf{P}$-a.s.

The conditional probability $\mathbf{P}(B|\mathscr{G})(\omega)$ is well defined as a random variable (a function in ω) for each fixed $B \in \mathscr{F}$. Since it is defined up to null sets, it is impossible in general to regard it as a function of B for a fixed ω. This inconvenience is overcome by the regular conditional probability, which was used in a proof of the fundamental theorem on the existence of solutions for stochastic differential equations and that of the strong Markov property of solutions of martingale problems.

Definition A.4.2 If a mapping $p(\cdot,\cdot|\mathscr{G}) : \Omega\times\mathscr{F} \ni (\omega, A) \mapsto p(\omega, A|\mathscr{G}) \in [0, 1]$ satisfies the conditions:

(i) for any $\omega \in \Omega$, $\mathscr{F} \ni A \mapsto p(\omega, A|\mathscr{G})$ is a probability measure on $(\Omega, \mathscr{F})$;
(ii) for any $A \in \mathscr{F}$, $\Omega \ni \omega \mapsto p(\omega, A|\mathscr{G})$ is $\mathscr{G}$-measurable;
(iii) for any $A \in \mathscr{F}$ and $B \in \mathscr{G}$,

$$\mathbf{P}(A \cap B) = \int_B p(\omega, A|\mathscr{G})\mathbf{P}(d\omega),$$

then it is called a **regular conditional probability** given the condition $\mathscr{G}$.

The condition (iii) can be written as
(iii)′ for any $X \in L^1(\Omega)$ and $B \in \mathscr{G}$,

$$\mathbf{E}[X\mathbf{1}_B] = \mathbf{E}\Big[\int_\Omega X(\omega')p(\cdot, d\omega'|\mathscr{G})\mathbf{1}_B\Big].$$

In other words, $\int_\Omega X(\omega')p(\omega, d\omega'|\mathscr{G}) = \mathbf{E}[X|\mathscr{G}](\omega)$, $\mathbf{P}$-a.s.

An example of probability spaces which do not admit regular conditional probabilities can be found in [99].

Definition A.4.3 Two measurable spaces $(\Omega, \mathscr{F})$ and $(\Omega', \mathscr{F}')$ are called **Borel isomorphic** if there exists an $\mathscr{F}/\mathscr{F}'$-measurable and one-to-one mapping f from Ω onto Ω' whose inverse f^{-1} is $\mathscr{F}'/\mathscr{F}$-measurable. If a measurable space $(\Omega, \mathscr{F})$ is Borel isomorphic with a Borel subset of $\mathbb{R}^1$, it is called a **standard measurable space**. Moreover, if $\mathbf{P}$ is a probability measure on $(\Omega, \mathscr{F})$, $(\Omega, \mathscr{F}, \mathbf{P})$ is called a **standard probability space**.

It is known that a standard probability space is Borel isomorphic with one of $\{1, 2, \ldots, n\}$ $(n = 1, 2, \ldots)$, $\mathbb{N}$ and $[0, 1]$. It is also known that a complete separable metric space endowed with the topological σ-field is a standard probability space.

Theorem A.4.4 *A standard probability space* $(\Omega, \mathscr{F}, \mathbf{P})$ *admits the regular conditional probability given a sub-*σ*-field* $\mathscr{G}$ *of* $\mathscr{F}$.

We refer the reader to [17, 45, 47] about standard probability space and regular conditional probability.

Next let $(S, \mathscr{B})$ be a measurable space and ξ be an S-valued random variable defined on Ω. For $X \in L^1(\Omega)$, $C \in \mathscr{B}$, define $\mathbf{R}$ by

$$\mathbf{R}(C) = \mathbf{E}[X\mathbf{1}_C(\xi)].$$

$\mathbf{R}$ is a set function on $\mathscr{B}$. Since $\mathbf{R}$ is absolutely continuous with respect to the probability law ν of ξ, there exists a measurable function Z on S such that

$$\mathbf{R}(C) = \int_C Z(s)\nu(\mathrm{d}s)$$

by the Radon–Nikodym theorem. The function Z is denoted by $\mathbf{E}[X|\xi = s]$ $(s \in S)$ and called the conditional expectation of X given $\xi = s$. By definition we have

$$\mathbf{E}[X\mathbf{1}_C(\xi)] = \int_C \mathbf{E}[X|\xi = s]\mathbf{P}(\xi \in \mathrm{d}s) \qquad (C \in \mathscr{B}).$$

Definition A.4.5 Let $(\Omega, \mathscr{F})$ be a measurable space. $\mathscr{F}$ is said to be **countably generated** if it possesses a countable subset $\mathscr{F}_0$ of $\mathscr{F}$ such that any probability measures $\mathbf{P}_1$ and $\mathbf{P}_2$ coincide provided that they are equal on $\mathscr{F}_0$.

If $(\Omega, \mathscr{F})$ is a standard measurable space, then $\mathscr{F}$ is countably generated.

Theorem A.4.6 *Let $(\Omega, \mathscr{F}, \mathbf{P})$ be a standard probability space, $\mathscr{G}$ be a sub-σ-field of $\mathscr{F}$, $p(\omega, \mathrm{d}\omega'|\mathscr{G})$ be the regular conditional probability given $\mathscr{G}$ and $\mathscr{H}$ be a countably generated sub-σ-field of $\mathscr{G}$. Then there exists a* $\mathbf{P}$*-null set $N \in \mathscr{G}$ such that, if $\omega \notin N$, $p(\omega, A|\mathscr{G}) = \mathbf{1}_A(\omega)$ for $A \in \mathscr{H}$.*

Corollary A.4.7 *Under the same framework as the theorem, let ξ be an S-valued random variable, $(S, \mathscr{B})$ being a measurable space. If $\mathscr{B}$ is countably generated and $\{x\} \in \mathscr{B}$ for all $x \in S$, then $p(\omega, \{\omega'; \xi(\omega') = \xi(\omega)\}|\mathscr{G}) = 1$ almost surely.*

A.5 Kolmogorov Continuity Theorem

Let D be a domain in $\mathbb{R}^d$ and B be a Banach space. A family $\{X(x)\}_{x\in D}$ of B-valued random variables is called a **random field**. The stochastic flow defined by the solution of a stochastic differential equation is a typical example.

Kolmogorov's continuity theorem gives a sufficient condition for a random field to have a continuous modification. It is also called the Kolmogorov–Centsov theorem or the Kolmogorov–Totoki theorem.

Theorem A.5.1 *Let $X = \{X(x)\}_{x\in D}$ be a random field. Suppose that there exist positive constants $\alpha_1, \alpha_2, \ldots, \alpha_d, \gamma, C$ with $\sum_{i=1}^d \alpha_i^{-1} < 1$ such that*

$$\mathbf{E}[\|X(x) - X(y)\|^\gamma] \leqq C \sum_{i=1}^d |x^i - y^i|^{\alpha_i} \qquad (x, y \in D),$$

where $\|\cdot\|$ is the norm of B. Then X has a continuous modification $\{\widetilde{X}(x)\}_{x\in D}$.

Moreover, for any cube E contained in D, there exists a positive random variable K with $\mathbf{E}[K^\gamma] < \infty$ such that

$$\|\widetilde{X}(x) - \widetilde{X}(y)\| \leqq K \sum_{i=1}^d |x^i - y^i|^{\beta_i} \qquad (x, y \in E),$$

where $\beta_i = \frac{\alpha_i(\alpha_0 - d)}{\alpha_0\gamma}$ and $\alpha_0 = d(\sum_{i=1}^d \frac{1}{\alpha_i})^{-1}$.

For a proof, see Kunita [62].

A.6 Laplace Transforms and the Tauberian Theorem

For a non-negative measure $F(\mathrm{d}\xi)$ on $[0, \infty)$, the function L defined by

$$L(\lambda) \equiv L_F(\lambda) = \int_0^\infty \mathrm{e}^{-\lambda\xi} F(\mathrm{d}\xi)$$

is the **Laplace transform** of F. For example, if $F(\mathrm{d}\xi) = \gamma\xi^{\gamma-1}\mathrm{d}\xi$ for $\gamma > 0$, then

$$L_F(\lambda) = \gamma \int_0^\infty \mathrm{e}^{-\lambda\xi}\xi^{\gamma-1}\mathrm{d}\xi = \frac{\Gamma(\gamma+1)}{\lambda^\gamma}, \quad \lambda > 0.$$

We mention some properties of Laplace transforms and the Tauberian theorem which was used in Chapter 7. For details, see [23, 125].

Theorem A.6.1 (Uniqueness) *Let F and G be non-negative measures on $[0,\infty)$ and assume that there exists an $a > 0$ such that $L_F(\lambda) = L_G(\lambda) < \infty$ $(\lambda > a)$. Then $F = G$.*

Theorem A.6.2 (Continuity) *Let F_n $(n \in \mathbb{N})$ be a sequence of non-negative measures on $[0,\infty)$ which have the Laplace transforms $L_{F_n}(\lambda)$ for $\lambda > a$ and assume that $L_{F_n}(\lambda)$ converges to some function $L(\lambda)$ for all $\lambda > a$ as $n \to \infty$. Then $L(\lambda)$ $(\lambda > a)$ is the Laplace transform of some measure F on $[0,\infty)$ and, for every bounded and continuous interval*[1] *$I = (\alpha,\beta)$ of F, $F_n(I) \to F(I)$ as $n \to \infty$.*

The Tauberian theorem is formulated by means of regularly varying functions.

Definition A.6.3 A function $\ell(\xi)$ on $[0,\infty)$ is called **slowing varying** as $\xi \to \infty$ if

$$\lim_{\xi\to\infty} \frac{\ell(\kappa\xi)}{\ell(\xi)} = 1$$

for any $\kappa > 0$. A function of the form $k(\xi) = \xi^\gamma \ell(\xi)$ $(\gamma > 0)$ is called **regularly varying**.

For $c, \delta > 0$, the function $\ell(\xi) = c(\log(1+\xi))^\delta$ is slowing varying as $\xi \to \infty$.

Theorem A.6.4 *Let γ be a positive number and $\ell(\xi)$ be a slowly varying function as $\xi \to \infty$. Then, for a right-continuous non-decreasing function F, F has an asymptotic behavior*

$$F(\xi) = \xi^\gamma \ell(\xi)(1 + o(1)) \quad (\xi \to \infty)$$

if and only if the Laplace transform $L(\lambda)$ of the corresponding Stieltjes measure $F(\mathrm{d}\xi)$ has the asymptotic behavior

$$L(\lambda) = \Gamma(\gamma+1)\lambda^{-\gamma}\ell\Big(\frac{1}{\lambda}\Big) \quad (\lambda \downarrow 0).$$

[1] $I = (\alpha,\beta)$ is said to be a continuous interval of a measure F if $F(\{\alpha\}) = F(\{\beta\}) = 0$.

That the asymptotic behavior of the Laplace transform L of F follows from that of F is called the **Abelian theorem** and the converse assertion is the **Tauberian theorem**. If we define the slow varying near 0, the assertion of Theorem A.6.4 holds, replacing $\xi \downarrow 0$ and $\lambda \to \infty$ for $\xi \to \infty$ and $\lambda \downarrow 0$, respectively.

References

[1] R. Adams and J. Fournier, *Sobolev Spaces*, 2nd edn., Academic Press, 2003.

[2] M. Aizenman and B. Simon, Brownian motion and Harnack's inequality for Schrödinger operators, *Comm. Pure Appl. Math.*, **35** (1982), 209–273.

[3] V. I. Arnold, *Mathematical Methods of Classical Mechanics*, 2nd edn., Springer-Verlag, 1989.

[4] J. Avron, I. Herbst, and B. Simon, Schrödinger operators with magnetic fields, I. general interactions, *Duke Math. J.*, **45** (1978), 847–883.

[5] P. Billingsley, *Probability and Measure*, 3rd edn., John Wiley & Sons, 1995.

[6] R. M. Blumenthal and R. K. Getoor, *Markov Processes and Potential Theory*, Academic Press, 1968.

[7] V. Bogachev, *Gaussian Measures*, Amer. Math. Soc., 1998.

[8] N. Bouleau and F. Hirsch, *Dirichlet Forms and Analysis on Wiener Space*, Walter de Gruyter, 1991.

[9] C. Cocozza and M. Yor, Démonstration d'un théorème de Knight à l'aide de martingales exponentielles, *Séminaire de Probabilités, XIV*, eds. J. Azama and M. Yor, Lecture Notes in Math., **784**, 496–499, Springer-Verlag, 1980.

[10] H. L. Cycon, R. G. Froese, W. Kirsch, and B. Simon, *Schrödinger Operators, with Application to Quantum Mechanics and Global Geometry*, Springer-Verlag, 1987.

[11] E. B. Davies, *Heat Kernels and Spectral Theory*, Cambridge University Press, 1989.

[12] B. Davis, Picard's theorem and Brownian motion, *Trans. Amer. Math. Soc.*, **213** (1975), 353–362.

[13] C. Dellacherie, *Capacités et Processus Stochastiques*, Springer-Verlag, 1971.

[14] D. Deuschel and D. Stroock, *Large Deviations*, Academic Press, 1989.

[15] J. L. Doob, *Stochastic Processes*, John Wiley & Sons, 1953.

[16] H. Doss, Liens entre équations différentielles stochastiques et ordinaires, *Ann. Inst. H. Poincaré Sect. B (N.S.)*, **13** (1977), 99–125.

[17] R. M. Dudley, *Real Analysis and Probability*, 2nd edn., Cambridge University Press, 2002.

[18] D. Duffie, *Dynamic Asset Pricing Theory*, 2nd edn., Princeton University Press, 1996.

[19] N. Dunford and J. Schwartz, *Linear Operators*, **II**, Interscience, 1963.

[20] R. Durrett, *Brownian Motion and Martingales in Analysis*, Wadsworth, 1984.
[21] R. Elliot and P. Kopp, *Mathematics of Financial Markets*, Springer-Verlag, 1999.
[22] K. D. Elworthy, *Stochastic Differential Equations on Manifolds*, Cambridge University Press, 1982.
[23] W. Feller, *An Introduction to Probability Theory and Its Applications*, Vol. II, John Wiley & Sons, 1966.
[24] E. Fournié, J.-M. Lasry, J. Lebuchoux, P.-L. Lions, and N. Touzi, Applications of Malliavin calculus to Monte Carlo methods in finance, *Finance Stoch.* **3** (1999), 391–412.
[25] M. Fukushima, Y. Oshima, and M. Takeda, *Dirichlet Forms and Symmetric Markov Processes*, 2nd edn., Walter de Gruyter, 2010.
[26] P. Friz and M. Hairer, *A Course on Rough Paths*, Springer-Verlag, 2014.
[27] P. Friz and N. Victoir, *Multidimensional Stochastic Processes as Rough Paths*, Cambridge University Press, 2010.
[28] B. Gaveau and P. Trauber, L'intégrale stochastique comme opérateur de divergence dans l'space fonctionnel, *J. Func. Anal.*, **46** (1982), 230–238.
[29] R. K. Getoor and M. J. Sharpe, Conformal martingales, *Invent. Math.*, **16** (1972), 271–308.
[30] E. Getzler, Degree theory for Wiener maps, *J. Func. Anal.*, **68** (1986), 388–403.
[31] I. S. Gradshteyn and I. M. Ryzhik, *Tables of Integrals, Series, and Products*, 7th edn., Academic Press, 2007.
[32] J.-C. Gruet, Semi-groupe du mouvement Brownien hyperbolique, Stochastics *Stochastic Rep.*, **56** (1996), 53–61.
[33] D. Hejhal, *The Selberg Trace Formula for PSL*$(2, \mathbb{R})$, Vol.1, Vol.2, Lecture Notes in Math., **548**, **1001**, Springer-Verlag, 1976, 1983.
[34] B. Helffer and J. Sjöstrand, Multiple wells in the semiclassical limit, I, *Comm. PDE*, **9** (1984), 337–408.
[35] B. Helffer and J. Sjöstrand, Puits multiples en limite semi-classique, II. Interaction moléculaire. Symétries. Perturbation., *Ann. Inst. H. Poincaré Phys. Théor.*, **42** (1985), 127–212.
[36] L. Hörmander, *The Analysis of Linear Partial Differential Operators*, I, *Distribution Theory and Fourier Analysis*, 2nd edn., Springer-Verlag, 1990.
[37] L. Hörmander, Hypoelliptic second order differential equations, *Acta Math.*, **119** (1967), 147–171.
[38] E. P. Hsu, *Stochastic Analysis on Manifolds*, Amer. Math. Soc., 2002.
[39] K. Ichihara, Explosion problem for symmetric diffusion processes, *Trans. Amer. Math. Soc.*, **298** (1986), 515–536.
[40] N. Ikeda, S. Kusuoka, and S. Manabe, Lévy's stochastic area formula and related problems, in *Stochastic Analysis*, eds. M. Cranston and M. Pinsky, 281–305, Proc. Sympos. Pure Math., **57**, Amer. Math. Soc., 1995.
[41] N. Ikeda and S. Manabe, Van Vleck–Pauli formula for Wiener integrals and Jacobi fields, in *Itô's Stochastic Calculus and Probability Theory*, eds. N. Ikeda, S. Watanabe, M. Fukushima, and H. Kunita, 141–156, Springer-Verlag, 1996.
[42] N. Ikeda and H. Matsumoto, Brownian motion on the hyperbolic plane and Selberg trace formula, *J. Funct. Anal.*, **163** (1999), 63–110.

[43] N. Ikeda and H. Matsumoto, The Kolmogorov operator and classical mechanics, *Séminaire de Probabilités XLVII*, eds. C. Donati-Martin, A. Lejay, and A. Rouault, Lecture Notes in Math., **2137**, 497–504, Springer-Verlag, 2015.

[44] N. Ikeda and S. Taniguchi, Quadratic Wiener functionals, Kalman-Bucy filters, and the KdV equation, in *Stochastic Analysis and Related Topics in Kyoto, in honor of Kiyosi Itô*, eds., H. Kunita, S. Watanabe, and Y. Takahashi, Adv. Studies Pure Math. **41**, 167–187, Math. Soc. Japan, Tokyo, 2004.

[45] N. Ikeda and S. Watanabe, *Stochastic Differential Equations and Diffusion Processes*, 2nd edn., North Holland/Kodansha, 1989.

[46] K. Itô, *Essentials of Stochastic Processes* (translated by Y. Ito), Amer Math. Soc., 2006. (Originally published in Japanese from Iwanami Shoten, 1957, 2006)

[47] K. Itô, *Introduction to Probability Theory*, Cambridge University Press, 1984. (Originally published in Japanese from Iwanami Shoten, 1978)

[48] K. Itô, Differential equations determining Markov processes, *Zenkoku Shijō Sūgaku Danwakai*, **244** (1942), 1352–1400, (in Japanese). English translation in *Kiyosi Itô, Selected Papers*, eds. D. Stroock and S. R. S. Varadhan, Springer-Verlag, 1987.

[49] K. Itô, On stochastic differential equations, *Mem. Amer. Math. Soc.*, **4** (1951).

[50] K. Itô and H. P. McKean, Jr., *Diffusion Processes and Their Sample Paths*, Springer-Verlag, 1974.

[51] K. Itô and M. Nisio, On the convergence of sums of independent Banach space valued random variables, *Osaka J. Math.*, **5** (1968), 35–48.

[52] M. Kac, *Integration in Function Spaces and Some of Its Applications*, Fermi Lectures, Accademia Nazionale dei Lincei, Scuola Normale Superiore, 1980.

[53] M. Kac, On distributions of certain Wiener functionals, *Trans. Amer. Math. Soc.*, **65** (1949), 1–13.

[54] M. Kac, On some connections between probability theory and differential and integral equations, *Proceedings of 2nd Berkeley Symp. on Math. Stat. and Probability*, 189–215, University of California Press, 1951.

[55] M. Kac, Can one hear the shape of a drum?, *Amer. Math. Monthly*, **73** (1966), 1–23.

[56] I. Karatzas and S. E. Shreve, *Brownian Motion and Stochastic Calculus*, 2nd edn., Springer-Verlag, 1991.

[57] I. Karatzas and S. Shreve, *Methods of Mathematical Finance*, Springer-Verlag, 1998.

[58] T. Kato, *Perturbation Theory for Linear Operators*, 2nd edn., Springer-Verlag, 1995.

[59] N. Kazamaki, The equivalence of two conditions on weighted norm inequalities for martingales, *Proc. Intern. Symp. SDE Kyoto 1976* (ed. K. Itô), 141–152, Kinokuniya, 1978.

[60] F. B. Knight, A reduction of continuous square-integrable martingales to Brownian motion, in *Martingales*, ed. H. Dinges, Lecture Notes in Math., **190**, 19–31, Springer-Verlag, 1971.

[61] H. Kunita, *Estimation of Stochastic Processes* (in Japanese), Sangyou Tosho, 1976.

[62] H. Kunita, *Stochastic Flows and Stochastic Differential Equations*, Cambridge University Press, 1990.
[63] H. Kunita, Supports of diffusion processes and controllability problems, *Proc. Intern. Symp. SDE Kyoto 1976* (ed. K. Itô), 163–185, Kinokuniya, 1978.
[64] H. Kunita, On the decomposition of solutions of stochastic differential equations, in *Stochastic Integrals*, ed. D. Williams, Lecture Notes in Math., **851**, 213–255, Springer-Verlag, 1981.
[65] H. Kunita and S. Watanabe, On square integrable martingales, *Nagoya Math. J.*, **30** (1967), 209–245.
[66] S. Kusuoka, The nonlinear transformation of Gaussian measure on Banach space and its absolute continuity, *J. Fac. Sci. Tokyo Univ.*, Sect. 1.A., **29** (1982), 567–590.
[67] N. N. Lebedev, *Special Functions and their Applications*, translated by R. R. Silverman, Dover, 1972.
[68] M. Ledoux, Isoperimetry and Gaussian analysis, in *Lectures on Probability Theory and Statistics*, Ecole d'Eté de Probabilités de Saint-Flour XXIV – 1994, ed. P. Bernard, Lecture Notes in Math., **1648**, 165–294, Springer-Verlag, 1996.
[69] J.-F. LeGall, Applications du temps local aux equations différentielle stochastiques unidimensionalles, *Séminaire de Probabilités XVII*, edn. J. Azema and M. Yor, Lecture Notes in Math., **986**, 15–31, Springer-Verlag, 1983.
[70] P. Lévy, Wiener's random function, and other Laplacian random functions, *Proceedings of 2nd Berkeley Symp. on Math. Stat. and Probability*, 171–186, University of California Press, 1951.
[71] T. Lyons, M. Caruana, and T. Lévy, Differential equations driven by rough paths, *École d'Été de Probabilités de Saint-Flour XXXIV - 2004*, Lecture Notes in Math., **1908**, Springer, 2007.
[72] T. Lyons and Z. Qian, *System Control and Rough Paths*, Oxford University Press, 2002.
[73] P. Malliavin, *Stochastic Analysis*, Springer-Verlag, 1997.
[74] P. Malliavin, Stochastic calculus of variation and hypoelliptic operators, *Proc. Intern. Symp. SDE Kyoto 1976* ed. K. Itô, 195–263, Kinokuniya, 1978.
[75] P. Malliavin, C^k-hypoellipticity with degeneracy, in *Stochastic Analysis*, eds. A. Friedman and M. Pinsky, 199–214, 327–340, Academic Press, 1978.
[76] P. Malliavin and A. Thalmaier, *Stochastic Calculus of Variations in Mathematical Finance*, Springer-Verlag, 2006.
[77] G. Maruyama, *Selected Papers*, eds. N. Ikeda and H. Tanaka, Kaigai Publications, 1988.
[78] G. Maruyama, On the transition probability functions of the Markov process, *Nat. Sci. Rep. Ochanomizu Univ.*, **5** (1954), 10–20.
[79] G. Maruyama, Continuous Markov processes and stochastic equations, *Rend. Circ. Mate. Palermo.*, **4** (1955), 48–90.
[80] H. Matsumoto, Semiclassical asymptotics of eigenvalues for Schrödinger operators with magnetic fields, *J. Funct. Anal.*, **129** (1995), 168–190.
[81] H. Matsumoto, L. Nguyen, and M. Yor, Subordinators related to the exponential functionals of Brownian bridges and explicit formulae for the semigroups of hyperbolic Brownian motions, in *Stochastic Processes and Related Topics*, eds. R. Buckdahn, E. Engelbert and M. Yor, 213–235, Gordon and Breach, 2001.

[82] H. Matsumoto and S. Taniguchi, Wiener functionals of second order and their Lévy measures, *Elect. Jour. Probab.*, **7**, No.14 (2002), 1–30.
[83] H. Matsumoto and M. Yor, A version of Pitman's $2M{-}X$ theorem for geometric Brownian motions, *C.R. Acad. Sc. Paris Série I*, **328** (1999), 1067–1074.
[84] H. P. McKean, Jr., *Stochastic Integrals*, Academic Press, 1969.
[85] H. P. McKean, Jr., Selberg's trace formula as applied to a compact Riemannian surface, *Comm. Pure Appl. Math.*, **101** (1972), 225–246.
[86] P. A. Meyer, *Probabilités et Potentiel*, Hermann, 1966.
[87] T. Miwa, E. Date, and M. Jimbo, *Solitons: Differential Equations, Symmetries and Infinite Dimensional Algebras* (translated by M. Reid), Cambridge University Press, 2000. (Originally published in Japanese from Iwanami Shoten, 1993)
[88] M. Musiela and M. Rutkowski, *Martingale Methods in Financial Modeling*, Springer-Verlag, 2003.
[89] S. Nakao, On the pathwise uniqueness of solutions of one-dimensional stochastic differential equations, *Osaka J. Math.*, **9** (1972), 513–518.
[90] A. A. Novikov, On an identity for stochastic integrals, *Theory Prob. Appl.*, **17** (1972), 717–720.
[91] A. A. Novikov, On moment inequalities and identities for stochastic integrals, *Proc. Second Japan–USSR Symp. Prob. Theor.*, eds. G. Maruyama and J. V. Prokhorov, Lecture Notes in Math., **330**, 333–339, Springer-Verlag, 1973.
[92] D. Nualart, *The Malliavin Calculus and Related Topics*, 2nd edn., Springer-Verlag, 2006.
[93] B. Øksendal, *Stochastic Differential Equations, an Introduction with Applications*, 6th edn., Springer-Verlag, 2003.
[94] S. Port and C. Stone, *Brownian Motion and Classical Potential Theory*, Academic Press, 1978.
[95] K. M. Rao, On the decomposition theorem of Meyer, *Math. Scand.*, **24** (1969), 66–78.
[96] D. Ray, On spectra of second order differential operators, *Trans. Amer. Math. Soc.*, **77** (1954), 299–321.
[97] L. Richardson, *Measure and Integration: a Concise Introduction to Real Analysis*, John Wiley & Sons, 2009.
[98] D. Revuz and M. Yor, *Continuous Martingales and Brownian Motion*, 3rd edn., Springer-Verlag, 1999.
[99] L. C. G. Rogers and D. Williams, *Diffusions, Markov Processes, and Martingales*, Vol.1, *Foundations*, 2nd edn., John Wiley & Sons, New York, 1994.
[100] L. C. G. Rogers and D. Williams, *Diffusions, Markov Processes, and Martingales*, Vol.2, *Itô Calculus*, 2nd edn., John Wiley & Sons, New York, 1994.
[101] K. Sato, *Lévy Processes and Infinitely Divisible Distributions*, Cambridge University Press, 1999.
[102] M. Schilder, Some asymptotic formulae for Wiener integrals, *Trans. Amer. Math. Soc.*, **125** (1966), 63–85.
[103] A. Selberg, Harmonic analysis and discontinuous groups in weakly symmetric Riemannian spaces with applications to Dirichlet series, *J. Indian Math. Soc.*, **20** (1956), 47–87.
[104] I. Shigekawa, *Stochastic Analysis*, Amer. Math. Soc., 2004. (Originally published in Japanese from Iwanami Shoten, 1998)

[105] S. Shreve, *Stochastic Calculus for Finance*, I, II, Springer-Verlag, 2004.
[106] B. Simon, *Functional Integration and Quantum Physics*, Academic Press, 1979.
[107] B. Simon, *Trace Ideals and Their Applications*, 2nd edn., Amer. Math. Soc., 2005.
[108] B. Simon, Schrödinger semigroups, *Bull. Amer. Math. Soc.*, **7** (1982), 447–526.
[109] B. Simon, Semiclassical analysis of low lying eigenvalues I, Non-degenerate minima: Asymptotic expansions, *Ann. Inst. Henri-Poincaré, Sect. A*, **38** (1983), 295–307.
[110] B. Simon, Semiclassical analysis of low lying eigenvalues II, Tunneling, *Ann. Math.*, **120** (1984), 89–118.
[111] D. W. Stroock, *Lectures on Topics in Stochastic Differential Equations*, Tata Insitute of Fundamental Research, 1982.
[112] D. W. Stroock, *Probability Theory: an Analytic View*, 2nd edn., Cambridge University Press, 2010.
[113] D. W. Stroock, An exercise in Malliavin calculus, *J. Math. Soc. Japan*, **67** (2015), 1785–1799.
[114] D. W. Stroock and S. R. S. Varadhan, *Multidimensional Diffusion Processes*, Springer-Verlag, 1979.
[115] D. W. Stroock and S. R. S. Varadhan, On the support of diffusion processes with applications to the strong maximum principle, *Proc. Sixth Berkeley Symp. Math. Statist. Prob. III.*, 361–368, University of California Press, 1972.
[116] H. Sugita, Positive generalized Wiener functions and potential theory over abstract Wiener spaces, *Osaka J. Math.*, **25** (1988), 665–696.
[117] H. J. Sussmann, On the gap between deterministic and stochastic ordinary differential equations, *Ann. Probab.*, **6** (1978), 19–41.
[118] S. Taniguchi, Brownian sheet and reflectionless potentials, *Stoch. Pro. Appl.*, **116** (2006), 293–309.
[119] H. Trotter, A property of Brownian motion paths, *Illinois J. Math.*, **2** (1958), 425–433.
[120] A.S. Üstünel and M. Zakai, *Transformation of Measure on Wiener Space*, Springer-Verlag, 2000.
[121] J. H. Van Vleck, The correspondence principle in the statistical interpretation of quantum mechanics, *Proc. Nat. Acad. Sci. U.S.A.*, **14** (1928), 178–188.
[122] S. Watanabe, Analysis of Wiener functionals (Malliavin calculus) and its applications to heat kernels, *Ann. Probab.*, **15** (1987), 1–39.
[123] S. Watanabe, Generalized Wiener functionals and their applications, Probability theory and mathematical statistics, *Proceedings of the Fifth Japan–USSR Symposium*, Kyoto, 1986, eds. S. Watanabe and Y. V. Prokhorov, 541–548, Lecture Notes in Math., **1299**, Springer-Verlag, Berlin, 1988.
[124] H. Weyl, Das asymptotische Verteilungsgesetz der Eigenschwingungen eines beliebig gestalteten elastischen Körpers, *Rend. Cir. Mat. Palermo*, **39** (1915), 1–50.
[125] D. V. Widder, *The Laplace Transform*, Princeton University Press, 1941.
[126] D. Williams, *Probability with Martingales*, Cambridge University Press, 1991.
[127] E. Wong and M. Zakai, On the relation between ordinary and stochastic differential equations, *Intern. J. Engng. Sci.*, **3** (1965), 213–229.

[128] T. Yamada and S. Watanabe, On the uniqueness of solutions of stochastic differential equations, *J. Math. Kyoyo Univ.*, **11** (1971), 155–167.
[129] Y. Yamato, Stochastic differential equations and nilpotent Lie algebras, *Z. Wahr. verw. Geb.*, **47** (1979), 213–229.
[130] M. Yor, *Exponential Functionals of Brownian Motion and Related Processes*, Springer-Verlag, 2001.
[131] M. Yor, Sur la continuité des temps locaux associés à certaines semimartingales, *Astérisque* **52–53** (1978), 23–35.
[132] M. Yor, On some exponential functionals of Brownian motion, *Adv. Appl. Prob.*, **24** (1992), 509–531. (Also in [130])
[133] K. Yoshida, *Functional Analysis*, 6th edn., Springer-Verlag, 1980.

Index